Allelopathy in Agroecosystems

Allelopathy in Agroecosystems has been co-published simultaneously as *Journal of Crop Production*, Volume 4, Number 2 (#8) 2001.

The *Journal of Crop Production* Monographic "Separates"

Below is a list of "separates," which in serials librarianship means a special issue simultaneously published as a special journal issue or double-issue *and* as a "separate" hardbound monograph. (This is a format which we also call a "DocuSerial.")

"Separates" are published because specialized libraries or professionals may wish to purchase a specific thematic issue by itself in a format which can be separately cataloged and shelved, as opposed to purchasing the journal on an on-going basis. Faculty members may also more easily consider a "separate" for classroom adoption.

"Separates" are carefully classified separately with the major book jobbers so that the journal tie-in can be noted on new book order slips to avoid duplicate purchasing.

Allelopathy in Agroecosystems, edited by Ravinder K. Kohli, PhD, Harminder Pal Singh, PhD, and Daizy R. Batish, PhD (Vol. 4, No. 2 #8, 2001). *Explains how the natural biochemical interactions among plants and microbes can be used as an environmentally safe method of weed and pest management.*

The Rice-Wheat Cropping System of South Asia: Efficient Production Management, edited by Palit K. Kataki, PhD (Vol. 4, No. 1 #7, 2001). *This book critically analyzes and discusses production issues for the rice-wheat cropping system of South Asia, focusing on the questions of soil depletion, pest control, and irrigation. It compiles information gathered from research institutions, government organizations, and farmer surveys to analyze the condition of this regional system, suggest policy changes, and predict directions for future growth.*

The Rice-Wheat Cropping System of South Asia: Trends, Constraints, Productivity and Policy, edited by Palit K. Kataki, PhD (Vol. 3, No. 2 #6, 2001). *This book critically analyzes and discusses available options for all aspects of the rice-wheat cropping system of South Asia, addressing the question, "Are the sustainability and productivity of this system in a state of decline/stagnation?" This volume compiles information gathered from research institutions, government organizations, and farmer surveys to analyze the impact of this regional system.*

Nature Farming and Microbial Applications, edited by Hui-lian Xu, PhD, James F. Parr, PhD, and Hiroshi Umemura, PhD (Vol. 3, No. 1 #5, 2000). *"Of great interest to agriculture specialists, plant physiologists, microbiologists, and entomologists as well as soil scientists and evnironmentalists. . . . very original and innovative data on organic farming." (Dr. André Gosselin, Professor, Department of Phytology, Center for Research in Horticulture, Université Laval, Quebec, Canada)*

Water Use in Crop Production, edited by M.B. Kirkham, BA, MS, PhD (Vol. 2, No. 2 #4, 1999). *Provides scientists and graduate students with an understanding of the advancements in the understanding of water use in crop production around the world. You will discover that by utilizing good management, such as avoiding excessive deep percolation or reducing runoff by increased infiltration, that even under dryland or irrigated conditions you can achieve improved use of water for greater crop production. Through this informative book, you will discover how to make the most efficient use of water for crops to help feed the earth's expanding population.*

Expanding the Context of Weed Management, edited by Douglas D. Buhler, PhD (Vol. 2, No. 1 #3, 1999). *Presents innovative approaches to weeds and weed management.*

Nutrient Use in Crop Production, edited by Zdenko Rengel, PhD (Vol. 1, No. 2 #2, 1998). *"Raises immensely important issues and makes sensible suggestions about where research and agricultural extension work needs to be focused." (Professor David Clarkson, Department of Agricultural Sciences, AFRC Institute Arable Crops Research, University of Bristol, United Kingdom)*

Crop Sciences: Recent Advances, Amarjit S. Basra, PhD (Vol. 1, No. 1 #1, 1997). *Presents relevant research findings and practical guidance to help improve crop vield and stability, product quality, and environmental sustainability.*

Allelopathy in Agroecosystems

Ravinder K. Kohli, PhD
Harminder Pal Singh, PhD
Daizy R. Batish, PhD
Editors

Allelopathy in Agroecosystems has been co-published simultaneously as *Journal of Crop Production*, Volume 4, Number 2 (#8) 2001.

CRC Press
Taylor & Francis Group
Boca Raton London New York

CRC Press is an imprint of the
Taylor & Francis Group, an **informa** business

CRC Press
Taylor & Francis Group
6000 Broken Sound Parkway NW, Suite 300
Boca Raton, FL 33487-2742

ISBN 13: 978-1-56022-090-9 (hbk)

Visit the Taylor & Francis Web site at
http://www.taylorandfrancis.com

and the CRC Press Web site at
http://www.crcpress.com

Allelopathy in Agroecosystems

CONTENTS

ABOUT THE EDITORS

Ravinder K. Kohli, PhD, an eco-physiologist of distinction, is Professor in the Department of Botany, Panjab University, Chandigarh, India. His research interests include plantation ecology, weed science and biochemical interactions. Recently, the Government of Chandigarh honored this recipient of the National Young Scientist Award of India, for his contributions in environment research and teaching. He is the leader and deputy leader of the IUFRO working parties 4.02.02 (Multipurpose Inventory) and 8.01.04 (Tropical and Sub-Tropical Forest Ecosystems), respectively. He has published over 100 research papers, apart from co-authoring 2 and co-editing 7 books on environment related themes. He has guided 14 PhD and 24 MPhil theses in experimental ecology and successfully completed 6 R & D projects. He is the General Secretary, Dayanand National Academy of Environmental Sciences and Vice-President, Society of Environmental Scientists. He is credited with organizing and delivering invited and keynote lectures besides chairing sessions in international and national scientific meetings in India and abroad.

Harminder Pal Singh, PhD, a promising young chemical ecologist, is Research Associate of the Council of Scientific and Industrial Research, India, working in the Department of Botany, Panjab University, Chandigarh. He specializes in tree-crop and weed-crop interactions in agroecosystems. He is the first recipient of the Young Scientist Award in Life Sciences of the Punjab Academy of Sciences, India. He has co-authored and co-edited one book each, apart from publishing 24 research papers.

Daizy R. Batish, PhD, a weed ecologist with a strong academic record, is Reader (Associate Professor) in the Department of Botany, Panjab University, Chandigarh, India. A dedicated and popular teacher, she teaches ecology and environmental botany to the university students. She also guides MPhil and doctoral research in weed allelopathy and forest dynamics. She has published 35 research papers and 7 articles in addition to co-editing 4 books.

Foreword

Biochemical interactions between plants (allelopathy) and from plants to other organisms influence growth and development in biotic and biotic communities. These biochemical activities have been recognized by scientists worldwide because they offer alternative uses in agriculture, including decreased reliance on synthetic herbicides, insecticides, and nematocides. Allelopathic scientists need to be aware of the important metabolic changes in plants, i.e., diurnal variation, growth, onset of flowering, fruit development, seeds for the future, as well as those that occur throughout the different periods of the life span. More than thirty different classes of secondary or natural products have been identified as allelopathic compounds. These compounds either individually or combined into an array of compounds such as are found in natural circumstances mediate these growth effects.

The joint occurrence of allelopathy and application of herbicides is common to many situations in modern agriculture. The potential for co-operative action from natural products and synthetic chemicals should be a consideration in crop management strategies.

Allelopathy arises because growth inhibiting or stimulating plant and microbial produced biochemicals are released into the environment. These allelopathic chemicals are one more influence in an array of potential stress conditions such as moisture, temperature, biotic factors, etc., that alter the growth of higher plants and microorganism. Because it is almost impossible to separate these factors in field situations, much of our understanding of allelopathy comes from studies which rely on bioassays.

Allelopathy is a component of most natural communities and agroecosystems, but frequently it is unrecognized. The adverse effects from allelochemicals arising from certain weeds and crops may reduce production in agricultural fields and managed forest systems. There is a need for evaluating the allelochemical effects of prior plants, residues, associated plants, and autotoxicity that may occur in cropping systems. It is possible to manage allelopa-

[Haworth co-indexing entry note]: "Foreword." Waller G. R. Co-published simultaneously in *Journal of Crop Production* (Food Products Press, an imprint of The Haworth Press, Inc.) Vol. 4, No. 2 (#8), 2001, pp. xxi-xxii; and: *Allelopathy in Agroecosystems* (ed: Ravinder K. Kohli, Harminder Pal Singh, and Daizy R. Batish) Food Products Press, an imprint of The Haworth Press, Inc., 2001, pp. xiii-xiv. Single or multiple copies of this article are available for a fee from The Haworth Document Delivery Service [1-800-342-9678, 9:00 a.m. - 5:00 p.m. (EST). E-mail address: getinfo@haworthpressinc.com].

thy as a mechanism of weed control in agriculture and forestry. This is currently being accomplished in some countries/regions through management practices that include residue manipulation, use of cover crops, crop rotations, desirable planting of cover crops, and adjusting herbicide applications to complement allelopathic weed control. Sunflower, grain sorghum, and rye are effective weed control crops. Evidence accumulated shows that allelochemicals and herbicides have an additive negative impact on sensitive species. Management of allelopathy will be applied in efforts which focus on reducing herbicide and pesticide usage to enable sustainable agricultural practices that have less deleterious effects on the environment. Allelopathic scientists anticipate that the tools of biotechnology will increasingly be applied to create agricultural and forestry plants that produce allelochemicals that aid in their own pest and weed control.

In spite of much optimism and some progress on unraveling the complexities of physiological and biochemical interactions between species, a firmer foundation describing the existence and functions of allelopathy is needed, if we are to maximize opportunities that capitalize on allelopathy in agroecosystems and explain its multiple roles in ecology.

G. R. Waller
Professor Emeritus
Department of Biochemistry & Molecular Biology,
Oklahoma State University, OK, USA
Founder President
International Allelopathy Society

Preface

The enigmatic, silent and complex biochemical interactions among plants, including microbes, generally go unnoticed in nature but lead to changes in vegetation structure, composition and species diversity. Such biochemical interactions, termed as Allelopathy, have emerged as a full fledged scientific discipline falling within the framework of chemical ecology. Allelochemicals are produced in plants as defensive compounds and upon release influence the growth and development of biological systems in different ecosystems. Allelopathy is now viewed with multifaceted approach and is recognized as a key factor in vegetation patterning, succession, forest regeneration and influencing growth and yield of other plants.

In agroecosystems, the influence of allelopathy on crop plants was probably realized from the beginning of the civilization when agriculture began. Generally, it is viewed as a causal factor in decreasing yield of crop plants. Accumulation of allelochemicals in the fields causes soil sickness and thereby the growth and yield of crop plants particularly when the same crop is grown continuously in monocultures. The problem of soil sickness is attributed to autotoxicity (autoallelopathy). During the last few years, this science has attracted the attention of scientists world over because of the pesticidal potential of allelochemicals. Allelopathic interactions in agricultural fields also influence soil microbial ecology, nutrient dynamics and various biotic and abiotic factors.

Several research papers, reviews, compendia and books on the theme of allelopathy have been published from time to time, but its comprehensive status especially in relation to agroecosystems, has remained neglected. Further, the progress in this field has been so much that it requires regular update inviting thoughts on current issues of improving cropping pattern, tillage, crop protection and preservation of genetic diversity. This volume being simultaneously published as a special issue of the *Journal of Crop Production* includes 20 chapters presenting a comprehensive treatise on the topic.

[Haworth co-indexing entry note]: "Preface." Kohli, Ravinder K., Harminder Pal Singh, and Daizy R. Batish. Co-published simultaneously in *Journal of Crop Production* (Food Products Press, an imprint of The Haworth Press, Inc.) Vol. 4, No. 2 (#8), 2001, pp. xxiii-xxv; and: *Allelopathy in Agroecosystems* (ed: Ravinder K. Kohli, Harminder Pal Singh, and Daizy R. Batish) Food Products Press, an imprint of The Haworth Press, Inc., 2001, pp. xv-xvii. Single or multiple copies of this article are available for a fee from The Haworth Document Delivery Service [1-800-342-9678, 9:00 a.m. - 5:00 p.m. (EST). E-mail address: getinfo@haworthpressinc.com].

The first chapter highlights the various aspects of allelopathy in agroecosystems presenting a global perspective on the topic. Chapters 2 and 3 provide an in-depth knowledge on the role, ecology and allelopathic potential of crops and weeds in agroecosystems. The management and problems associated with crop residues have been discussed in Chapters 4 and 5. Chapter 6 highlights the significance of crop rotation in diluting the allelopathic effects of the crops and non-crops.

The use of allelochemicals as natural pesticides is currently the hot topic catching the attention of scientists' world over. As many as 6 chapters aim to discuss the prospects of allelochemicals of plant and microbe origin in combating weeds, plant pathogens and other harmful microbes. This includes the pesticidal potential of microbes that are the key factors in synthesizing the variety of chemicals and also capable of bio-transforming the chemicals produced by other plants (Chapter 7). The screening and testing of allelopathic plants for weed and pest management are covered in Chapter 8 and 9. One of the fast emerging trends in the sustainable weed management is the use of plant pathogens. This has been discussed in detail in Chapter 10. It is being increasingly realized that modern day crop cultivars have less allelopathic potential compared to their wild relatives owing the selection pressure for the high yields. Efforts are, therefore, being made in several parts of the world to identify crop cultivars with high allelopathic potential and then incorporate their genes into the cultivars in-use through breeding methods and other molecular genetic techniques. This has been discussed in Chapter 11, taking rice as model crop. Chapter 12 discusses the potential of alfalfa for the management of weeds. The allelopathy of tomato plant that involves the volatile inhibitors has been covered in Chapter 13 whereas that of a weed *Cyperus esculentus* indicating the growth stage with maximum allelopathic potential has been discussed in Chapter 14. Another aspect of allelopathy in agroecosystems is soil sickness due to autotoxicity. The mechanism and means to overcome autotoxicity have been discussed in Chapter 15 with special reference to cucurbit crops. The activity, persistence and role of specific allelochemicals (sorgoleone and benzoxazinones), identified from the crop plants have been discussed in Chapters 17 and 18. The chapters 19 and 20 deal with the case studies with reference to the pastures in Australia and agroecosystems in Spain.

In preparing this publication, we had to seek the help from a number of sources. We are thankful to all contributors without whose hard work, dedication and expertise this work would have never completed. Thanks are due to all reviewers who inspite of their busy schedule and commitments contributed immensely in reviewing and giving their expert comments for improving the manuscripts.

Further, we owe our gratitude to Dr. Amarjit Singh Basra, Editor, *Journal of Crop Production*, and Editor-in-Chief, Food Products Press, The Haworth Press, Inc., USA for giving us this opportunity to invite contributions from the leading research groups and edit this publication.

Ravinder K. Kohli
Harminder Pal Singh
Daizy R. Batish

Allelopathy in Agroecosystems: An Overview

H. P. Singh
Daizy R. Batish
R. K. Kohli

SUMMARY. Allelopathy plays an important role in the agroecosystems leading to a wide array of interactions between crop-crop, crop-weed and tree-crops. Generally, these interactions are harmful to the receiver plants but provide a selective benefit to the donor. Soil microbes play a key role in determining such interactions as they not only alter the nature of allelopathic interactions but also modify the expression of allelochemicals. Soil sickness problem in the croplands could also be attributed to the allelopathic property or even the autotoxicity. The allelochemicals released largely by the plant residues that are left in the fields after the harvest of crops add to the multifarious problems. If properly managed, these residues could be used for controlling weeds and pests. As is true for any chemical based response, allelopathic interactions are also concentration specific. The promotory functions that are inbuilt need to be worked out and exploited. Now a days allelopathic interactions, in general, and the allelochemicals, in particular, are viewed as an important tool for sustainable weed and pest management, and disease control. In this direction, a number of strate-

---

H. P. Singh is Research Associate, Daizy R. Batish is Reader, and R. K. Kohli is Professor, Department of Botany, Panjab University, Chandigarh-160 014, India.

Address correspondence to: R. K. Kohli at the above address (E-mail: rkkohli45@yahoo.com).

H. P. Singh is thankful to Council of Scientific and Industrial Research (CSIR), India for financial assistance.

[Haworth co-indexing entry note]: "Allelopathy in Agroecosystems: An Overview." Singh, H. P., Daizy R. Batish, and R. K. Kohli. Co-published simultaneously in *Journal of Crop Production* (Food Products Press, an imprint of The Haworth Press, Inc.) Vol. 4, No. 2 (#8), 2001, pp. 1-41; and: *Allelopathy in Agroecosystems* (ed: Ravinder K. Kohli, Harminder Pal Singh, and Daizy R. Batish) Food Products Press, an imprint of The Haworth Press, Inc., 2001, pp. 1-41. Single or multiple copies of this article are available for a fee from The Haworth Document Delivery Service [1-800-342-9678, 9:00 a.m. - 5:00 p.m. (EST). E-mail address: getinfo@haworthpressinc. com].

gies like use of cover or smother or companion crops for weed management, direct use of allelochemicals as natural pesticides, and even the transfer of allelopathic traits/principles to modern day cultivars are being adopted. The purified allelochemicals and/or their derivatives and even the compounds synthesized on their chemistry can be used as novel agrochemicals for sustainable management in an eco-friendly manner. The present paper aims to discuss all the above mentioned roles and aspects of allelopathy in the agroecosystems.

KEYWORDS. Allelopathy, sustainable agriculture, role of microbes, soil environment, novel agrochemicals, eco-friendly herbicides, pest management

INTRODUCTION

Plants produce an array of secondary metabolites which apart from playing an important physiological function for the self, result in an orderly interaction between them and consequential impact on the neighboring environment. The chemical interactions are diverse and complex and provide the donor plant a number of selective advantages (Reigosa, Sánchez-Moreiras, and González, 1999). There are several historical references in the literature that point to the inhibitory effects of one plant on the other through the release of chemical substances in the environment (Theophrastus, 300 BC; Pliny II, 1 AD; Culpeper, 1633; Young, 1804; de Candolle, 1832). Since all such references were based on simple observations or were merely statements/anecdotes that lacked scientific proofs and experimentation, these failed to attract the attention of scientific community. Moreover, the lack of coherence regarding the appropriate and meaningful terminology for chemical based interactions added to the confusion, as different workers used different terms for the chemicals responsible for the interactions.

The term allelopathy was coined in 1937 by Hans Molisch choosing two Greek words '*Allelo*' and '*Pathos*' literally meaning *mutual sufferings*. Based on the Molisch's concept, Rice (1984) defined *Allelopathy as any direct or indirect positive or negative effect of one plant on the other (including the microbes) through the release of chemicals into the environment*.

The progress in the field of allelopathy started rather slowly in the beginning of Post-Molisch period until 1960's when Muller and his co-workers did some classic and pioneer work on the Californian Chaparral (Muller, Muller, and Haines, 1964; Muller, 1965, 1966; del Moral and Muller, 1969; Chou and Muller, 1972). Later, Rice revolutionized this field by his excellent research work and exhaustive writings. Now, allelopathy is a well-established interdisciplinary science with multifaceted approach (Inderjit, Dakshini, and Foy, 1999). This is evident from the fact that International Allelopathy Society has already organized two International Allelopathy Congresses–one in Spain and the second in Canada. During the last two decades, the science of allelopathy has attracted a number of scientists from the diverse fields and there has been a spurt in the number of research papers, reports, articles, reviews and books in this field. To mention a few–Grodzinsky (1970-1974, 1977); Rice (1984, 1995); Thompson (1985); Einhellig (1985b); Chou and Waller (1989); Kohli (1990, 1993, 1997); Rizvi and Rizvi (1992); Putnam and Tang (1986); Waller (1987); Narwal and Tauro (1994); Inderjit, Dakshini, and Einhellig (1995); Seigler (1996); Kohli, Batish, and Singh (1998a); Chou, Waller, and Reinhardt (1999); Cutler and Cutler (1999a); Inderjit, Dakshini, and Foy (1999); Macias et al. (1999a); and Narwal (1999a).

HISTORICAL BACKGROUND

In the historical perspective of allelopathy, the first statement related to it came from none other than Theophrastus (300 BC), who observed that chickpea plants exert adverse effect on the other by exhausting the soil. Years later, Pliny II (Plinius Secundus, 1 A.D.) made almost similar observations. Likewise, a few more observations and references were made by some other scientists like Culpeper, Young and de Candolle. However, none of these was based on experimentation (Rice, 1984). In the beginning of the 19th Century, de Candolle–a well-known plant taxonomist generated some interest in this field when he observed that root exudations of some plants are the cause of *Soil Sickness* and that this can be overcome by the suitable crop rotation (de Candolle, 1832). His observations were based on simple experiments. Later, nevertheless, his hypothesis was rejected by a majority of the scientists. After a lag period, in the beginning of the 20th century, some interest in this field was again generated by the studies of Schreiner and his co-workers in USA (Schreiner and Reed, 1907, 1908) and Pickering and his team in UK (Willis, 1997). To summarize, the history of allelopathy could be divided into 3 phases of its development:

i. *de Candolle Phase:* The period of late 18th and early 19th century, especially between 1785 and 1845,
ii. *Pre-Molisch Phase:* The period in the beginning of the 20th century (from 1900-1920) known by the work of Pickering and Schreiner, and

iii. *Post-Molisch Phase:* 1937 onwards which actually could progress since 1960 (Willis, 1997).

However, as a result of better techniques and methodologies, collaborative approach among the scientists and availability of suitable bioassays, this science picked up the momentum during the last two decades or so.

ALLELOPATHY–PERSPECTIVES AND PROBLEMS

The field of allelopathy offers an exciting, inter-disciplinary, complex and challenging studies and, therefore, invites the attention of the botanists, phytochemists, microbiologist, soil scientists, agronomists, ecologists and foresters from world over. The current focus of allelopathic research is to understand various ecological problems in agriculture, forestry/plantations, wastelands, aquatic and other ecosystems. The increase in the literature and experimental improvements in the subject, apart from the better computational facilities has provided this science a firm footing. In spite of all this, criticism from a section of scientists like Harper (1977); Connell (1990); Begon, Harper, and Townsend (1995); and Watkinson (1998) on the science can not be ignored as such. The critics base their arguments on the following:

- Lack of adequate field studies; as most of the studies reported are based on laboratory work
- Difficulties in determining critical concentrations of allelochemicals in soil
- Inappropriate bioassay models
- Difficulty in establishing the role of microbes in litter decomposition, release of allelochemicals and subsequent transformation
- Problems in determining the mechanism of action, particularly when the chemical nature of allelochemicals is unknown.

In view of the above apprehensions and difficulties, the proofs required for establishing an allelopathic phenomenon, than any other aspect of ecology, is usually put to rigorous evaluation (Williamson, 1990). Perhaps for such reasons, the traditional ecologists either prefer to omit the phenomenon or just make a brief mention of it in their writings/books. In spite of the lack of legitimate support, the probing of the science of allelopathy on one hand and experimental handicaps that it suffers from, on the other, it has now become an integral component of chemical ecology (Anaya, 1999). In terms of practical significance, the allelochemicals are now considered as an alternative to synthetic herbicides and pesticides for controlling weeds and pests in the agroecosystems for the protection of the quality of environment (Macias, 1995; Macias et al., 1997; Anaya, 1999; Rizvi et al., 1999). Allelochemicals are also used to manage the biotic resources (Anaya et al., 1995). Further, in several instances the synthetic chemicals fail to provide desired results owing to development of resistance among the pests. Under such circumstances, allelochemicals, being the natural plant products, provide an excellent source of novel agrochemicals for the sustainable management of pests and weeds in the agroecosystems (Cutler, 1988; Macias, 1995; Duke et al., 1997; Kohli, Batish, and Singh, 1998a,b; Cutler and Cutler, 1999a,b; Rizvi et al., 1999; Duke et al., 2000).

ALLELOCHEMICALS-DIVERSITY, NATURE, MODE OF RELEASE AND SIGNIFICANCE

The chemicals responsible for the phenomenon of allelopathy are generally referred to as *Allelochemicals* or *Allelochemics* (Whittaker and Feeny, 1971). Since the inception of the term allelopathy, these chemicals have been given different names by the different workers. To name a few, there are terms like *Antibiotic*, *Koline*, *Marasmin* and *Phytoncide* based on the type of donor and the recipient (Grümmer, 1955); *Saproinhibitins and Phytoinhibitins* based on the type of donor (Fuerst and Putnam, 1983). These being non-descriptive and non-specific nor have received much acceptability (Putnam and Weston, 1986). Of late, these chemicals have also been named as Plant Eco-Chemicals (Mizutani, 1991, 1999). These phytochemicals are synthesized in plants as secondary metabolites that appear to have no direct function in the growth and development but serve as a defensive adaptation. The significance of their synthesis seems to be an interaction between the plant and its environment (Harborne, 1989; Berenbaum, 1995a,b; Seigler, 1996). Plant allelochemicals/phytotoxins are generally localized and sequestered in certain specialized organs which may be glandular or sub-epidermal (Duke et al., 1999). The distribution of particular secondary metabolites, is restricted within a group of taxonomically related species. For instance, 'salicin,' a phenol glucoside is characteristic of the members of the family Salicaceae while sesquiterpene lactones are found in the members of family Asteraceae. Contrary to the earlier belief that secondary plant products are the metabolic wastes and thus, functionless (Ellis, 1997), their ecological significance is now being realized. They not only dispel harmful insects, pests and pathogens, but also increase reproductive fitness of the plants (Taiz and Zeiger, 1998). Perhaps, their presence represents an evolutionary change through heritable mutations and natural selection. Depending upon the nature of metabolite, there is, however, insufficient data that selection pressure has favored plants with allelopathic phenomenon (Newman, 1978). Allelochemicals are also now being considered even for their role in biological life support system for long duration space missions (Stutte, 1999).

The allelochemicals broadly classified as plant phenolics and terpenoids, show a great chemical diversity and are involved in a number of metabolic and ecological processes. The mode of release from the plant could be through leachation, volatilization, root exudation and the death and decay of the fallen plant parts either through biotic or abiotic means (Anaya, 1999). Upon release, these are involved in a number of metabolic and physico-chemical processes (Rice, 1984; Einhellig, 1985a; Waller, Feng, and Fujii, 1999; Mizutani, 1999). The toxicity of released allelochemicals in the environment is a function of concentration, flux rates, age and metabolic stage of the plant, prevailing climate and season, and environmental conditions (Rice, 1984; Daizy, 1990; Wyman-Simpson et al., 1991; Wardle, Nicholson, and Rahman, 1993; Weidenhamer, 1996; Gallet and Pellissier, 1997; Nilsson, Gallet, and Wallstedt, 1998). Besides, their production not only varies in quantity and quality throughout the year (Devi, Pellissier, and Prasad, 1997) but is also dependent upon the age and the cultivar of the donor plant (Argandona, Niemeyer, and Corcuera, 1981; Wyman-Simpson et al., 1991; Hanson et al., 1981; Burgos, Talbert, and Mattice, 1999). A density dependent effect of allelochemicals has also been demonstrated wherein the effect has been shown to be more at low plant densities compared to the high densities (Thijs, Shann, and Weidenhamer, 1994). This provides a strong evidence for the chemical interference (Weidenhamer, 1996). Under natural conditions, the released allelochemicals act synergistically and exert inhibitory effect as complex mixtures and therefore the inhibitory effects are observed at concentrations well below their individual inhibitory levels (Blum, 1995, 1996).

ALLELOPATHY AND SOIL ENVIRONMENT

Allelopathic interactions in soil environment may be a function of interacting substances that may be neutral, promoters and/or even inhibitors (Blum et al., 1993). The importance of the allelopathic interactions in natural and managed systems greatly depend upon turn over rate of allelochemicals and their interaction with clay, organic matter and other competitive sinks which change physico-chemical and biotic characteristics of soil (Blum, 1995; Blum, Shafer, and Lehman, 1999). A number of reports suggest that various soil factors affect allelopathic compounds and hence the phenomenon (Rice, 1984; Dalton, Blum, and Weed, 1989; Cheng, 1995; Blum, 1996). Since the soil particles adsorb the allelochemicals, their degree of sorption is greatly influenced by the soil texture. Generally, the loose and sandy soils exhibit more inhibitory effect than the heavy and loamy soil, as allelochemicals fail to be sorbed on the former (Oleszek and Jurzysta, 1987). The soil factors which affect the allelopathic expression of a plant are soil pH, organic carbon, organic matter and available nitrogen (Blum, 1996). The sorption of the allelochemicals is greatly influenced by slightly basic pH and it increases with organic matter and multivalent cation content (Dalton, Blum and Weed, 1983). These characteristics, in turn, influence the uptake and immobilization of inorganic ions and thus accumulation of nutrients (Facelli and Pickett, 1991; Blum, 1996). Low organic matter in soil induces those microbes which are able to utilize phenolics as carbon source (Blum and Shafer, 1988). The allelopathic activities are also influenced by the soil nutrients (Koeppe, Southwick, and Bittell, 1976; Rice, 1984). Other abiotic factors which influence the allelopathic phenomenon or the accumulation of allelochemicals are temperature (Fischer et al., 1994), irradiance (Bhowmik and Doll, 1983), moisture deficit (Einhellig, 1989), etc. Allelochemicals, particularly the phenolic acids, interfere with the mineral uptake (Einhellig, 1995) and under stress conditions the amount of allelochemicals is substantially increased (Hall, Blum, and Fites, 1982) and the allelopathic effects are observed at less than optimal concentrations of allelochemicals (Einhellig, 1996). Allelochemicals also influence the availability of the nutrients to plants. For instance, some phenolic acids are known to bind with the minerals like iron, manganese and aluminum, and thus increase the availability of phosphate which otherwise forms complex with these metal ions (Appel, 1993).

One of the important aspects of the allelochemicals in the soil environment can be the demonstration of their active concentration. Most of the bioassays are based on single toxin effect and do not provide adequate information on the concentration required for plant growth inhibition, since in soil toxins occur in complex organic mixtures. Much depends upon their differential utilization by the soil microbes and associated rate of this uptake, metabolic conversion and creation of anaerobic conditions in the medium (Pue et al., 1995). Magnitude of allelopathic effect is also governed by distribution of allelochemicals and plant roots in the rhizosphere (Lehman, Blum, and Gerig, 1994).

ALLELOPATHIC INTERACTIONS IN AGROECOSYSTEMS

An agroecosystem, a man-made distinct unit in agricultural fields, is regulated by ecological principles where all the biotic and abiotic factors play an important role. Modern agriculture is an outcome of long process of domestication, improvement of new techniques with the prime objective of increasing crop production and requires a high degree of inputs and modern technology (Anaya, 1999). This has led to spectacular increase in food production and self-sufficiency in several countries. In spite of all this progress, scientists world over are worried that this increase in food production has not been able to keep the desired pace with the unabated growth in population especially in the tropics and subtropics. Moreover, agriculture is a fragile resource and is shrinking very fast. Increasing production on the existing land, therefore, seems the only alternative. In order to accomplish this target, sectoral monocultural practice is being undertaken where any other biological component (even if it is the normal seral stage of the succession) is considered non-crop and hence removed to minimize competition for the crop plants. To eliminate the non-crop biological components such as weeds, insects and pathogens, the synthetic chemicals are being used indiscriminately to the extent of even making the agroecosystems unsustainable. Their excessive use has not only affected the human health and impaired the quality of environment but also, in some circumstances failed to provide expected protection due to development of resistance against synthetic chemicals. For enhancing crop productivity, new high yielding varieties are also being selected/evolved and so are the pests, insects and pathogens. This has become an unending process. The modern agroecosystems with monocultures of high yielding crop varieties are characterized by the presence of residues of synthetic agrochemicals (herbicides/fungicides/pesticides/fertilizers), less diversity and resistant pests which makes it ecologically unsustainable. Sustainable agroecosystems on the other hand are organic, regenerative, biodynamic and resource conserving (Anaya, 1999). Hardly any crop rotation practice, cover cropping, companion cropping or polyculture cropping practice, as was used to be done in our traditional systems, is followed in the present agricultural set up. Fast deteriorating agroecosystems have become a serious global problem and to improve them a holistic approach, though slow, needs to be followed (Allen and van Dusen, 1988).

In the modern agriculture, allelopathy has a great potential in improving crop productivity, genetic diversity, maintenance of ecosystem stability, nutrient cycling and nutrient conservation, weed control, disease management and pest control (Jordan, 1993; Einhellig, 1996; Halbrendt, 1996; Swanton and Murphy, 1996; Weston, 1996; Kohli, Batish, and Singh, 1998a; Anaya, 1999; Chou, 1999; Caldiz and Fernandez, 1999; Cutler, 1999a,b). Before discussing the application part, it is important to understand the allelopathic interactions among major interacting components of agroecosystems.

Allelopathic Weeds

Out of 30,000 plant species identified as weeds, 250 are really important and about 80 are known to reduce crop yields (Sauerborn, 1999). They are the part of dynamic agroecosystems created by man with the aim of producing useful plants. Weeds, which have been described as unwanted plants at a given place, are integral component of agroecosystems and have co-evolved with the crop plants. They have a number of physiological, agronomic and reproductive characteristics which make them successful compared to the other plants (Cobb, 1992). In the agroecosystems, they compete with the crop plants for resources, interfere in crop handling, reduce crop yield and deteriorate their quality and thus the financial loss. They are rather more harmful than the insects and pathogens, as they are static, co-evolved with the crop plants and often develop resistance to the herbicides. Nearly 12% of the total loss of crop yields has been attributed to the weeds alone (Anaya, 1999). A number of allelopathic weeds have been reported to affect crop plants in agroecosystem (Putnam and Weston, 1986; Kohli, Batish, and Singh, 1998a). Some of the important weeds exhibiting allelopathy and crop losses are *Agropyron repens* (Lynch and Penn, 1980), *Ageratum conyzoides* (Kohli, 1997), *Amaranthus retroflexus* (Bhowmik and Doll, 1983), *Avena fatua* (Schumacher, Thill, and Lee, 1983), *Brassica napus* (Mason-Sedun, Jessop, and Lovett, 1986; Choesin and Boerner, 1991), *Chenopodium album* (Qasem and Hill, 1989), *Chenopodium murale* (Qasem, 1993), *Cyperus rotundus* (William and Warren, 1975), *Datura stramonium* (Levitt, Lovett, and Garlick, 1984), *Echinochloa crus-galli* (Bhowmik and Doll, 1979), *Lantana camara* (Singh, Tamma, and Nigg, 1989; Arora and Kohli, 1993), *Parthenium hysterophorus* (Kohli and Batish, 1994), and *Sicyos deppei* (Cruz-Ortega et al., 1998; Anaya, 1999).

Though weeds interfere with the crop plants and cause huge economic losses, yet they are an integral component of the agricultural system (Altieri and Liebman, 1988; Zimdahl, 1999). The herbicidal control of weeds is not only a costly affair but also deteriorates the quality of soil, water, other life support systems, human health and food. Therefore, their management in an eco-friendly and low-input manner requires alternative management practices which involve both short and long-term strategies and emphasize greatly on prevention of weed emergence so as to minimize the competition with crops (Buhler, 1996, 1999). Some of the proposed suggestions for weed management in agroecosystems are:

- Understanding the biology, ecology and life cycle pattern of weeds in order to devise the suitable means of their control.
- Understanding the exact mode of their interference in the croplands, i.e., whether density dependent or density independent, competitive or allelopathic, or both.
- Changing cropping patterns like increased rotation, new tillage and improved resource utilization (Wyse, 1992; Buhler, 1999).
- Finding alternate control methods like using competitive cultivars or allelopathic crop plants or allelochemical/natural plant products for weed management (Buhler, 1999).

Allelopathic Crops

A number of crops have also been known to exhibit allelopathic property on other crops growing in succession or simultaneously or may even exhibit autotoxicity (Einhellig, 1985a; Putnam and Weston, 1986; Chou, 1999; Anaya, 1999). Among these, especially the cover crops may be exploited for the purpose of weed management (Teasdale, 1996; Weston, 1996; Foley, 1999). Crop autotoxicity provides an interesting dimension to allelopathy in agroecosystems where a crop is inhibitory to the self, particularly in the successive years (Einhellig, 1985a; Singh, Batish, and Kohli, 1999a; Reigosa, Sánchez-Moreiras, and González, 1999). For the common man and farmers, the commonly encountered problem of *Soil Sickness* or *Soil Fatigue* is attributed to crop autotoxicity as it is not even rectified by the addition of fertilizers (Yu, 1999a). The principal causes of crop autotoxicity include the deliberate leaving of crop residues or old roots in soil that release phytotoxins which may directly affect the succeeding crops, cause microbial imbalance, changes in organic matter of soil, increase ion leakage, disturb nutrient uptake and immobilization (Waksman, 1937; Katznelson, 1972; Kimber, 1973; Yu and Matsui, 1997). Some of the highly worked out important crops exhibiting crop autotoxicity include rice (Chou, 1995), wheat (Kimber, 1973), maize (Yakle and Cruse, 1983, 1984), sugarcane (Chou, 1995), alfalfa (Miller, 1983, 1996), sunflower (Kohli, 1993) and a few others (Rice, 1995; Singh, Batish, and Kohli, 1999a). Several vegetable crops like cucumber, carrot, fennel, watermelon, eggplant, tomato and even pea are known to exhibit autotoxicity (Singh, Batish, and Kohli, 1999a; Yu, 1999a). The problem of crop autotoxicity is particularly acute in croplands where tillage is not practiced.

There are a number of instances where crops are known to affect even the other crops growing in vicinity leading to serious implications to agricultural practices like stubble mulch farming or reduced or no-tilled farming where crop residues are left on the soil surface to avoid soil erosion, preserve moisture, increase nutrient balance and a number of other benefits (Einhellig, 1985a; Rice, 1984, 1995; Chou, 1995, 1999). The residues of preceding crops affect the performance of other crops through the release of allelochemicals (Kimber, 1973; Guenzi and McCalla, 1966; Lodhi, Bilal, and Malik, 1987; Thorne et al., 1990; Kohli, 1993). Kalburtji and Gagianas (1997) reported the poor performance of cotton following the sugar beet crop due to allelochemicals released from the residues of sugar beet. However, a number of water-soluble phytotoxins/allelochemicals released by the residues accumulate in the soil and the crop roots may have a chance encounter with these chemicals leading to serious repercussions on the quality and quantity of crop yields (Einhellig, 1985a). Due to this reason a number of traditional cropping practices like cover cropping, companion cropping, polyculture and green manuring, etc., need to be thoroughly revised. Some studies have shown that crop residues, if properly managed, may even lead to enhancement in yield (Chou, 1999). Different approaches that need to be looked into are:

- Understanding the mechanism of crop-weed interference in agroecosystems.
- Revival of traditional practices that are organic, biodynamic and sustainable (Anaya, 1999).
- Shift in cropping pattern from mono- to poly-cultures and effectively reviving the old practices of cover cropping, companion cropping and crop rotation with new and scientific approach.
- Improving crops through genetic manipulations using conventional breeding techniques and molecular genetic recombinant technology so that they have ability to protect themselves from the pests (Foley, 1999).

Allelopathic Agroforestry Trees

Trees are fast becoming an integral part of the agriculture under various intensive and extensive agroforestry programmes. Several studies show that this practice increases productivity, improves soil quality, microclimate and nutrient cycling, conserves soil, manages weeds and increases overall sustainability, even though a number of negative interactions (including allelopathic) have also been recognized (Ong, 1991; Burch and Parker, 1992; Harris and Natrajan, 1987; Rao, Nair, and Ong, 1997). A number of tree species like *Acacia* spp., *Albizzia lebbeck*, *Eucalyptus* spp., *Grewia optiva*, *Glircidia sepium*, *Leucaena leucocephala*, *Moringa oleifera*, *Populus deltoides*, are known to affect the performance of crops through the phenomenon of allelopathy (Rice, 1984; Kohli, 1990; Malik and Sharma, 1990; Kohli, Singh, and Verma, 1990; Ralhan, Singh, and Dhanda, 1992; Singh and Kohli, 1992; Rice, 1995; Singh, Batish, and Kohli, 1999b; Singh, Kohli, and Batish, 1998, 1999; Rizvi et al., 1999). In most of the instances, the litter from the tree interferes with the growth and establishment of the adjoining crop plants. Unfortunately, very little has been done to understand this phenomenon in agroforestry systems, though it is an important factor in determining the success of trees (Kuo, Chou, and Hu, 1983; Kohli, Batish, and Singh, 1998a; Rizvi et al., 1999). Therefore, understanding allelopathic mechanism operative in agroforestry systems can enhance the crop productivity through control of weeds, nematodes, pathogens and insects. Allelochemicals from agroforestry trees can be used for controlling the weeds (Kohli, Batish, and Singh, 1998b). Agroforestry can, therefore, be manipulated to make agroecosystems sustainable through proper management and/or mulching of the litter of the trees growing in agroecosystems to improve the soil quality, conserve moisture and bring about the cooling effect.

Role of Microbes

Microbes have a profound effect on the allelopathic activity by altering and/or transforming the amount of allelochemicals, particularly the phenolic acids, in the soil depending upon the available carbon source and other environmental factors (Pellissier, 1998; Blum, Shafer, and Lehman, 1999; Pellissier and Souto, 1999). The microbes may metabolize the released phenolic acids by addition or deletion of side groups, polymerization, production of other organic molecules and/or incorporation of carbon from other phenolic acids into microbial biomass (Martin and Haider, 1976; Haider, Martin, and Rietz, 1977; Blum, 1998; Blum, Shafer, and Lehman, 1999). The transformed or newly synthesized phenolics may differ in their phytotoxicity from the original ones that entered the soil (Blum, 1998). Further, in the soil the preferential utilization of carbon sources may also affect the plant-microbe-soil system and the allelopathic phenomenon (Pue et al., 1995). Some of the bacteria like *Streptomyces sagononensis*, *S. hygroscopicus*, *Pseudomonas fluorescencs* and many others are allelopathic and inhibit the growth of the nearby plant (Suslow and Schroth, 1982; Hoagland, 1990; Alstrom, 1991; Heisey and Putnam, 1986; Barazani and Friedman, 1999). The allelochemicals from microorganisms are generally non-specific and inhibit the growth of several annual and perennial species (Cutler, 1988; Hoagland, 1990). They may be effective at very low concentration, i.e., cycloheximide at 1 μg/l (Heisey et al., 1988) and have variable effect on the different cultivars (Alstrom, 1991).

On the other hand, allelochemicals may influence the growth of microbes positively or negatively and, thereby indirectly interfere with the availability of nutrients, particularly nitrogen and phosphorus, in the soil (Pellissier, 1993; Wardle and Nilsson, 1997; Anaya, 1999). Crop residues of several crops like corn, barley, and potato in the continuous single crop cultivation support several non-pathogenic microbial population which may considerably harm the growth of plants (Bakker and Schippers, 1987; Schippers, Bakker, and Bakker, 1987; Fredrickson, Elliot, Engibous, 1987; Turco et al., 1990; Alstrom, 1992). Phenolic compounds released in soil from decomposing residues may cause microbial imbalance (Chou, 1995). However, flooding may eliminate phytotoxins leading to improved microbial balance and restoring the yield (Wang, Kao, and Li, 1984).

In orchards microbes play an important role in replant problem besides the autotoxicity (Singh, Kohli, and Batish, 1999a). Presence of *Penicillium expansum* in apple orchards facilitates the release of allelochemicals (Börner, 1963; Berestetsky, 1972). Vesicular Arbuscular Mycorhizae (VAM) also helps in changing rhizosphere microflora and help in increasing biomass of apple seedling (Catska, 1994). In case of peach orchards, nematodes are known to play an important role in releasing and hydrolyzing amygdalin–a cynogenic compound causing autotoxicity and replant problem. In case of *Asparagus* it has been shown that allelochemicals synergise with fungal pathogens thereby markedly increasing the disease incidence (Peirce and Colby, 1987).

ROLE OF ALLELOPATHY IN SUSTAINABLE AGRICULTURE

The modern agriculture is commercial and target oriented and hence governed by the market laws and not the sustainable ecological principles. Thus, in order to maintain the sustainability of the agroecosystems it is essential to determine the factor(s) disturbing it and adopt appropriate conservation strategies. Integrated biological weed control in croplands follow two approaches–the technological and ecological (Müller-Schärer, Scheepens, and Greaves, 2000). There is, therefore, an urgent need to explore the eco-friendly means of alternative weed management for profitability and environmental sustainability. There is also need to revive our traditional system practices of polyculture, crop rotation, cover crops and use of green manure. For the management of the weeds, allelopathy particularly that of crops, can play an important role, if suitably understood, managed and applied (Weston, 1996; Anaya, 1999; Chou, 1999; Wu et al., 1999).

Use of Allelopathic Cover/Smother/Rotational/Companion Crops

Use of cover crops and smother crops has been an age-old practice. Cover crops are usually grown during the fallow period to reduce soil erosion, conserve moisture, improve nutrient status and manage weeds besides providing biomass (Gallandt, Liebman, and Huggins, 1999; Foley, 1999). Likewise, smother crops are grown in rotation or as catch crop and shade out the weeds due to their quick growth and thick stands. Some of the noteworthy cover, smother or companion or rotational crops are buckwheat (*Fagopyrum esculentum*), foxtail millet (*Setaria italica*), rye (*Secale cereale*) and *Sorghum* spp., alfalfa (*Medicago sativa*), sunflower and cruciferous plants (Putnam, DeFrank, and Barnes, 1983; Leather, 1983, 1987; Oleszek, 1987; Schreiber, 1992; Kohli, 1993; Foley, 1999; Narwal, 1999b; Vaughn, 1999; Weston, Nimbal, and Jeandet, 1999). It has been seen that allelopathic cover and smother crops can considerably reduce the weed species in the croplands if suitably manipulated and may provide a non-herbicidal method of weed control (Robbins, Crafts, and Raynor, 1942; Altieri and Doll, 1978; Barnes and Putnam, 1983, Liebel et al., 1992; Swanton and Murphy, 1996; Weston, 1996; Anaya, 1999; Chou, 1999). *Mucuna pruriens* var. *utilis*, commonly known as velvetbean and used as green manure to smother weeds, has a potential to suppress the weeds through the release of chemical substance L-3,4-dihydroxy phenylalanine (L–DOPA) present in this legume (Fujii, Shibuya, and Yasuda, 1991; Fujii, 1999). They not only help in enhancing the biomass but also possess the weed suppressing abilities. Genetic approaches should, therefore, be explored for crop improvement and development of molecular markers to aid in breeding cover crops so as to bring about weed suppression (Foley, 1999). Putnam, DeFrank, and Barnes (1983) reported the use of mulches and cover crops in order to achieve weed management naturally. The residues of cover crops like rye, wheat, barley killed by contact herbicides and desiccants reduced the density and biomass of broad-leaved weeds (Barnes and Putnam, 1983). Later, from the rye and wheat mulch phytotoxins like β-phenyl lactic acid and β-hydroxybutyric acid were identified (Shilling, Liebl, and Worsham, 1985). Rye residues release 2,4-dihydroxy-1,4 (2H)-benzoxasin-3-one (DIBOA) which along with its break down product 2(3H)-benzoxazolinone (BOA) suppresses the growth of weeds (Barnes and Putnam, 1987). Wheat straw has been found to be an excellent mulch crop due to the presence of triterpenoids and other phenolic acids in no-till farming (Gaspar, Neves, and Pereira, 1999).

Screening of Crops and Transfer of Allelopathic Traits Using Modern Techniques

Several wild accessions of modern day crops plants are found to possess allelopathic traits that impart them resistance against weeds and pests (Hoult and Lovett, 1993). During the process of cultivation and selection of high yielding varieties these traits were never preferred and hence gradually got eliminated from the modern day plants. Since accessions are of different origin and stage of improvement, this combined with other factors show that allelopathic potential is a polygenic character and weakly correlated with the yield and other characters (Olofsdotter, Navarez, and Moody, 1995). However, they can be incorporated in the present day cultivars through several modern techniques as a plant protection strategy for major food crops. Even the different cultivars of the present day crop plants differ in their allelopathic property and superior genotypes with allelopathic potential can be selected for Integrated Weed Management (IWM) programmes (Wu et al., 1999). Some of the crops which have been screened for allelopathic traits in the wild accessions are *Avena* sp. (Fay and Duke, 1977), *Cucumis sativa* (Putnam and Duke, 1978), *Glycine max* (Massantini, Coporali, and Zellini, 1977), *Brassica* spp. (Sarmah, Narwal, and Yadav, 1992; Narwal, 1999b) and *Oryza sativa* (Olofsdotter, Navarez and Moody, 1995; Dilday et al., 1998; Olofsdotter et al., 1999). Dilday et al. (1998) used 16,000 rice accessions collected from different countries out of which many exhibited allelopathic suppression of ducksalad (*Heteranthera limosa*) and redstem (*Ammenia coccinea*). Even an hybrid between allelopathic and non-allelopathic accessions through breeding techniques revealed fewer weed species in the rice fields and the workers found that allelopathic activity in rice is inherited quantitatively (Dilday et al., 1998). There is thus a great potential for breeding of allelopathic high yielding varieties (Olofsdotter, 1998; Olofsdotter et al., 1999).

These allelopathic crops have also been studied at the genetic and molecular levels. The genes responsible for synthesis of allelopathic chemical in these plants can be transferred through DNA Recombinant Technology or/and even through conventional breeding methods. The selected phenotypes can be analyzed by genetic and molecular genetic techniques like progeny analysis, Polymer Chain Reaction (PCR), Random Amplified Polymorphic DNA (RADP), Restriction Fragment Length Polymorphism (RFLP), and Near Isogenic Lines (NIL) (Williams et al., 1990; McCloskey and Holt, 1990), cloning genes (Lovett, 1994; Jansen, 1996) and also by preparing genetic maps of the higher plants. In case of maize, 5 genes have been identified for the biosynthesis of DIBOA (Frey et al., 1997). The genes responsible for biosynthesis of protective chemicals thus are important tools in agriculture. In an attempt to improve the crop yield, these got eliminated from the crop plants during the centuries long domestication process, as the phytochemicals for which they encoded were not palatable.

Allelochemicals as Novel Agrochemicals

The overuse of the synthetic chemicals for the control of pests is posing a serious threat to the environment (Chou, 1995; Cutler and Cutler, 1999b; Kohli et al., 1999). It is strongly realized that the natural plant products, being biodegradable are eco-friendly and can be relied upon more to enhance the crop productivity in a sustainable way. Allelochemicals–the natural plant products from the higher plants and microbes can, therefore, prove to be ideal agrochemicals (Cutler, 1988, 1999a; Putnam, 1988; Rice, 1995; Dayan et al., 1999; Duke, Dayan, and Rimando, 1998; Duke et al., 1996a,b, 1997, 2000; Hoagland, 1990, 1999; Kohli et al., 1999; Macias et al., 1997, 1999b). Three categories of the agrochemicals and pharmaceuticals such as benzodiazepines, phenoxy compounds and organo-phosphates with potential use as fungicides, insecticides or growth regulators have been identified (Cutler and Cutler, 1999b). They are normally broad-spectrum molecules that are bio-efficaceous, economical and environmentally safe. Besides, they have greater shelf life, require less space, have wide range of storage condition and ease of application. Even the changes in the structure of the natural plant product can improve the bio-efficacy for the desired properties. In case of parthenin–a sesquiterpene lactone and a known allelochemical of *Parthenium hysterophorus* the changes in the structure made by various photo-chemical reactions considerably altered the phytotoxicity of the chemical (Saxena et al., 1991) and the analogues were found to have better growth regulatory effect compared to parthenin itself (Batish et al., 1997).

Some of the allelochemicals from higher plants with known weed suppressing ability are given below in the Table 1.

TABLE 1. Allelochemicals from higher plants with weed suppressing ability.

Allelochemical	Plant Source	Chemical Nature	Reference
Ailanthone	*Ailanthus altissisima*	Quassinoid	Heisey, 1996
Artemisinin	*Artemisia annua*	Sesquiterpene lactone	Duke et al., 1987
Azadirachtin	*Azadirachta indica*	Sesquiterpene lactone	Koul, Isman, and Ketkar, 1990
Allyl- and Benzyl-Isothiocyanate (AITC, BITC)	*Brassica* spp.	-	Wolf, Spencer, and Kwolek, 1984; Boydston and Hang, 1995; Krishnan, Holshouser, and Nissen, 1998.
Caffeine	*Coffea arabica*	Alkaloid	Rizvi, Mukerji, and Mathur, 1980
1,4- and 1,8-Cineole	*Eucalyptus* spp., *Salvia* spp.,	Monoterpenes	Kohli, Batish, and Singh, 1998b; Romagni, Allen, and Dayan, 2000
Cnicin	*Centaurea maculosa*	Sesquiterpene lactone	Kelsey and Locken, 1987
DIBOA and BOA	*Secale cereale*	Benzoxazinones	Barnes and Putnam, 1987
Juglone	*Juglans nigra*	Quinone	Rietveld, 1983
Leptospermone	*Callistemon* spp.	Triketone	Lee et al., 1997
L-DOPA	*Mucuna pruriens*	-	Fujii, 1999
Mimosine	*Leucaena leucocephala*	Non-protein amino acid	Rizvi, Sinha, and Rizvi, 1990
Parthenin	*Parthenium hysterophorus*	Sesquiterpene lactone	Pandey, 1996; Batish et al., 1997
Saponins	*Medicago sativa*	Saponins	Waller, Jurzysta, and Thorne, 1993
Sorgoleone	*Sorghum bicolor*	Quinoline	Einhellig and Souza, 1992
α-Terthienyl (α-T)	Members of Asteraceae	-	Lambert et al., 1991

Some of the natural compounds like cineole, benzoxazinones, quinolinic acid and leptospermones from higher plants have been marketed under different trade names by companies based in Germany, USA and Japan, e.g., Benzoxazinones and Quinolinic acid by BASF, Germany; Cineole as Cinmethylene by Shell (USA); Leptospermones as Triketones by Zeneca (Dayan et al., 1999; Kohli, Batish, and Singh, 1998a). Though cinmethylene has attracted the attention of scientists worldwide, yet there has been a little success with this herbicide (Grayson et al., 1987).

Microbes (fungi, bacteria and actinomycetes) too provide an excellent source of natural herbicide products. Some of the potent phytotoxins from microbes with potential herbicidal properties are Anisomycin from *Streptomyces* spp., AAL-toxin from *Alternaria alternata*, Bialaphos from *Streptomyces viridochromogenes* and *S. hygroscopicus*, Coformycin from *Actinomycete*, Cornexistin from *Paecilomyces variotii*, Cyperine from *Ascochyta cypericola* and *Phoma sorghina*, (+)-epiepoformin from *Scropulariopsis brumptii*, (−)-dehydropyrenophorin from *Drechslera avenae*, gougerotin from *Streptomyces* spp., Herbicidins A, B, E, F, G from *S. saganonensis*, Herbimycins from *S. hygroscopicus*, Hydantocidin from *Streptomyces hygroscopicus*, Phosphinothricin from *S. hygroscopicus*, Pironetin from *Streptomyces* spp., Rhizobitoxine from *Bradyrhizobium japonicum* and *Pseudomonas andropogonis*, Tentoxin from *Alternaria alternata* f. sp. *tenuis* and other *Alternaria* spp., Ustloxins A, B, C, D from *Ustilaginoidea virens*, and vulgamycin from *Streptomyces hygroscopicus* (Owens et al., 1972; Sugawara and Strobel, 1986; Huang et al., 1989; Mitchell, 1989; Babczinski et al., 1991; Duke et al., 1996a,b; Duke, Dayan, and Rimando, 1998; Cutler, 1999a; Hoagland, 1999). Although hundreds of such compounds have been patented, yet only two, *viz.*, bialaphos and phosphinothricin have been successfully commercialized (Duke et al., 1996b; Dayan et al., 1999). Both have antibiotic and herbicidal properties and even the transgenic plants resistant to these chemicals have been prepared (Hoagland, 1999). A metabolite, 6-pentyl-2H-pyrane-2-one isolated from *Trichoderma* species, is reported to have antifungal activity and based on this a natural and soft fungicide has been developed (Parker, Hill, and Cutler, 1999). Recently, a compound, trans-4-aminoproline identified from a fungus *Ascochyta caulina*, has been tested to be a promising mycoherbicide (Evidente et al., 2000).

Though the commercialized products of natural products represent a small fraction, yet they offer novel herbicide target sites (Duke et al., 1997, 2000). The use of such products can restore the beneficial balance of natural environment that is often lost in the agroecosystems (Hill et al., 1999). Some of the marketing strategies for the natural plant products need careful revision (Devine, Duke, and Fedtke, 1993). Further, attention to the following may also prove to be useful for their future exploitation in agroecosystems.

- Carefully monitored patent search.
- Protocols on synthesizing compounds for large-scale production should be developed.
- Use of modern genetic and molecular techniques such as cell culture, genetic engineering and biotechnology should be considered.
- Structural modifications of the compound that can improve bio-efficacy and other properties should be considered.
- Plant derived chemicals could be used as template for commercial pesticides.
- Compounds must be thoroughly tested for toxicological and environmental properties.

Although natural products are normally considered ideal compounds, yet they may have some limitations too (Duke, Dayan, and Rimando, 1998; Dayan et al., 1999). These are:

- Their structures are normally too complex and therefore, the process of simplification may lead to loss of activity.
- In some cases, they too have short environment life which fails to provide the desired effects, because for a compound to act some persistence is required.
- The purification process of natural plant products is slow and cumbersome.
- The molecular sites of action of these compounds may not be known.
- The allelochemicals may not be always safe. Sometimes they may cause mammalian toxicity.

Allelopathy for Pathogen and Disease Management

World wide plant diseases are estimated to cause about 20% yield reductions of major food and cash crops (Oerke et al., 1994). The problems of pollution and hazardous effects of synthetic chemicals on the non-target plants have led scientists to find alternate methods such as use of natural plant products and their analogues (Cutler, 1988; Al-Abed, Qasem and Abu-Blan, 1993; Qasem and Abu-Blan, 1995, 1996; Amadioha, 1998; Cutler, 1999b; Cutler and Cutler, 1999a). The phenomenon of allelopathy which involves natural plant products offers one of the best biological means of disease control. Further, the disease resistance and defense mechanism is quicker in plants than the inter-specific interactions that take a lot of time. Natural products from the higher plants and microorganisms are an excellent source of antifungal agents. A number of studies have reported the use of pure or crude forms of allelochemicals, besides the role played by allelopathic rotational and companion crops for the control of plant diseases (Rice, 1995;

Rizvi et al., 1999). Allelopathic crops can also greatly reduce the incidence of soil-borne diseases through crop residues releasing inhibitory substances (Yu, 1999b). Some allelochemicals also play an important role in plant defense by acting as signal transducers (Gaspar, Neves, and Pereira, 1999).

DIMBOA and DIBOA commonly found in the cereals play an important role against the harmful predators such as insects, fungi and even bacteria (Kutchan, 1997). The allelochemicals from the weeds also inhibit the growth and development of many plant pathogenic fungi and bacterial species (Qasem, 1996 a,b). Under some circumstances, even enhancement of pathogen susceptibility by allelochemicals has also been reported (Lynch and Penn, 1980). Several species of *Brassica* and even mustard oil are reported to inhibit the growth and development of a large number of soil-borne phytopathogenic fungi, and this affect is attributed to the presence of glucosinolates which degrade enzymatically to isothiocyanates (Wilcoxon and McCallan, 1935; Walker, Morell, and Foster, 1937; Pryor, Walker, and Stahmann, 1940; Papavizas and Lewis, 1971; Angus et al., 1994; Kirkegaard, Wong, and Desmarchelier, 1996; Olivier et al., 1999; Vaughn, 1999). Saponins from some higher plants like roots of alfalfa (*Medicago sativa*), unripe berries of *Phytolacca dodecandra*, fruits of *Swartzia medagascariensis* have fungicidal properties against a number of phytopathogenic fungi (Oleszek, 1999; Oleszek, Hoagland, and Zablotowicz, 1999; Waller, 1999). A large number of flavonoids have been reported to have fungicidal property against a number of seed borne fungi (Weidenborner et al., 1990, 1992; Weidenborner and Jha, 1993; Dixon, 1999).

Some of the higher plants with the potential to suppress plant pathogens are listed in Table 2.

A number of allelopathic compounds from microbes like *Polyangium* spp., *Paecilomyces variotii*, *Streptomyces albospinus* and *Fusicoccum* spp. have a fungicidal potential against a number of phytopathogens (Cutler, 1999a). Several bacterial secondary metabolites like Phenazines and Phloroglucinols active in suppression of root diseases are known. These can be transferred to plants through DNA recombinant technology to produce transgenic plants which will have natural protection against pathogens and this method is environmentally safe (Fujimoto, Weller, and Thomashow, 1995; Weller and Thomashow, 1999). Neem (*Azadirachta indica*), an Indian tree, possesses a number of active principles like azadirachtin with potential to kill pathogens (Ghewande, 1989). Its seed-cake, seed and fruit extract, seed kernel powder and seed oil have been reported to control a wide spectrum of fungal pathogens (Gunasekaran et al., 1986; Jeyarajan et al., 1987; Srivastava et al., 1997). The biological activity of neem against pathogens is attributed to the presence of sulphurous compounds in its seed oil. Moreover, neem products also act as deterrent to pathogen carrying insects, thereby decreas-

TABLE 2. Allelochemicals from higher plants with ability to suppress plant pathogens.

Source	Compound	Plant Pathogens	Reference
Allium tuberosum	–	*Pseudomonas solanacearum*	Yu, 1999
A. sativum	Ajoene	*Phytophthora drechsleri* f. sp. *cajana*	Singh et al., 1992
Artemisia borealis	Polyacetylenes	*Cladosporium cucumerinum*	Wang et al., 1990
A. nelagrica	Essential oils	*Pythium aphanidermatum*	Kishore and Dwivedi, 1991
Brassica nigra	Allyisothiocyanate	*Helminthosporium solani* *Verticillium dahliae*	Olivier et al., 1999
Callistemon lanceolatus	Essential oils	*Pythium aphanidermatum*	Kishore and Dwivedi, 1991
Coffea arabica	1,3,7-trimethylxanthine	*Drechslera maydis*	Rizvi et al., 1980
Eucalyptus spp.	Volatile oils	*Sclerotium rolfsii*	Singh and Dwivedi, 1990
E. rostrata	Essential oil, leaf powder	*Sclerotium cepivorum*	Salama et al., 1988
Ferula assafoetida	Asafoetida	*Rhizoctonia solani*, *Sclerotinia sclerotiorum*	Chaurasia and Ram, 1992
Juniperus communis	Essential oil	*Pythium aphanidermatum*	Kishore and Dwivedi, 1991
Medicago sativa	Medicagenic acid 3-O-glucoside (Root saponins)	*Trichoderma viride*	Oleszek et al., 1990
Nicotiana tabacum	α- and β-duvatriene diols	*Peronospora tabacina*	Menetrez et al., 1990
Phellodendron amurense	Berberine	*Botrytis cinerea*, *Alternaria alternata* and some bacteria	Park and Choi, 1999
Pinus spp.	Essential oils	*Pythium aphanidermatum*	Kishore and Dwivedi, 1991
P. taeda	Essential oils	*Pythium aphanidermatum*	Kishore and Dwivedi, 1991
Ranunculus arvensis	Aqueous extracts	*Fusarium oxysporium* f. sp. *lycopersici*, *Helminthosporium sativa*	Qasem, 1999
Salvia splendens	Root extracts	Bacteria	Qureshi, Ahmed, and Kapadia,1989
Tagetes erecta	Essential oils	*Pythium aphanidermatum*	Kishore and Dwivedi, 1991
Zingiber officinale	Gingerenone A	*Pyricularia oryzae*	Endo, Kanno, and Oshima, 1990

ing the disease incidence (Saxena, Khan, and Bajet, 1985). Greenhouse studies have shown that water and ethanol extracts from the leaves and oil extracts from the seeds of *Azadirachta indica* significantly reduced the radial growth of *Pyricularia oryzae* causing blast disease of rice under *in vitro* conditions. This effect was comparable to commercial fungicide cabendazim at the concentration of 0.1% used *in vivo* (Amadioha, 2000).

Though a large number of plants and microorganisms with antifungal activity have been tested, yet only a few fungicides based on their chemistry have been marketed. Two such compounds are Pyrrolnitrin from *Pseudomonas pyrrocinia* and Strobilurin A from *Strobilurus tenacellus* (Gullino, Leroux, and Smith, 2000).

Allelopathy for Insect and Nematode Control

Allelopathy can also be exploited for the control of harmful insects (Anaya, 1999). Being non-motile, plants adopt either structural changes or biochemical strategies and the presence of large amount of secondary products that deter insects. These may be flavonoids, saponins, terpenoids, alkaloids, phenolics and limonoids. Over 2000 plant species have been reported to possess insecticidal activity (Devakumar and Parmar, 1993). However, only a few have been commercialized as yet. In addition to such chemicals, a number of other synthetic insecticides based on their chemistry have been synthesized (Khambay et al., 1997a, b). Allelochemicals deter insects by acting as metabolic poisons or as feeding deterrents (Brattsten, 1986). A number of flavonols, flavones and their derivatives are reported to deter a wide spectrum of insects (Berhow and Vaughn, 1999). 2-phenylethylisothiocyanate (PEITC) isolated from turnip roots has been reported to be a natural insecticide (Lichtenstein, Morgan, and Mueller, 1964). Rice extracts have been reported to be active against brown planthopper *Nilaparvata lugens* (Zhang et al., 1999). Saponins, one of the important allelochemicals, have not shown much promising results regarding their insecticidal activity. Though they have been tested against a number of alfalfa pests, but they lack species specificity (Oleszek, 1999; Oleszek, Hoagland, and Zablotowicz, 1999). Azadirachtin, a tetra triterpenoid from neem (*Azadirachta indica*), is known to be effective against nearly 300 insect pests (Devakumar and Parmar, 1993). It acts as a phage and oviposition deterrent, repellent, anti-feedant, growth retardant, etc. (Schumutterer, 1995). Remarkably, it is relatively non-toxic to warm-blooded animals including human beings.

Nematodes possess thick wall that is resistant to chemicals and attack all kinds of plants either as endo-parasites or as ecto-parasites on the roots. These are even more harmful when they form associations with the pathogens. To control them is really a tedious job. Though, they have a number of natural enemies, yet effective control measures are required to check their growth. Allelochemicals from the plants have the potential to control them. Some of the important ones include *Azadirachta indica* (Alam and Khan, 1975), *Coffea* species (Tronocorn et al., 1986), *Leucaena leucocephala* (Jain and Hassan, 1985) and *Mucuna pruriens* (Reddy et al., 1986). Green manure of rapeseed (*Brassica napus*), extracts from black mustard (*Brassica nigra*) and glucosinolates (mainly AITC) have been reported to be effective against a number of nematodes like *Meloidogyne chitwoodi*, *Caenorhabditis elegans*, *Heterodera schachtii*, *H. rostochiensis* (Mojtahedi et al., 1991, 1993; Lazzeri, Tacconi, and Palmieri, 1993; Vaughan, 1999).

CONCLUSION AND FUTURE DIRECTIONS

The science of allelopathy exhibit complexities of biochemical interactions among plants including microbes plays an important role in agroecosystems where more often a decline in crop productivity, problem of soil sickness and depletion of biodiversity are noticed. The allelochemicals responsible for bringing about biochemical interactions are products of secondary metabolism and exhibit diversity in chemical nature and functions as a group. The phenomenon of allelopathy in agroecosystems can be exploited or manipulated for enhancing crop production, if its mechanism is well understood. Some of the major approaches by which the allelopathy in agroecosystems can be utilized are:

- Revival of traditional cropping pattern with new and scientific approaches, *viz.*, use of allelopathic cover crops, smother crops, companion crops and crop rotation.
- Changing the cropping sequence and cultural practices such as crops with divergent life cycle, and selective allelopathic potential in rotational sequence to check the build-up of weeds in the fields.
- Selecting crop cultivars with greater weed-suppressing/allelopathic activity to inhibit the pests–weeds, pathogens, diseases, and nematodes.
- Identification and transfer of allelopathic genes from wild relatives to modern day crop cultivars using traditional plant breeding practice and modern recombinant DNA technology.
- Employing metabolic engineering strategies for quantitative and qualitative alteration of plants ability to produce bioactive natural products/ allelochemicals.

The allelochemicals from crops as well as from other plants can be utilized for the purpose of managing weeds, pathogens, diseases, insect and nematodes under Integrated Pest and Weed Management Programmes (IPM, IWM) in sustainable agriculture. Such natural products/chemicals have short environmental life, low environmental impact and mammalian toxicity, and are more site specific and target oriented. Screening and testing of such allelochemicals/phytochemicals is, therefore, a useful area of research. Even the chemical compounds can be synthesized based on their novel chemistry. Biotechnology can play an important role in the use of allelochemicals and the phenomenon of allelopathy for sustainable enhancement of crop production. For this, several genetical and molecular-genetical approaches like RAPD, RFLP, etc., should therefore be followed to improve the allelopathic crop plants so as to get the desired results of managing weeds and pests as well as enhancing crop yields. Very little molecular and genetic evidences regarding the evolutionary origin of metabolic pathways leading to the synthesis of secondary metabolites are available. Proliferation of cloned genes for natural product biosynthetic enzymes and systematic sequencing of *Arabidopsis* genome may, however, provide an insight into several aspects of secondary metabolism (Facchini, 1999).

Microorganisms which are efficient competitors and inhibitors of soil pathogens should be promoted rather than killing them. Induced resistance is an important tool in protecting crops from disease. Attempts should made to induce natural resistance in crop plants by using agrochemicals and microbes antagonistic to plant pathogens. Efforts should be made to explore the allelochemicals whose mode of action involves the regulation of physiological and biochemical aspects of host-pathogen interactions, and not killing the pathogen (Hoagland, 1999).

REFERENCES

Al-Abed, A.S., J.R. Qasem and H.A. Abu-Blan. (1993). Antifungal effects of some common wild plant species on certain plant pathogenic fungi. *Dirasat (Pure and Applied Science)* 20: 149-158.

Alam, M.M. and A.M. Khan. (1974). Control of phytonematodes with oil cake amendments in spinach fields. *Indian Journal of Nematology* 4: 239-240.

Allen, P. and D. van Dusen. (1988). Sustainable agriculture: choosing the future. In *Global Perspectives on Agroecology and Sustainable Agriculture. Proceedings of the 6th International Scientific Conference of the International Federation of Organic Movements (IFOAM)*. California, USA: University of California. pp. 1-13.

Alström, B. (1991). Role of bacterial cyanide production in differential reaction of plant cultivars to deleterious rhizosphere pseudomonads. *Plant and Soil* 133: 93-100.

Alström, S. (1992). Saprophytic soil microflora in relation to yield reductions in soil repeatedly cropped with barley (*Hordeum vulgare* L.). *Biology and Fertility of Soils* 14: 145-150.

Altieri, M.A. and J.D. Doll. (1978). The potential of allelopathy as a tool for weed management in crops. *Proceedings of National Academy of Sciences* 24: 495-502.

Altieri, M.A. and M. Liebman. (1988). *Weed Management in Agroecosystems: Ecological Implications*. Florida: CRC Press.

Amadioha, A.C. (1998). Control of powdery mildew of pepper (*Capsicum annuum* L.) by leaf extracts of papaya (*Carica papaya* L.). *Journal of Herbs, Spices and Medicinal Plants* 6(2): 41-47.

Amadioha, A.C. (2000). Controlling rice blast *in vitro* and *in vivo* with extracts of *Azadirachta indica*. *Crop Protection* 19: 287-290.

Anaya, A.L. (1999). Allelopathy as a tool in the management of biotic resources. *Critical Reviews in Plant Sciences* 18: 697-739.

Anaya, A.L., B.E. Hernandez-Bautisian, H.R. Pelayo-Benavides, M. Calera, and E. Fernandez-Luiselli. (1995). Allelopathy in Mexican plants–more recent studies. In *Allelopathy, Organisms, Processes, and Applications*, eds. Inderjit, K.M.M. Dakshini, and F.A. Einhellig. ACS Symposium Series 582. Washington, DC: American Chemical Society, pp. 224-241.

Angus, G.F., P.A. Gardner, J.A. Kirkegaard, and J.M. Desmarchelier. (1994). Biofumigation: Isothiocyanates released from *Brassica* roots inhibit the growth of take-all fungus. *Plant and Soil* 162: 107-112.

Appel, H.M. (1993). Phenolics in Ecological interactions: the importance of oxidation. *Journal of Chemical Ecology* 19: 1521-1552.

Argandona, V.H., H.M. Niemeyer, and L.J. Corcuera. (1981). Effect of content and distribution of hydroxamic in wheat on infestation by aphid *Schizaphis graminum*. *Phytochemistry* 20: 673-676.

Arora, R.K. and R.K. Kohli. (1993). Autotoxic impact of essential oils extracted from *Lantana camara* L. *Biologia Plantarum* 35: 293-297.

Babczinski, P., M. Dorgerioh, A. Lobberding, H.-J. Santel, R.R. Schmidt, P. Schmitt, and C. Wunsche. (1991). Herbicidal activity and mode of action of vulgamycin. *Pesticide Science* 33: 439-446.

Bakker, A.W. and B. Schippers. (1987). Microbial cyanide production in the rhizosphere in relation to potato yield reduction and *Pseudomonas* spp. mediated plant growth stimulation. *Soil Biology and Biochemistry* 19: 451-457.

Barazani, O. and J. Friedman. (1999). Allelopathic bacteria and their impact on higher plants. *Critical Reviews in Plant Sciences* 18: 741-755.

Barnes, J.P. and A.R. Putnam. (1983). Rye residues contribute weed suppression in no-tillage cropping systems. *Journal of Chemical Ecology* 9: 1045-1057.

Barnes, J.P. and A.R. Putnam. (1987). Role of benzoxazinones in allelopathy by rye (*Secale cereale* L.). *Journal of Chemical Ecology* 13: 889-906.

Batish, D.R., R.K. Kohli, D.B. Saxena, and H.P. Singh. (1997). Growth regulatory response of parthenin and its derivatives. *Plant Growth Regulation* 21: 189-194.

Begon, M., J.L. Harper, and C.R. Townsend (1995). *Ecology Individuals, Populations and Communities*. UK: Blackwell Science Publishers.

Berenbaum, M.R. (1995a). The chemistry of defense: the theory and practice. *Proceeding National Academy of Sciences, USA* 92: 2-8.

Berenbaum, M.R. (1995b). Turnabout for fair play: secondary roles for primary compounds. *Journal of Chemical Ecology* 21: 925-940.

Berestetsky, O.A. (1972). Formation of phytotoxic substances by soil micro-organisms on root residues of fruit trees. In *Physiological-Biochemical Basis of Plant*

Interactions in Phytocenoses, ed. A.M. Grodzinsky. Vol. 3. Kiev: Naukova-Dumka, pp. 121-124 (In Russian with English summary).

Berhow, M.A. and S.F. Vaughn. (1999). Higher plant flavonoids: Biosynthesis and chemical ecology. In *Principles and Practices in Plant Ecology: Allelochemical Interactions*, eds. Inderjit, K.M.M. Dakshini, and C.L. Foy. Boca Raton, USA: CRC Press, pp. 423-438.

Bhowmik, P.C. and J.D. Doll. (1979). Evaluation of allelopathic effects of selected weed species on corn and soybeans. *Proceedings of North Central Weed Control Conference* 34: 43-45.

Bhowmik, P.C. and J.D. Doll. (1983). Growth analysis of corn and soybean response to allelopathic effects of weed residues at various temperatures and photosynthetic photon flux densities. *Journal of Chemical Ecology* 9: 1263-1280.

Blum, U. (1995). The value of model plant-microbe-soil system for understanding processes associated with allelopathic interactions: One example. In *Allelopathy, Organisms, Processes, and Applications*, eds. Inderjit, K.M.M. Dakshini, and F.A. Einhellig. ACS Symposium Series 582. Washington, DC: American Chemical Society, pp. 127-131.

Blum, U. (1996). Allelopathic interactions involving phenolic acids. *Journal of Nematology* 28: 259-267.

Blum, U. (1998). Effects of microbial utilization of phenolic acids and their phenolic acid breakdown products on allelopathic interactions. *Journal of Chemical Ecology* 24: 685-708.

Blum, U. and S.R. Shafer (1988). Microbial populations and phenolic acids in soil. *Soil Biology and Biochemistry* 20: 793-800.

Blum, U., S.R. Shafer, and M.E. Lehman. (1999). Evidence for inhibitory allelopathic interactions involving phenolic acids in field soils: Concepts *vs.* an experimental model. *Critical Reviews in Plant Sciences* 18: 673-693.

Blum, U., T.M. Gerig, A.D. Worsham, and L.D. King (1993). Modification of allelopathic effects of *p*-coumaric acid on morning glory seedling biomass by glucose, methionine and nitrate. *Journal of Chemical Ecology* 19: 2791-2811.

Börner, H. (1963). Der Einflutz verschie-dener facteren auf die Bildung von patulin und einer phenolischen Verbindung *Penicillium* expansum auf Blatt-und Wurzebückstanden des Apfels. *Phytopathologische Zeitschrift* 49: 1-28.

Boydston, R.A. and A. Hang. (1995). Rapeseed (*Brassica napus*) green manure crop suppresses weeds in potato (*Solanum tuberosum*). *Weed Technology* 9: 669-675.

Brattsten, L.B. (1986). Fate of ingested plant allelochemicals in herbivorous insects. In *Molecular Aspects of Insect-Plant Interactions*, ed. L.B. Brattsten and S. Ahmand. New York: John Wiley and Sons, pp. 211.

Buhler, D.D. (1996). Development of alternative weed management strategies. *Journal of Production Agriculture* 9: 501-505.

Buhler, D.D. (1999). Expanding the context of weed management. *Journal of Crop Production* 2: 1-7.

Burch, W.R., Jr. and J.K. Parker. (1992). *Social Science Application in Asian Agroforestry*. USA: Winrock International.

Burgos, N.R., Talbert, R.E., and Mattice, J.D. (1999). Cultivar and age differences in the production of allelochemicals by *Secale cereale*. *Weed Science* 47: 481-485.

Caldiz, D.O. and L.V. Fernandez. (1999). Allelopathy as a possible strategy for weed control in agriculture and forestry systems. In *Recent Advances in Allelopathy Vol. 1. A Science for the Future*, eds. F.A. Macías, J.C.B. Galindo, J.M.G. Molinillo, and H.G. Cutler. Universidad de Cádiz, Spain: Servicio de Publicaciones, pp. 451-462.

Catska, V. (1994). Interrelationships between vesicular arbuscular mycorrhiza and rhizosphere microflora in apple replant disease. *Biologia Plantarum* 36: 99-104.

Chaurasia, S.N.P. and D. Ram. (1992). Inhibition of mycelial growth and sclerotium formation in *Rhizoctonia solani* Kuhn and *Sclerotinia sclerotiorum* (Lib) de Bary by asafoetida. *Tropical Science* 30: 15-19.

Cheng, H.H. (1995). Characterization of the mechanism of allelopathy: modeling and experimental approaches. In *Allelopathy, Organisms, Processes and Applications*, eds. Inderjit, K.M.M. Dakshini, and F.A. Einhellig. ACS Symposium Series 582, Washington, DC: American Chemical Society, pp. 132-141.

Choesin, D.N. and R.E. Boerner. (1991). Allyl isothiocyanate release and allelopathic potential of *Brassica napus* (Brassicaceae). *American Journal of Botany* 78: 1083-1090.

Chou, C.H. (1995). Allelopathy and sustainable agriculture. In *Allelopathy: Organisms, Processes and Applications*, eds. Inderjit, K.M.M. Dakshini, and F.A. Einhellig. ACS Symposium Series 582. Washington, DC: American Chemical Society, pp. 211-223.

Chou, C.H. (1999). Roles of allelopathy in plant biodiversity and sustainable agriculture. *Critical Reviews in Plant Sciences* 18: 609-636.

Chou, C.H. and C.H. Muller. (1972). Allelopathic mechanism of *Arctostaphylos glandulosa* var. Zacaensis. *American Midland Naturalist* 88: 324-347.

Chou, C.H. and G.R. Waller. (1989). *Phytochemical Ecology: Allelochemicals, Mycotoxins, and Insect Pheromones and Allomones*. Academia Sinica Monograph Series No. 9. Taipei, Taiwan: Institute of Botany, Academia Sinica, 489 p.

Chou, C.H., G.R. Waller, and C.F. Reinhardt. (1999). *Biodiversity and Allelopathy: From Organisms to Ecosystems in the Pacific*. Taipei, Taiwan, ROC: Academia Sinica.

Cobb, A. (1992). *Herbicides and Plant Physiology*. London: Chapman and Hall.

Connell, J.H. (1990). Apparent versus real competition plants. In *Perspectives in Plant Competition*, eds. J.B. Grace and D. Tilman. New York, USA: Academic Press, pp. 9-26.

Cruz-Ortega, R., A.L. Anaya, B.E. Hernández-Bautista, and G. Laguna-Hernández. (1998). Effect of allelochemical stress produced by *Sicyos deppei* on the root ultrastructure of *Cucurbita ficifolia* and *Phaseolus vulgaris*. *Journal of Chemical Ecology* 24: 2039-2057.

Culpeper, N. (1633). *English Physitian and Complete Herball*. London: Foulsham. (Reprinted, 1955).

Cutler, H.G. (1988). Perspectives on discovery of microbial phytotoxins with herbicidal activity. *Weed Technology* 2: 525-532.

Cutler, H.G. (1999a). Potentially useful natural product herbicides from microorganisms. In *Principles and Practices in Plant Ecology: Allelochemical Interactions*, eds. Inderjit, K.M.M. Dakshini, and C.L. Foy. Boca Raton, USA: CRC Press, pp. 497-516.

Cutler, H.G. (1999b). Allelopathy in the biological control of plant diseases. In *Recent Advances in Allelopathy Vol. 1. A Science for the Future*, eds. F.A. Macías, J.C.B. Galindo, J.M.G. Molinillo, and H.G. Cutler. Universidad de Cádiz, Spain: Servicio de Publicaciones, pp. 397-414.

Cutler, H.G. and S.J. Cutler. (eds.). (1999a). *Biologically Active Natural Products: Agrochemicals*. Boca Raton, USA: CRC Press, 299 p.

Cutler, H.G. and S.J. Cutler. (1999b). Agrochemicals and pharmaceuticals. In *Biologically Active Natural Products: Agrochemicals*, eds. H.G. Cutler and S.J. Cutler. Boca Raton, USA: CRC Press, pp. 1-14.

Daizy Rani. (1990). Phytotoxic Properties of *Parthenium hysterophorus* L. Ph.D. Thesis, Panjab University, Chandigarh, India.

Dalton, B.R., U. Blum, and S.B. Weed. (1983). Allelopathic substances in ecosystem: effectiveness of sterile soil components in altering the recovery of ferulic acid. *Journal of Chemical Ecology* 9: 1185-1201.

Dalton, B.R., U. Blum, and S.B. Weed. (1989). Differential sorption of exogenously applied ferulic, *p*-coumaric, *p*-hydroxybenzoic and vanillic acids in soil. *Soil Science Society of America Journal* 53: 757-762.

Dayan, F., J. Romagni, M. Tellez, A. Rimando, and S. Duke. (1999). Managing weeds with natural products. *Pesticide Outlook* (October): 185-188.

de Candolle, M.A.P. (1832). *Physiologie Vegetale*. Tome-III. Béchet Jeune, Lib., Fac. Méd. Paris, pp. 1474-1475.

del Moral R. and C.H. Muller. (1969). Fogdrip : a mechanism of toxin transport from *Eucalyptus camaldulensis*. *American Midland Naturalist* 83: 254-282.

Devakumar, C. and B.S. Parmar. (1993). Pesticides of higher plant and microbial origin. In *Botanical and Pesticides*, eds. B.S. Parmar and C. Devakumar. SPS Publication No. 4. New Delhi: Society of Pesticide Science, India and Westvill Publishing House, pp. 1-73.

Devi, S.R., F. Pellissier, and M.N.V. Prasad. (1997). Allelochemicals. In *Plant Ecophysiology*, ed. M.N.V. Prasad. New York: John Wiley & Sons, pp. 253-303.

Devine, M., S.O. Duke, and C. Fedtke. (1993). *Physiology of Herbicide Action*. Englewood Cliffs, New Jersey: Prentice Hall.

Dilday, R.H., W.G. Yan, K.A.K. Moldenhauer, and K.A. Gravois. (1998). Allelopathic activity of rice for controlling major aquatic weeds. In *Allelopathy in Rice*, ed. M. Olofsdotter. Proceedings of the workshop on Allelopathy in rice. Manila, Philippines: International Rice Research Institute (IRRI), pp. 7-26.

Dixon, R.A. (1999). Isoflavonoids: biochemistry, molecular biology, and biological functions. In *Comprehensive Natural Products Chemistry*, ed. U. Sankawa. Vol. 1. Netherlands, Elsevier Sci., pp. 773-823.

Duke, S.O., A.M. Rimando, M.V. Duke, R.N. Paul, J.E.S. Ferreira, and R.J. Smeda. (1999). Sequestration of phytotoxins by plants: implications for biosynthetic production. In *Biologically Active Natural Products: Agrochemicals*, eds. H.G. Cutler and S.J. Cutler. Boca Raton, USA: CRC Press, pp. 127-136.

Duke, S.O., F.E. Dayan, A. Hernández, M.V. Duke, and H.K. Abbas. (1997). Natural products as leads for new herbicide modes of action. In *Proceedings of The 1997 Brighton Crop Protection Conference–Weeds. Vol. 2*, eds. K.E. Pallett. Farnham, UK: British Crop Protection Council, pp. 579-586.

Duke, S.O., F.E. Dayan, and A. Rimando (1998). Natural products as tools for weed management. In *Proceedings of Special Lecture Meeting on "Recent Topics of Weed Science and Weed Technology,"* pp. 1-11.

Duke, S.O., F.E. Dayan, J.G. Romagni, and A.M. Rimando. (2000). Natural products as sources of herbicides: current status and future trends. *Weed Research* 40: 99-111.

Duke, S.O., H.K. Abbas, M.V. Duke, H.J. Lee, H.J., K.C. Vaughn, T. Amagasa, and T. Tanaka. (1996a). Microbial phytotoxins as potential herbicides. *Journal of Environment Science and Health* B31: 427-434.

Duke, S.O., H.K. Abbas, T. Amagasa, and T. Tanaka. (1996b). Phytotoxins of microbial origin with the potential for use as herbicides. In *Crop Protection Agents from Nature: Natural Products and Analogues. Critical Reviews on Applied Chemistry,* ed. L.G. Copping. Vol. 35. Cambridge, UK: Society for Chemical Industries, pp. 82-113.

Duke, S.O., K.C. Vaughn, E.M. Groom, and H.N. Elsholy. (1987). Artemisinin, a constituent of annual wormwood (*Artemisia annua*) is a selective phytotoxin. *Weed Science* 35: 499-505.

Einhellig, F.A. (1985a). Effect of allelopathic chemicals on crop productivity. In *Bioregulators for Pest Control,* ed. P.A. Hedin. ACS Symposium Series 276. Washington, DC: American Chemical Society, pp. 109-132.

Einhellig, F.A. (1985b). Allelopathy–a natural protection, allelochemicals. In *Handbook of Natural Pesticides: Methods. Vol. 1. Theory, Protection and Detection,* ed. N.B. Mandava. Boca Raton, FL: CRC Press, Inc., pp. 161-200.

Einhellig, F.A. (1989). Interactive effects of allelochemicals and environmental stress. In *Phytochemical Ecology: Allelochemicals, Mycotoxins, Insect Pheromones and Allomones,* eds. C.H. Chou and G.R. Waller. Academia Sinica Monograph No. 9. Taipei, Taiwan: Institute of Botany, Academia Sinica, pp. 101-118.

Einhellig, F.A. (1995). Allelopathy : Current status and future goals. In *Allelopathy: Organisms Processes and Applications,* eds. Inderjit, K.M.M. Dakshini, and F.A. Einhellig. ACS Symposium Series 582. Washington, DC: American Chemical Society, Washington, pp. 1-24.

Einhellig, F.A. (1996). Interactions involving allelopathy in cropping systems. *Agronomy Journal* 88: 886-893.

Einhellig, F.A. and I.F. Souza. (1992). Phytotoxicity of sorgoleone found in grain sorghum root exudates. *Journal of Chemical Ecology* 18: 1-11.

Ellis, B.E. (1997). Metabolism of defense and communication. In *Plant Metabolism,* eds. D.T. Dennis, D.H. Turpin, D.D. Lefebvere, and D.B. Layzell. Essex, UK: Addison Wesley Longman, pp. 148-160.

Endo, K., E. Kanno, and Y. Oshima. (1990). Structures of antifungal diarylheptenones, gingerenones A, B, C and isogingerenone B, isolated from the rhizomes of *Zingiber officinale*. *Phytochemistry* 29: 797-799.

Evidente, A., A. Andolfi, M. Vurro, M.C. Zonno, and A. Motto. (2000). Trans-4-aminoproline, a phytotoxic metabolite with herbicide activity produced by *Ascochyta Caulina*. *Phytochemistry* 53: 231-237.

Facchini, P.J. (1999). Plant secondary metabolism out of evolutionary abyss. *Trends in Plant Science* 4: 381-418.

Facelli, J.M. and S.T.A. Pickett. (1991). Plant litter: Its dynamics and effects on plant community structure. *Botanical Review* 57: 1-32.

Fay, P.K. and W.B. Duke. (1977). An assessment of allelopathic potential in *Avena germplasm*. *Weed Science* 25: 224-228.

Fischer, N.H., G.B. Williamson, J.D. Weidenhamer, and D.R. Richardson. (1994). In search of allelopathy in the Florida scrub: the role of terpenoids. *Journal of Chemical Ecology* 20: 1355-1380.

Foley, M.C. (1999). Genetic approach to the development of cover crops for weed management. *Journal of Crop Production* 2(1): 77-93.

Fredrickson, J.K., L.F. Elliot, and J.C. Engibous. (1987). Crops residues as substrates for host-specific inhibitory pseudomonads. *Soil Biology and Biochemistry* 19: 127-134.

Frey, M., P. Chomet, E. Glawischnig, C. Stettner, S. Grün, A. Winklmair, W. Eisenreich, A. Bacher, R.B. Meeley, S.P. Briggs, K. Simcox, and G. Alfons. (1997). Analysis of a chemical plant defense mechanism in grasses. *Science* 277: 696-699.

Fuerst, E.P. and A.R. Putnam. (1983). Separating the competitive and allelopathic components of interference: theoretical principles. *Journal of Chemical Ecology* 9: 937-944.

Fujii, Y. (1999). Allelopathy of velvetbean: Determination and identification of L-DOPA as a candidate of allelopathic substances. In *Biologically Active Natural Products: Agrochemicals*, eds. H.G. Cutler and S.J. Cutler. Boca Raton, USA: CRC Press, pp. 33-47.

Fujii, Y., T. Shibuya, and T. Yasuda. (1991). Discrimination of allelopathy of velvetbean (*Mucuna pruriens*) with stairstep experiment and rotary greenhouse experiments. *Japanese Journal of Soil Science and Plant Nutrition* 62: 258-264 (In Japanese with English summary).

Fujimoto, D.K., D.M. Weller, and L.S. Thomashow. (1995). Role of secondary metabolites in root disease suppression. In *Allelopathy: Organisms, processes and Applications*, eds. Inderjit, K.M.M. Dakshini, and F.A. Einhellig. ACS Symposium Series 582. Washington, DC: American Chemical Society, pp. 330-347.

Gallandt, E.R., M. Liebman, and D.R. Huggins. (1999). Improving soil quality: implications for weed management. *Journal of Crop Production* 2(1): 95-121.

Gallet, C. and F. Pellissier. (1997). Phenolic compounds in natural solutions of coniferous forests. *Journal of Chemical Ecology* 23: 2401-2412.

Gaspar, E.M.M.S.M., H.J.C. Neves, and M.M.A. Pereira. (1999). Triterpenoids and other potentially active compounds from wheat straw: isolation, identification, and synthesis. In *Biologically Active Natural Products: Agrochemicals*, eds. H.G. Cutler and S.J. Cutler. Boca Raton, USA: CRC Press, pp. 69-80.

Ghewande, M.P. (1989). Management of foliar diseases of groundnut (*Arachis hypogaea*) using plant extracts. *Indian Journal of Agricultural Science* 59: 133-134.

Grayson, P.T., K.S. Williams, P.A. Freehauf, R.R. Pease, W.T. Ziesel, R.L. Sereno, and R.E. Reinsfelder. (1987). The physical and chemical properties of herbicide cinmethylin (SD 95481). *Pesticide Science* 21: 143-151.

Grodzinsky, A.M. (ed.). (1970-1974). *Physiological-Biochemical Basis of Plant Interactions in Phytocenoses*. Vol. 1-5. Kiev: Naukova Dumka (In Russian, with English Summary).

Grodzinsky, A.M. (ed.). (1977). *Interactions of Plants and Microorganisms in Phytocenoses*. Kiev: Naukova Dumka (In Russian, with English Summary).

Grümmer, G. (1955). *Die gegenseitige Beeinflussung höherer Pflanzen-Allelopathie*. Fisher, Jena.

Guenzi, W.D. and T.M. McCalla. (1966). Phenolic acids in oats, wheat, sorghum, and corn residues and their phytotoxicity. *Agronomy Journal* 58: 303-304.

Gullino, M.L., P. Leroux, and C.M. Smith. (2000). Uses and challenges of novel compounds for plant disease control. *Crop Protection* 19: 1-11.

Gunasekaran, M., N. Ramadoss, M. Ramiajh, P. Bhaskaran, and T. Ramananthan. (1986). Role of neem cake in the control of Thanjavaran wilt of coconut. *Indian Coconut Journal* 17: 7-12.

Haider, K., J.P. Martin, and E. Rietz. (1977). Decomposition in soil of ^{14}C-labeled coumaryl alcohols: free and linked into dehydropolymer and plant lignins and model humic acids. *Soil Science Society of America Journal* 41: 556-562.

Halbrendt, J.M. (1996). Allelopathy in the management of plant-parasitic nematodes. *Journal of Nematology* 28: 8-14.

Hall, A.B., U. Blum, and R.C. Fites (1982). Stress modification of *Helianthus annuus* L. debris on seed germination. *American Journal of Botany* 69: 776-783.

Hanson, A.D., P.L. Traynor, K.M. Ditz, and D.A. Reicosky. (1981). Gramine in barley forage–effects of genotype and environment. *Crop Science* 21: 726-730.

Harborne, J.B. (1989). *Introduction to Biochemical Ecology*. New York: Academic Press.

Harper, J.L. (1977). *Population Biology of Plants*. London: Academic Press.

Harris, D. and M. Natrajan. (1987). Physiological basis for yield advantages in a sorghum/groundnut intercrop exposed to drought. II. Plant temperature, water stress and components of field. *Field Crop Research* 17: 213-288.

Heisey, R.M. (1996). Identification of an allelopathic compound from *Ailanthus altissima* (Simaroubaceae) and characterization of its herbicidal activity. *American Journal of Botany* 83: 192-200.

Heisey, R.M. and A.R. Putnam. (1986). Herbicidal effects of geldanamycin and nigericin, antibiotics from *Streptomyces hygroscopicus*. *Journal of Natural Products* 49: 859-865.

Heisey, R.M., S.K. Mishra, A.R. Putnam, J.R. Miller, C.J. Whitenack, J.E. Keller, and J. Huang. (1988). Production of herbicidal and insecticidal metabolites by soil microorganisms. In *Biologically Active Natural Products, Potential Use in Agriculture*, ed. H.G. Cutler. ACS Symposium Series 380. Washington DC: American Chemical Society, pp. 65-78.

Hill, R.A., M.A. Eden, H.G. Cutler, P.A.G. Elmer, T. Reglinski, and S.R. Parker. (1999). Practical natural solutions for plant disease control. In *Biologically Active Natural Products: Agrochemicals*, eds. H.G. Cutler and S.J. Cutler. Boca Raton, USA: CRC Press, pp. 201-210.

Hoagland, R.E. (1990). Microbes and microbial products as herbicides: an overview. In *Microbes and Microbial Products as Herbicides*, ed. R.E. Hoagland. ACS Symposium Series 439. Washington, DC: American Chemical Society, pp. 2-52.

Hoagland, R.E. (1999). Biochemical interactions of the microbial phytotoxin phosphinothricin and analogs with plants and animals. In *Biologically Active Natural*

Products: Agrochemicals, eds. H.G. Cutler and S.J. Cutler. Boca Raton, USA: CRC Press, pp. 107-125.

Hoult, A.H.C. and J.V. Lovett. (1993). Biologically active secondary metabolites of barley. III. A method for identification and quantification of hordenine and gramine in barley by high-performance liquid chromatography. *Journal of Chemical Ecology* 19: 2245-2254.

Huang, J., A.R. Putnam, G.M. Werner, S.K. Mishra, and C. Whitenack. (1989). Herbicidal metabolites from a soil dwelling fungus (*Scopulariopsis brumptii*). *Weed Science* 37: 123-128.

Inderjit, K.M.M. Dakshini, and C.L. Foy (1999). *Principles and Practices in Plant Ecology: Allelochemical Interactions*. Boca Raton, USA: CRC Press, 589 p.

Inderjit, K.M.M. Dakshini, and F.A. Einhellig. (1995). *Allelopathy: Organisms, Processes and Applications*. ACS Symposium Series 582. Washington, DC: American Chemical Society. 389 pp.

Jain, R.K. and M. Hassan. (1985). Toxicity of koo-babool (*Leucaena leucocephala* L.) extracts to *Meloidogyne incognita* and *Helicotyulenchus dihystera*. *Indian Journal of Nematology* 14: 179-189.

Jansen, R.C. (1996). Complex plant traits: time for polygenic analysis. *Trends in Plant Science* 1: 89-94.

Jeyarajan, R., S. Doraiswamy, R. Bhaskaran, and S. Jayaraj. (1987). Effect of neem and other products in the management of plant diseases in India. In *natural Pesticides from the Neem Tree and Other Tropical Plants*, eds. H. Schumutterer and K.R.S. Ascher. Germany: GTZ, Eschborn, pp. 635-644.

Jordan, N. (1993). Prospects for weed control through crop interference. *Ecological Applications* 3: 84-91.

Kalburtji, K.L. and A. Gagianas. (1997). Effect of sugarbeet as a preceding crop on cotton. *Journal of Agronomy and Crop Science* 178: 59-63.

Katznelson, J. (1972). Studies on clover soil sickness. 1. The phenomenon of soil sickness in berseem and Persian clover. *Plant and Soil* 36: 379-393.

Kelsey, R.G. and I.J. Locken. (1987). Phytotoxic properties of cnicin, a sesquiterpene lactone from *Centaurea maculosa* (spotted knapweed). *Journal of Chemical Ecology* 13: 19-33.

Khambay, B.P.S., D. Batty, D.G. Beddie, I. Denholm, and M.R. Cahill. (1997a). A new group of plant-derived naphthoquinone pesticides. *Pesticide Science* 50: 291-296.

Khambay, B.P.S., D.G. Beddie, and M.S.J. Simmonds. (1997b). In pursuit of insecticidal compounds from plants. In *Phytochemicals Diversity-A Source of New Industrial Products*, eds. S. Wrigley, M. Hayes, R. Thomas, and E. Crystal. Cambridge, UK: The Royal Society of Chemistry, pp. 158-169.

Kimber, R.W.L. (1973). Phytotoxicity from plant residues. III. The relative effect of toxins and nitrogen immobilization on the germination and growth of wheat. *Plant and Soil* 38: 543-555.

Kirkegaard, J.A., P.T.W. Wong, and J.M. Desmarchelier, (1996). *In vitro* suppression of fungal root pathogen of cereals by *Brassica* tissues. *Plant Pathology* 45: 593-603.

Kishore, N. and R.S. Dwivedi. (1991). Fungitoxicity of the essential oil of *Tagetes erecta* L. against *Pythium aphanidermatum* Fitz. the damping off pathogen. *Flavour and Fragrance Journal* 6: 291-294.

Koeppe, D.E., L.M. Southwick, and J.E. Bittell (1976). The relationship of tissue chlorogenic acid concentrations and leaching of phenolics from sunflowers grown under varying phosphate nutrient conditions. *Canadian Journal of Botany* 54: 593-599.

Kohli, R.K. (1990). *Allelopathic Potential of Eucalyptus*. Project Report, MAB-DOEn Project, India.

Kohli, R.K. (1993). *Allelopathic Impact of Sunflower on Weeds and Crops of Punjab*. Project report, Punjab State Council for Science and Technology, India.

Kohli, R.K. (1997). *Ageratum Caused Threat to Plant Diversity*. Project Report, MAB-DOEn Project, India.

Kohli, R.K., D. Batish, and H.P. Singh (1998a). Allelopathy and its implications in Agroecosystems. *Journal of Crop Production* 1: 169-202.

Kohli, R.K., D. Singh, and R.C. Verma. (1990). Influence of eucalypt shelterbelt on winter season agroecosystems. *Agriculture Ecosystems and Environment* 33: 23-31.

Kohli, R.K. and D.R. Batish. (1994). Exhibition of allelopathy by *Parthenium hysterophorus* in agroecosystems. *Tropical Ecology* 35: 295-307.

Kohli, R.K., D.R. Batish, and H.P. Singh. (1998b). Eucalypt oils for the control of parthenium (*Parthenium hysterophorus* L.). *Crop Protection* 17: 119-122

Kohli, R.K., D.R. Batish, H.P. Singh, and V. Arora. (1999). Allelopathy as a tool for weed and pest management. *Journal of Punjab Academy of Sciences* 1: 127-131.

Koul, O., M.B. Isman, and M. Ketkar. (1990). Properties and uses of neem, *Azadirachta indica*. *Canadian Journal of Botany* 68: 1-11.

Krishnan, G., D.L. Holshouser, and S.J. Nissen. (1998). Weed control in soybean (*Soybean max*) with green manure crops. *Weed Technology* 12: 97-102.

Kuo, Y.L., C.H. Chou, and T.W. Hu. (1983). Allelopathic potential of *Leucaena leucocephala*. In *Allelochemicals and Pheromones*, eds. C.H. Chou and G.R. Waller. Academia Sinica Monograph Series no. 5. Taipei, Taiwan, ROC: Academia Sinica, pp. 107-119.

Kutchan, T.M. (1997). Natural product munitions-new prospects for plant protection. *Research News* 2: 449-450.

Lambert, J.D.H., G. Campbell, J.T. Arnason, and W. Majak. (1991). Herbicidal properties of alpha-terthienyl, a naturally occurring phytotoxin. *Canadian Journal of Plant Science* 71: 215-218.

Lazzeri, L., R. Tacconi, and S. Palmieri. (1993). *In vitro* activity of some glucosinolates and their reaction products toward the population of nematode *Heterodera schachtii*. *Journal of Agricultural and Food Chemistry* 41: 825-829.

Leather, G.R. (1983). Sunflowers (*Helianthus annuus*) are allelopathic to weeds. *Weed Science* 31: 37-42.

Leather, G.R. (1987). Weed control using allelopathic sunflowers and herbicide. *Plant and Soil* 98: 17-23.

Lee, D.L., M.P. Prisbylla, T.H. Cromartie, D.P. Dagarin, S.W. Howard, W.M. Provan, M.K. Ellis, T. Fraser, and L.C. Mutter. (1997). The discovery and structural requirements of inhibitors of *p*-hydroxypyruvate dioxygenase. *Weed Science* 45: 601-609.

Lehman, M.E., U. Blum, and T.M. Gerig. (1994). Simultaneous effects of ferulic and *p*-coumaric acids on cucumber leaf expansion in split-root experiments. *Journal of Chemical Ecology* 20: 1773-1782.

Levitt, J., J.V. Lovett, and P.R. Garlick. (1984). Effect of allelochemicals of *Datura stramonium* L. (Thorn-apple) on root tip ultrastructure of *Helianthus annuus* L. *New Phytologist* 97: 217-218.

Lichtenstein, E.P., D.G. Morgan, and C.H. Mueller. (1964). Naturally occurring insecticides in cruciferous tissue. *Journal of Agricultural and Food Chemistry* 12: 158-161.

Liebl, R., F.W. Simmons, L.M. Wax, and E.W. Stoller. (1992). Effect of rye (*Secale cereale*) mulch on weed control and soil moisture in soybeans (*Glycine max*). *Weed Technology* 6: 838-846.

Lodhi, M.A.K., R. Bilal, and K.A. Malik. (1987). Allelopathy in agroecosystems: wheat phytotoxicity and its possible roles in crop rotation. *Journal of Chemical Ecology* 13: 1881-1891.

Lovett, M. (1994). Fishing for complements: Finding genes by direct selection. *Trends in Genetics* 10: 352-357.

Lynch, J.M. and D.J. Penn. (1980). Damaged to cereals caused by decaying weed residues. *Journal of Science Food and Agriculture* 310: 321-324.

Macias, F.A. (1995). Allelopathy in the search for natural herbicide models. In *Allelopathy: Organisms, Processes, and Applications*, eds. Inderjit, K.M.M. Dakshini, and F.A. Einhellig. ACS Symposium No. 582. Washington, D.C.: American Chemical Society, pp. 310-329.

Macías, F.A., D. Castellano, R.M. Oliva, P. Cross, and A. Torres (1997). Potential use of allelopathic agents as natural agrochemical. In *Proceedings of the 1997 Brighton Crop Protection Conference–Weeds. Vol. 1*, ed. L. Copping. Farnham, UK: British Crop Protection Council, pp. 33-38.

Macías, F.A., J.C.B. Galindo, J.M.G. Molinillo, and H.G. Cutler. (eds.). (1999a). *Recent Advances in Allelopathy Vol. 1. A Science for the Future*. Universidad de Cádiz, Spain: Servicio de Publicaciones, 514p.

Macias, F.A., J.M.G. Molinillo, J.C.G. Galindo, R.M. Varela, A. Torres, and A.M. Simonet (1999b). Terpenoids as potential use as natural herbicide templates. In *Biologically Active Natural Products: Agrochemicals*, eds. H.G. Cutler and S.J. Cutler. Boca Raton, USA: CRC Press, pp. 15-31.

Malik, R.S. and K. Sharma. (1990). Moisture extraction and crop yield as function of distance from a row of *Eucalyptus tereticornis*. *Agroforestry Systems* 12: 187-195.

Martin, J.P. and K. Haider. (1976). Decomposition of specifically carbob-14-labeled ferulic acid: free and linked into model humic acid type polymers. *Soil Science Society of America Journal* 40: 377-380.

Mason-Sedun, W., R.S. Jessop, and J.V. Lovett. (1986). Differential phytotoxicity among species and cultivars of the genus *Brassica* to wheat. 1. Laboratory and field screening of species. *Plant and Soil* 93: 3-16.

Massantini, F., F. Caporali, and G. Zellini. (1977). Evidence for allelopathic control of weed in lines of soybeans. *Proceedings of the EWRS Symposium Methods of Weed Control and Their Integration* 1: 20-32.

McCloskey, W.B. and J.S. Holt. (1990). Triazine resistance in *Senecio vulgaris* parental and nearly isonuclear backcrossed biotypes is correlated with reduced productivity. *Plant Physiology* 92: 954-962.

Menetrez, M.L., H.W. Spurr, Jr., D.A. Danehower, and D.R. Lawson. (1990). Influence of tobacco leaf surface chemicals on *Peronospora tabacina* Adam sporangia. *Journal of Chemical Ecology* 16: 1565-1576.

Miller, D.A. (1983). Allelopathic effects of alfalfa. *Journal of Chemical Ecology* 9: 1059-1072.

Miller, D.A. (1996). Allelopathy in forage crop systems. *Agronomy Journal* 88: 854-859.

Mitchell, R.E. (1989). Biosynthesis of rhizobitoxine from L-aspartic acid and L-threohydroxythreonine by *Pseudomonas andropogonis*. *Phytochemistry* 28: 1617-1620.

Mizutani, J. (1991). Plant ecochemicals. *Journal of Pesticide Science* 16: 679-686.

Mizutani, J. (1999). Selected allelochemicals. *Critical Reviews in Plant Sciences* 18: 653-671.

Mojtahedi, H., G.S. Santo, A.N. Hang, and J.H. Wilson. (1991). Suppression of root-knot nematode populations with selected rapeseed cultivars as green manure. *Journal of Nematology* 23: 170-174.

Mojtahedi, H., G.S. Santo, J.H. Wilson, and A.N. Hang. (1993). Managing *Meloidogyne chitwoodi* on potato with rapeseed as green manure. *Plant Disease* 77: 42-46.

Molisch, H. (1937). *Der Einfluss einer Pflanze auf die andere–Allelopathie*. Fischer, Jena.

Muller, C.H. (1965). Inhibitory terpenes volatilized from *Salvia* shrubs. *Bulletin of the Torrey Botanical Club* 92: 38-45.

Muller, C.H. (1966). The role of chemical inhibition (allelopathy) in vegetational composition. *Bulletin of the Torrey Botanical Club* 93: 332-351.

Muller, C.H., W.H. Muller, and B.L. Haines. (1964). Volatile growth inhibitors produced by shrubs. *Science* 143: 471-473.

Müller-Schärer, H., P.C. Scheepens, and M.P. Greaves. (2000). Biological control of weeds in European crops: recent achievements and future works. *Weed Research* 40: 83-98.

Narwal, S.S. (ed.). (1999a). *Allelopathy Update: Basic and Applied Aspects*. NY: Science Publishers, Inc.

Narwal, S.S. (1999b). Smothering crops as in weeds management: preliminary studies. In *Recent Advances in Allelopathy Vol. 1. A Science for the Future*, eds. F.A. Macías, J.C.B. Galindo, J.M.G. Molinillo, and H.G. Cutler. Universidad de Cádiz, Spain: Servicio de Publicaciones, pp. 365-382.

Narwal, S.S. and P. Tauro. (1994). *Allelopathy in Agriculture and Forestry*. Jodhpur, India: Scientific Publishers.

Newman, E.I. (1978). Allelopathy: Adaptation or Accident? In *Biochemical Aspects of Plant and Animal Co-evolution*, ed. J.B. Harborne. London: Academic Press, pp. 327-342.

Nilsson, M.C., C. Gallet, and A. Wallstedt. (1998). Temporal variability of phenolics and batatasin. III in *Empetrum hermaphroditum* leaves over an eight-year period: interpretations of ecological function. *Oikos* 81: 6-16.

Oerke, E.C., H.W. Dhene, F. Schonbeck, and A. Weber. (1994). *Crop Protection and Crop Production*. Amsterdam: Elsevier.

Oleszek, W. (1987). Allelopathic effects of volatiles from some cruciferae species on lettuce, barnyard grass and wheat growth. *Plant and Soil* 102: 271-273.

Oleszek, W. (1999). Allelopathic significance of plant saponins. In *Recent Advances in Allelopathy Vol. 1. A Science for the Future*, eds. F.A. Macías, J.C.B. Galindo, J.M.G. Molinillo, and H.G. Cutler. Universidad de Cádiz, Spain: Servicio de Publicaciones, pp. 167-178.

Oleszek, W., K.R. Price, I.J. Colquhoun, M. Jurzysta, M. Polszynski, and G.R. Fenwick. (1990). Isolation and identification of alfalfa root saponins: their activity in relation to fungal bioassay. *Journal of Agricultural and Food Chemistry* 38: 1810-1817.

Oleszek, W. and M. Jurzysta. (1987). The allelopathic potential of alfalfa root medicagenic acid glycosides and their fate in soil environment. *Plant and Soil* 98: 67-80.

Oleszek, W., R.E. Hoagland, and R.M. Zablotowicz. (1999). Ecological significance of plant saponins. In *Principles and Practices in Plant Ecology: Allelochemical Interactions*, eds. Inderjit, K.M.M. Dakshini, and C.L. Foy. Boca Raton, USA: CRC Press, pp. 451-465.

Olivier, C., S.F. Vaughn, E.S.G. Mizubuti, and R. Loria. (1999). Variation in allyl isothiocyanate production within *Brassica* species and correlation with fungicidal activity. *Journal of Chemical Ecology* 25: 2687-2701.

Olofsdotter, M. (1998). Allelopathy for weed control in organic farming. In *Sustainable Agriculture for Food, Energy and Industry–Strategies Towards Achievement*, eds. N. El-Bassam, R.K. Behl, and B. Prochnow. London, UK: James and James (Science Publishers) Ltd., pp. 453-457.

Olofsdotter, M., D. Navarez, and K. Moody. (1995). Allelopathic potential in rice (*Oryza sativa* L.) germplasm. *Annals of Applied Biology* 127: 543-560.

Olofsdotter, M., D. Navarez, M. Rebulanan, and J.C. Streibig (1999). Weed suppressing rice cultivars–does allelopathy play a role. *Weed Research* 39: 441-454.

Ong, C.K. (1991). A framework for quantifying the various effects of tree-crop interactions. In *Tree-Crop Interactions*, eds. C.K. Ong and P. Huxley. Wallingford, UK: CAB International, pp. 1-23.

Owens, L.D., J.F. Thompson, and P.V. Fennessey. (1972). Dihydrorhizobitoxine, a new ether amino-acid from *Rhizobium japonicum*. *Journal of Chemical Society, Chemical Communications* 1972: 715.

Pandey, D.K. (1996). Phytotoxicity of sesquiterpene lactone parthenin on aquatic weeds. *Journal of Chemical Ecology* 22: 151-160.

Papavizas, G.C. and J.A. Lewis. (1971). Effect of amendments on fungicides on *Aphanomyces* root rot of peas. *Phytopathology* 61: 215-220.

Park, Y.G. and M.S. Choi. (1999). Allelopathic effects for cork tree (*Phellodendron amurense* Rupr.) on microorganisms growth. In *Recent Advances in Allelopathy Vol. 1. A Science for the Future*, eds. F.A. Macías, J.C.B. Galindo, J.M.G. Molinillo, and H.G. Cutler. Universidad de Cádiz, Spain: Servicio de Publicaciones, pp. 81-88.

Parker, S.R., R.A. Hill, and H.G. Cutler. (1999). Spectrum of activity of antifungal natural products and their analogues. In *Biologically Active Natural Products: Agrochemicals*, eds. H.G. Cutler and S.J. Cutler. Boca Raton, USA: CRC Press, pp. 175-183.

Peirce, L.C. and L.W. Colby. (1987). Interaction of asparagus-root filtrate with *Fusarium oxysporium* f. sp. *asparagi*. *Journal of American Society of Horticultural Science* 112: 35-40.

Pellissier, F. (1993). Allelopathic effect of phenolic acids from humic solutions on two spruce mycorrhizal fungi *Cenococcum graniforme* and *Laccaria laccata*. *Journal of Chemical Ecology* 19: 2105-2114.

Pellissier. F. (1998). The role of soil community in plant population dynamics: is allelopathy a key component? *Trends in Ecology and Evolution* 13: 407.

Pellissier, F. and X.C. Souto. (1999). Allelopathy in northern temperate and boreal semi-natural woodland. *Critical Reviews in Plant Sciences* 18: 637-652.

Plinius Secundus, C. (I.A.D.). *Natural History*. 10 Vols. Translated to English by H. Rackam, W.H.S. Jones, and D.E. Eichholz. Cambridge, Massachusetts: Harvard University Press, pp. 1938-1963.

Pryor, D.E., J.C. Walker, and M.A. Stahmann. (1940). Toxicity of allyl isothiocyanate vapor to certain fungi. *American Journal of Botany* 27:30-38.

Pue, K.J., U. Blum, T.M. Gerig, and R. Shafer (1995). Mechanism by which non-inhibitory concentrations of glucose increase inhibitory activity of *p*-coumaric acid on morning-glory seedling biomass accumulation. *Journal of Chemical Ecology* 21: 833-847.

Putnam, A.R. (1988). Allelochemicals from plants as herbicides. *Weed Technology* 2: 510-518.

Putnam, A.R. and C.S. Tang. (1986). *The Science of Allelopathy*. USA: John Wiley and Sons, Inc., 317 p.

Putnam, A.R. and L.A. Weston. (1986). Adverse impacts of allelopathy in agricultural systems. In *The Science of Allelopathy*, ed. A.R. Putnam and C.S. Tang. USA: John Wiley and Sons. Inc., pp. 43-56.

Putnam, A.R. and W.B. Duke. (1978). Allelopathy in agroecosystems. *Annual review of Phytopathology* 16: 431-451.

Putnam, A.R., J. DeFrank, and J.P. Barnes. (1983). Exploitation of allelopathy for weed control in annual and perennial cropping systems. *Journal of Chemical Ecology* 9: 1001-1010.

Qasem, J.R. (1993). Allelopathic effect of nettle-leaved goosefoot (*Chenopodium murale*) on wheat and barley. *Dirasat* 20B: 80-94.

Qasem, J.R. (1996a). Fungicidal activity of *Ranunculus asiaticus* and other weeds against *Fusarium oxysporium* f. sp. *lycopersici*. *Annals of Applied Biology* 128: 533-540.

Qasem, J.R. (1996b). Aqueous extract effects of some common weed species against certain plant pathogenic fungi. *ACTES, Institute of Agronomy and Veterinary, (Maroc)* 16: 11-19.

Qasem, J.R. (1999). Biological activity of corn buttercup (*Ranunculus arvensis* L.). In *Recent Advances in Allelopathy Vol. 1. A Science for the Future*, eds. F.A. Macías, J.C.B. Galindo, J.M.G. Molinillo, and H.G. Cutler. Universidad de Cádiz, Spain: Servicio de Publicaciones, pp. 287-300.

Qasem, J.R. and H.A. Abu-Blan. (1995). Antifungal activity of aqueous extracts from some common weed species. *Annals of Applied Biology* 127: 215-219.

Qasem, J.R. and H.A. Abu-Blan. (1996). Fungicidal activity of some common weed extract against different plant pathogenic fungi. *Journal of Phytopathology* 144: 157-161.

Qasem, J.R. and T.A. Hill. (1989). Possible role of allelopathy in the competition between tomato, *Senecio vulgaris* L. and *Chenopodium album* L. *Weed Research* 29: 349-356.

Qureshi, I.H., S. Ahmed, and Z. Kapadia. (1989). Antimicrobial activity of *Salvia splendens*. *Pakistan Journal of Scientific and Industrial Research* 32: 597-599.

Ralhan, P.K., A. Singh, and R.S. Dhanda. (1992). Performance of wheat as intercrop under poplar (*Populus deltoides* Bartr.) plantations in Punjab (India). *Agroforestry Systems* 19: 217-222.

Rao, M.R., P.K.R. Nair, and C.K. Ong. (1997). Biophysical interactions in tropical agroforestry system. *Agroforestry Systems* 38: 3-50.

Reigosa, M.J., A. Sánchez-Moreiras, and L. González. (1999). Ecophysiological approach in allelopathy. *Critical Reviews in Plant Sciences* 18: 577-608.

Reddy, K.C., A.R. Soffes, G.M. Prine, and R.A. Dunn. (1986). Tropical legumes for green manure. II. Nematode populations and their effects on succeeding crop yields. *Agronomy Journal* 78: 5-10.

Rietveld, W.J. (1983). Allelopathic effects of juglone on germination and growth of several herbaceous and woody species. *Journal of Chemical Ecology* 9: 295-308

Rice, E.L. (1984). *Allelopathy*. 2nd edition. Orlando, Florida: Academic Press Inc., 422 p.

Rice, E.L. (1995). *Biological Control of Weeds and Plant Diseases–Advances in Applied Allelopathy*. Norman, USA: University of Oklahoma Press, 439 p.

Rizvi, S.J.H., D. Mukerji, and S.N. Mathur. (1980). A new report on a possible source of natural herbicide. *Indian Journal of Experimental Biology* 18: 777-778.

Rizvi, S.J.H., M. Tahir, V. Rizvi, R.K. Kohli, and A. Ansari (1999). Allelopathic interaction Agroforestry systems. *Critical Reviews in Plant Sciences* 18: 773-796.

Rizvi, S.J.H., R.C. Sinha, and V. Rizvi. (1990). Implications of mimosine allelopathy in agroforestry. *Proceedings IUFRO XIX World Congress of Forestry* 2: 22-27.

Rizvi, S.J.H., V. Jaiswal, D. Mukerji, and S.N. Mathur. (1980). Antifungal properties of 1,3,7-trimethylxanthine isolated from *Coffea arabica*. *Naturwissenschaften* 67: 459-460.

Rizvi, S.J.H. and V. Rizvi. (eds.). (1992). *Allelopathy: Basic and Applied Aspects*. London: Chapman and Hall.

Robbins, W.W., A.S. Crafts, and R.N. Raynor. (1942). *Weed Control–A Textbook and Manual*. New York, NY: McGraw-Hill Book Company, Inc.

Romagni, J.G., S.N. Allen, and F.E. Dayan. (2000). Allelopathic effects of volatile cineoles on two weedy plant species. *Journal of Chemical Ecology* 26: 303-313.

Salama, A.-A.M., I.M.K. Ismail, M.I.A. Ali, and S.A.-E. Ouf. (1988). Possible control of white rot disease of onions caused by *Sclerotium cepivorum* through soil amendments with *Eucalyptus rostrata* leaves. *Revista Ecological Biological Sciences* 25: 305-314.

Sarmah, M.K., S.S. Narwal, and J.S. Yadav. (1992). Smothering effect of *Brassica* species on weeds. In *proceedings of First National Symposium on Allelopathy in Agroecosystems*, eds. P. Tauro and S.S. Narwal. Hisar: Indian Society of Allelopathy, pp. 51-55.

Sauerborn, J. (1999). Legumes used for weed control in agro-ecosystems in the tropics. *Plant Research and Development* 50: 74-82.

Saxena, D.B., P. Dureja, B. Kumar, D. Rani, and R.K. Kohli. (1991). Modification of parthenin. *Indian Journal of Chemistry* 30B: 849-858.

Saxena, R.C., Z.R. Khan, and N.B. Bajet. (1985). Neem seed derivatives for preventing ice tungro virus transmission by the green leafhopper *Nephotettix virescens* (Distant). Pihl. *Phytopathology* 21: 88-102.

Schippers, B., A.W. Bakker, and P.A.H.M. Bakker. (1987). Interactions of deleterious and beneficial rhizosphere microorganisms and the effect of cropping practices. *Annual Review of Phytopathology* 25: 339-358.

Schreiber, M.M. (1992). Influence of tillage, crop rotation, and weed management on giant foxtail (*Setaria faberi*) population dynamics and corn yield. *Weed Science* 40: 645-653.

Schreiner, O. and H.S. Reed. (1907). The production of deleterious excretions by roots *Bulletin of the Torrey Botanical Club* 34: 279-303.

Schreiner, O. and H.S. Reed. (1908). The toxic action of certain organic plant constituents. *Botanical Gazette* 45: 73-102.

Schumacher, W., D. Thill, and G. Lee. (1983). Allelopathic potential of wild oat (*Avena fatua*) on spring wheat. *Journal of Chemical Ecology* 9: 1235-1246.

Schumutterer, H. (ed.) (1995). *The Neem Tree.* Weinheim, FRG: VCH Verlags-gesellschaft.

Seigler, D.S. (1996). Chemistry and mechanism of allelopathic interactions. *Agronomy Journal* 88: 876-885.

Shilling, D.G., R.A. Liebl, and A.D. Worsham. (1985). Rye (*Secale cereale* L.) and wheat (*Triticum aestivum* L.) mulch: The suppression of certain broadleaf weeds and the isolation and identification of phytotoxins. In *The Chemistry of Allelopathy*, ed. A.C. Thompson. ACS Symposium Series 268. Washington DC: American Chemical Society, pp. 243-271.

Singh, H.P., D.R. Batish, and R.K. Kohli. (1999a). Autotoxicity: concept, organisms and ecological significance. *Critical Reviews in Plant Sciences* 18: 757-772.

Singh, H.P., D.R. Batish, and R.K. Kohli. (1999b). Allelopathic effects of *Leucaena leucocephala* on *Zea mays*. *Journal of Tropical Forest Science* 11: 801-808.

Singh, D. and R.K. Kohli. (1992). Impact of *Eucalyptus tereticornis* Sm. shelterbelts. *Agroforestry Systems* 20: 253-266.

Singh, H.P., R.K. Kohli, and D.R. Batish. (1998). Shelter effects of poplar (*Populus deltoides*) on growth and yield of wheat in North India. *Agroforestry Systems* 40: 207-213.

Singh, H.P., R.K. Kohli, and D.R. Batish. (1999). Impact of *Populus* deltoides and *Dalbergia sissoo* shelterbelts on wheat–a comparative study. *International Tree Crops Journal* 10: 51-60.

Singh, M., R.V. Tamma, and H.N. Nigg. (1989). HPLC identification of allelopathic compounds from *Lantana camara*. *Journal of Chemical Ecology* 15: 81-89.

Singh, R.K. and R.S. Dwivedi. (1990). Fungicidal properties of neem and blue gum against *Sclerotium rolfsii* Sacc., a foot-rot pathogen of barley. *Acta Botanica Indica* 18: 260-262.

Singh, U.P., V.B. Chauhan, K.G. Wagner, and A. Kumar. (1992). Effect of ajoene, a compound derived from garlic (*Allium sativum*), on *Phytophthora drechsleri* f. sp. *cajana*. *Mycologia* 84: 105-108.

Srivastava, K.K., P.K. Gupta, Y.C. Tripathi, and R. Sarvate. (1997). Antifungal activity of plant products on spermoplane fungi of *Azadirachta indica* (Neem) seeds. *Indian Forester* 1223: 157-161.

Stutte, G.W. (1999). Phytochemicals. Implications for long-duration space missions. In *Biologically Active Natural Products: Agrochemicals*, eds. H.G. Cutler and S.J. Cutler). Boca Raton, Florida: CRC Press, pp. 275-286.

Sugawara, F. and G. Strobel. (1986). (−)-Dihydroperenophorin, a novel and selective phytotoxin produced by *Drechslera avenae*. *Plant Science* 43: 1-5.

Suslow, T.V. and M.N. Schroth. (1982). Role of deleterious rhizobacteria as minor pathogens in reducing crop growth. *Phytopathology* 72: 111-115.

Swanton, C.J. and S.D. Murphy. (1996). Weed Science beyond the weeds: the role of integrated weed management (IWM) in agroecosystem health. *Weed Science* 44: 437-445.

Taiz, L. and E. Zeiger. (1998). *Plant Physiology*. 2nd edition. Sunderland, Massachusetts: Sinauer Associates, Inc.

Teasdale, J.R. (1996). Contributions of cover crops to weed management in sustainable agricultural systems. *Journal of Production Agriculture* 9: 475-479.

Theophrastus. (ca 300BC). *Enquiry into Plants and Minor Works on Odours and Weather Signs*. 2 Vols. Translated to English by A. Hort. and W. Heinemann. London, 1916.

Thijs, H., J.R. Shann, and J.D. Weidenhamer. (1994). The effect of phytotoxins on competitive out come in a model system. *Ecology* 75: 1959-1964.

Thompson, A.C. (ed.) (1985). *The Chemistry of Allelopathy*. ACS Symposium Series 268. Washington, DC: American Chemical Society.

Thorne, R.L.Z., G.R. Waller, J.K. McPherson, D.G. Krezer, Jr., and C.C. Young. (1990). Autotoxic effects of old and new wheat straw in conventional-tillage and no-tillage wheat. *Botanical Bulletin of the Academia Sinica* 31: 35-49.

Tronocorn, N.M, S. Ferraz, J.M. Santos, and A.J. Regazzy. (1986). Evaluation of effect of coffee bean straw castings mixed with the soil on the development of *Meloidogyne exigua* Goeldi 1887 in coffee seedlings. *Nematologica Brasilica* 10: 85-102.

Turco, R.F., M. Bischoff, D.P. Breakwell, and D.R. Griffith. (1990). Contribution of soil-borne bacteria to the rotation effect in corn. *Plant and Soil* 122: 115-120.

Vaughn, S.F. (1999). Glucosinolates as natural pesticides. In *Biologically Active Natural Products: Agrochemicals*, eds. H.G. Cutler and S.J. Cutler. Boca Raton, USA: CRC Press, pp. 81-94.

Waksman, S.A. (1937). Soil deterioration and soil conservation from the viewpoint of soil microbiology. *Journal of American Society of Agronomy* 29: 113-122.

Walker, J.C., S. Morell, and H.H. Foster. (1937). Toxicity of mustard oils and related sulphur compounds to certain fungi. *American Journal of Botany* 24: 536-541.

Waller, G.R. (1987). *Allelochemicals: Role in Agriculture and Forestry.* ACS Symposium Series No. 330. Washington, DC: American Chemical Society, 606 p.

Waller, G.R. (1999). Recent advances in saponins used in foods, agriculture, and medicine. In *Biologically Active Natural Products: Agrochemicals*, eds. H.G. Cutler and S.J. Cutler. Boca Raton, USA: CRC Press, pp. 243-274.

Waller, G.R., M. Jurzysta, and R.L.Z. Thorne. (1993). Allelopathic activity of root saponins from alfalfa (*Medicago sativa*) against weeds and wheat. *Botanical Bulletin of the Academia Sinica* 34: 1-11.

Waller, G.R., M.-C. Feng, and Y. Fujii. (1999). Biochemical analysis of allelopathic compounds: Plants, microorganisms and soil secondary metabolites. In *Principles and Practices in Plant Ecology: Allelochemical Interactions*, eds. Inderjit, K.M.M. Dakshini, and C.L. Foy. CRC Press: Boca Raton, Florida, pp. 75-98.

Wang, T.S.C., M.M. Kao, and B.W. Li. (1984). The exploration and improvement of the yield of monoculture sugarcane in Taiwan. In *Tropical Plants*, ed. C.H. Chou. Taipei, Taiwan, ROC: Academia Sinica, pp. 1-9.

Wang, Y., M. Toyota, F. Krause, M. Hamburger, and K. Hostettmann. (1990). Polyacetylenes from *Artemisia borealis* and their biological activities. *Phytochemistry* 29: 3101-3105.

Wardle, D.A., K.S. Nicholson, and A. Rahman. (1993). Influence of plant age on the allelopathic potential of nodding thistle (*Carduus nutans* L.) against pasture grasses and legumes. *Weed Research* 33: 69-78.

Wardle, D.A. and M.C. Nilsson. (1997). Microbe-plant competition, allelopathy and arctic plants. *Oecologia* 109: 291-293.

Watkinson, A.R. (1998). Reply from A.R. Watkinson. *Trends in Ecology and Evolution* 13: 407.

Weidenborner, M., H. Hindorf, H.C. Jha, and P. Tsotsonos. (1990). Antifungal activity of the flavonoids against storage fungi of the genus *Aspergillus*. *Phytochemistry* 29: 1103.

Weidenborner, M., H. Hindorf, H.C. Weltzien, and H.C. Jha. (1992). An effective treatment of legume seeds with flavonoids and isoflavonoids against storage fungi of the genus *Aspergillus*. *Seed Science and Technology* 20: 447.

Weidenborner, M. and H.C. Jha. (1993). Antifungal activity of flavonoids and their mixtures against different fungi occurring on grain. *Pesticide Science* 38: 347.

Weidenhamer, J.D. (1996). Distinguishing resource competition and chemical interference: Overcoming the methodological impasse. *Agronomy Journal* 88: 866-875.

Weller, D.M. and L.S. Thomashow. (1999). The role of microbial metabolites in biological control of plant pathogens. In *Recent Advances in Allelopathy Vol. 1. A Science for the Future*, eds. F.A. Macías, J.C.B. Galindo, J.M.G. Molinillo, and H.G. Cutler. Universidad de Cádiz, Spain: Servicio de Publicaciones, pp. 415-422.

Weston, L.A. (1996). Utilization of allelopathy for weed management in agroecosystems. *Agronomy Journal* 88: 860-866.

Weston, L.A., C.I. Nimbal, and P. Jeandet. (1999). Allelopathic potential of grain sorghum [*Sorghum bicolor* (L.) Moench] and related species. In *Principles and Practices in Plant Ecology: Allelochemical Interactions*, eds. Inderjit, K.M.M. Dakshini, and C.L. Foy. Boca Raton, USA: CRC Press, pp. 467-477.

Whittaker, R.H. and P.P. Feeny. (1971). Allelochemics: Chemical interactions between species. *Science* 171: 757-770.

Wilcoxon, F. and S.E.A. McCallan. (1935). Fungicidal action of organic thiocyanates, resorcinol derivatives and other organic compound. *Contributions of Boyce Thompson Institute* 7: 333-339.

William, R.D. and G.F. Warren. (1975). Competition between purple nutsedge and vegetables. *Weed Science* 23: 317-323.

Williams, J.G.K., A.R. Kubelik, J. Livak, J.A. Rafalski, and S.V. Tingey. (1990). DNA polymorphisms amplified by arbitrary primers are useful as genetic markers. *Nucleic Acids Research* 18: 6531-6535.

Williamson, G.B. (1990). Allelopathy, Koch's postulates and neck riddles. In *Perspectives in Plant Competition*, eds. J.B. Grace and D. Tilman. New York, USA: Academic Press, pp. 143-162.

Willis, R.J. (1997). The history of allelopathy. 2. The second phase (1900-1920): The era of S.U. Pickering and the U.S.D.A. bureau of Soils. *Allelopathy Journal* 4: 7-56.

Wolf, R.B., G.S. Spencer, and W.F. Kwolek. (1984). Inhibition of velvet leaf (*Abutilon theophrasti*) germination and growth by Benzyl isothiocyanate, a natural toxicant. *Weed Science* 32: 612-615.

Wu, H., J. Pratley, D. Lemerle, and T. Haig. (1999). Crop cultivars with allelopathic capability. *Weed Research* 39: 171-180.

Wyman-Simpson, C.L., G.R. Waller, M. Jurzysta, J.K. McPherson, and C.C. Young. (1991). Biological activity and chemical isolation of root saponins of six cultivars of alfalfa (*Medicago sativa* L.). *Plant and Soil* 135: 83-94.

Wyse, D.L. (1992). Future of weed science research. *Weed Technology* 6: 162-165.

Yakle, G.A. and R.M. Cruse. (1983). Corn plant residue age and placement effects upon early corn growth. *Canadian Journal of Plant Science* 63: 871-877.

Yakle, G.A. and R.M. Cruse. (1984). Effect of fresh and decomposing corn plant residue extracts on corn seedling development. *Soil Science Society of America Journal* 48: 1143-1146.

Young, A. (1804). *The Farmers Calender.* London.

Yu, J.Q. (1999a). Autotoxic potential of vegetable crops. In *Allelopathy Update: Basic and Applied Aspects*, ed. S.S. Narwal. NY: Science Publishers, Inc., pp. 149-162.

Yu, J.Q. (1999b). Allelopathic control of soil-borne diseases in vegetable crops. In *Allelopathy Update: Basic and Applied Aspects*, ed. S.S. Narwal. NY: Science Publishers, Inc., pp. 335-344.

Yu, J.Q. (1999c). Allelopathic suppression of *Pseudomonas solanacearum* infection of tomato (*Lycopersicon esculentum*) in a tomato-Chinese chive (*Allium tuberosum*) intercropping system. *Journal of Chemical Ecology* 25: 2409-2418.

Yu, J.Q. and Y. Matsui. (1997). Effects of root exudates of cucumber (*Cucumis sativus* L.) and allelochemicals on ion uptake by cucumber seedling. *Journal of Chemical Ecology* 23: 817-827.

Zhang, G., Zhang, W., Lian, B., Gu, L., Zhou, Q., and Liu, T.-X. (1999). Insecticidal effects of extracts from two rice varieties to brown planthopper, *Nilaparvata lugens*. *Journal of Chemical Ecology* 25: 1843-1855.

Zimdahl, R.L. (1999). *Fundamentals of Weed Science*. 2nd ed. San Diego: Academic Press.

Weed Allelopathy, Its Ecological Impacts and Future Prospects: A Review

J. R. Qasem
C. L. Foy

SUMMARY. The importance, characteristics, positive and negative impacts, and future role of weeds as an integral part of the natural and agroecosystems are evaluated and discussed. Interference between plants in nature and the importance of differentiating between competition and allelopathy are interpreted. Allelopathy as one component of weed/crop interference, allelochemicals from weed species and their possible mechanism of action are listed and discussed. Weed species with inhibitory action against cultivated crops, other weed species, and plant pathogens, as well as self-inhibitory (autopathic) species are reviewed. Stimulatory or inhibitory allelopathic effects of different crop

---

J. R. Qasem is Visiting Professor of Plant Physiology/Weed Science and Fulbright Scholar, on sabbatical leave from the Department of Agricultural Resources and Environment, Faculty of Agriculture, University of Jordan, Amman, Jordan.

C. L. Foy is Professor of Plant Physiology/Weed Science, Department of Plant Pathology, Physiology and Weed Science, Virginia Polytechnic Institute and State University, Blacksburg, VA 24061 USA.

Address correspondence to: J. R. Qasem, Professor, Department of Agricultural Resources and Environment, Faculty of Agriculture, University of Jordan, Amman, Jordan (E-mail: jrqasem@agr.ju.edu.jo).

The authors express gratitude to the University of Jordan, Virginia Polytechnic Institute and State University, and Fulbright Foundation for offering this opportunity to the first author to share this work, through a sabbatical leave program funded through a Fulbright Award in 1999-2000.

[Haworth co-indexing entry note]: "Weed Allelopathy, Its Ecological Impacts and Future Prospects: A Review." Qasem, J. R., and C. L. Foy. Co-published simultaneously in *Journal of Crop Production* (Food Products Press, an imprint of The Haworth Press, Inc.) Vol. 4, No. 2 (#8), 2001, pp. 43-119; and: ***Allelopathy in Agroecosystems*** (ed: Ravinder K. Kohli, Harminder Pal Singh, and Daizy R. Batish) Food Products Press, an imprint of The Haworth Press, Inc., 2001, pp. 43-119. Single or multiple copies of this article are available for a fee from The Haworth Document Delivery Service [1-800-342-9678, 9:00 a.m. - 5:00 p.m. (EST). E-mail address: getinfo@haworthpressinc.com].

plants, trapping and catching species, and the potential of allelopathic weeds in inhibiting or stimulating certain parasitic weed species are discussed and evaluated. Allelopathy as a mechanism and future strategy for agricultural pest control and farm management and the potential use and development of some allelochemicals as natural pesticides or plant growth regulators are also considered and discussed.

KEYWORDS. Allelopathy, weeds, weed allelochemicals, autotoxicity, antifungal chemicals, parasitic weeds, ecological implications, genetic engineering

INTRODUCTION

Weeds and Their Significance in the Agroecosystem

Weeds are defined in anthropocentric terms by agriculturists, including weed scientists and crop growers as certain groups of plants that are considered objectionable under certain given situations. They are diverse in their habit and habitats throughout the world. Although they account for not more than 1% of the total plant species on the earth, nevertheless they cause great problems to humankind by interfering in food production, health, economic stability, and welfare. The term *Weed* is not a natural one but rather one given to those plants that are deemed to be out of place, growing where farmers do not desire them. They are heavy consumers of environmental resources and plant growth factors, undesirable, aggressive, and highly competitive, invaders, which appear by themselves without being sown by farmers, diverting energy to a direction undesired by man. They are uniquely tolerant to unfavorable environmental conditions either themselves or by their propagules, and tolerate well older methods of control. Many of them even resist modern synthetic herbicides. Their ultimate values in natural and agroecosystems have not been completely discovered, but many of their negative and positive impacts on both systems are very familiar to growers, weed scientists, ecologists, and other specialists in agricultural and environmental sciences.

Weeds as an agricultural problem would not be recognized without human activities and interference in the environment. They did not exist as a problem in the balanced natural ecosystem before agriculture was developed. Therefore, man has created this problem and he is the only concern in finding a solution to the problems they cause to agriculture, the environment, and human life and activities. Thus, weeds are as old as agriculture itself and have substantially adapted characteristics that enable them to grow, flourish, invade, and dominate an important part of the natural system. Absolutely they represent a significant component of the agroecosystem resulting from the continuous selection pressure imposed on them by man, his tools, agricultural practices, and methods of weed control. They have long been exposed to both natural and human breeding and selection systems that enabled them to survive and insure their future generations. They have also proved to be excellent invaders into human made habitats, taking advantage of all measures that man imposed to eliminate or to keep them fully under his control. This is most likely the reason behind their existence and excellent adaptation, performance, and abundance (in certain cases) in the environment.

Traditionally, weeds are characterized and described by their abundant seed production (annuals, biennials and many perennials), vegetative reproduction, ease of dissemination of their reproductive organs, long period of seed dormancy which may extend to thousands of years enabling them to avoid and escape environmental and human hazards, and seed modifications which adapt them for short and long distance dispersal to invade and colonize a new area. Seed resistance to animal digestion and other morphological structures facilitate their movement through animals and manure and dissemination to different geographical and environmental sites. Seed polymorphism, is an excellent indicator of the effect of environmental conditions and the response, modification and behavior of certain weed species to changes in the surrounding environment. This trait affects growth and survival, therefore it is another character that contributes positively to the survival of this group of plants. This character appears to have been important in helping certain weed species and their future generations to exist in the environment, especially if it is combined with a weed's ability to reproduce vegetatively and to tolerate adverse environmental conditions.

At present, with the rapid development in agriculture technology through genetic engineering, molecular biology, gene replication and transfer, herbicide technology, and the aid of cell and tissue culture techniques, it is expected that the near future will witness new methods and characteristics of weeds in response to all or some of these advances in biotechnology. Herbicide resistance of weed species had not occurred at the time when much of this technology was developed; whereas nowadays, more than 110 weed species have been reported to resist chemical treatments of different individual or groups of herbicides and some of these species are also resistant to the newly developed herbicides (Ross and Lembi, 1999).

Certainly, weeds will continue to exist in nature as an integral part of the natural ecosystem and a significant element in our agroecosystem. Considering this fact, the best way to counteract their problems is to live with them, but through remediation and rebalancing the environment for better accommodating our primary interest. This may be achieved through a proper revision of our understanding toward these organisms from all aspects. Subsequent reconsideration of our control measures adopted against them seems important. These may be replaced by safer, more reliable, and better-managed methods in both natural and agroecosystems. We strongly believe that weed scientists in all of their fields of specialization (biology, ecology, physiology and chemistry) understand and are aware of nature's laws. Therefore, increasingly, they are respecting and embracing the recently emerged approach in weed science titled "weed management" as an alternative for chemical weed control *per se*. While this approach is not new to farmers, and has long been traditionally and partially practiced in different parts of the world, its scientific implication is presently witnessing a wider acceptance and acknowledgement as a strategy for weed control.

Achievement of 100% control of weeds probably is an unrealistic objective in a balanced ecosystem if resistance, allelopathic potential, and other noxious characteristics of weeds are to be avoided and a less polluted environment is our final and ultimate goal. Less numbers of weeds or low dense weed populations may have some positive effects. Better crop growth through chemical stimulation or competition response has been reported and this may be important for weed species serving as hosts for certain parasitic weeds, to disturb their germination and attachment to host crops or by serving as catch or parasite trapping species.

Weed Impact in Agro- and Natural Ecosystems

As stated before, weeds represent highly successful organisms that form an important part of the natural and agroecosystems. From the evolutionary point of view they are considered as pioneers of secondary succession, colonizers, and competitors, taking advantage of human-made habitats, and highly responsive to changes in environmental conditions in such a beneficial and advantageous way enabling them to survive and grow in nature (Grime, 1986). Therefore, they represent a good example of the ability of organisms to adapt to the continued disturbance that humans create in their environment. Understanding the mechanisms through which weeds respond and behave to changes in the environment may open the way toward their control.

Clearly there is much concern by farmers and food producers toward weeds and their direct and indirect harmful effects on agriculture that are to be expected under different environmental conditions (Patterson, 1985). This may also be intensified in the future, since weeds are opportunistic species, rapidly modifying their behavior with a high phenological plasticity to suit better any newly prevailing conditions. For this reason, concern is growing among farmers and weed scientists regarding recent developments in weed control technology. There is great fear about the future situation and what solutions will be required if some species of weeds became well adapted to this technology. How will the new problem impact food and feed production? Taking into consideration that different weeds and crop plants are genetically related, the potential for gene exchange between these is quite high and well demonstrated. However, until now weed scientists and farmers have been more concerned with the disadvantages and negative effects rather than advantages or positive impacts of weeds on agricultural and environmental systems. This concern, however, may be justified since reflecting the high degree of awareness in the relative success that weeds have achieved and may attain in the future, and the fear from declining or failure in controlling or counteracting their problems.

Concentration on the negative impact of species often referred to as weeds not negated some positive impacts that weeds may have in nature. Therefore, another more balanced term "*Anthropomorphic characteristics of weeds*" has emerged. This term is frequently used in weed science to describe both positive and negative roles that weeds may have in natural and agroecosystems. It includes regarding many weed species as wild flowers, that prevent soil erosion, balance soil nutrients, supply parasite preferred food in absence of crop plants, provide suitable habitats for predators, supply nutrients to the soil especially legume weeds, and in this regard the positive impact of weed competition on protein content of the competing species is not to be ignored. Some additional sources of genetic materials for plant breeding programs, soil and climatic indicators, medicinal, food or feed species, or forage and rangeland species, traps for insects and plant diseases, and probably many other features yet may be discovered. However, in spite of all of the above mentioned advantages of these species to humans and agriculture, most attention continues to be directed toward their negative impacts, and therefore different terms have emerged and are used in this regard.

In agriculture, weeds are of concern because they compete with cultivated crop plants for growth factors including water, light, nutrients, space and perhaps under certain conditions, carbon dioxide (essential for photosynthesis). Weeds harbor insects and plant pathogens and serve as hosts for parasitic weeds, adding more complications to the control of these important pests. Weed interference in agriculture deserves attention because of its effect on quantity and quality of agricultural (plant and animal) products and increasing the cost of farm operations and subsequently production, and processing costs. They interfere with water resources and management either through direct utilization of water for growth and abundance or by blocking waterways and directly or indirectly spoiling and polluting water resources. They impair human and animal health, create fire hazards, and are aesthetically unappealing (Zimdahl, 1999).

ALLELOPATHY AS ONE COMPONENT OF WEED/CROP INTERFERENCE

The concept of interference was first proposed by Harper (1964) to include both competition for growth factors, and allelopathy resulting from the release of toxic chemicals that are added to the surrounding environment. The adverse effect of a neighboring plant in an association is termed interference by Muller (1969) who postulated that two components, competition and allelopathy, contribute to plant interference. More recently, Zimdahl (1999) defined interference as "the total adverse effect that plants exert on each other when growing in a common ecosystem and it includes competition and allelopathy." However, all of the above definitions highlighted the negative side of the term and ignored any possible positive impact of it on the associated organisms. The concept of the term, however, should include all kinds of relationships (positive, negative, or neutral) among and between species and populations. We strongly believe that widening the concept of the term to include both positive and negative interactions should better define the recent revision of the term allelopathy (Rice, 1984) and the definition given by the International Allelopathy Society (IAS, 1996) based on Molisch's understanding of the term. Otherwise any stimulatory or positive impact of allelopathy should not be regarded and included under this title.

The main cause of interference may include consumption and depletion of limited resources, production of toxic chemical compounds that are released into the environment, parasitism, and protection. However, negative interference includes competition and other deleterious effects of one plant on another, including allelopathy and parasitism.

Competition, as one component of interference, arises from the interaction between individuals of the same or different species in response to limited supply in one or more of the environmental factors. It occurs when the supply of certain environmental factors fall below the actual total demand for them by competing individuals, and often leads to the relative success of one individual or plant species and relative failure of another (Qasem, 1987). Clements (1904, 1928) was the first who tried to explain the cause, nature, and extent of competition between plants and concluded that competition is a purely physical process, and with few exceptions, results from plants, interaction over physical environmental factors. Pavlychenko and Harrington (1934) defined competition as "a powerful natural force tending toward limitation or extinction of the weak competitor, it occurs over physical factors." Grime (1986) defined competition as "tendency of two neighboring plants to utilize the same quantum of light, molecule of water, ion of mineral, or volume of space," and by this he specified the factors and the conditions under which competition may occur. However, Grime intelligently avoided any indication of competition between plants under high supply of these factors, since the actual requirements of each competing individual is to be determined and luxury accumulation of certain factors (e.g., nutrients) by certain species may not respect the actual need of these species. However, word selection in Grime's definition of competition allowed definition of the term according to mechanism rather than its effect and, based on that, competition is solely due to resource acquisition which is only one part of the mechanism through which one plant affects location and stability of its neighbors through manipulation or modification of the environment. According to Zimdahl (1999), competition is the active acquisition of limited resources by an organism that results in reduced supply and consequently reduced growth of other organisms. All of the above definitions are ultimately in agreement, in that limitated availability of one or more growth factors is required for competition to occur. However, no logical and convincing explanation has been given for the luxury acquisition of resources by certain species and its relation to conditions under which competition occurs. This, however, may have its own impact and interpretation on the relationship and possible interaction effects of allelopathy and competition in nature, since a high supply of environmental factors should not overcome allelopathic effects of one species on another but most likely should lessen the influence of competition up to a certain stage of development. This may appear in conflict with the conditions under which allelopathy can operate.

While competition between plants has been observed for hundreds of years, it is only recently that man discovered that competition was only one component of plant relationships. Allelopathy, is another term introduced to the science of plant ecology to describe the addition of chemical compounds (toxic or non-toxic) from a plant into the environment that affects germination, growth, health, development, and population biology or behavior of another plant species.

The allelopathy phenomenon in nature has concerned farmers and scientists through ages, whereas allelopathy mechanisms of plant interaction are an important component of the aggressive nature of many species in both natural and agroecosystems. The history of this science may go back to Theophrastus (300 BC) who speculated that there might be chemical interaction between weeds and plants but provided little firm evidence to substantiate this claim. De Candolle (1832), a pioneer in allelopathy research of weeds on crop plants, concluded that exudates of certain weed species injured specific crop plants. However, Rice (1974, 1983, 1984, 1995) gave comprehensive documentary references on the history of allelopathy. These reviews revealed that allelopathy between plants in nature noticed in pre-historic time and some of our common weeds were the first to inflict or afflict by this mechanism (Theophrastus ca. 300 B.C; De Candolle, 1832).

The radical idea put forward by De Candolle (1832) through his research stimulated interest in the chemical ecology of plants, but it was not until 1937 that Hans Molisch coined the term allelopathy and defined it as "stimulatory or inhibitory biochemical interactions between and among plants of all levels of complexity including microorganisms." However, this was a somewhat erroneous definition because the Greek words from which allelopathy was derived *allelo* or "each other," similar to the Greek word *allelon* means "one another" and the suffix *patho* or *pathose* means suffering, disease or intense feeling (Zimdahl, 1999), infer a different meaning; therefore, the literal meaning of the term negates any possible positive action of allelopathic chemicals which agrees with neither the term, nor with the effects of most concern to and farmers and nonfarmers (Rice, 1984).

Other definitions of the term allelopathy were given by different authors (Bridges, 1995; Cardina, 1995; Anderson, 1996; Aldrich and Kramer, 1997; Zimdahl, 1999); some were consistent with the general concept of the term and the spirit of Molisch's definition of allelopathy, while others referred to it as a negative process. However, all emphasized, as a condition, the release of the chemicals involved into the environment. This clearly demonstrates that allelopathy is a natural phenomenon and chemicals involved in this process should be naturally released in amounts enough to exert the effect. Any deviation from this understanding of the term's spirit should be regarded as an artificial approach.

It is worth mentioning, that this science has undergone a great deal of development in the last 63 years (1937-1999), since allelopathy research in various fields has shown that it could be applied in the fields of agriculture and forestry, viz., plant production, plant protection, agroforestry systems, and other fields of science. In addition it may become one of the strategies to reduce environmental pollution and increase agricultural production in sustainable agriculture in the future. Thus, it is receiving more attention from agriculturists and bioscientists.

In the last two decades there has been an increase in publications on allelopathy and a considerable amount of literature is available that implicates allelopathy as an important form of plant interference (Rice, 1983, 1984, 1995; Putnam and Tang, 1986; Waller, 1987; Rizvi and Rizvi, 1992; Narwal, 1994; Kohli, Batish, and Singh, 1998; Inderjit and Keating, 1999; Inderjit, Dakshini, and Foy, 1999; Macias et al., 1999). In relation to weed allelopathy, Putnam (1983) estimated about 50 weed species that possess allelopathic properties; then three years later, Putnam and Tang (1986) and Patterson (1986) cited about 90 weed species with allelopathic potential.

More recently, Kohli, Batish, and Singh (1998) listed about 50 weeds; Inderjit and Keating (1999) listed 112 common weed species in agroecosystems with allelopathic activity. However, in the present review 240 weed species are reported to have inhibitory action on crop plants only. Therefore, major progress in the science has recently occurred and the phenomenon is of worldwide importance, so that now allelopathy impacts virtually all plant science and pestology disciplines as well as microbiology and natural product chemistry.

Competition and Allelopathy in Context

The previously given information on the components of interference clearly showed that two operating active mechanisms are mainly involved in plant-plant interactions at all levels of complexity. However, a feature that distinguishes allelopathy from competition is that some factors are being added to the environment in allelopathy as opposed to some factors being removed from it in competition. Density dependent effect is another major factor differentiating between allelopathy and competition. Competition increases with density up to certain limits after which no more reduction in crop yield occurs, however, yield per plant basis continues to be decreased. In the case of allelopathy, increasing plant density has been shown to reduce the effects of toxins (Weidenhamer, Morton, and Romeo, 1987; Weidenhamer, Hartnett, and Romeo, 1989; Thijs, Shann, and Weidenhamer, 1994). Possible allelopathic effects are most likely occurring when competition for resources does not seem adequate to explain the observation. Therefore, a dramatic decrease in plant biomass or mortality is usually evident for one species but not the other. Some other features differentiating the two components are listed in Table 1.

Allelopathy Techniques: Problems and Proofs

One of the main challenges facing researchers in the field of allelopathy is the lack of suitable experimental techniques to enable separating allelopathy from other mechanisms of plant interference, and mainly competition. This issue has questioned the occurrence and the significance of allelopathy in nature and its impact on the agroecosystem, and created an intense debate among scientists in different disciplines. While allelopathy researchers argue that the subject has a significant ecological role, many ecologists are merely tolerant to this opinion and claim that allelopathic observations are purely experimental and have no ecological impact. The skepticism arises from the experimental techniques used, and the major problem encountered in designing a convincing research program has been the subtle nature of allelopathy

TABLE 1. Some of the differences in nature and mechanism of action of allelopathy and competition.

Allelopathy	Competition
Mainly due to chemical compounds added to the environment	Results from shortage in supply of environmental factors
No specific plant characteristics for allelopathy	There are certain plant characteristics of the competitors
The effect can occur at any growth stage of the donor species and toxic chemical is most concentrated at flowering and late-growth stages	The effect is more intense at early growth stages and during the critical period of weed competition
Can't be overcome or reversed by higher supply of growth factors	Can be totally reversed at certain growth stages by increasing the levels of growth factors in the environment, or reducing its effect at other growth stages.
Results from the interaction between environmental factors but not competition for them	Occurs due to vying for environmental factors
Characterizing certain plant species and most likely observed with perennials and aggressive annual weeds	All species can cause competition but at certain densities
Donor density increases the effect but receiver density decreases it	Density of all competing species increases the effect through inter- and intra-specific competition
Can occur in absence of the donor species in many cases or without being with the receiver plant in time and place (volatilization, leaching, debris, etc.)	Competing species must be found together in place and time
Principally, occurs naturally but there are external factors affecting it	Can be created by farmers and their agricultural interferences
It is an outcome of competition	Not resulting from allelopathy influence
Greatly influenced by soil microorganisms and other soil factors can modify its action	Less affected by soil factors especially microorganisms
Microorganisms may intensify or nullify the effect	No role of microorganisms in preventing its occurrence but may reduce its effect by supplying more nutrient, especially from nitrogen fixing-bacteria
Does not require a certain density of the donor species to occur	It requires certain density to start (critical density)
Afflicted species show morphological, physiological, and anatomical symptoms	The effect is mainly reflected and observed on the general appearance of all species involved
Only receiver plants suffer	All competing plants are negatively affected but at various degrees depending on their competitive abilities
Influence of environmental conditions (e.g., light quality, temperature, day length) is important for its occurrence	Influence of these factors is perhaps some what less appreciated
May be caused by living or dead plants	Occurs only between living organisms
With reduction in growth or plant size, toxic chemicals become more concentrated	With reduction in growth or plant size, the competitive ability decreases

Allelopathy	Competition
All plant parts may be involved in the process	Inter-specific competition only occurs between shoots or roots
Donor plants tend to be found in patches or colonies in nature	No spatial growth arrangement
Root and shoot exudates have a significant role in the process	They are not a factor in extenuating or intensifying the effect if they are toxins
Symptoms appear in the form of toxicity, inhibition of growth, subsequent death of cells and tissues, or total unrecoverable collapse of the whole plant	In most cases, mineral deficiency symptoms are common, but mostly plants can complete their life cycles and produce
Some allelopathic species have an aromatic or special odor or emanate volatile materials	Odor and volatility are not important factors
Excessive amounts of growth factors are neither a causal factor nor do they overcome the effect	Excessive amounts of growth factors, lead to toxicity or negative effect
Can affect seed germination	Does not affect this process but does affect growth, development, and yield
Positively influenced by herbicides	Mostly all are negatively affected by herbicides at certain stage of development
Depends on chemical compounds in plants or the environment that can be isolated, identified, and then synthesized	Depending on chemicals, and their presence in the environment or in plants is not a condition
Agricultural practices can't negate the influence after removal of the donor	Agricultural practices may deny completely its effect when the competitor is removed
May be more selective in its action	Not selective at critical density and all species can be affected
Can't be artificially induced, by definition	Can be induced artificially under laboratory or glasshouse conditions
Affects internally and externally or internal effect reflected externally	The effect is mainly external but reflected internally
Can be demonstrated even on one individual plant	There should be more than one individual plant involved to show the effect
Many of allelopathic species have herbicidal or pesticidal action	The harmful effect is mainly internal and a physiological one
Mostly found in wild species	Strong and found in cultivated species

in certain species and the complexity of plant interactions. However, the large number of publications on plant allelopathy and allelopathic chemicals, and the progress in research has not been associated with similar progress in development of allelopathy techniques used in these studies. In spite of the diverse number of the techniques suggested (Einhellig, 1986; Qasem and Hill, 1989a), finding a single publication confirming results using different experimental techniques is rare, thus these may remain far from differentiating between the effect of allelopathy and that of competition and hardly isolating the effect of the former mechanism on the target species. Researchers apparently have a wide choice of extraction techniques that they can use. It is critical, however, that natural conditions should be simulated in these methods. Extraction techniques may be useful for the first test to explore allelopathic plants, but the problem remains in widely regarding and solely using this method as the main basis to judge allelopathic activity of plant species (see Tables 2, 3, 4). It is preliminary, perhaps misleading, and an artificial method. It doesn't reflect the natural conditions of the phenomenon and the free release of allelochemicals in nature. The solvents used sometimes are toxic to plants, and may be hot or cool. Plant tissues, chopped or non-chopped, boiled or non-boiled, the amount used for extraction and volume of solvent used, are usually varied. Sterilizing or not of plant tissues is another problem added to the complications of plant age, growth stage, and location from where they collected. Maceration or not of plant tissues, extract pH and osmotic potential, extracting of toxic and non-toxic, releasable or not of chemicals in plant cell, all account for more difficulties in full reliance on this method. Moreover, conditions of the experiment and the lay of target seeds or seedlings, either direct soaked in extracts or sown in/on sand soil, agar, or cotton media in Petri-dishes, all affect the results obtained and are artificial conditions and materials irrelevant to the nature of allelopathy phenomenon. In addition, many experiments conducted under laboratory or glasshouse conditions can be precisely controlled. This may raise questions as to the ideal (natural) conditions under which an experiment should be performed. Many researchers have shown that they can extract an allelochemical but unfortunately failed to elucidate its path to a receiver species. Therefore, the mere presence of allelopathic compounds in plant tissues does not demonstrate allelopathy (Heisey, 1990). Altieri and Doll (1978) emphasized the problem in having sufficient concentrations of toxins to influence the growth of a receiver plant by concluding that it is easy to extract toxins and easy to demonstrate plant interactions, but difficult to show that the two are naturally related, and if so how. Therefore, the choice of the bioassay used to assess the importance of allelopathy is a major determinant of the conclusions. For example it is insufficient to conclude that a plant will be influenced when it is mature just because its seed germination or seedling development was inhibited by allelochemicals from a donor plant. Problems associated with allelopathy techniques were referred to by different researchers (Qasem and Hill, 1989b; Blum, 1999; Foy, 1999).

Allelopathy has been reported for many crop and weed species (Rice, 1974, 1984; Kohli, Batish, and Singh, 1998; Inderjit and Keating 1999) but strong evidence concerning its occurrence in nature is lacking. Putnam and Duke (1978) suggest that several reasons account for the difficulty in separating allelopathy from competition. These include: a general lack of nomenclature to adequately describe plant responses that occur in this manner, a lack of suitable

TABLE 2. Weed species with allelopathic inhibitory activity on crop plants.

Weed species	Source of toxin	Afflicted crop species	Reference
Abutilon theophrasti	Residue, seeds	Radish, corn, soybean, tomato	Gressel and Holm, 1964; Bhowmik and Doll, 1984
Acalypha indica	Extract	Bajra	Murthy and Zakharia, 1980
Ageratum conyzoides	Extract, leachate, volatile	Wheat	Sugha, 1979; Chuihua, Fei, and Shoushan, 1999
Agropyron repens	Extract, residues	Bean, soybean, alfalfa	Weston and Putnam, 1986
Allium vineale	–	Oat	Osvald, 1950
Alternanthera triandra	Root washings	Soybean, groundnut, blackgram	Tiwari et al., 1985
Amaranthus blitoides	Extract, residue	Barley, wheat	Qasem, 1994b, 1995b
Amaranthus dubius	Extract, leachate, residue	Beans, corn, sorghum	Altieri and Doll, 1978
Amaranthus gracilis	Extract	Wheat, barley	Qasem, 1994b, 1995b
Amaranthus palmeri	Extract, residue, vapor	Cabbage, carrot, onion, sorghum	Bradow and Connick, 1987; Menges, 1988
Amaranthus retroflexus	Extract, exudate, leachate, residue	Barley, cabbage, carrot, cauliflower, corn, cotton, eggplant, pepper, squash, soybean, tobacco, tomato, wheat	Bhowmik and Doll, 1979, 1983; Menges, 1987, 1988; Qasem, 1994b, 1995b,c
Amaranthus spinosus	Living plant	Coffee	Vander Veen, 1935
Ambrosia artemisifolia	Extract, exudate, residue	Alfalfa, corn, rye, sorghum, tomato	Jackson and Willemsen, 1976; Bhowmik and Doll, 1984
Ambrosia elatior	Living plant	–	Toth, Benecs-Bardi, and Beres, 1999
Ambrosia psilostachya	Extract, leachate	*Andropogon ternarius*, several	Neill and Rice, 1971
Ambrosia trifida	Extract	Peas, radish, sorghum, wheat	Le Tourneau, Failes, and Heggeness, 1956; Rasmussen and Einhellig, 1979
Ammi majus	Extract, residue	Wheat	Qasem, 1995a
Anagallis foemina	Extract	Wheat	Qasem, 1995a
Anaphalis araneosa	Extract	Wheat	Tripathi, Singh, and Rai, 1981
Antennaria microphylla	Extract	Lettuce, wheat	Selleck, 1972; Manners and Galitz, 1986
Antennaria neglecta	Extract	Wheat	Selleck, 1972
Anthemis cotula	Residue	Wheat	Qasem, 1995a
Argemone mexicana	Living plants	Bajra, mustard, radish, wheat	Sharma and Nathawat, 1987
Aristolochia clematitis	Extract	Lettuce	Toma et al., 1999
Artemisia absinthium	Residue	Bean, fennel, pea	Bode, 1940; Funke, 1943

TABLE 2 (continued)

Weed species	Source of toxin	Afflicted crop species	Reference
Artemesia annua	Extract	Grasses, corn	Lydon, Teasdale, and Chen, 1997
Artemisia californica	Volatile	Cucumber	Muller, Muller, and Haines, 1964
Artemisia princeps	Volatile	Apple	Ito et al., 1981
Artemisia vulgaris	Leachate, volatile	Cucumber, peppergrass, red clover	Mann and Barnes, 1945; Melkania, Singh, and Bisht, 1982; Inderjit and Foy, 1999
Asclepias syriaca	Extract	Sorghum, wheat	Le Tourneau, Failes, and Heggeness, 1956; Rasmussen and Einhellig, 1975
Asphodelus tenuifolius	Living plants	Wheat	Porwal and Gupta, 1986
Aster sp.	Extract	Black cherry, red pine	Horsley, 1977; Norby and Kozlowski, 1980
Avena fatua	Exudate	Wheat, several	Tinnin and Muller, 1971; Schumacher, Thill, and Lee, 1983
Berteroa incana	Extract, residue	Corn, soybean	Bhowmik and Doll, 1979
Bidens bipinnata	Soil extracts	7 crops	Meissner, Nel, and Beyers, 1986
Bidens pilosa	Root exudate	Bean, corn, lettuce, sorghum	Stevens and Tang, 1985
Boerhavia diffusa	Extract	Bajra, til	Sen, 1976
Borreria articularis	Extract	Jute, mustard	Biswas and Chakraborti, 1984
Bothriochloa pertusa	Extract, exudate, residue	Corn, lettuce, pepper, tomato	Rice, 1984
Brachiaria mutica	Leachate	Lettuce, rice, wheat	Chou, 1989
Brassica campestris	Extract	Barley, radish, wheat	Jimenez-Osornio and Gliessman, 1987
Brassica napus	Extract, volatile	Lettuce, wheat, several	Muller, 1969, Mason-Sedun, Jessop, and Lovett, 1986
Brassica nigra	Extract	Wheat, several	Muller, 1969, Mason-Sedun, Jessop, and Lovett, 1986
Bromus japonicus	Extract	Nitrogen-fixing bacteria	Rice, 1964
Bromus tectorum	Extract	Nitrogen-fixing bacteria	Rice, 1964
Calendula officinalis	Residue	Wheat	Qasem,1995a
Calluna vulgaris	Exudate	Oat	Salas and Vieitez, 1972
Camelina alyssum	Leachate	Flax	Grummer and Beyer, 1960
Camelina sativa	Leachate	Flax	Grummer and Beyer, 1960; Lovett and Duffield, 1981
Cannabis sativa	Leachates, volatiles, residues	Mustard, sorghum, blackgram, fenugreek	Inam, Hussain, and Bano, 1989
Carthamus nitidus	Extract	Wheat	Qasem, 1995a

Weed species	Source of toxin	Afflicted crop species	Reference
Carduus nutans	Extract, leachate, residue	Pasture grasses and legumes	Wardle, Nicholson, and Rahman, 1993
Cassia sophora	Leachate	Bean, lettuce, mustard, wheat	Datta and Dasmahapatra, 1984
Celosia argentea	Extract, exudate	Bajra, til	Pandya, 1975; Ashraf and Sen, 1978
Cenchrus biflorus	Extract	Bajra, til	Sen, 1976
Cenchrus pauciflorus	-	-	Rice, 1964
Centaurea diffusa	Residue	Barley, ryegrass, tomato	Fletcher and Renney, 1963; Muir and Majak, 1983
Centaurea maculosa	Residue	Barley, tomato	Fletcher and Renney, 1963; Locken and Kelsey, 1987
Centaurea nigra	Residue	Tomato	Vezina and Doyon, 1983
Centaurea repens	Residue	Barley	Fletcher and Renney, 1963
Centaurea solstitialis	Toxins	Alfalfa, cucumber, lettuce	Merril, 1989
Cerastium decotomum	Extract	Wheat	Qasem, 1995a
Chenopodium album	Extract, exudate, residue	Corn, cucumber, oat, soybean, tomato, wheat, maize	Bhowmik and Doll, 1979; Caussanel, 1979; Porwal and Gupta, 1986; Qasem and Hill, 1989a
Chenopodium ambrosioides	Extracts, volatile oils	Mothbean, *Amaranthus hypochondriacus*	Jimenez-Osornio, Kumamoto, and Wasser, 1996
Chenopodium murale	Extract, leachate, residue	Barley, cabbage, carrot, eggplant, pepper, wheat	Qasem, 1990, 1993, 1995c
Chromolaena odorata	Exudate, leachate, residue	Chinese cabbage, cucumber, pepper, rape, spinach, teak	Sahid and Sugua, 1993; Ambika, 1999
Cirsium arvense	Extract, leachate, residue	Alfalfa, barley, wheat	Stachon and Zimdahl, 1980; Hussain, Khan, and Hussain, 1987
Cirsium discolor	Extract	Wheat	Le Tourneau, Failes, and Heggeness, 1956
Citrullus colocynthis	Extract	Bajra	Bhandari and Sen, 1971
Citrullus lanatus	-	-	Bhandari and Sen, 1972
Clerodendrum viscosum	Leachate	*Brassica nigra*	Datta and Chakrabarti, 1982
Convolvulus arvensis	Extract	Wheat	Qasem, 1995a
Convolvulus sepium	Leachate	*Danthonia sericea*	Quinn, 1974
Crotolaria pallida	Leachate	Bean, lettuce, mustard, wheat	Datta and Dasmahapatra, 1984
Croton bonplandianum	Extract, leachate	Barley, lettuce, mustard, rice	Datta and Sinha-Roy, 1975, 1983
Cucumis cellosus	Extract	Bajra, til	Sen, 1976

TABLE 2 (continued)

Weed species	Source of toxin	Afflicted crop species	Reference
Cynodon dactylon	Extract, residue	Barley, blum, coffee, orange, peach	Vander Veen, 1935; Horowitz and Friedman, 1971;Weller, Skroch and Monaco, 1985; Meissner, Nel, and Beyers,1989
Cyperus esculentus	Extract, residue	Corn, lettuce, oat, sorghum, orange	Tames, Gesto, and Vieitez, 1973; Horowitz and Friedman, 1971; Drost and Doll, 1980; Bezuidenhout and Reinhardt, 1999
Cyperus kyllingia	Essential oil	Grassland, lawns	Komai and Tang, 1989
Cyperus rotundus	Extract	Barley, radish, rice, sorghum, soybean	Das and Pal, 1970; William and Warren, 1975; Tang et al., 1995
Daboecia polifolia	Extract	Oat	Salas and Vietez, 1972
Datura innoxia	Extract, volatile	Bajra, lettuce, pepper, pea, tomato	Hussain et al.,1979
Datura stramonium	Leachate	Flax, sunflower	Lovett et al., 1981; Levitt, Lovett, and Garlick,1984
Delphinium ajacis	Toxin	Pea	Waller and Burston, 1969
Dichanthium annulatum	Extract	Lettuce, mustard, bajra	Dirvi and Hussain, 1979
Digera alternifolia	Leachate	Cotton, eggplant, onion, pepper, pea, tomato	Dubey, 1973; Ashraf and Sen, 1980
Digera arvenis	Exudate	Bajra	Sarma, 1974
Digera muricata	Extract	Legume horsegram	Suseelamma and Raju, 1992
Digitaria decumbens	Extract	Lettuce	Chou, 1987; Chou and Young, 1975
Digitaria sanguinalis	Extract, residue	Apple, corn, several	Parenti and Rice, 1969; Ito et al., 1981
Diplotaxis erocoides	Extract, residue	Wheat, tomato	Qasem, 1995a; Obeid, 1996
Echinochloa crus-galli	Extract	Corn, soybean, others	Bhowmik and Doll, 1979
Echinops echinatus	Extract	Pearl millet, til	Jha and Sen, 1981
Eichornia crassipes	Extract	Radish	Ahmed, Ito, and Hussain, 1982; Sun, Yu, and Yu, 1988
Elytrigia repens	Extract, leachate, residue	Alfalfa, barley, bean, flax, legumes, wheat	Kommedahl, Kotheimer, and Bernardini, 1959; Weston and Putnam, 1986
Encelia farinosa	Extract	Corn, pepper, tomato	Gray and Bonner, 1948
Eragrostis poaeoides	Extract, residue	Bajra, mustard	Hussain, Zaidi, and Chughtai, 1984
Erica arborea	Extract	Oat	Salas and Vieitez, 1972
Erica australis	Extract	Alfalfa, ryegrass	Carballeira and Cuervo, 1980
Erica ciliaris	Exudate	Oat	Salas and Vieitez, 1972
Erica cinerea	Exudate	Oat	Salas and Vieitez, 1972

Weed species	Source of toxin	Afflicted crop species	Reference
Erica mediterranea	Exudate	Oat	Salas and Vieitez, 1972
Erica scoparia	Extract	Red clover, oat	Salas and Vieitez, 1972; Ballester, Albo, and Vieitez, 1977
Erica tetralix	Exudate	Oat	Salas and Vieitez, 1972
Erica umbellata	Exudate	Oat	Salas and Vieitez, 1972
Erigeron spp.	Extract	Eulalia (*Miscanthus sinensis*)	Kobayashi et al., 1980
Eruca sativa	Extract	Wheat	Qasem, 1995a
Erucaria hespanica	Extract	Wheat	Qasem, 1995a
Eugenia unifora	Extract	Chicory	Aquila, Silva and Laufer, 1999
Eupatorium adenophorum	Extract, residue	Garden pea, wheat	Tripathi, Singh, and Rai, 1981; Song et al., 1999
Eupatorium euppillifolium	–	Barley	Cited by Zimdahl, 1999
Eupatorium maximiliani	Extract	Common bean, lettuce, rice	Correa et al., 1999
Eupatorum odoratum	Leachate, residue	Fenugreek, cowpea	Fadayomi and Oyebade, 1984; Ambika and Jayachandra, 1980
Eupatorum riparium	Extract	Wheat	Tripathi, Singh, and Rai, 1981
Euphorbia corollata	Toxin	–	Rice, 1964
Euphorbia esula	Extract, residue	Pea, tomato, wheat	Le Tourneau and Heggeness, 1957; Manners, 1987
Euphorbia geniculata	Extract	Wheat	Sugha, 1979
Euphorbia granulata	Leachate	Lettuce	Hussain, 1980
Euphorbia helioscopia	Residue	Wheat	Qasem, 1995a
Euphorbia hirta	Root secretions	Groundnut, soybean	Tiwari et al., 1985
Euphorbia prostrata	Extract	Alfalfa, cotton	Alsaadawi, Sakeri, and Dulaimy, 1990
Euphorbia supina	Extract, exudate	Several	Brown, 1968; Rice, 1969
Falcaria vulgaris	Extract	Wheat	Qasem, 1995a
Festuca arundinacea	Leachates	Lettuce	Buta, Spaulding, and Reed, 1987
Fumaria densiflora	Extract	Wheat	Qasem, 1995a
Fumaria indica	Extract	Wheat	Chaghtai, Sadiq, and Ibrar, 1986
Galinsoga ciliata	Extract	Wheat	Tripathi, Singh, and Rai, 1981
Galium aparine	Toxin	Alfalfa, lettuce	Komai, Iwamura, and Ueki, 1983
Galium mollugo	Extract	Onion, radish, sunflower, wheat	Kohmuenzer, 1965
Gendelia tournefortii	Residue	Wheat	Qasem, 1995a
Gliricidia sepium	Extract	Sweet potato, *Hypetus capital*	Alan and Barrantes, 1988

TABLE 2 (continued)

Weed species	Source of toxin	Afflicted crop species	Reference
Gloriosa superb	Extract	Pea	Sugha, 1978
Gomphrena decumbens	Leachate	Sorghum, til	Solomon and Bhandari, 1981
Helenium amarum	Extract, residue	Alfalfa, ryegrass	Smith, 1989
Helianthus annuus	Toxin, extract	Wheat, mustard, several	Spring and Hager, 1982; Leather, 1987
Helianthus mollis	–	Radish, wheat	Anderson, Katz, and Anderson, 1978
Heliotropeum eichwaldi	Extract	Bean	Srivastava, 1969
Hemarthia altissima	Exudate, residue	*Desmodium intotum*	Young and Bartholomew, 1981; Tang and Young, 1982
Hieracium umbellatum	Extract, leachate	Flax, radish, red fescue, wheat	Becker and Guyot, 1951
Hieracium vulgatum	Extract, leachate	Flax, radish, red fescue, wheat	Becker and Guyot, 1951
Holcus mollis	–	–	Mann and Barnes, 1947
Hydrocotyl sibthorpioides	Extract	Radish, rape, tomato	Tsuzuki, Watanabe, and Shida, 1978
Imperata cylindrica	Leachate, residue	Cucumber, fenugreek, radish, rice, tomato	Eussen, 1978; Inderjit and Dakshini, 1991a
Indigofera cordifolia	Extract	Bajra, til	Sen, 1976
Ipomoea lucunosa	Extract, leachate	Soybean	LaBonte and Darding, 1988
Iva xanthifolia	Extract	Wheat	Le Tourneau, Failes, and Heggeness, 1956
Kochia scoparia	Extract, seeds	Radish	Wali and Iverson, 1978
Lactuca scariola	–	–	Rice, 1964
Lactuca serriola	Extracts, residue	Vegetable crops	Obeid, 1996
Lantana camara	Volatile, residues	Chinese cabbage, chilli, cucumber, pepper, rape, rice, spinach, wheat, maize, soybean	Das and Pal, 1970; Mersie and Singh, 1987a; Sahid and Sugau, 1993
Leersia hexandra	Extract, residue	Rice	Chou, Lee, and Oka,1984
Lepidium draba	Extract, exudate, leachate, residue	Barley, cabbage, carrot, cucumber, onion, pepper, squash, tomato, wheat	Qasem, 1994a, Qasem, 1999b
Lepidium virginicum	Extract	*Coronilla veria*	Bieber and Hoveland, 1968
Leucas cephalotus	Extract	Bajra, brome, clover, lettuce	Murthy and Zakharia, 1980
Lolium multiflorum	Leachate, residue	Alfalfa, lettuce	Naqvi and Muller, 1975
Lolium perenne	Extract, leachate	Lettuce, white clover	Buta, Spaulding, and Reed, 1987; Mattner and Parbery, 1999

Weed species	**Source of toxin**	**Afflicted crop species**	**Reference**
Lychnis alba	Extract	Wheat	Bhowmik and Doll, 1979
Madia glomerata	Leachate, residue	*Agropyron intermediate*	Carnahan and Hull, 1962
Matricaria inodora	Living Plant	Barley	Mann and Barnes, 1945
Melilotus alba	Extract	Wheat	Oudhia and Tripathi, 1999
Mercurialis annua	Residue	Wheat	Qasem, 1995a
Miscanthus floridulus	Extract, leachate	Lettuce	Chou and Chung, 1974
Nepeta cataria	Extract	Peas, wheat	Le Tourneau, Failes, and Heggeness, 1956
Notobasis syriaca	Residue	Wheat	Qasem, 1995a
Ocimum sanctum	Volatile	Rice	Das and Pal, 1970
Oenothera biennis	Extract	*Coronilla varia*	Bieber and Hoveland, 1968
Onopordon jordanicolum	Residue	Wheat	Qasem, 1995a
Oryza perennis	Extract, residue	Chinese cabbage	Chou, Chang, and Oka, 1991
Oxalis corniculata	Extract	Wheat	Leela, 1984
Panicum maximum	Extract	Lettuce	Chou and Young, 1975
Parthenium hysterophorus	Extract, leachate, residue	Bean, cabbage, tomato, cauliflower, cowpea, lentil, peanut, wheat, mung bean, soybean, chilly, clover, brinjal, pigeon pea, brinjal	Sarma, Giri and Subramanyam, 1976; Kanchan and Jayachandra, 1979b; Mersie and Singh, 1987b, 1988; Kohli and Batish, 1994; Kohli et al., 1996
Peganum harmala	Residue	Bajra	Sharma, Sidana, and Singhvi, 1982
Phalaris minor	Extract, exudate	Wheat	Porwal and Gupta, 1986; Oudhia and Tripathi, 1999
Piqueria trinervia	Extract	Several	Gonzalez et al., 1981
Plantago lanceolata	Extract, residue	Wheat	Sugha, 1979; Qasem, 1995a
Plantago purshii	–	–	Rice, 1964
Pluchea lanceolata	Leachate	Carrot, Indian mustard, okra, tomato	Inderjit and Dakshini, 1994
Poa pratensis	Residue	Tomato, wheat	Alderman and Middleton, 1925; Cochran, Elliott, and Papandick, 1977
Polygonum aviculare	Exudate, leachate, residue	Cotton, medic, sorghum	Alsaadawi and Rice, 1982; Kloot and Boyce, 1982
Polygonum orientale	Extract	Lettuce, mustard, rice	Datta and Chatterjee, 1978, 1980a
Polygonum pensylvanicum	Extract	Alfalfa, wheat	Le Tourneau, Failes, and Heggeness, 1956
Polygonum persicaria	–	Flax, potato	Martin and Rademacher, 1960
Polygonum sachalinense	Exudate	Lettuce	Inoue et al., 1992
Polygonella myriophylla	Infested soil	*Paspalum notatum*	Weidenhamer and Romeo, 1989

TABLE 2 (continued)

Weed species	Source of toxin	Afflicted crop species	Reference
Polypogon monspeliensis	–	–	Inderjit and Dakshini, 1995
Portulaca oleracea	Extract	Alfalfa, pea, radish, wheat	Le Tourneau, Failes, and Heggeness, 1956; Gressel and Holm, 1964
Proboscidea louisianica	Extract	Cotton	Mercer, Murry, and Verhalen, 1987; Riffle et al., 1990
Prosopis glandulosa	Extract, leachate	Chrysanthemum, cicer, lettuce, wheat	Sankhla, Baxi, and Chatterji, 1965; Alam and Azmi, 1989
Prosopis juliflora	Extract	Cassia, radish	Goel, Sexena, and Kumar, 1989
Pteridium aquilinum	Humus	Aspen, lettuce	Li et al., 1993; Dolling, 1996
Pulicaria wightiana	Extract	Bajra	Murthy and Zakharia, 1980
Ranunculus arvensis	Extract, leachate, residue, volatile	Wheat	Qasem, 1999a
Ranunculus asiaticus	Extract	Wheat	Qasem, 1995a
Ranunculus testiculatus	Extract	Grasses, herbs	Buchanan, Harper, and Frischknecht, 1978
Rorippa sylvestris	Extract	Lettuce	Misutani and Yamane, 1991
Rorippa indica	Extract	Lettuce	Yamane et al., 1992
Rumex crispus	Extract, residue	Corn, radish, sorghum, wheat	Einhellig and Rasmussen, 1973; Qasem, 1995a
Ruta graveolens	Leachate	Radish	Aliotta and Cafiero, 1999
Saccharum spontaneum	Leachate	Wheat	Amritphale and Mall, 1978
Salsola kali	Leachate	Radish, several	Lodhi, 1979
Salvadora oleoides	Extract	Sorghum	Mohnot and Soni, 1976
Salvia apiana	Volatile	Oat	Muller, Muller, and Haines, 1964
Salvia leucophylla	Volatile	Cucumber, oat	Muller, Muller, and Haines, 1964; Muller, 1969, Muller and Muller,1964
Salvia millifera	Inhibitory zones	Cucumber, oat	Muller, 1965, Muller and Muller,1964
Salvia reflexa	Leachate, volatile	Sorghum, wheat	Lovett and Lynch, 1979
Salvia syriaca	Extract,exudate, leachate, residue, volatile	Barley, cabbage, carrot, chickpea, cucumber, lentil, onion, pepper, squash, tomato, wheat	Qasem and Abu-Irmaileh, 1985; Abu-Irmaileh and Qasem, 1986; Qasem, 1999b
Sasa cernua	Soil	Lettuce, wheat	Li et al., 1992
Scandix pecten-veneris	Extract	Wheat	Qasem, 1995a
Schinus molle	Extract	Cucumber, wheat	Anaya and Gomez-Pompa, 1971
Senecio vulgaris	Extract, residue	Norway spruce	Schutt et al., 1975

Weed species	Source of toxin	Afflicted crop species	Reference
Setaria faberii	Extract, residue	Corn, soybean	Schreiber and Williams, 1967; Bell and Koeppe, 1972; Bhowmik and Doll, 1979
Setaria glauca	Extract, residue	Corn, soybean, several	Gressel and Holm, 1964; Bhowmik and Doll, 1984
Setaria viridis	Extract	Corn, soybean	Rice, 1964; Bhowmik and Doll, 1979
Sisymbrium irio		Onion	Menges and Tamez, 1981
Solanum surattense	Extract	Bajra, sorghum, til	Sharma and Sen, 1971; Mohnot and Soni, 1977
Solidago altissima	Extract	Eulalia (*Miscanthus sinensis*)	Kobayashi et al., 1980
Solidago sp.	-	Rice, several	Le Tourneau, Failes, and Heggeness, 1956
Solidago virgaurea	–	Flax, radish, wheat	Becker, Guyot, and Montegut, 1951
Sonchus arvensis	Toxin	Garden cress, rye, wheat	Bukolova, 1971
Sorghum halepense	Extract, exudate, residue	Barley, cabbage, onion, orange, rice, sunflower	Abdul-Wahab and Rice, 1967; Horowitz and Friedman, 1971; Menges, 1987
Spergula arvensis	Residue, natural population	Broccoli, collard, millet, pea	Harrison, Peterson, and Snook, 1998
Sphaeranthus indicus	Extract	Wheat	Oudhia and Tripathi, 1999
Sphenoclea zeylanica	Toxin	Rice	Sakashita et al., 1999
Stachys parviflora	Extracts, residue, volatile	Mustard, chinese cabbage, lentil, bean, radish	Hussain et al., 1986
Stellaria media	Living plant	Barley	Mann and Barnes, 1950; Inderjit and Dakshini, 1998
Stevia eupatoria	Volatile	White clover	Lovett, 1982
Striga densiflora	Extract, leachate	Canola	Zuberi et al., 1989
Tagetes minuta	Soil extract	7 crop species	Meissner, Nel, and Beyers, 1986
Tagetes patula	Residue	Bean, corn, sorghum	Altieri and Doll, 1978
Tephrosia purpurea	Leachate, volatile, soil	Cluster bean, pearl millet, til	Sundaramoorthy and Sen, 1990
Trianthema portulacastrum	Extract	Moth bean	Sethi and Mohnot, 1988
Trichodesma amplexicaula	Extract	Bajra, til	Sen, 1976
Trichodesma sedgwickianum	Residue	Bajra, til	Bansal and Sen, 1982
Trichostema lancealotum	Extract, volatile	Barley, soft brome	Heisey and Delwiche, 1983
Tridax procumbens	Volatile	Rice	Das and Pal, 1970

TABLE 2 (continued)

Weed species	Source of toxin	Afflicted crop species	Reference
Urgenia indica	Leachate	Corn, mung bean, pigeon pea, sorghum	Khare, 1980
Urtica urens	Residue	Wheat	Qasem, 1995a
Vulpia spp.	Extract, residue	Pasture spp.	An, Pratley, and Haig, 1999
Xanthium pennsylvanicum	-	-	Rice, 1964
Xanthium strumarium	Extract, residue	Lettuce, corn	Inam, Hussain, and Bano, 1987

and precise techniques to separate allelopathic interaction from competition, and a failure to prove the existence of direct versus indirect influence via other organisms or micro-environmental modification. There is a real need to demonstrate that an allelopathic compound is released into the environment at a concentration high enough to cause allelopathic influence or its modification. This however, depends on a host of factors which would be very difficult, if not impossible, to duplicate in laboratory or glasshouse work. Therefore, obtaining a chemical from plant tissues and identifying it as an allelopathic agent is also a misleading tool in identifying allelopathic species, since it neither agrees with the general term of allelopathy nor represents the natural situation.

On the other hand, failure to overcome the interference effects of weed species on crop plants by adding more resources (e.g., fertilizer) has been considered by many researchers as a strong evidence on allelopathy occurrence. This, however, may not hold true and could be a misleading explanation for the observed effect which should not justify such a claim. It is evident that species response as well as their requirements for growth factors are different, in quantities, qualities and time or growth stages at which these are critically required. In addition certain species are well known by their abilities to capture more resources, and exceed their actual need of these (luxury accumulation), thus leading to their limitation of these in the growing habitat and to severe competition effects.

Many workers (Newman, 1978; Aldrich, 1984; Willis, 1985; Putnam and Tang, 1986; Qasem and Hill, 1989b; Horsley, 1991; Blum, 1999; Foy, 1999) have suggested conditions for the existence of allelopathy or protocols similar to Koch's postulates for proof of disease. Proof of allelopathy requires certain consequent steps (Aldrich, 1984; Putnam, 1985; Putnam and Tang, 1986) including:

- Demonstrating interference by using suitable controls, observing, and describing symptomology, quantifying growth reduction and the degree of interference in a natural system.

- Isolating, characterizing, and assaying the suspected chemical against the species that were previously affected, then synthesizing the suspected toxin.
- Identifying the chemical.
- Demonstrating that the isolated chemical causes the observed effect when added back to the system in which the plant is growing at appropriate rates and times in nature.
- Monitoring release, movement, and uptake of the chemical from the donor plant and detecting it in the environment around the recipient.

WEED ALLELOPATHY (WEED SPECIES AND THEIR TARGETS)

Weeds vs. Crop Plants

A huge body of literature has been accumulated on the allelopathic effects of weeds on cultivated crops (see Rice, 1974, 1984, 1995). However, the influence of weeds through this mechanism of plant interference can be positive (stimulatory) or negative (inhibitory). The first is not much concern to agriculturists, since it is to be expected that some nutrient accumulated in the dead plant tissues will finally be released to the soil during the decay process. Also, nitrogen fixation by legume weeds may promote growth of crop species. Certain chemicals may be stimulatory in nature either before or after been modified by soil microorganisms. Dealing with the stimulatory action of allelochemicals may later necessiate separation between the factors causing this effect, leading to more complications in separating the effect of allelochemicals from that of nutrients. This is probably the reason behind the low number of publications on positive aspect of allelopathic interaction between plants, although some findings in this regard such as agrostemmin, allantoin, and gibberellin produced by *Agrostemma githago* are extremely important and agrostemmin application to wheat field at extremely low rate (1.2 g/ha) resulted in a significant increase in grain yield of this crop (Gajic, and Nikocevic, 1973; Gajic et al., 1976). Positive effects may be accomplished by other alternatives to achieve this goal in crop production. However, the harmful effects of allelochemicals on crop plants are difficult to alleviate. Therefore, it is of more concern to workers in agricultural and environmental sciences in general. In the present review 240 weed species of different growth habit and life form have been reported to exhibit harmful allelopathic effects on a wide range of crop species and through different mechanisms. Such deleterious effects are alarming since all allelopathic species are mainly either aggressive annual or perennial weeds and many world's worst weeds are involved (Table 2). Evidence for allelopathy by many of the species

TABLE 3. Weeds with allelopathic inhibitory effects on other weeds.

Weed species (Donor)	Source of toxin	Weed species (Reciever)	References
Agrostemma githago	Extract	*Ranunculus acer*	Gajic, 1973
Amaranthus retroflexus	Residue	*Cyperus esculentus*	Simkins and Doll, 1982
Ambrosia artemisiifolia	Extract	*Amaranthus retroflexus*, *Amaranthus chlorostachis*, *Chenopodium album*	Bruckner, Molnar, and Szabo, 1999
Ambrosia cumanensis	Extract, leachate	Different weed sopecies	Anaya et al., 1978
Ambrosia psilostachya	Exudate, leachate, volatile	*Amaranthus retroflexus*, *Aristida oligantha*, *Bromus japonca*, *Erigeron canadensis*, *Haplopappus ciliatus*	Neill and Rice, 1971
Anastatica heirochuntica	Extract	*Cloeme desertifolia*, *Diplotaxis harra*, *Farsetia aegyptiaca*, *Rumex cyperius*, *Trigonella stellata*	Hegazy, Mansour, and Abdel-Hady, 1990
Antennaria microphylla	Extract, toxin	*Euphorbia esula*	Selleck, 1972; Manners and Galtiz, 1986
Argemone mexicana	Extract	*Cyperus rotundus*	Srivastava and Das, 1974
Artemisia annua	Extract	*Amaranthus retroflexus*, *Chenopodium album*	Lydon, Teasdale, and Chen, 1997
Artemisia californica	Volatile	*Avena fatua*	Muller, Muller, and Haines, 1964
Artemisia herba-alba	Volatile	*Helianthemum ledifolium*, *Stipa capensis*, *Zygophyllum dumosum*	Friedman, 1987
Avena fatua	Residue	Different herbs	Tinnin and Muller, 1971
Brassica nigra	Volatile	*Bromus rigidus*, *Echinochloa crus-galli*	Bell and Muller, 1973; Oleszek, 1987
Cassia uniflora	Leachate	*Parthenium hysterophorus*	Joshi, 1991
Cenchrus ciliaris	Exudate in soil	*Calotropis procera*	Cheam, 1984
Centaurea solstitialis	-	*Echinochloa crus-galli*, *Lolium perenne*	Merril, 1989
Chenopodium album	Residue	*Cyperus esculentus*	Simkins and Doll, 1982
Chenopodium ambrosioides	Extract, leachate, residue	*Abutilon indicum*, *Cassia sophera*, *Cassia tora*, *Evolvolus numularius*, *Tephrosia hamiltonii*	Datta and Ghosh, 1987
Chenopodium murale	Extract	*Brassica juncea*, *Cortularia pallida*, different weeds	Datta and Ghosh, 1982; Biswas and Ghosh, 1984
Cirsium arvense	Extract, residue	*Amaranthus retroflexus*, *Carduus pycnocephalus*, *Circium vulgaris*, *Lolium perenne*, *Setaria viridis*, *Silybum marianum*	Bendall, 1975
Clerodendrum viscosum	Residue	*Abutilon indicum*, *Amaranthus spinosus*, *Cassia tora*, *Tephrosia hamiltonii*	Datta and Chakrabarti, 1982
Datura innoxia	Extract	*Brassica campestris*, *Setaria italica*	Mubarak and Hussain, 1978

Weed species (Donor)	Source of toxin	Weed species (Reciever)	References
Dichanthium annulatum	Extract	*Brassica campestris*, *Pennisetum typhoideum*, *Setaria italica*	Dirvi and Hussain, 1979
Digitaria sanguinalis	Exudate	*Aristida oligantha*	Parenti and Rice, 1969
Echinops echinatus	Extract	*Argemone mexicana*	Sarma, 1974
Eleusine indica	Residue	Weed species	Altieri and Doll, 1978
Eragrostis curvula	Living plants	*Cenchrus longispinus*	Johnston, 1989
Eragrostis poaeoides	–	*Searia italica*	Hussain, Zaidi, and Chughtai, 1984
Eupatorium adenophorum	Extract	*Lantana camara var. aculeata*, *Rumex nepalensis*	Dhyani, 1978; Tripathi, Singh, and Rai, 1981
Eupatorium maximiliani	Extract	*Amaranthus retroflexus*, *Bidens pilosa*	Correa et al., 1999
Eupatorium odoratum	Residue	*Euphorbia heterophylla*	Fadayomi and Oyebade, 1984
Euphorbia acalyphoides	Extract	*Striga hermonthica*	Ibrahim et al., 1985
Euphorbia esula	Residue	*Digitaria sanguinalis*	Steenhagen and Zimdahl, 1979
Euphorbia granulata	Leachate	*Cynodon dactylon*, *Dicanthium annulatum*, *Euphorbia pilulifera*, *Oxalis corniculata*	Hussain, 1980
Euphorbia pilulifera	Extract	*Striga hermonthica*	Ibrahim et al., 1985
Euphorbia prostrata	Extract	*Amaranthus retroflexus*, *Cynodon dactylon*	Alsaadawi, Sakeri, and Dulaimy, 1990
Euphorbia supina	Extract, exudate, residue	*Aristida oligantha*, *Bromus japonicus*, *Bromus tectorum*, *Erigeron canadensis*	Brown, 1968
Festuca arundinacea	-	*Digitaria sanguinalis*	Peters and Mohammed Zam, 1981
Galium aparine	Toxin	*Digitaria sanguinalis*	Komai, Iwamura, and Ueki, 1983
Gliricidia sapium	Extract	*Hypetis capitata*, *Ipomoea sp.*	Alan and Barrantes, 1988
Helianthus annuus	Living plants	*Bromus japonicus*, *Erigeron canadensis*, *Haplopappus ciliatus*, *Rudbeckia hirta*	Wilson and Rice, 1968
Imperata cylindrica	-	*Melilotus parviflora*	Inderjit and Dakshini, 1991a
Lantana camara	Extract, residue	*Cyclosorus dentatus*, *Echinochloa crassipes*, *Lolium* spp., *Morrenia odorata*, *Lantana camara*	Archhireddy and Singh, 1984; Wadhawani and Bhardwaja, 1981; Saxena, 1992; Arora and Kohli, 1993
Leptochloa filiformis	Residue	Weed species	Altieri and Doll, 1978
Lolium temulentum	Leachate, residue	*Bromus* sp.	Naqvi and Muller, 1975

TABLE 3 (continued)

Weed species (Donor)	Source of toxin	Weed species (Reciever)	References
Lycoris radiata	Extract	Different weeds	Ueki and Takahashi, 1984
Parthenium hysteropherus	Toxin	*Ageratum conyzoides*	Batish et al., 1997
Polygonum aviculare	Exudate, leachate, residue	*Chenopodium album*, *Cynodon dactylon*	Alsaadawi and Rice, 1982
Polygonum orientale	Extract, leachate	*Brassica juncea*, different weeds	Datta and Chatterjee, 1978,1980b
Polygonum sachalinese	-	*Amaranthus viridis*, *Phleum pratense*	Inoue et al., 1992
Prosopis juliflora	Extract	*Euphorbia caducifolia*	Sen and Chawan, 1970
Ranunculus bulbosus	Extract, residue	*Echinochloa crus-galli*, *Ipomoea lacunosa*, *Sida spinosa*	Gander and Oliver, 1998
Rumex crispus	Extract	*Amaranthus retroflexus*	Einhellig and Rasmussen, 1973
Ruta graveolens	Extract	*Amaranthus retroflexus*, *Chenopodium album*, *Cynodon dactylon*, *Cyperus rotundus*, *Euphorbia exigua*, *Portulaca oleracea*	Aliotta and Cafiero, 1999
Salsola kali	Leachate, residue	*Kochia scoparia*	Lodhi, 1979
Salvia leocophylla	Volatile	*Avena fatua*	Muller, Muller, and Haines, 1964
Sasa cernua	Volatile, soil	*Amaranthus viridis*, *Echinochloa crus-galli*	Li et al., 1992
Senecio jacobaea	Extract	*Lolium perenne*	Ahmed and Wardle, 1994
Solanum surattense	Extract	*Agremone mexicana*	Sarma, 1974
Sorghum halepense	Residue	*Amaranthus retroflexus*, *Aristida oliginata*, *Bromus tectorum*, *Digitaria sanguinalis*, *Setaria viridis*, *Sorghum halepense*	Abdul-Wahab and Rice, 1967
Sporobolus pyramidatus	Leachate, residue	*Amaranthus retroflexus*, *Aristida oliginata*, *Bromus tectorum*, *Cynodon dactylon*	Rasmussen and Rice, 1971
Tribulus terrestris	Leachate	*Amaranthus lividus*, *Brassica tournefortii*, *Bromus tectorum*, *Chenopodium murale*, *Conyza bonariensis*, *Eruca sativa*, *Hordeum murinum*, *Solanum nigrum*	El-Ghareeb, 1991
Typha latifolia	-	*Phragmatis communis*	Szezepanska, 1971
Vitex negundo	Leachate	*Digitaria decumbens*	Chou and Yao, 1983

TABLE 4. Weed species with autotoxic (autopathy) effect.

Weed species	Source of inhibition	Reference
Amaranthus palmeri	Vapor	Connick et al., 1987
Ambrosia cumanensis	Extract, leachate	Anaya and Del Amo, 1978; Anaya et al., 1987
Ammi majus	Toxin	Friedman, Rushkin and Waller, 1982
Bromus inermis	Leachate, residue	Benedict, 1941
Cenchris ciliaris	Toxin	Lahiri and Kharabanda, 1962
Cenchris setigerus	Toxin	Lahiri and Kharabanda, 1962
Cirsium arvense	Extract	Bendall, 1975
Cirsium palustre	Extract	Ballegaard and Warncke, 1985
Convolvulus sepium	Leachate	Quinn, 1974
Datura innoxia	Extract	Mubarak and Hussain, 1978
Echinops echinatus	Toxin	Soni and Mohnot, 1988
Erigeron canadensis	Toxin	Keever, 1950
Helianthus rigidus	Decayed material	Curtis and Cottam, 1950
Kochia scoparia	Decayed leaves	Wali and Iverson, 1978; Lodhi, 1979
Lasiurus sindicus	Toxin	Lahiri and Kharabanda, 1962
Parthenium argentatum	Leachate	Bonner and Galston, 1944
Parthenium hysterophorus	Leachate	Kumari and Kohli, 1987
Phytolacca americana	Extract, leachate	Edwards et al., 1988
Plantago lanceolata	Leachate	Newman and Rovira, 1975
Prosopis juliflora	Extract	Warrag, 1995
Salsola kali	Residue	Lodhi, 1979
Sorghum halepense	Decayed material	Abdul-Wahab and Rice, 1967
Solanum rostratum	Decayed material	Ahshapanek, 1962
Sporobolus pyramidatus	Leachate	Rasmussen and Rice, 1971
Typha latifolia	Extract	McNaughton, 1968

reported, however, remains weak and may not hold true under closer scrutiny (Putnam and Tang, 1986), because of the techniques followed in detecting this action. For some species, the evidence is quite convincing and results have been confirmed through different experimental techniques or in nature. The number listed in this category will most likely increase in the future because research on this type of allelopathy interaction is progressing rapidly through the efforts of researchers in different fields of science worldover.

Weeds vs. Weeds

The concept of allelopathy interaction between/among weeds is interesting. Certain weed species have the potential to be used in solving problems of other weed species and represents an excellent strategic source of natural chemicals that may be involved in developing natural herbicides. The significance of this effect may be better demonstrated where other methods of weed control may fail, or are not possible or available (e.g., aquatic weeds). Replacing certain aquatic weed species by less important weeds through allelopathy may partially or totally diminish such an effect on the long-term, especially when the effect is a selective one. On the other hand, in certain environments, herbicide use may be restricted for reasons such as pollution. Natural weed control would be the obvious answer in such a situation. Szezpanski (1977) suggested that in aquatic environments (where pollution is a problem) aquatic plants could be used to control weeds. Weeds with high potential of herbicidal action and their target species are shown in Table 3. These include many of the world's worst weeds that are negatively affected by allelopathic weeds, confirming that struggles between these species and between these and crop plants are not only due to competition but also to significant allelopathic effects.

Weed Autopathy (Autotoxicity)

Weed species having self-inhibitory activity are listed in Table 4. Autoxicity is different from allelopathy in that it is self-toxic and can't be stimulating. This phenomenon is well documented for many crop species, including certain fruit trees in a phenomenon called replanting problem (Rice, 1974, 1984). Many weed species show such self-toxic chemicals in their seeds or tissues. Recently, Singh, Batish, and Kohli (1999) have reviewed the phenomenon of autotoxicity among plants, its ecological role and significance. The nature, cause, and purpose of this kind of toxicity are still questionable and not fully understood. Newman (1982) questioned the purpose of autotoxicity and Whittaker and Feeny (1971) described it as "an evolutionary paradox." It is expected that autotoxicity has a significant role in the survival strategy of certain species in nature through regulating their natural populations, especially under adverse environmental conditions and short supply of germination growth factors under which this mechanism is most likely naturally operating. In addition, it is possible that the allelochemicals have a multifunctional role and may, for example, deter grazing animals (Singh, Batish, and Kohli, 1999). Friedman and Waller (1985) concluded that some experimental results were ambiguous due to imprecise techniques and they cited many ways in which plants avoid autotoxicity, yet these mechanisms may be disturbed or destroyed during research and, hence, the apparently positive results inferring that autotoxicity exists. On the other hand, the autotoxic action of the plant residues added to the soil is most likely due to liberation of some stored toxic materials from specific storage plant cells or tissues, highly regulated and released in living plants, and act against other species or individuals but become toxic to the same species when leached out of the dead tissues into the surrounding environment. However, population regulation most likely occurs at early growth stages, when seedlings are expected to be more sensitive to allelochemicals. In general, there has been limited research on autotoxicity and on weeds in particular. Theoretically, autotoxic species would be ecologically best suited in mixed stands. However, probably the positive exploitation of this phenomenon remains in managing population sizes of certain noxious weed species through application of their residues on or in the soil and thus preventing some of the negative effects on desired plants. Another possible uses through providing a source for highly selective natural herbicides that may be used to regulate their populations in cultivated land or controlling their abundance in nature (Singh, Batish, and Kohli, 1999).

Weeds vs. Plant Pathogens

One of the negative characteristics and impacts of weeds on agricultural systems is that they provide suitable and alternative hosts for many of agricultural pests including insects and plant pathogens. While the role of weeds is to be added to their overall negative effects on crop growth and yield, beneficial effects of certain allelopathic weeds have been long observed and known in nature through their effects on infection, growth, and development of these pests in general and plant pathogens in particular. However, literature on the role of higher plants in general as a source of fungi or bacteria toxins and their importance in controlling different plant pathogens is excessive and varied (Tripathi et al., 1983; Frange, 1984; Qasem, Al-Abed, and Abu-Blan, 1995; Qasem, 1996 a,b). Many workers reported antimicrobial activity of plant extracts (Frange, 1984; Qasem, 1996a,b,c) and recent studies on the subject emphasized the importance of natural chemicals as an alternative to synthetic pesticides in any future strategy for pest control (Macias et al., 1994; Singh, 1994).

Antipathogenic weeds have been regarded as natural renewable sources of allelochemicals that inhibit growth and development of many plant pathogenic fungi and bacteria species (Qasem, 1996 a,b). Weed species reported to have inhibitory effects against many plant pathogens are shown in Table 5. However, the number of weeds having such an effect here certainly is not a final one and may well exceed this number especially if those affecting plant pathogenic nematodes, rectsia, and other pathogens are considered. It is concluded that many weed species may serve as an excellent source of

TABLE 5. Weeds with antifungal or antibacterial properties.

Weed species	Source of toxin	Pathogen affected	Reference
Adenocalyamma allicea	Oil	*Drechslera oryzae*	Chaturvedie et al., 1987
Anagallis arvensis	Extract, residue	*Fusarium oxysporum, Graphium ulmi, Helminthosporium carbonum, Helminthosporium oryzae, Helminthosporium sativum, Helminthosporium turicum, Penicillium digitatum, Sclerotinia sclerotiorum, Verticillium dhaliae*	Rafiq, Nasir, and Bhatti, 1984; Desevedavy et al., 1987; Al-Abed, Qasem, and Abu-Blan, 1993; Qasem, 1996a,b,c; Qasem and Abu-Blan, 1996
Anemone pulsatilla	Extract	*Mycobacterium tuberculosis*	Seegal and Holden, 1945
Artemisia borealis	Toxin	*Cladosperorium cucumerinum*	Wang et al., 1990
Atriplex leucoclada	Extract	*Helminthosporium sativum*	Qasem and Abu-Blan, 1996
Calendula arvensis	Extract	*Fusarium oxysporum*	Qasem, 1996a
Cerastium dicotatum	Extract	*Fusarium oxysporum*	Qasem, 1996a
Chenopodium album	Extract	*Coletotrichum falcatum, Helminthosporium carbonum*	Rafiq, Nasir, and Bhatti, 1984
Chenopodium murale	Extract	*Alternaria solani, Penicillium digitatum, Sclerotinia sclerotiorum, Verticillium dhaliae*	Qasem, 1996b; Qasem and Abu-Blan, 1995; Qasem and Abu-Blan, 1996
Convolvulus arvensis	Extract	*Fusarium oxysporum*	Qasem, 1996c
Crepis aspera	Extract	*Helminthosporium sativum, Penicillium digitatum, Sclerotinia sclerotiorum, Verticillium dhaliae*	Qasem, 1996b; Qasem and Abu-Blan, 1996
Cuscuta reflexa	Extract	*Helminthosporium turicum*	Rafiq, Nasir, and Bhatti, 1984
Cyperus rotundus	Tuber oil	*Aspergillus niger, Aspergillus orizae, Bacillus subtilis, Candida didensii, Escherichia coli, Saccharomyces cerevisiae*	Komai, Sato, and Ueki, 1982
Ecballium elaterium	Extract	*Botrytis cinerea*	Bar-Nun and Mayer, 1990
Echinops echinatus	Phenolic compounds	*Alternaria tenuissima*	Singh et al., 1988
Erodium cruciatum	Extract	*Sclerotinia sclerotiorum Verticillium dahliae*	Qasem, 1996b; Qasem and Abu-Blan, 1995
Euphorbia helioscopia	Extract	*Fusarium oxysporum, Sclerotinia sclerotiorum, Verticillium dhaliae*	Qasem, 1996a, b,c
Falcaria vulgaris	Extract	*Alternaria solani, Fusarium oxysporum*	Qasem, 1996a,c; Qasem and Abu-Blan, 1996
Foeniculum vulgare	Extract	*Fusarium oxysporum*	Qasem, 1996c
Galium tricornutum	Extract	*Verticillium dhaliae*	Qasem, 1996b
Helianthus annuus	Sesquiterpenes	*Bacillus brevis*	Spring, Rodon, and Macias, 1992
Helichrysum decumbens	Phloroglucinol, acetophenone	*Cladosporium herbarum*	Tomas-Barberan et al., 1990
Helichrysum italicum	Phloroglucinol, acetophenone	*Cladosporium cucumerinum, Escherichia coli*	Tomas-Barberan et al., 1990
Helichrysum nitens	Toxin	*Cladosporium cucumerinum*	Tomas-Barberan, Msonthi, and Hostettmann, 1988

Weed species	Source of toxin	Pathogen affected	Reference
Helichrysum stoechas	Phloroglucinol, acetophenone	*Cladosporium herbarum*	Tomas-Barberan et al., 1990
Inula viscosa	Extract, residue, volatile	*Fusarium oxysporum, Helminthosporium sativum*	Qasem, 1996c; Qasem, Al-Abed, and Abu-Blan, 1995
Larrea tridentata	Leachate	*Escherichia coli, Saccharomyces cerevisiae*	Downum et al., 1989
Mercurialis annua	Extract	*Rhizoctonia solani*	Qasem and Abu-Blan, 1996
Micanthus sinensis	-	*Pyricularia oryzae*	Mori et al., 1987
Notobasis syriaca	Extract	*Helminthosporium sativum*	Qasem and Abu-Blan, 1996
Parthenium hysterophorus	Leachate	*Synechococcus elongatus*	Megharaj et al., 1987
Polygonum aviculare	Extract	*Fusaruim oxysporum*	Qasem, 1996a
Posidonia oceanica	-	Bacteria	Bernard and Pesando, 1989
Ranunculus asiaticus	Extract, residue, leachate, volatile	*Alternaria solani, Helminthosporium sativum, Penicillium digitatum, Rhizoctonia solani, Sclerotinia sclerotiorum, Verticillium dhaliae*	Qasem and Abu-Blan, 1995, 1996; Qasem, 1996 a,b,c
Ranunculus bulbosus	-	*Epidermophyton floccosum, Rhodotorula glutinis*	Mares, 1987
Ranunculus clematis	-	*Alternaria tenius, Curvularia lunata, Fusarium nivale, Helminthosporium graminicum*	Misra, Misra, and Dixit, 1974
Ranunculus sp.	Extract	*Monilia albicans*	Seegal and Holden, 1945
Ranunculus sceleratus	Extract	*Alternaria tenius, Curvularia lunata, Fusarium nivale, Helminthosporium graminicum*	Misra and Dixit, 1978
Robus idaeus	Leachate	Mycorrhizal fungi	Cote and Thibault, 1988
Rumex crispus	Extract	*Helminthosporium sativum*	Qasem and Abu-Blan, 1996
Salvia canariensis	Diterpenes	*Bacillus subtilis, Staphylococcus aureus*	Gonzalez et al., 1989
Salvia cardiophylla	Diterpenes	*Bacillus subtilis, Staphylococcus aureus*	Gonzalez et al., 1989
Salvia texana	Diterpenes	*Bacillus subtilis, Staphylococcus aureus*	Gonzalez et al., 1989
Scorpiurus muricatus	Extract	*Fusarium oxysporum*	Qasem, 1996a,c
Sisymbrium irio	Extract	*Alternaria solani, Fusarium oxysporum, Helminthosporium sativum, Verticillium dhaliae*	Qasem and Abu-Blan, 1995; Qasem, 1996c
Solanum nigrum	Extract	*Fusarium oxysporum, Trichoderma lignorum*	Singh, 1977; Qasem, 1996a
Sonchus oleraceous	Extract	*Helminthosporium sativum, Rhizoctonia solani*	Qasem and Abu-Blan,1996
Stellaria media	Extract	*Fusarium oxysporum*	Qasem, 1996a
Trachypogon pulmosus	Extract	*Bacillus subtilis, Escherichia coli, Staphylococcus aureus, Staphylococcus haemolyticus*	Stiven, 1952
Urtica urens	Extract	*Fusarium oxysporum*	Qasem, 1996c
Vicia narbonensis	Extract	*Helminthosporium sativum*	Qasem and Abu-Blan, 1996

beneficial allelochemicals important for the agrichemical industry or may be used as such in different ways leading to more balanced and pest managed populations in nature. A feature could be positively added to some beneficial characteristics of many weed species and offer a good possibility for inclusion into integrated pest management programs.

Weeds vs. Parasitic Weeds: A Possible Approach for Parasitic Weed Control, Indication and Some Thoughts

Parasitic weeds or parasitic flowering plants represent a major threat to agriculture because they are difficult to control with available weed control methods, except by soil fumigation which is not economically feasible on a wide scale. The huge number of seeds these species produce accompanied with long seed viability add more complications to their control. In addition, certain species such as *Orobanche* have a wide host range from both wild and cultivated plant species that facilitate its spread and infestation of new areas worldwide.

Parasitic weeds may be classified according to the presence or absence of chlorophyll pigments or to their degree of dependence on host plants for food and/or water, to two main categories, hemi-parasites and holo-parasites. While certain species (e.g., *Orobanche* and *Striga*) require preconditioning periods and host stimulants for seed germination, others (e.g., *Cuscuta* and *Osyris*) do not (Parker and Rachis, 1993).

Stimulants are mainly natural chemicals released through the host root system into the surrounding environment and received by seeds of parasitic species. Stimulants may be individual chemicals or a complex of these, working individually or jointly in promoting parasite seed germination. These chemicals encourage seed germination of the parasite probably through their effects on the hormonal balance of the seeds themselves and thus overcoming their special chemical-based dormancy. They may also act as hormones (e.g., strigols and their analogs of GA-like chemicals) affecting embryo germination and growth of germination tube of the parasite. However, identified natural germination stimulants of host root exudates with known chemical structures are strigol (isolated from maize root exudate), alectrol (from cowpea root exudate) and sorgolactone (from sorghum root exudate) all coming from *Striga* host plants (Wegmann, 1998). Strigol is a powerful germination stimulant of *Striga* seeds that was first isolated and identified from the root exudate of cotton plants (a non-host crop) and stimulates seed germination of *Striga* by 85 and 100% at concentrations of 10^{-4} to 10^{-6} (Hsiao, Worsham, and Moreland, 1981). However, a number of synthetic and natural germination stimulants of *Striga* species are lactonic or contain one or more unsaturated lactone rings (Worsham, Klingman and Moreland, 1962; Johnson, Rosebery, and Parker, 1976; Musselman, 1980). In addition, some of these chemicals, such as coumarin-type compounds are well known allelochemicals (Worsham, Klingman, and Moreland, 1962; Rice, 1984). On the other hand, no structure of germination stimulant has been reported from *Orobanche* host plants (Wegmann, 1998), while different species have shown strong ability to stimulate seed germination of different *Orobanche* species by more than 90%. It is noteworthy that the structures of the germination stimulants exuded from roots of these host species remain unknown although several haustoria-inducing compounds have now been characterized including xenognosin A and B, identified in an exudate of *Astragalus gummifer*, and soyasapogenol B that was identified in roots of *Lespedeza sericea*.

Germination stimulants are produced in small amounts and concentrations enough to promote germination but probably at higher amounts or concentrations different actions may be obtained. However, in most cases enough and optimal production, concentration, and release of these stimulants may encourage a higher number of parasite seeds to germinate and to attach themselves on the root system of host species. Therefore, root health, growth, and size may be critical factors in determining tolerance or susceptibility of different species or cultivars to the parasite attack, and subsequently the magnitude of the negative effects on yield of these species or cultivars. Plants differ in growth and size of their root systems, and abilities to produce chemical stimulants which leads to differences in parasite infestation of different hosts or even different cultivars of the same host species (Qasem and Kasrawi, 1995). Tolerant species with small or less developed root systems or above-ground canopies may produce low amounts of stimulatory chemicals with less release of these through the root systems. They have clearly low photosynthetic leaf areas and growth rate in general, and therefore less competitive ability leading to low yield.

Considering Molisch's concept of the term allelopathy, seed germination stimulants as well as chemicals that induce dormancy are both allelochemicals effective at minute amounts and concentrations but at high concentration they may become toxic, and have an inhibitory or herbicidal action. The origin of their synthesis may be roots and thus absent in shoots, or may be produced in shoots and transported to the root system and throughout to the surrounding environment. The mechanism of release and concentration of germination stimulants is under precise control and well-calibrated in place and time.

It is well established that cultivated host species are different in their abilities to stimulate seed germination of certain parasitic weeds (e.g., *Orobanche*) which may be considered as a criterion differentiating host species into preferable, intermediate, or less preferred hosts. This may be based on the amount of germination stimulants these species release from their roots (genetically and/or environmentally controlled) and received by seeds of the parasite. Differences in the chemical nature of germination stimulants and their susceptibility to microbial degradation, in addition to the number and nature of the chemicals involved in the complex of germination stimulants, may determine their stimulatory or inhibitory action and the amount received by parasite seeds.

The ability of root exudates to stimulate germination of the parasite is not restricted to host species but is of wide occurrence in the plant kingdom (Musselman, Riopel, and Eplee, 1978). Non-host species may produce toxic chemicals that inhibit or kill seeds in the soil. They may produce stimulants, but attachment may fail due to plant mechanical or physical resistance mechanisms. Stimulants produced at low amounts and/or concentrations may not be sufficient to overcome seed dormancy or to balance the promoters/inhibitors ratio in seeds. These species may not produce any stimulants enhancing seed germination of the parasite. This may have blocked and negated any work on others and thus kept researchers far away from dealing with other possible mechanisms that non-host species may have specifically through allelochemicals.

Hosts of different parasitic species from cultivated plants may be wide enough to include a significant number of crops of different plant families, but differences in susceptibility of these or their cultivars are quite varied (Parker and Rachis, 1993). Therefore, preferable, non-preferable, or tolerant host species or cultivars are most likely found and identified for each parasitic species. Differences between species may be due to mechanical or physical, physiological, or chemical factors. In this regard, two main groups of plants in relation to stimulatory action can be easily distinguished from other host species and considered extremely important. These are trapping and catching species. Although the number known from these in general and for certain species in particular (*Orobanche*, *Striga*) is extremely low compared with host species, they are promising and have a great potential in solving, partially or totally, problems that certain parasitic weeds exert on agriculture.

Many cultivated species or cultivars have been reported to suppress weeds in general including cucumber (Putnam and Duke, 1974), oats (Fay and Duke, 1977), sunflower (Leather, 1983), soybean (Massantini, Caporalli and Zellini, 1977), rice (Dilday, Lin, and Yan, 1994) and rye (Perez and Ormeno-Nunze, 1993) through allelopathic mechanism. Stimulation of crop growth by root exudates of certain weed species also has been reported (Gajic and Nikocevic, 1973; Stevens and Merril, 1985). On the other hand, different cultivated species including plants producing ethylene or strigoles with great potential to stimulate seed germination of certain parasitic weeds, including different species of *Orobanche* and *Striga*, without being themselves attacked, have been reported. Many of these crops have been studied and identified through laboratory or glasshouse work. *Orobanche* stimulating species include: sorghum, barley, and *Vicia dasycarpa* spp. *villosa* (Kasasian, 1973; Linke, Schnell and Saxena, 1991) for *O. crenata*; bean, sorghum, maize, and cucumber for *O. ramosa* (Labrada and Perez, 1988) and sorghum, cowpea, chili, hemp, mung bean, flax, lucerne, soybean and chickpea for *O. cernua* (Krishnamurthy and Rao, 1976; Krishnamurthy, Lal, and Nagarajan, 1977). Linke, Schnell and Saxena (1991) reported significant reduction in *O. crenata* infestation after three years of growing *Vicia dasycarpa* spp. *villosa*. More recently, Al-Menoufi and Adam (1998) concluded the possible use of fenugreek (*Trigonella foenum-graecum*), lupine (*Lupinus termis*), coriander (*Coriandrum sativum*), and turnip (*Brassica rapa*) in an intercropping system with faba bean or tomato to reduce the infection rate of *Orobanche* plants attached to the host root systems. Intercropping was also suggested by Megahed (Cited by Al-Menoufi, Safwani and Adam,1996) and Al-Menoufi (1991, 1992) who found significant reduction in *O. crenata* infestation in faba bean intercropped with fenugreek. The use of a variety of mixtures of plant species is considered as a possible strategy to affect *Orobanche*-host interaction by combining certain host species with different degrees of susceptibility (Bouhatous and Jacquard, 1994). The non-suitable host crops of *O. cerenua* including chili, sorghum, cowpea, mothbean (*Phaseolus aconitifolius*) and *Hibiscus sabdariffa* stimulated germination of the parasite at a high level, while sesame stimulated germination but without offering it any chance for further growth and development; instead filamentous infection type of the parasite started shriveling and dying. Root diffusate of flax (*Linum usitatissimum*), pepper (*Capsicum annuum*), sesame (*Sesamum indicum*) and Egyptian clover (*Trifolium alexandrinum*) enhanced *Orobanche* seed germination, but the germinated seeds normally failed to attach to the root systems of these crops (Abu-Shakra, Maih, and Saghir, 1977; Ibrahim et al., 1973; Krishnamurthy, Lal, and Nagarajan, 1977; Musselman, 1980; Al-Menoufi, 1991). Dhanapal, Struik, and ter Borg (1998) reported strong induced germination of *O. cernua* seeds by trap crops especially green gram (*Vigna radiata*) and sunhemp (*Crotalaria juncea*). *Striga* seed germination can be induced by root exudates of some non-host plants including cotton root exudates that stimulate *S. asiatica* seed germination without serving as a host plant for this parasite.

Catch crops are normal hosts of the parasites that only stimulate germination but may arrest growth and accommodate the parasite to maturity stage. They are rarely as economically favorable as trap crops because of the lack of production, but can sometimes be grown during a short, unreliable, rainy season unsuitable for normal cropping (Parker and Rachis, 1993). However, normal crop hosts are likely to be the most effective catch crop of all, provided the parasite can be controlled by herbicides or other means. It seems likely that catch species are not offering enough support either through growth hormones or availability of food and/or water necessary for further growth and development of the parasite, although these are potentially rich in their contents from stimulatory chemicals. Another possibility is the lack of compatibility between the parasite and host root tissues and their internal chemicals. Therefore, parasite development may not be possible and arrested shortly after attachment. Certainly the availability of germination or growth stimulants in host, trap, or catch species is as important as the non-stimulant, toxic, or inhibitory allelochemicals in non-host species regardless of any mechanical resistance of the host tissues. The amount of allelochemicals either stimulating or inhibiting parasite seed germination and development in host tissues may determine how long the parasite can remain connected to host tissues and the ability of the host to accommodate parasite's needs for water and nutrients. Heavy stimulation may result in intraspecific competition between shoots of the parasite and those with the host root system (interspecific competition) for a limited supply of growth material by the host, which may be the reason behind failure of parasite growth and development after attachment to the host root system. Whether stimulant or suppressant allelochemicals are found in shoot tissues of trapping or catching species is a matter of controversy, but it is to be expected that allelopathic inhibitory chemicals may also be found in shoot or root tissues of different host and non-host species and can affect parasite germination, growth, and development in a way similar to their effects on any other plant species.

The possibility of wild species to inhibit or stimulate totally or partially, parasite seed germination remains high and exists in nature since they have a high allelopathic potential compared with cultivated species, although studies on this fertile area of allelopathic research are either scarce or lacking. In one study, Ibrahim et al. (1985) reported that extracts of several *Euphorbia* species were most active in inducing germination of *Striga hermointhica*. Nevertheless, in one of the most recent and extensive studies conducted to investigate the role of allelopathy in controlling *Orobanche ramosa*, Qasem (unpublished data) screened 118 plant species, including summer and winter species, traditionally known as non-host crops, medicinal herbs, and a large number of common weed species for the effect of their root exudates on germination, growth, and development of the parasite on their root systems and the effect of their root exudates released to the soil on the parasite's ability to attack subsequently sown tomato plants as a preferable host crop. Great variation was detected between species in their abilities to stimulate *Orobanche* seed germination and attachment to their root systems. Certain species were highly stimulatory; others prevented or showed low infestation. The differences were better demonstrated when the number of emerged *Orobanche* shoots was considered. *Orobanche* size ranged between a pinhead-sized infection to fully developed mature shoots.

Results showed that some of the previously reported crop species as trapping crops (e.g., flax, fenugreek, chili, and sweet pepper) were attacked by the parasite. Flax showed an extremely low infestation but other species were highly infested. More differences were detected in *Orobanche* infesting tomato plants sown directly in the same soil after the tested species were removed. Results were impressive and may reflect differences between species in amount, concentration, and chemical nature of germination stimulants they released to the soil and the ability of these in stimulating or inhibiting *Orobanche* seed germination, attachment, and development on their root systems. In another extensive screening study, Qasem (unpublished data) investigated the allelopathic effects of dried shoot residues of 137 weed species and certain residue mixtures of these incorporated into an *Orobanche ramosa* infested soil mixture, on germination, growth, and development of the parasite on tomato plants sown in these pots. Striking results were obtained. Residues of certain weed species highly stimulated parasite germination and infestation; others reduced or almost completely prevented its infestation. Work using different experimental techniques is in progress to quantify the role of inhibitory allelochemicals on seed germination of various *Orobanche* species. Results clearly showed great variation in the effect of allelochemicals from different sources on *Orobanche* infestation of the host crop. Certain species, however, were tested for the first time and found as an excellent catching species of *Orobanche ramosa* while others proved to be strong inhibitors to seed germination of the parasite in the presence of the host root exudates and through their shoot residues incorporated in the soil. The results indicated that like any other plant species affected by allelochemicals, parasitic weeds are not exceptional and may also suffer or be highly promoted by allelochemicals from other species added to their environment. In all cases, however, there is a great potential for exploitation of allelochemicals in control programs of these species by offering stimulation action in absence of host crops, therefore offering a suicidal killing of the parasite or inhibitory action through their herbicidal effect in preventing seed germination or destroying them in the soil through the effect on their viability or killing the germinated seeds prior to their attachment to host crop. Such species offer a novel approach for development of natural herbicides for *Orobanche* in the future.

WEED ALLELOCHEMICALS

Production, Parts, Release, Factors, and Transport

Allelochemicals are present in virtually all plant parts and tissues including stems, leaves, flowers, buds, bark, pollen grains, seeds, fruits, roots, and rhizomes (Rice, 1974, 1984; Putnam and Tang, 1986). However, their concentration is different from one part to another. Many studies have shown more concentration of allelochemicals in shoots as opposed to roots. Other studies reported high concentrations of allelopathic agents in seeds of different weed species, a problem that may be in part responsible for seed longevity of certain weed species by protecting seeds from attack and decay by microorganisms. In general, and for maximal effect against target plant species, allelochemicals need to be concentrated in roots, leaves, or stems rather than flowers and seeds. Leaves, roots, and seeds are the common source (Rice, 1974). Toxins may be localized in cell vacuoles away from metabolic centers or may be deposited in specialized storage structures such as trichomes of *Salvia reflexa* (Lovett and Speak, 1979). Some of the chemicals are released in toxic form to the receiver plants while others such as juglone (Norrington-Davis, 1980) become toxic only on release and after being modified in soil. The exact mechanisms of allelochemical storage and release merit further investigation for different weed species under various conditions. Some known mechanisms of allelochemical release in weed species are summarized in Table 6.

Rice (1983) hypothesized that allelochemicals in certain cases may not even be released, but they could move directly from donor to the receiver through underground bridges such as root grafts or mychorrhiza. On the other hand, it appears that certain mechanisms of allelochemical release are more likely occurring than others under specific conditions since, in general, volatile inhibitors may be most significant ecologically in arid and semi-arid

TABLE 6. Some mechanisms of allechemical release in weeds.

Mechanism of Release	Donor Weed	Receiver Species	Reference
Volatilization from stems and leaves	*Salvia reflexa*	Wheat seedlings	Lovett and Levitt, 1981
Volatilization from roots	*Raphanus raphnistrum*	Maize seeds	Cited by Fletcher and Kirkwood, 1982
Leaching from shoots	*Lepidium draba*	Wheat	Qasem, 1994a
Leaching or exudation from roots	*Chenopodium album*	Tomato	Qasem and Hill, 1989a
Pollen exudation	*Parthenium hysterophorous*	*Crotalaria pallida*	Kanchan and Jayachandra, 1980
Fruit leachates	*Chenopodium murale*	Wheat and Barley	Qasem, 1990
Seed	*Datura stramonium*	Sunflower seeds	Levitt, Lovett, and Garlick, 1984
Flowers	*Erica vagrans*	Red clover	Arnes, Vieitez, and Mantilla, 1974
Residue decay	*Setaria faberii*	Maize	Bell and Koeppe, 1972

regions, while root exudates may be the principal mechanism in aquatic habitats. Once released, allelochemicals must reach the receiver plant in sufficient amounts to affect its growth (Newman, 1978). Water in form of rain or dew has been reported as the major transporting agent except in case of volatile inhibitors. Most allelochemicals reach the receiver plant via the soil (although a few are adsorbed directly onto the foliage). In the soil, allelochemicals may be modified, accumulated, leached, or taken up by the receiver plant. However, both possible allelopathic actions, either stimulatory or inhibitory, may be expected, depending on the concentration of the chemical received by the target plant. The importance of soil-water-allelochemical interaction can't be overemphasized, although studies on this aspect are scanty.

Quantities of allelochemicals produced are widely influenced by genetic as well as environmental factors including light (quality, intensity and duration) since more allelochemicals are produced upon plant exposure to ultraviolet light, ionizing radiation, red and far-red light and also during the peak of the growing season, and their interactions. It is long known that different plant species and even different cultivars are widely different in production of various allelochemicals and in their abilities to produce certain toxins in different parts. More differences in allelochemical production may be observed at different growth and development stages. Mineral (Bo, Ca, Mg, N, P, K, S) deficiency, drought stress, temperature (may be greater at cool temperature), application of plant growth regulators (2,4-D, maleic hydrazide, and other allelochemicals), infection by pathogens, or attack by insects may greatly increase the quantities of allelopathic chemicals produced. In this regard more research is needed on the effect of some agricultural practices and farm operations on the production of certain allelochemicals and if any differences may be obtained in response to specific operation (e.g., till vs. no till; high fertility vs. low fertility; moisture stress or accessibility). In addition, studies aimed to clarify the role of planting density and design, agrochemicals, and certain soil factors such as salinity are worth pursuing.

Nature, Mechanisms of Action and Future Development

Allelochemicals form a part of the natural ecosystem within which plants grow and are an important factor in ecological relationships between weeds and crop plants. Therefore, understanding the general nature of these chemicals is essential in recognizing the significant role they have in the interaction between plants and with other organisms such as pathogens and insects. However, allelopathic agents are usually thought to be secondary chemicals or waste products resulting from the primary metabolic pathways in plants (Levin, 1976) and do not appear to play a role in the primary metabolism essential for plant survival (Swain, 1977). They would logically include chemicals produced by microbial plants and those that enhance or inhibit growth. Swain (1977) estimated over 10,000 known low molecular weight secondary metabolites in higher plants and fungi. However, the number estimated for these secondary chemicals in nature is approaching 400,000.

Acetate and amino acids are basic components from which secondary compounds are derived while allelochemicals represent a wide spectrum of chemical classes ranging from simple molecules such as ammonia, hydrocarbons, and aliphatic acids to more complex quinones juglone, and the terpenes, camphor and cineole, to very complex conjugated flavonoids and polycyclic structures. They represent hydrolyzable tannins, terpenes, steroids, organic acids, low molecule carbohydrate and aldehydes, aromatic acids, simple unsaturated lactones, long chain fatty acids, quinones, flavonoids, cinnamic acid derivatives (coumarins, phenols, benzoic acids), alkaloids, cyanogenic glucosides and sulphides, cyanohydrins, amino acids and derivatives, polypeptides, purines necleosides, toxic gases (HCN, NH_3), and a variety of other chemicals (Rice, 1984). However, only certain compounds have been identified to be allelopathic in plant tissues, including juglone, scopoletin, hydroxamic acid, and sorgoleone (Putnam and Tang, 1986). Some allelochemicals identified in different weed species are shown in Table 7.

Limited work on the mode of action of allelochemicals suggests that they affect a variety of sites and biochemical processes, many of which are familiar to those affected by synthetic herbicides. Probably other novel sites and mechanisms remain to be discovered. The major effect of some allelochemicals have been discussed and summarized by different workers (Rice, 1979, 1984; Muller, 1986; Einhellig, 1986, 1995; Waller, 1989; Inderjit, Dakshini, and Foy, 1999). Some known mechanism of action of allelopathic weed species are given in Table 8.

Different workers referred to allelopathic agents as the future natural pesticides or nature's herbicides in action (Putnam, 1983; Rice, 1995). These chemicals are becoming more attractive to replace partially or totally the synthetic chemicals. Persistence, toxicity, pollution, and public concern create a need to reduce chemical inputs (Wickliffe et al., 1997). Undesirable pesticide mobility has resulted in recent rate adjustments and often label changes for several herbicides (Yenish, Worsham, and York, 1996). Because of their biodegradability, natural products are considered safer than synthetic chemicals to the environment They are fairly selective at low doses and they have no proven long-term environmental effects due to the existence of natural decomposing microorganisms. Allelochemicals could be produced using the donor plant in a factory system and could be also used as a basic framework from which to formulate new synthetic herbicides. However, the allelochemicals isolated from plants and soils have been less active than synthetic herbicides by an order or two of magnitude (Putnam, 1988). Devel-

TABLE 7. Allelochemicals of some common weed species.

Weed species	Allelochemical	Reference
Abutilon theophrasti	Flavonoides, free amino acids	Gressel and Holm, 1964
Ageratum conyzoides	Ageratochromene and its derivatives, monoterpenes, sesquiterpenes	Chuihua, Fei, and Shoushan, 1999
Agrostema githago	Agrostemin	Gajic et al., 1976
Amaranthus palmeri	Methyl ketones, pentanone, alcohols, aldehydes	Bradow and Connick, 1988a,b
Ambrosia artemisiifolia	Glucose ester of caffeic acid, sesquiterpenes	Rice, 1965
Ambrosia cumanensis	Sesquiterpenes lactones	Del Amo and Anaya, 1978
Ambrosia psilostachya	Glucose ester of caffeic acid, sesquiterpenes	Rice, 1965
Ammi majus	Furanocoumarins (Xanthotoxin)	Friedman, Rushkin, and Waller, 1982
Antennaria microphylla	Hydroquinone, arbutin, caffeic acid	Manners and Galitz, 1986
Artemisia absinthium	Quercitrin, sesquiterpene	Grummer, 1961
Artemesia annua	Sesquiterpenes lactones (artemisinin)	Lydon, Teasdale, and Chen, 1997
Artemesia herba-alba	Terpenes, terpenoides	Friedman, 1987
Artemesia princeps	Terpenoids	Kil et al., 1991
Avena fatua	Scopoletin, vanillic acid, chlorogenic, ellagic, caffeic, ferulic, *p*-coumaric, *p*-hydroxy-benzoic acids	Schumacher, Thill, and Lee, 1983; Qureshie, Hussain, and Shaukat, 1987
Bidens pilosa	Phenylheptatriyne	Campbell et al., 1982
Brassica nigra	Allyl isothiocyanate	Bell and Muller, 1973
Brassica spp.	Mustard oil glycosides	Choesin and Boerner, 1991
Bromus japonicus	Phenolics (Glucose ester of ferulic acid)	Rice and Parenti, 1967
Camelina alyssum	*p*-hydroxy benzoic acid, vanillic acid, ferulic acid	Grummer and Beyer, 1960
Camelina sativa	Benzyl isothiocyanate	Lovett and Duffield, 1981
Cassia auriculata	Proanthocyanidin	Bhatia and Chawan, 1976
Centaurea maculosa	Sesquiterpene lactone cnicin	Locken and Kelsey, 1987
Centaurea repens	Sesquiterpene lactone chromenes	Stevens and Merrill, 1985
Centaurea silastitialis	Eupatoriochromene, encecalin	Merril, 1989
Chenopodium album	Chlorogenic acid	Mallik, Puchala, and Grosz, 1994
Chenopodium ambrosioides	Terpenes (*p*-cymenen, ascaridole, aritazone)	Datta and Ghosh, 1987
Chenopodium murale	Oxalic acid	Datta and Ghosh, 1987
Chromolaena odorata	Phenolic acids, alkaloids	Ambika, 1999
Cirsium arvense	Ferulic, caffeic, chlorogenic, *p*-coumaric, *p*-hydroxy benzoic, vanillic acids	Hussain, Khan, and Hussain, 1987

TABLE 7 (continued)

Weed species	Allelochemical	Reference
Clerodendrum viscosum	Terpene (clerodin)	Datta and Chakrabarti, 1982
Croton bonplandianum	Abscisic acid, phaseic acid	Datta and Sinha-Roy, 1975
Cyperus esculentus	*p*-hydroxy benzoic, vanillic, syringic, ferulic, *p*-coumaric acid	Tames, Gesto, and Vieitez, 1973
Cyperus rotundus	Sesquiterpenes, catechol tannins	Komai and Ueki, 1975
Datura stramonium	Alkaloids (scopolamine, hyoscamine, tropane)	Lovett et al., 1981; Lovett and Potts, 1987
Delphinium ajacis	Diterpenoid alkaloids, delcosine, ajacocnine	Waller and Burstrom, 1969
Digitaria sanguinalis	Chlorogenic, isochlorogenic, sulfosalicylic acids	Rice, Penfound, and Rohrbaugh, 1960; Parenti and Rice, 1969
Echinops echinatus	Phenolics, terpenoids	Soni and Mohnot, 1988
Elytrigia repens	Organic acid, flavone tricin, polyacetylene ragropyrene, 5-hydroxy indole-3-acetic acid (5-HIAA), 5-hydroxy tryptophan (5-HTP)	Penn and Lynch, 1982; Waller, 1987; Hagen, 1989
Erica australis	Protocatchecuic, vanillic, *p*-hydroxybenzoic, *p*-coumaric	Carballeira and Cuervo, 1980
Erica sp.	Phenolics (orcinol, orcinolglucoside)	Ballester and Vieitez, 1971; Salas and Vieitez, 1972
Erigeron spp.	C16-polyacetylenes, C10-polyacetylenes	Kobayashi et al., 1980
Euphorbia corrollata	Gallic and tannic acid	Rice, 1965, 1969
Euphorbia esula	Kaemferol 3-glucuronide, 1-hexacosanol	Manners, 1987
Euphorbia supina	Gallic and tannic acids	Rice, 1965, 1969
Galium aparine	Asperuloside of a monoterpene lacton glucoside	Komai, Iwamura, and Ueki, 1983
Galium mollugo	Chlorogenic acid, scopolin, flavonoid, diosmetin trioside	Kohmuenzer, 1965
Imperata cylindrica	Phenolic, vanillic, p-coumaric, syringic acid, scopolin, scopoletin, chlorogenic, isochlorogenic acid	Abdul-Wahab and Al-Naib, 1972; Eussen, 1978; Eussen and Niemann, 1981
Kochia scoparia	Phenolic compounds	Wali and Iverson, 1978
Lantana camara	Phenolic compounds	Singh, Tamma, and Nigg, 1989
Melilotus alba	Coumarin	Winter, 1961
Miscanthus floridulus	Phenolic acids	Chou and Chung, 1974
Parthenium hysterophorus	Sesquiterpenes lactones, phenolics	Kanchan and Jayachandra, 1980
Pteridium aquilinum	Phenolic acids, p-coumaric	Glass, 1976; Whitehead, Dibb, and Hartley, 1982
Phleum pratense	Glycoside of esculetin	Avers and Goodwin, 1956

Weed species	Allelochemical	Reference
Plantago occidentalis	Chlorogenic acid, isochlorogenic acid, neochlorogenic acid, band-510, *o*-coumaric acid	Al-Naib and Rice, 1971
Pluchea lanceolata	Flavonone, dihydroflavonol, isoflavonoid	Inderjit and Dakshini, 1991a,b, 1992
Polygonum aviculare	Long chain fatty acid	Alsaadawi and Rice, 1982
Polygonum orientalis	Luteolin, apigenin, glycosides	Datta and Chatterjee, 1978, 1980a
Polygonum sachalinense	Anthraquinone (emodin, physcion)	Inoue et al., 1992
Proboscidea louisianica	Piperitenone, *p*-cymen-9-ol, vanillin, alpha-bisabolol, phenethyl alcohol	Riffle, 1988
Rorippa sylvestris	Salicylic, *p*-hydroxybenzoic, vanillic, syringic acid, hirsutin, pyrocatechol, isothiocyanates	Kawabata et al., 1989; Yamane et al., 1992
Rumex crispus	Phenolic compounds	Einhellig and Rasmussen, 1973
Ruta graveolens	Flavonoides, furocoumarins	Aliotta et al., 1999
Salsola kali	Ferulic acid	Lodhi, 1979
Salvia apiana *Salvia leucophylla* *Salvia millifera*	α-pinene, β-pinene, camphore, camphene, cineole, dipentene	Muller and Muller, !964; Muller, Muller, and Haines, 1964
Salvia reflexa	Aromatic compounds	Lovett and Lynch, 1979
Sasa cernua	*p*-coumaric, ferulic, vanillic, *p*-hydroxybenzoic acids	Li et al., 1992
Solidago altissima	Polyacetylenes, cis-dehydromatricaria	Numata, Kobayashi and Ohga, 1973; Kobayashi et al., 1980
Sporobolus pyramidatus	*p*-coumaric acid, ferulic acid	Rasmussen and Rice, 1971
Sorghum halepense	Phenolic compounds, chlorogenic, *p*-coumaric acids, *p*-hydroxybenzeldehyde	Rice, 1965; Abdul-Wahab and Rice, 1967
Tagetes erecta	α-terthienyl	Campbell et al., 1982
Tagetes minuta	α-terthienyl	Meissner, Nel, and Beyers, 1986
Tagetes patula	Thiophenes, benzofurans	Tang et al., 1987
Vulpia spp.	Phenolic compounds	An, Pratley, and Haig, 1999
Xanthium strumarium	Caffeic, *p*-coumaric, *p*-hydroxybenzoic, chlorogenic, ellagic acids	Inam, Hussain, and Bano, 1987

opment of allelochemicals as herbicides is hampered by the chemical complexity of many natural compounds, making synthesis difficult, and by the problem of working with small quantities available from natural sources (Duke and Abbas, 1995). Patterson (1986) reported that no synthetic herbicides had evolved from allelopathic background. Abbas et al. (1995) reported that various broadleaf crop and weed species showed susceptibility to AAL-toxin, a natural metabolite produced by *Alternaria alternata* f. sp. *lycopersi-*

TABLE 8. Some known mechanisms of action of certain allelopathic weeds.

Effect	Donor weed or toxin	Receiver crop	Reference
Interference with cell division	*Salvia leucophylla*	Cucumber	Muller, 1965
Effect on hormone induced growth	*Aegilops kotschyi*	–	Wurzoburger and Leshem, 1969
Effect on membrane permeability	*Salvia leucophylla*	–	Muller, 1969
Reduced photosynthesis	*Abutilon theophrasti*	Soybean	Colton and Einhellig, 1980
Effect on mineral uptake	*Chenopodium album*	Tomato	Qasem and Hill, 1989a
Reduced soil phosphorus availability	*Plantago lanceolata*	–	Newman and Miller, 1977
Effect on respiration	*Artemisia tridentata*	Wheat	Weaver and Klarich, 1977
Inhibition of respiratory activity	*Parthenium hysterophorus*, parthenin		
Inhibition of water conductance	*Abutilon theophrasti*	Soybean	Colton and Einhellig, 1980
Inhibition of soil microflora	*Polygonum aviculare*	Soil microflora	Alsaadawi and Rice, 1982
Interference with protein and nucleic acid metabolism	Cinnamic acid	Lettuce	Cameron and Julian, 1980
Effect on stomata opening	*Agropyron repens*	Flax	Vikherkova, 1970

ci, the pathogen that causes stem canker of tomato. Monocots are generally immune to AAL-toxin. Because of differential selectivity of species to AAL-toxin, Abbas et al. (1995) proposed that it could be exploited for selective weed control. Likewise, Batish et al. (1997) have reported that parthenin, a sesquiterpene lactone from *Parthenium hysterophorus* severely affects the germination and seedling growth of bill-goat weed but no effect was observed on wheat. These workers further reproted that parthenin has a growth regulatory activity similar to that of IAA (Batish et al., 1997). However, despite the number of natural products used commercially or under commercial development as pesticides and plant growth regulators, two main facts stand behind these advances. The first reflects the better effectiveness and ease of developing allelochemicals as insecticides, reflecting probably the more concern on public health and finding safer chemicals to replace the extremely toxic synthetic insecticides rather than less toxic herbicides. The second showed that most of these developed pesticides are of microbial (fungi or bacteria origin) derivatives or products (Putnam and Tang, 1986), although recent studies on allelopathy clearly indicate toward weeds as potentially high sources of natural allelochemicals of all types and with different modes and mechanisms of action.

SOME ECOLOGICAL IMPLICATIONS AND GENETIC ENGINEERING OF ALLELOPATHIC WEED SPECIES

Public concern about synthetic pesticides in general, and herbicides in particular, is exacerbated by lack of proper knowledge, awareness, and technical training of herbicide handling, application, and sprayer operation. Sustainable concern also results from the high cost of herbicides, persistence of some, and weed resistance to many of these chemicals or their groups. All urge searching for alternative methods of weed control that are safer to humans and the environment. Weed management seems to be the feasible substitute and more convenient than use of herbicides exclusively. Integration of allelopathy into this package may have its significant implication. However, allelochemicals pose both a problem and an opportunity. Considering their negative impacts, certainly they cause great yield losses to agriculture worldwide. Understanding factors controlling their production, and release, and affecting their activation, inactivation, nature, and mechanism of action, facilitates possible exploitation of natural chemicals in a beneficial way toward preserving and/or improving crop yield and quality. Some important developments have already found their way into commercial use and are available to farmers, while others are now underway.

Allelopathy is significant for weed ecologists as another factor affecting changes in weed populations and composition, as an important mechanism of weed interference with germination, growth, development, and yield of many economic crops and other species of ecological importance. Its positive impact remains in regarding this tool as potentially significant in reducing losses of desired plants from weeds. However, the effect on germination may be the easiest to identify, but that on growth may be widely spread under natural conditions. Weed seed allelopathy and its bioecological importance in preventing seed decay and attack by soil microorganisms, and in regulating seed germination, is another role of allelopathy mechanism providing an additional factor facilitating survival strategies of many weed species. It is well known that weed problems in agroecosystems result not only from the huge number of propagules, but also from the long extended dormancy period and the lack of uniformity in their germination. Allelopathy may play a significant role in this regard. In addition, it can be effectively exploited in several ways for weed management purposes through the use of allelopathic cover or smother crops, allelopathic rotational or companion crops (Putnam and Duke, 1978), trap or catch species to control certain parasitic weeds, crude toxic extracts directly or indirectly applied to target species (Habib and Rahman, 1988), crop growth stimulants to enhance the competitive ability of desired species (Gajic et al., 1976), crop straw or debris soil cover (Putnam, 1988), and naturally produced herbicides (Putnam, 1988; Lovett, 1990). Breeding of allelopathic cultivars with weed suppressing potential is another important factor. The possible significant role that allelopathy might have in agriculture has been reviewed by An, Partley, and Haig (1998).

Recently, scientists in different disciplines have attempted to exploit allelopathy as a pest management strategy either through isolating, identifying, and synthesizing certain allelochemicals as natural pesticides (Putnam and Tang, 1986), or by direct application of allelopathic plant parts into the field. Much work is underway worldwide to adopt this natural mechanism of plant interference as a significant tool for a weed management strategy in the agroecosystem. Screening a large number of germplasm collections of different crop species for allelopathic influence on various weed species, and searching for superior genotypes with allelopathic influence has already commenced. Available evidence revealed that crop cultivars differ significantly in their abilities to suppress certain weed species and indicates possible development of crop cultivars able to inhibit growth of the principal weeds in a given area through allelopathic action and thus decrease the need for synthetic herbicides. Encouraging results were obtained and effective weed-suppressing types have been recommended in cucumber, oat, sorghum, soybean, sunflower, rice, and probably others (Putnam and Duke, 1974; Fay and Duke, 1977; Massantini, Caporalli, and Zellini, 1977; Leather, 1983). This character may be incorporated into a cultivar by conventional plant breeding methods or other genetic recombination strategies. The transfer of genetic material from one organism to another genetically unrelated is now commonplace while concentrated efforts using advances in plant biotechnology (e.g., DNA recombination techniques) facilitate gene identification, characterization, splicing, replication, regulation, and transfer. Genetic markers have allowed researchers to systematically map and characterize genes controlling quantitative character. In addition, better understanding and significant advances of genetics of allelopathic crops and genetic manipulation of this trait have been achieved (Wu et al., 1999). No doubt new techniques available in genetic engineering, molecular biology, tissue culture, etc., offer a great potential for identifying allelochemicals and their roles. This information may help to clarify the genetics of this trait and subsequently allow gene transfer to certain cultivars, aimed towards improving their competitive efficiency. Advanced technology offers opportunities to readily incorporate the capacity to produce specific allelochemicals into given plants for a desired effect, which may be essentail in the case of parasitic weed control, stress tolerance, or disease resistance. However, much work on this aspect is still waiting if better understanding of allelopathy genetics is our ultimate goal of useful exploitation of this bioecological tool.

Crops are widely different in their abilities of intra- and interspecific interference, and also in their tolerance to different environmental factors. Many studies have clearly demonstrated genetic variability of the allelopathic character within crops (Wu et al., 1999), which may be considered as an important genetic reservoir for selection of allelopathic cultivars. Mason-Sedun, (1986) and Putnam and Tang (1986) have suggested that allelopathy is a genetically controlled mechanism. Therefore, progress in developing allelopathic cultivars may be highly dependent on our deeper knowledge and understanding of the genetic control mechanism of allelopathy. However, isolating and quantifying allelochemicals in a genetically diverse population is a real challenge added to another probable constraint of having multiple rather than single genes involved in this process. Moreover, the type of genes controlling this mechanism, either dominant or recessive may add more complication in revealing this mechanism. Olofsdotter, Navarez, and Moody (1995) speculated that allelopathic mechanism might be a polygenic trait weakly correlated with yield or other important agronomic characteristics. Regulatory genes may be also involved in the process of releasing allelochemicals into the soil environment (Wu et al., 1999).

On the other hand, a strong relationship between the genetics of allelopathy mechanism and herbicide resistance may be expected. Both are very well expressed in response to stress conditions. Genes controlling each mechanism may not be activated or effectively expressed unless interaction with external stresses has occurred. It has been established that allelopathic compounds are released in greater amounts when plants are under stress (Rice, 1984; Einhellig, 1996). Continuous exposure of a plant to stress factors may encourage more responses and expression of recessive genes controlling either mechanism in a form of chemical resistance or allelopathic defense or adaptation mechanism. The genes conferring these characters may become more dominating and active compared with other dominant genes controlling other beneficial characters such as high yielding or vigorous growth. However, plant self-expression through allelopathy or chemical resistance is varied according to species genetic differences, and quantity, and quality of the environmental factors imposed and causing these mechanisms to operate. This may imply that probably all plant species have an allelochemical defense mechanism and are able to produce allelochemicals, but the strength of this argument or position may be greatly dependent on the degree of species/environment interaction, genetics, and whether the genes responsible are dominant or recessive. All these factors may work jointly to determine amounts, and may be the types of allelochemicals produced or excreted to the environment. It is well known that different environmental factors affect production of different allelochemicals in different species. On the other hand, the similarity in mechanism of action of herbicides and allelochemicals suggests that herbicide resistance and allelopathy may be more or less genetically related mechanisms. However, the significant role that stresses exert on both mechanisms, implies that external conditions may regulate the genes, promote the biosynthesis of allelochemicals or inactivate the herbicides within plant tissues and stimulate degradation of the herbicides in herbicide resistant plants, or release allelopathic agents into the surrounding environment. Similarities between allelopathy mechanism of action and herbicide resistant or susceptible traits may facilitate use of the same techniques to study the genetic gain for the allelopathic activity through marker-assisted selection in crop breeding programs. When an allelopathy inheritance character becomes known, a breeding program could be initiated to select novel secondary metabolites, to enhance the biosyntheses of these biochemicals and to regulate the release of allelochemicals for biological management of weeds. Possible transfer of allelopathic agents from wild species or crop progenitors to crop cultivars appeared highly promising with successful transfer of allelochemical production from *Agropyron glaucum* to wheat (*Triticum aestivum*) (Panchuk and Prutinskaya, 1973; Grodzinsky and Panchuk, 1974). In this regard, it may be important to ensure that genetic manipulation of allelochemicals of a crop does not have adverse effects on non-target organisms or the environment (Einhellig and Leather, 1988), and potential gene flow to weed species needs to be kept under control.

Genetic engineering for herbicide resistance has been successfully achieved with cultivars of different economic crops (soybean, cotton, corn, and others), which are now commercially available to farmers. However, in the future, a possibly feasible method of pest control is the genetic engineering for herbicide resistance aimed to control parasitic weeds and/or developing parasitic weed resistant cultivars. The use of natural plant chemicals offers an additional avenue for alleviating the problem. Overall, allelochemicals should serve as a model for future herbicides if environmental compatibility is a required feature. Incorporation of allelopathic traits together with other plant interference potentials into commercial cultivars could be a major step toward further development of sustainable crop production systems with less herbicide dependence. Methods of cell and tissue culture would provide continuous production processes for industrial scale operation, and plant-derived chemicals can be expected to play an increasingly significant role in agricultural and ecological sciences. At present, allelopathy may also become available and considered as one of the viable weed control methods in the third world where herbicide technology is limited.

CONCLUSION AND FUTURE PROSPECTS

Despite the complexity in research methodology, allelopathy as a science is rapidly growing and the phenomenon does exist as one component of plant interference in nature. Also, its significant role in nature is now fairly well acknowledged. Although complications in plant-plant interference proved to be a significant barrier in understanding how allelopathy operates, research on allelopathy in recent years has increased in both scale and diversity. However, the void in our knowledge on the subject is still not fulfilled by experimental field evidence, and a good deal of more intensive precise work is still required.

No doubt allelopathy research has advanced a long way since the era of De Candolle, but it seems that we are still far from completely answering the critical questions concerning its exact ecological role due to many factors involved that may affect this process, either positively or negatively. Allelochemicals are added and removed from the environment by numerous processes, the dynamics of which are poorly understood. If allelopathy is to be regarded as a scientifically credible hypothesis, evidence should be provided on its ecological occurrence, and its separation from competition must be achieved. Reaching this advanced level, chemical warfare in the plant community can be extended to encompass other organisms (mainly agricultural pests) and probably many others of primary importance to humans and the environment. Allelochemicals are diverse in their chemical nature, concentration in different plant parts and for the same plant species under different environmental conditions, method of release, nature of effect, selectivity, and susceptibility to degradation by different environmental factors. Concentrations of these allelochemicals within plant tissues or after their release to the environment are greatly influenced by prevailing environmental conditions determining their release, activation (stimulatory or inhibitory), or inactivation. Allelochemicals may be toxic or nontoxic at the time of release by donor species, but their natural effect may be completely reversed by environmental factors or receiver species. However, the stimulatory (especially at low concentration) or inhibitory (at high concentration) effects of allelochemicals may offer a wide array of possible beneficial uses for a better-managed environment.

Although studies to measure differences between species in their responses to allelopathic plants or chemicals are relatively small compared with the great bulk of literature on the subject, studies devoted mainly to this aspect aimed to evaluate the effect of different allelopathic species against a standard common species are lacking. Such studies may determine the extent of allelopathic effects that different allelopathic species may have on a common target plant and differences between donors in production of allelochemicals and its relation to the overall growth characteristics of these species. In addition, such studies may determine the range of chemicals to which this species is susceptible and whether all species producing the same allelochemical are able to influence the target similarly, or not. Exposing all species to the same stress conditions and measuring the amount of allelochemicals they release when grown in artificial conditions and separately may be also a useful approach. In fact, we urge a protocol similar to but expanded from Koch's postulates be followed to develop proofs. There is a clear need for better understanding of plant succession and mutualism, and identification of soil organisms and their role in activating or inactivating allelochemicals in nature. Differentiation between allelochemicals and other chemicals of natural origin and the extent of the effect of the former group would seem to be important for understanding the ecological role of allelochemicals, in natural as well as in agricultural ecosystems. The genetic mechanism and its relation to prevailing environmental conditions, the types of genes responsible, and differences between species relative to allelopathic traits appeares crucial in providing some useful information in this realm and extending our understanding this complex area of plant interference in nature.

Major research work should be directed toward finding methods to increase weed seed decay, overcome dormancy, or inhibit weed seed germination. In this regard, and in spite of the considerable amount of information relative to allelopathy that has accumulated in the scientific literature, its significance to agriculture and in particular to weed science has yet to be fully recognized and appreciated (Anderson, 1996). There is little doubt that allelochemicals inhibit plant growth, but a problem still exists in separating allelopathic effects from competition. The alleopathic potentials of certain perennial and annual weeds that influence the growth and yield of desired plants has been well documented (Rice, 1984, 1995; Inderjit and Foy, 1998). However, more available knowledge of life cycles, especially of annual weeds, is important in understanding the role of allelopathy. Establishing the role of allelopathy interference in the field is essential before using allelopathy in successful weed management, although different mechanisms of interference such as resource competition, allelopathy, microbial nutrient immobilization, etc., cannot be separated under field conditions.

REFERENCES

Abbas, H.K., T. Tanaka, S.O. Duke and C.D. Boyette. (1995). Susceptibility of various crop and weed species to AAL-toxin, a natural herbicide. *Weed Technology* 9: 125-130.

Abdul-Wahab, A.S. and E.L. Rice. (1967). Plant inhibition by johnsongrass and its possible significance in old field succession. *Bulletin of the Torrey Botanical Club* 94: 486-497.

Abdul-Wahab, A.S. and F.A.G. Al-Naib. (1972). Inhibitional effects of *Imperata cylindrica* (L.). *Bulletin of the Iraqi Natural History Museum Publication* 5: 17-24.

Abu-Irmaileh, B.E. and J.R. Qasem. (1986). Aqueous extract effects of *Salvia syriaca* L. in various lines of four crops. *Dirasat* 13: 147-170.

Abu-Shakra, S., A.A. Maih, and A.R. Saghir. (1970). Germination of seeds of branched broomrape (*Orobanche ramosa* L.). *Horticultural Research* 10: 119-124.

Ahmed, M. and D.A. Wardle. (1994). Allelopathic potential of vegetative and flowering ragwort (*Senecio jacobaea* L.) plants against associated pasture species. *Plant and Soil* 164: 61-68.

Ahmed, S.A., M. Ito, and F. Hussain. (1982). Phytotoxic effect of water hyacinth water extract and decayed residue. *Weed Research. (Japan)* 27: 177-182.

Ahshapanek, D.C. (1962). *Ecological Studies on Plant Inhibition of Solanum rostratum*. Dissertation, University of Oklahoma, Norman, USA.

Al-Abed, A.S., J.R. Qasem and H.A. Abu-Blan. (1993). Antifungal effects of some common wild plant species on certain plant pathogenic fungi. *Dirasat (Pure and Applied Science)* 20: 149-158.

Alam, S.M. and A.R. Azmi. (1989). Influence of *Prosopis glandulosa* water extract on seedling growth of wheat cultivars. *Pakistan Journal of Scientific and Industrial Research* 32: 708.

Alan, E. and U. Barrantes. (1988). Effecto alelopatico del madero negro (*Gliricidia sepium*) en la germinacion y crecimiento inicial de algunas malezas topicales. *Turrialba* 38: 271-278.

Alderman, W.H. and J.A. Middleton. (1925). Proceedings of the American Society of Horticultural Science 22: 307. Cited by A.R. Putnam and L.A. Weston. 1986. Adverse impact of allelopathy in agricultural systems. In *The Science of Allelopathy*, eds. A.R. Putnam and C.S. Tang. New York: John Wiley and Sons, pp. 43-50.

Aldrich, R.J. (1984). *Weed Crop Ecology. Principles in Weed Management*. MA: Breton Publisher.

Aldrich, R.J. and R.J. Kramer. (1997). *Principles in Weed Management*. 2nd ed., Ames, IA: Iowa State University Press, pp. 455.

Aliota, G. and G. Cafiero. (1999). Biological properties of rue (*Ruta graveolens* L.): Potential use in sustainable agricultural systems. In *Principles and Practices in Plant Ecology: Allelochemical Interactions*, eds. Inderjit, K.M.M. Dakshini and C.L. Foy. Boca Raton, CRC Press LLC, pp. 551-563.

Aliota, G., G. Cafiero, S. Conti, M.M. Corsaro, and A. Oliva. (1999). Germination of radish (*Raphanus sativus* L.) seeds in presence of rue (*Ruta graveolens* L.) infusion: a cellular perspective. In *Second World Congress on Allelopathy: Critical Analysis and Future Prospects*. Lakehead University, Canada. August 8-13. Abstract no. 47.

Al-Menoufi, A.O. (1991). Crop rotation as a control measure of *Orobanche crenata* in *Vicia faba* fields. In *Progress in Orobanche Research*, eds. K. Wegmann and L.J. Musselman. Proceedings of International Workshop on *Orobanche* Research. Obermarchtal, 1989., Tubingen: Eberhard-Karls-Universitat, pp. 241-247.

Al-Menoufi, A.O. (1992). *Cultural practices to control Orobanche crenata in faba bean field in Egypt*. Final Report, Project Financed by Alexandria University, Egypt (1991-1992).

Al-Menoufi, A.O. and M.A. Adam. (1998). Susceptibility/resistance of some wild *Lycopersicon* accessions to *Orobanche ramosa*. In *Current Problems of Orobanche Researches*, eds. K. Wegmann, L.J. Musselman, and D.M. Joel. Proceedings of the 4th International *Orobanche* Workshop, Sept. 23-26, 1998. Albena, Bulgaria: Institute for wheat and sunflower "Dobroudja," pp. 259-266.

Al-Menoufi, A.O., N.A. El-Safwani, and M.A. Adam. (1996). Biological and chemical inhibition of *Orobanche* seed germination. In *Advances in Parasitic Plant Research*, eds. M.T. Moreno and J.I. Cubero. Sixth International Parasitic Weed Symposium. 16-18 April, Cordoba, Spain, pp. 417-423.

Al-Naib, F.A. and E.L. Rice. (1971). Allelopathic effects of *Plantanus occidentalis*. *Bulletin of the Torrey Botanical Club* 98: 75-82.

Alsaadawi, I.S. and E.L. Rice. (1982). Allelopathic effects of *Polygonum aviculare* L. I. Vegetational patterning. *Journal of Chemical Ecology* 8: 993-1009.

Alsaadawi, I.S., I.A.K. Sakeri, and S.M. Al-Dulaimy. (1990). Allelopathic inhibition of *Cynodon dactylon* (L.) Pers. and other plant species by *Euphorbia prostrata* L. *Journal of Chemical Ecology* 16: 2747-2754.

Altieri, M.A. and J.D. Doll. (1978). The potential of allelopathy as a tool for weed management in crops. *Pest Articles and News Summaries* 24: 495-502.

Ambika, S.R. (1999). Allelopathic interference of *Chromolaena odorata* (L.) King and Robinson. In *Second World Congress on Allelopathy: Critical Analysis and Future Prospects*. Lakehead University, Thunder Bay, Ontario, Canada. August 8-13. Abstract no. 49.

Ambika, S.R. and Jayachandra. (1980). Suppression of plantation crops by *Eupatorium* weed. *Current Science* 49: 874-875.

Amritphale, D. and L.P. Mall. (1978). Allelopathic influence of *Saccharum spontaneum* L. on growth of three varieties of weed. *Science and Culture* 44: 28-30.

An, M., J.E. Pratley, and T. Haig. (1998). Allelopathy: From concept to reality. In *Proceedings of the 9th Australian Agronomy Conference*, Wagga Wagga, Australia, pp. 563-566.

An, M., J.E. Pratley and T. Haig. (1999). Allelochemical dynamic of decaying *Vulpia* residues, and their corresponding biological activity. In *Second World Congress on Allelopathy: Critical Analysis and Future Prospects*. Lakehead University, Thunder Bay, Ontario, Canada. August 8-13. Abstract no. 50.

Anaya, A.L. and A. Gomez-Pompa. (1971). Inhibition del crecimiento producida por el "piru" (*Schinus molle* L.). *Revista Sociedad Mexicana de Historia Natural* 32: 99-109.

Anaya, A.L. and S. Del Amo. (1978). Allelopathic potential of *Ambrosia cumanensis* H.B.K. (compositae) in a tropical zone of Mexico. *Journal of Chemical Ecology* 4: 289-304.

Anaya, A.L., S. Del Amo, G. Ruy-Ocotla, and L.M. Ortiz. (1978). Allelopathic potential of a coffee plantation. In *2nd International Congress of Ecology*. Jerusalem. Abstract no.8.

Anaya, A.L., L. Ramos, J.G. Hernandez, and R. Cruz. (1987). Allelopathy in Mexico. In *Allelochemicals Role in Agriculture and Forestry*, ed. G.R. Waller. ACS Symposium Series 330. Washington, DC: American Chemical Society, pp. 89-101.

Anderson, R.C., A.J. Katz, and M.R. Anderson (1978). Allelopathy as a factor in the success of *Helianthus mollis* Lam. *Journal of Chemical Ecology* 4: 9-16.

Anderson, W.P. (1996). *Weed Science: Principles and Applications*. 3rd ed. St. Paul, MN: West Publishing Company, pp. 388.

Aquila, M.E.A., L.P. da Silva, and M.P. Laufer. (1999). Screening bioassay to detect allelopathy in a natural Brazilian woodland: Evaluation of *Eugenia uniflora* (Myrtaceae). In *Second World Congress on Allelopathy: Critical Analysis and Future Prospects*. Lakehead University, Thunder Bay, Ontario, Canada. August 8-13. Abstract no. 52.

Archhireddy, N.R. and M. Singh. (1984). Allelopathic effects of lantana (*Lantana camara*) on milkvine (*Morrenia odorata*). *Weed Science* 32: 757-761.

Arines, J., E. Vieitez and J.L.F. Mantilla. (1974). Cited by W.W. Fletcher and R.C. Kirkwood (1982), p. 96-97. In *Herbicides and Plant Growth Regulators*, London: Granada, pp. 93-99.

Arora, R.K. and R.K. Kohli. (1993). Autotoxic impact of essential oils extracted from *Lantana camara* L. *Biologia Plantarum* 35: 293-297.

Ashraf, N. and D.N. Sen. (1978). Allelopathic potential of *Celosia argentea* in arid land crop fields. *Oecologia Plantarum* 13: 331-338.

Ashraf, N. and D.N. Sen. (1980). Allelopathic influence of *Digera alterniflora* on cultivated crops in India arid zone. *Indian Journal of Weed Science* 12: 69-74.

Avers, C.J. and R.H. Goodwin. (1956). Studies on roots. IV. Effects of coumarin and scopoletin on the standard root growth pattern of *Phleum pratense*. *American Journal of Botany* 43: 612-620.

Ballegaard, T.K. and E. Warncke. (1985). Observations on antitoxic effects on seed germination and seedling growth in *Cirsium palustre* from a spring area in Jutland, Denmark. *Horticultural Ecology* 8: 63-65.

Ballester, A., J.M. Albo, and E. Vieitez. (1977). The allelopathic potential of *Erica scoparia* L. *Oecologia* 30: 55-61.

Bansal, R.P. and D.N. Sen. (1982). Effect of arid zone weed *Trichodesma sedgwikianum* on growth and yield of bajra and til crops. *Indian Journal of Botany* 5: 45-49.

Bar-Nun, N. and A.M. Mayer. (1990). Cucurbitacins protect cucumber tissue against infection by *Botrytis cinerea*. *Phytochemistry* 29: 787-791.

Batish, D.R., R.K. Kohli, H.P. Singh, and D.B. Sexena. (1997). Studies on herbicidal activity of parthenin–a constituent of *Parthenium hysterophorus* towards bill-goat weed. *Current Science* 73: 369-371.

Batish, D.R., R.K. Kohli, D.B. Sexena, and H.P. Singh. (1997). Growth regulatory response of parthenin and ita derivatives. *Plant Growth Regulation* 21: 189-194.

Becker, Y. and L. Guyot. (1951). Sur les toxines racinaires des sols incultes. *C.R. Academia des Sciences* 232: 105-107.

Becker, Y., L. Guyot and J. Montegut. (1951). Sur quelques incidences phytosociolgiques du probleme des excretions racinaires. *C. R. Academia des Sciences* 232: 2472-2474.

Bell, D.T. and C.H. Muller. (1973). Dominance of California annual grassland by *Brassica nigra*. *American Midland Naturalist* 90: 277-299.

Bell, D.T. and D.E. Koeppe. (1972). Noncompetitive effects of giant foxtail on the growth of corn. *Agronomy Journal* 64: 321-325.

Bendall, G.M. (1975). The allelopathic activity of California thistle (*Cirsium arvense*) in Tasmania. *Weed Research* 15: 77-81.

Benedict, H.M. (1941). The inhibitory effect of dead roots on the growth of broomgrass. *Journal of the American Society of Agronomy* 33: 1108-1109.

Bernard, P. and D. Pesando. (1989). Antibacterial and antifungal activity of extracts from rhizomes of the Mediterranean teagrass *Psidonia oceanica* (L.). Delile. *Botanica Marina* 32: 85-88.

Bezuidenhout, S.R. and C.F. Reinhardt. (1999). Growth stage of *Cyperus esculentus* influence its allelopathic effects. In *Second World Congress on Allelopathy: Critical Analysis and Future Prospects*. Lakehead University, Thunder Bay, Ontario, Canada. August 8-13. Abstract no. 59.

Bhandari, M.C. and D.N. Sen. (1971). Effect of *Citrullus colocynthis* (Linn) Schrad. On the seedling growth of *Pennisetum typhoideum* Rich. *Zeitschrift fuer Pflanzenphysiologie* 64: 466-469.

Bhandari, M.C. and D.N. Sen. (1972). Growth regulation specificity exhibited by substances present in the fruit pulp of *Citrullus lanatus* (Thunb.) Mansf. *Zeitschrift fuer Naturforschung* 27: 72-75.

Bhatia, R.K. and D.D. Chawan. (1976). Occurrence of phenolic substances in seed coat of *Cassia species* and their effect on early seedling growth. *Geobios* (Jodhpur) 3: 214-216.

Bhowmik, P.C. and J.D. Doll. (1979). Evaluation of allelopathic effects of selected weed species on corn and soybeans. *Proceedings of North Central Weed Control Conference* 34: 43-45.

Bhowmik, P.C. and J.D. Doll. (1983). Growth analysis of corn and soybean in response to allelopathic effects of weed residues at various temperature and photosynthetic flux densities. *Journal of Chemical Ecology* 9: 1263-1280.

Bhowmik, P.C. and J.D. Doll. (1984). Allelopathic effects of annual weed residues on growth and nutrient uptake of corn and soybeans. *Agronomy Journal* 76: 383-385.

Bieber, G.L. and C.S. Hoveland. (1968). Phytotoxicity of plant material on seed germination of crownvetch, *Coronilla varia* L. *Agronomy Journal* 60: 185-188.

Biswas, E. and K.N. Ghosh. (1984). Allelopathic effects of *Chenopodium murale*–An aspect of weed control for non-associated species. *Science and Culture* 50: 99-102.

Biswas, E. and N.K. Chakraborti. (1984). Allelopathy in *Borreria articularis*: Inhibition of seed germination and seedling growth. *Science and Culture* 50: 263-264.

Blum, U. (1999). Designing laboratory plant debris-soil bioassy: some reflection. In *Principles and Properties in Plant Ecology, Allelochemical Interactions*, eds. Inderjit, K.M.M. Dakshini, and C.L.Foy. Boca Raton: CRC Press LLC, pp. 17-23.

Bode, H.R. (1940). Uber die Blattauss cheidungen des wermuts und ihre Werkung auf andere. *Pflanzeno Planta* 30: 567-589.

Bonner, J. and A.W. Galston. (1944). Toxic substances from the culture media of guayule which may inhibit growth. *The Botanical Gazette (Chicago)* 106: 185-198.

Bouhatous, B. and P. Jacquard. (1994). The effect of combinations of hosts on infection capacity of *Orobanche crenata* Forsk. In *Biology and Management of Orobanche*, eds. A.H. Pieterse, J.A.C. Verklei and S.J. Borg. Proceedings of the Third International Workshop on *Orobanche* and Related *Striga* Research. Amesterdam, The Netherlands: Royal Tropical Institute, pp. 320-333.

Bradow, J.M. and W.J. Connick, Jr. (1987). Allelochemicals from palmer amaranth *Amaranthus palmeri* Wats. *Journal of Chemical Ecology* 13: 185-202.

Bradow, J.M. and W.J. Connick, Jr. (1988a). Volatile methyl ketone seed-germination inhibitors from *Amaranthus palmeri* S.Wats. residues. *Journal of Chemical Ecology* 14: 1617-1631.

Bradow, J.M. and W.J. Connick, Jr. (1988b). Seed germination inhibition by volatile alcohols and other compounds associated with *Amaranthus palmeri* residues. *Journal of Chemical Ecology* 14: 1633-1648.

Bridges, D.C. (1995). Ecology of Weeds. In *Handbook of Weed Management Systems*, ed. A.E. Smith. New York: Marcel Dekker, pp. 19-34.

Brown, D.D. (1968). *The Possible Ecological Significance of Inhibition by Euphorbia supina*. M.Sc. Thesis. University of Oklahoma, Norman, USA.

Bruckner, D.J., A. Molnar, and L.G. Szabo. (1999). The allelopathic effect of common ragweed (*Ambrosia artemisiifolia* L.) in tissue culture. In *Second World Congress on Allelopathy: Critical Analysis and Future Prospects*. Lakehead University, Thunder Bay, Ontario, Canada. August 8-13. Abstract no. 63.

Buchanan, B.A., K.T. Harper and N.C. Frischknecht. (1978). Allelopathic effect of bur buttercup tissue on germination and growth of various grasses and herbs *in vitro* and in soil. *Great Basin Naturalist* 38: 90-96.

Bukolova, T.P. (1971). A study of the mechanism of action of water-soluble substances of weeds on cultivated plants. In *Physiological-Biochemical Basis of Plant Interactions in Phytocenoses*, ed. A.M. Grodzinsky. Vol. 2. Kiev: Naukova Dumka, pp. 66-69 (In Russian, English summary).

Buta, J.G., D.W. Spauldin, and A.N. Reed. (1987). Diffferential growth responses of fractionated turfgrass seed leachates. *HortScience* 22: 1317-1319.

Cameron, H.J. and J.R. Julian. (1980). Inhibition of protein synthesis in lettuce (*Lactuca sativa* L.) by allelopathic compounds. *Journal of Chemical Ecology* 6: 989-995.

Campbell, G., J.D.H. Lambert, T. Arnason, and G.H.N. Towers. (1982). Allelopathic properties of α-terthienyl and phenylheptatriyne, naturally occurring compounds from species of Asteraceae. *Journal of Chemical Ecology* 8: 961-972.

Carballeira, A. and A. Cuervo. (1980). Seasonal variation in allelopathic potential of soils from *Erica australis* L. heathland. *Acta Oecologica* 1: 345-353.

Cardina, J. (1995). Biological weed management. In *Handbook of Weed Management Systems*, ed. A.E. Smith. New York: Marcel Dekker, pp. 279-341.

Carnahan, G. and A.C. Hull. (1962). The inhibition of seeded plants by tarweed. *Weeds* 10: 87-90.

Caussanel, J.P. (1979). Effect of non competitifs entre le chenopode blanc (*Chenopodium album*) de la mais (INRA 258). *Weed Research* 19: 123-125.

Chaghtai, S.M., A. Sadiq, and M. Ibrar. (1986). Phytotoxicity of *Fumaria indica* on wheat (*Triticum aestivum* var. Blue Silver). *Pakistan Journal of Botany* 18: 59-64.

Chaturvedi, R., A. Dikshit and S.N. Dixit. (1987). *Adenocalyamma allicea*, a new source of a natural fungitoxicant. *Tropical Agriculture* 64: 318-322.

Cheam, A.H. (1984). Journal of Agriculture of Western Australia 12: 69. Cited by J.V. Lovett, (1986). Allelopathy: The Australian experience. In *The Science of Allelopathy*, eds. A.R. Putnam and C. Tang. New York: John Wiley and Sons, pp. 75-99.

Choesin, D.N. and R.E.J. Boerner. (1991). Allyl isothiocyanate release and the allelopathic potential of *Brassica napus* (Brassicaceae). *American Journal of Botany* 78: 1083-1090.

Chou, C.H. (1989). Allelopathic research of subtropical vegetation in Taiwan. IV. Competitive phytotoxic nature of the leachate from four subtropical grasses. *Journal of Chemical Ecology* 15: 2149-2159.

Chou, C.H. (1987). Allelopathy in subtropical vegetation and soils in Taiwan. In *Allelochemicals: Role in Agriculture and Forestry*, ed. G.R.Waller. ACS Symposium Series 330, Washington, DC: American Chemical Society, pp. 102-117.

Chou, C.H. and C. Yao. (1983). Phytotochemical adaptation of coastal vegetation in Taiwan. I. Isolation, identification and biological activities of compounds in *Vitex negundo* L. *Botanical Bulletin of the Academia Sinica* 24: 155-168.

Chou, C.H. and C.C. Young. (1975). Phytotoxic substances in twelve subtropical grasses. *Journal of Chemical Ecology* 1: 183-193.

Chou, C.H. and Y.T. Chung. (1974). The allelopathic potential of *Miscanthus floridulus*. *Botanical Bulletin of the Academia Sinica* 15: 14-27.

Chou, C.H., F.J. Chang, and H.I. Oka. (1991). Allelopathic potential of wild rice, *Oryza perennis*. *Taiwania* 36: 201-210.

Chou, C.H., M.I. Lee and H.I. Oka. (1984). Possible allelopathic interaction between *Oryza perennis* and *Leersia hexandra*. *Botanical Bulletin of the Academia Sinica* 25: 1-19.

Chuihua, K., H. Fei, and H. Shoushan. (1999). Allelopathic potential of *Ageratum conyzoides* under different environmental conditions. In *Second World Congress on Allelopathy: Critical Analysis and Future Prospects*. Lakehead University, Thunder Bay, Ontario, Canada. August 8-13. Abstract no. 70.

Clements, F.E. (1904). Development and structure of vegetation. *Report of the Botanical Survey of Nebraska* 7: 5-175.

Clements, F.E. (1928). *Plant Succession and Indicators. A Definition Edition of Plant Succession and Plant Indicators*. New York: The H.W. Wilson Company, pp. 72-73.

Cochran, V.L., L.F. Elliott, and R.I. Papandick. (1977). The production of phytotoxins from surface crop residues. *Soil Science Society of America Journal* 41: 903-908.

Colton, C.E. and F.A. Einhellig. (1980). Allelopathic mechanisms of velvet leaf (*Abutilon theophrasti* Medic, Malvaceae) on soybean. *American Journal of Botany* 67: 1407-1413.

Connick, W.J. Jr., J.M. Bradow, M.G. Legenori, S.L. Vail, and R.M. Menges. (1987). Identification of volatile allelochemicals from *Amaranthus palmeri* S. Wats. *Journal of Chemical Ecology* 13: 463-472.

Correa, J.de.F., I.F. deSouza, A.M. Ladeira, M.C.M. Young, and M. Haragushi. (1999). Allelopathic potential and identification of substances of *Eupathorium maximiliani* leaves. In *Second World Congress on Allelopathy: Critical Analysis and Future Prospects*. Lakehead University, Thunder Bay, Ontario, Canada. August 8-13. Abstract no. 71.

Cote, J.F. and J.R. Thibault. (1988). Allelopathic potential of raspberry foliar leachates on growth of ectomycorrhizal fungi associated with black spurce. *American Journal of Botany* 75: 966-970.

Curtis, J.T. and G. Cottam. (1950). Antibiotic and autotoxic effects in prairie sunflower. *Bulletin of the Torrey Botanical Club* 77: 187-191.

Das, T.M. and S.K. Pal. (1970). Effects of volatile substances of aromatic weeds on germination and subsequent growth of rice embryos. *Bulletin of the Botanical Society of Bengal* 24: 101-103.

Datta, S.C. and A.K. Chatterjee. (1978). Some characteristics of an inhibitory factor in *Polygonum orientale* L. *Indian Journal of Weed Science* 10: 23-33.

Datta, S.C. and A.K. Chatterjee. (1980a). Allelopathy in *Polygonum orientale*: Inhibition of seed germination and seedling growth of mustard. *Comparative Physiology and Ecology* 5: 54-59.

Datta, S.C. and A.K. Chatterjee. (1980b). Allelopathic potential of *Polygonum orientale* L., in relation to germination and seedling growth of weeds. *Flora* 169: 456-465.

Datta, S.C. and K.N. Ghosh. (1982). Effect of pre-sowing treatment of mustard seeds with leaf and inflorescence extracts of *Chenopodium murale. Indian Journal of Weed Science* 14: 1-6.

Datta, S.C. and K.N. Ghosh. (1987). Allelopathy in 2 species of *Chenopodium*-inhibition of germination and seedling growth of certain weeds. *Acta Societatis Botanicorum Poloniae* 56: 257-270.

Datta, S.C. and S. Dasmahapatara. (1984). Allelopathy in two legumes: Inhibition of mustard (*Brassica*) germination and seedling growth. *Comparative Physiology and Ecology* 9: 285-289.

Datta, S.C. and S.D. Chakarbarti. (1982). Allelopathy in *Clerodendrum viscosum*: Inhibition of mustard (*Brassica*) germination and seedling growth. *Comparative Physiology and Ecology* 7: 1-7.

Datta, S.C. and S.P. Sinha-Roy (1983). Characteristics of an inhibitory factor from *Croton bonplandianum* Baill. *Acta Agronomica Academia Scientiarum Hungerica* 32: 124-129.

Datta, S.C. and S.P. Sinha-Roy. (1975). Phytotoxic effects of *Croton bonplandianum* Baill. on weedy associates. *Vegetatio* 30: 157-163.

De Candolle,M.A.P. (1832). *Physiologie Vegetale*. Tome III, pp. 1474-1475. Bechet Jeune, Lib, Fac. Med., Paris.

Del Amo, S. and A.L. Anaya. (1978). Effect of some sesquiterpene lactones on the growth of certain secondary topical species. *Journal of Chemical Ecology* 4: 305-313.

Desvedavy, C., M. Amoros, L. Girre, and P. Cmouard. (1987). *In vivo* antifungal properties of the daisy *Bellis perennis* and the scarlet pimpernel *Anagallis arvensis*, their use in the treatment of *Graphium ulmi. Fitoterapia* 58: 229-234.

Dhanapal, G.N., P.C. Struik, and S.J. terBorg. (1998). Effect of natural stimulants with and without GR 24 on broomrape germination. In *Current Problems of Orobanche Researches*, eds. K. Wegmann, L.J. Musselman, and D.M. Joel. Proceedings of the 4th International *Orobanche* Workshop. Sept. 23-26, 1998. Albena, Bulgaria, pp. 51-57.

Dhyani, S.K. (1978). Allelopathic potential of *Eupatorium adenophorum* on seed germination of *Lantana camara* var. *aculeata*. *Indian Journal of Forestry* 1: 311.

Dilday, R.H., J. Lin, and W. Yan. (1994). Identification of allelopathy in the USDA-ARS rice germplasm. *Australian Journal of Experimental Agriculture* 34: 907-910.

Dirvi, A.D. and F. Hussain. (1979). Allelopathic effects of *Dichanthium annulatum* (Forsk) Stapf on some cultivated plants. *Pakistan Journal of Scientific and Industrial Research* 22: 194-197.

Dolling, A. (1996). Changes in *Pteridium aquilinum* growth and phytotoxicity following treatments with lime, sulphuric acid, wood ash, glyphosate and ammonium nitrate. *Weed Research* 36: 293-301.

Downum, K.R., S. Villegas, E. Rodriguez, and D.J. Keil. (1989). Plant photosynsitizers, a survey of their occurrence in arid and semiarid plants from North America. *Journal of Chemical Ecology* 15: 345-355.

Drost, D.C. and J.D. Doll. (1980). The allelopathic effect of yellow nutsedge (*Cyperus esculentus*) on corn (*Zea mays*) and soybean (*Glycine max*). *Weed Science* 28: 229-233.

Dubey, P.S. (1973). Phytotoxicity of weeds to crops. I. Effect on germination. *Science and Culture* 39: 556-558.

Duke, S.O. and H.K. Abbas. (1995). Natural products with potential use as herbicides. In *Allelopathy: Organisms, Processes and Applications*, eds. Inderjit, K.M.M. Dakshini, and F.A. Einhellig. Washington, DC: American Chemical Society, pp. 348-362.

Edwards, M.E., E.M. Harris, F.H. Wagner, M.C. Cross, and G.S. Miller. (1988). Seed germination of American pokeweed (*Phytolacca americana*) I. Laboratory techniques and autotoxicity. *American Journal of Botany* 75: 1794-1802.

Einhellig, F.A. (1986). Mechanism and mode of action of allelochemicals. In *The Science of Allelopathy*, eds. A.R. Putnam and C.S. Tang. New York: John Wiley and Sons, pp. 171-188.

Einhellig, F.A. (1995). Mechanism of action of allelochemicals in allelopathy. In *Allelopathy: Organisms, Processes and Applications*, eds. Inderjit, K.M.M. Dakshini and F.A. Einhellig. American Chemical Symposium Series no. 582. Washington, DC: American Chemical Society, pp. 96-116.

Einhellig, F.A. (1996). Interactions involving allelopathy in cropping systems. *Agronomy Journal* 88: 886-893.

Einhellig, F.A. and G.R. Leather. (1988). Potential for exploiting allelopathy to enhance crop production. *Journal of Chemical Ecology* 4: 1829-1844.

Einhellig, F.A. and J.A. Rasmussen (1973). Allelopathic effects of *Rumex crispus* on *Amaranthus retroflexus*, grain sorghum and field corn. *American Midland Naturalist* 90: 79-86.

El-Ghareeb, R.M. (1991). Suppression of annuals by *Tribulus terrestris* in abandoned field in the sandy desert of Kuwait. *Journal of Vegetable Science* 2: 147-154.

Eussen, J.H.H. (1978). Isolation of growth inhibitory substances from alang-alang {*Imperata cylindrica* (L.) Beauve.}. In *Studies on the Tropical Weed Imperata cylindrica (L.) Beauv. var. major*, ed. J.H.H. Eussen. Paper No. 7. Utrecht: Drukkerij Elinkwijk Bv.

Eussen, J.H.H. and G.J.Niemann. (1981). *Zeitschrift fur Pflanzenphysiol* 102: 263. Cited by J.V. Lovett (1986). Allelopathy: The Australian experience. In *The*

Science of Allelopathy, eds. A.R. Putnam and C.S. Tang. New York: John Wiley and Sons, pp. 75-99.

Fadayomi, O. and E.O. Oyebade. (1989). An investigation of allelopathy in siamweed (*Eupatorium odoratum*). *Geobios* 11: 145-150.

Fay, P.K. and W.B. Duke. (1977). An assessment of allelopathy in *Avena* germplasm. *Weed Science* 25: 224-228.

Fletcher, A.R. and A.J. Renney. (1963). A growth inhibitor found in *Centaurea* spp. *Canadian Journal of Plant Science* 43: 475-481.

Fletcher, W.W. and R.C. Kirkwood. (1982). *Herbicides and Plant Growth Regulators*. London: Granada, pp. 93-99.

Foy, C.L. (1999). How to make bioassays for allelopathy more relevant to field conditions with particular reference to cropland weeds. In *Principles and Properties in Plant Ecology, Allelochemical Interactions*, eds. Inderjit, K.M.M. Dakshini and C.L. Foy. Boca Raton: CRC Press LLC, pp. 25-33.

Frange, N.S. (1984). *Evaluation of Medicinal Plant Extracts as Protectant and Therapeutant Against Legume Pathogens*. Thesis, University of Philippines. 192 p.

Friedman, J. (1987). Allelopathy in desert ecosystems. In *Allelochemicals: Role in Agriculture and Forestry*, ed. G.R. Waller, ACS Symposium Series 330. Washington, DC, American Chemical Society, pp. 53-68.

Friedman, J. and G.R. Waller. (1985). Allelopathy and autotoxicity. *Trends in Biochemical Sciences* 10: 47-50.

Friedman, J., E. Rushkin, and G.R. Waller. (1982). Highly potent germination inhibitor in aqueous elute of fruits of Bishop's weed (*Ammi majus*) and avoidance of autoinhibition. *Journal of Chemical Ecology* 8: 55-56.

Funke, G.L. (1943). The influence of *Artemisia absinthium* on neighboring plants. *Blumea* 5: 281-293.

Gajic, D. (1973). The effect of agrostemins as a means of improvement of the quality and quantity of the grass cover of the Zlatibor–as a preventive measure against the weeds. In *Yogoslav Symposium on Weed Control in Hilly and Mountainous Areas*. Sarajevo.

Gajic, D. and G. Nikocevic (1973). Chemical allelopathic effect of *Agrostemma githago* upon wheat. *Fragmenta Herbologica Jugoslavica* 18: 1-5.

Gajic, D., S. Malencic, M. Vrbaski, and S. Vrbaski. (1976). Study of the quantitative and qualitative improvement of wheat yield through agrostemin as an allelopathic factor. *Fragmenta Herbologica Jugoslavica* 63: 121-141.

Gander, J.R. and L.R. Oliver. (1998). The potential of allelopathic weeds for weed control in soybean. *Weed Science Society of America. Abstract 38*: 62.

Glass, A.D.M. (1976). The allelopathic potential of phenolic acids associated with the rhizosphere of *Pteridium aquilinum*. *Canadian Journal of Botany* 54: 2440-2444.

Goel, U., D.B. Saxena, and B. Kumar. (1989). Comparative study of allelopathy as exhibited by *Prosopis juliflora* Swartz and *Prosopis cineraria* (L.) Druce. *Journal of Chemical Ecology* 15: 591-600.

Gonzalez de la Parra, M., A.L. Anaya, F. Esinose, M. Jimenez, and R. Castillo. (1981). Allelopathic potential of *Piqueria trinerva* (compositae) and piquerols A and B. *Journal of Chemical Ecology* 7: 509-515.

Gonzalez, A.G., T. Abad, I.A. Jimenez, A.G. Ravelo, J.G. Luis, Z. Aguiar, L.S. Andres, M. Plasencia, J.R. Herrera and L. Moujir. (1989). A first study of autobacterial activity of diterpenes from some *Salvia* species (Lamiaceae). *Biochemical Systematics and Ecology* 17: 293-296.

Gray, R. and J. Bonner. (1948). Structure determination and synthesis of a plant growth inhibitor, 3-acetyl-6-methoxy benzaldehyde, found in the leaves of *Encelia farinosa*. *Journal of American Chemical Society* 70: 1249-1253.

Gressel, J.B. and L.G. Holm. (1964). Chemical inhibition of crop germination by weed seeds and nature of inhibition by *Abutilon theophrasti*. *Weed Research* 4: 44-53.

Grime, J.P. (1986). *Plant Strategies and Vegetation Processes*. Chichester, England: John Wiley and Sons.

Grodzinsky, A.M. and Panchuk, M.A. (1974). Allelopathic properties of crop residues of wheat-wheat grass hybrids. In *Physiological-Biochemical Basis of Plant Interactions in Phytocenosis*, ed. A.M. Grodzinsky. Kiev: Naukova Dumka, pp. 51-55 (In Russian with English summary).

Grummer, G. (1961). The role of toxic substances in the interrelationships between higher plants. In *Mechanisms in Biological Competition*, ed. F.L. Milthorpe. New York: Academic Press, pp. 219-228.

Grummer, G. and H. Beyer. (1960). The influence exerted by species of *Camelina* on flax by means of toxic substances. *Symposium of the British Ecological Society* 1: 153-157.

Habib, S.A. and A.A.A. Rahman. (1998). Evaluation of some weed extracts against field dodder on alfalfa (*Medicago sativa*). *Journal of Chemical Ecology* 14: 443-452.

Hagen, R.D. (1989). Isolation and identification of 5-hydroxyindole-3-acetic acid and 5-hydroxy-tryptophan, major allelopathic aglycons in quackgrass (*Agropyron repens* L. Beauv.). *Journal of Agricultural and Food Chemistry* 37: 1143-1149.

Harper, J.L. (1964). The nature and consequence of interference among plants. *Proceedings of the 11th International Conference of Genetics*. The Hague, 1963. pp. 465-481.

Harrison, H.F., J.K. Peterson, and M. Snook. (1998). Investigation of the allelopathic effect of corn spurry (*Spergula arvensis*) on cole crops and English pea. *Weed Science Society of America. Abstract no.* 61.

Hegazy, A.K., K.S. Mansour, and N.F. Abdel-Hady. (1990). Allelopathic and autotoxic effects of *Anastatica heirochuntica* L. *Journal of Chemical Ecology* 16: 2183-2193.

Heisey, R.M. (1990). Evidence of allelopathy by tree-of-heaven (*Ailanthus altissima*). *Journal of Chemical Ecology* 16: 2039-2055.

Heisey, R.M. and C.C. Delwiche. (1983). A survey of California plants for water extractable and volatile inhibitors. *The Botanical Gazette* 144: 382-390.

Horowitz, Z.M. and T. Friedman. (1971). Biological activity of subterranean residues of *Cynodon dactylon* L., *Sorghum halepense* L. and *Cyperus rotundus* L. *Weed Research* 11: 88-93.

Horsley, S.B. (1977). Allelopathic inhibition of black cherry by fern, grass, goldenrod, and aster. *Canadian Journal of Forestry Research* 7: 205-216.

Horsley, S.B. (1991). Allelopathy. In *Biophysical Research for Asian Agroforestry*, eds. M.E. Avery, M.G.R. Cannel, and C.K. Ong. Washington: Winrock International, pp. 167-183.

Hsiao, A.I., A.D. Worsham, and D.E.Moreland. (1981). Regulation of witchweed (*Striga asiatica*) seed conditioning and germination by dl-strigol. *Weed Science* 29: 101-104.

Hussain, F. (1980). Allelopathic effects of Pakistani weeds: *Euphorbia granulata* Forssk. *Oecologia* 45: 267-269.

Hussain, F., B. Mubarak, I.ul. Haq, and H.H. Naqvi. (1979). Allelopathic effects of *Datura innoxia* L. *Pakistan Journal of Botany* 11: 141-153.

Hussain, F., M.I. Zaidi, and S.R. Chughtai. (1984). Allelopathic effects of Pakistani weeds. *Eragrostis poaeoides* P. Beauv. *Pakistan Journal of Scientific and Industerial Research* 27: 159-164.

Hussain, F., N. Ahmed, and M. Akram. (1986). Allelopathy as experienced by *Stachys parviflora*. *Pakistan Journal of Scientific and Industerial Research* 29: 1458-460.

Hussain, F., T.W. Khan, and A. Hussain. (1987). Allelopathic effects of *Cirsium arvense* (L.) Scop. In *Modern Trends of Plant Science Research in Pakistan*, eds. I. Ilahi. and F. Hussain. Proceedings Third National Conference of Plant Scientists, Department of Botany. Peshawar: University of Peshawar, pp. 24-28.

Ibrahim, I.A., H. Elarosi, G.M. Hassouna, and O.A. Al-Menoufi. (1973). Studies on the parasitism of broomrapes in Egypt. Series 1: certain factors affecting seed germination of *Orobanche crenata* and *O. ramosa*. *Egyptian Journal of Plant Pathology* 5: 43-54.

Ibrahim, N.E., A.G.T. Babiker, W.G. Edwards, and C. Parker. (1985). Activity of extracts from *Euphorbia* species on the germination of *Striga* species. *Weed Research* 25: 135-140.

Inam, B., F. Hussain, and F. Bano. (1987). Allelopathic effects of Pakistani weeds: *Xanthium strumarium* L. *Pakistan Journal of Scientific and Industerial Research* 30: 530-533.

Inderjit and C.L. Foy. (1998). An ecological update on weed allelopathy. *Weed Science Society of America. Abstract* 38: 40.

Inderjit and C.L. Foy. (1999). Nature of the interference mechanism of mugwort (*Artemisia vulgaris*). *Weed Technology* 13: 176-182.

Inderjit and K.I. Keating. (1999). Allelopathy: Principles, procedure, processes, and promises for biological control. *Advances in Agronomy* 67: 141-231.

Inderjit and K.M.M. Dakshini. (1991a). Investigation on some aspects of chemical ecology of cogongrass, *Imperata cylindrica* (L.) Beauv. *Journal of Chemical Ecology* 17: 343-352.

Inderjit and K.M.M. Dakshini. (1991b). Hesperetin 7-rutinoside (hesperidin) and taxifolin 3-arabinoside as germination and growth inhibitors in the soil associated with the weed *Pluchea lanceolata* (DC.) C.B. Clark (Asteraceae). *Journal of Chemical Ecology* 17:1585-1591.

Inderjit and K.M.M. Dakshini. (1992). Formononetin 7-O-glycoside, an additional inhibitor from the soil associated with the weed, *Pluchea lanceolata* (DC.) C.B. Clark (Asteraceae). *Journal of Chemical Ecology* 18: 713-718.

Inderjit and K.M.M. Dakshini. (1994).Alleopathic effect of *Pluchea lancoelata* (Asteraceae) on characteristics of four soil and tomato and mustard growth. *American Journal of Botany* 81: 799-804.

Inderjit and K.M.M. Dakshini. (1995). Allelopathic potential of the annual weed *Polygonum monospeliensis*, in crops in India. *Plant and Soil* 173: 251-256.

Inderjit and K.M.M. Dakshini. (1998). Allelopathic interference of chickweed, *Stellaria media* with seedling growth of wheat (*Triticum aestivum*). *Canadian Journal of Botany* 76: 1317-1321.

Inderjit, K.M.M. Dakshini and C.L. Foy, eds. (1999). *Principles and Practices in Plant Ecology, Allelochemical Interactions*. Boca Raton: CRC Press LLC, pp. 589.

Inoue, M., H. Nishimura, H.H. Li, and J. Mizutani. (1992). Allelochemicals from *Polygonum sachalinese* FR. Schm. (Polygonaceae). *Journal of Chemical Ecology* 18: 1833-1840.

International Allelopathy Society (IAS). (1996). *Constitution and Bylaw of IAS*. IAS Newsletter, Cadiz-Spain.

Ito, M., Y. Matsushita, Y. Umeki, and K. Ueki. (1981). Determination of physiological effects of grass mulch in orchards: Bioassay of leachates from weed residues. *Weed Research (Japan)* 26: 221-227 (In Japanese with English summary).

Jackson, J.R. and R.W. Willemson. (1976). Allelopathy in the first stages of secondary succession on the piedmont of New Jersey. *American Journal of Botany* 63: 1015-1023.

Jha, P.K. and D.N. Sen. (1981). Allelopathic potential of *Echinops echinatus* Roxb. *Proceedings 8th Asian-Pacific Weed Science Society Conference* 2: 197-202.

Jimenez-Osornio, F.M.V.Z.J., J. Kumamoto, and C. Wasser. (1996). Allelopathic activity of *Chenopodium ambrosioides* L. *Biochemical Systematics and Ecology* 24: 195-205.

Jimenez-Osornio, J.J. and S.R. Gliessman. (1987). Allelopathic interference in a wild mustard (*Brassica campestris* L.) and broccoli (*Brassica oleracea* L. var. *italica*) intercrop agroecosystem. In *Allelochemicals: Role in Agriculture and Forestry*, ed. G.R. Waller. ACS Symposium Series 330. Washington, DC: American Chemical Society, pp. 262-274.

Johnson, A.W., G. Rosebery and C. Parker. (1976). A novel approach to *Striga* and *Orobanche* control using synthetic germination stimulants. *Weed Research* 16: 223-227.

Johnston, W.H. (1989). Consol lovegrass (*Eragrostis curvula* complex) controls spring burrgrass (*Cenchrus* spp.) in south-western New South Wales. *Australian Journal of Agriculture* 29: 37-42.

Joshi, S. (1991). Interference effects of *Cassia uniflora* Mill on *Parthenium hysterophorus* L. *Plant and Soil* 132: 213-218.

Kanchan, S.D. and Jayachandra. (1979a). Allelopathic effects of *Parthenium hysterophorus* L. I. Exudation of inhibitors through roots. *Plant and Soil* 53: 27-35.

Kanchan, S.D. and Jayachandra. (1979b). Allelopathic effects of *Parthenium hysterophorus* L. I. Inhibitory effect of weed residues. *Plant and Soil* 53: 37-47.

Kanchan, S.D. and Jayachandra. (1980). Pollen allelopathy–A new phenomenon. *New Phytologist* 84: 739-746.

Kasasian, L. (1973). Control of *Orobanche*. *Pest Articles and News Summaries* 19: 368-371.

Kawabata, J., Y. Fukushi, R. Hayashi, K. Suzuki, Y. Mishima, A.Yamane, and J. Mizutani. (1989). 8-methylsulfinyloctyl isothiocyanate as allelochemical candidate from *Rorippa sylvestris* Besser. *Agricultural and Biological Chemistry* 53: 3361-3362.

Keever, C. (1950). Causes of succession on old fields of the Piedmont, North Carolina. *Ecological Monographs* 20: 229-250.

Khare, L.J. (1980). Phytotoxicity of the weed *Urgenia indica* Kunth on the seed germination of associated crops. *Indian Journal of Botany* 3: 87-91.

Kil, B.S., D.Y. Kim, Y.S. Kim, and S.Y. Lee. (1991). Phytotoxic effects of naturally occurring chemicals from *Pinus koraiensis* on experimental species. *Korean Journal of Ecology* 14: 149-157.

Kloot, P.M. and K.G. Boyce. (1982). Australian Weeds 1: 11. Cited by Lovett, J.V. (1986). Allelopathy: The Australian experience. In *The Science of Allelopathy*, eds. A.R. Putnam and C. S. Tang. New York: John Wiley and Sons, pp. 75-99.

Kobayashi, A., S. Morimoto, Y. Shibata, K. Yamashita and M. Numata. (1980). C_{10}-poly-acetylenes as allelopathic substances in dominants in early stages of secondary succession. *Journal of Chemical Ecology* 6: 119-131.

Kohli, R.K. and D.R. Batish. (1994). Exhibition of allelopathy by *Parthenium hysterophorus* L. in agroecosystems. *Tropical Ecology* 35: 295-307.

Kohli, R.K., D. Batish, and H.P. Singh. (1998). Allelopathy and its implications in agroecosystems. *Journal of Crop Production* 1: 169-202.

Kohli, R.K., D. Rani, and R.C. Verma. (1993). A mathematical model to predict the tissue response to parthenin–an allelochemical. *Biologia Plantarum* 35: 567-576.

Kohli, R.K., D. Rani, H.P. Singh, and S. Gupta. (1996). Response of crop seeds towards the leaf leachates of *Parthenium hysterophorus* L. *Indian Journal of Weed Science* 28: 104-106.

Kohmuenzer, S. (1965). Botanical and chemical studies of the collective species *Gallium mollugo* with reference to karyotypes growing in Poland. VI. Effect of extracts and some other chemical components of *Galium mollugo* on the germination of seeds and growth of selected plants. *Dissertationes Pharmaceuticae et Pharmacologicae* 17: 369-379 (Via CA 64: 10085-10086g).

Komai, K. and C.S. Tang. (1989). A new chemotype of purple nutsedge (*Cyperus rotundus* L.) in Hawaii. *Phytochemistry* 28: 1883-1886.

Komai, K. and K. Ueki. (1975). Chemical properties and behavior of polyphenolic substances in purple nutsedge (*Cyperus rotundus* L.). *Weed Research (Japan)* 20: 66-71 (Japanese with English summary).

Komai, K., J. Iwamura, and K. Ueki. (1983). Plant growth inhibitor in the seed of catchweed. *Weed Research (Japan)* 28: 205-209.

Komai, K., S. Sato, and K. Ueki. (1982). Antimicroorganisms activities of essential oil in purple nutsedge. *Memoirs Faculty of Agriculture, Kinki University* 15: 33-41 (Japanese, with English summary).

Kommedahl, T., J.B. Kotheimer, and J.V. Bernardini. (1959). The effects of quackgrass on germination and seedling development of certain crop plants. *Weeds* 7: 1-12.

Krishnamurthy, G.V.G., R. Lal, and K. Nagarajan. (1977). Further studies on the effect of various crops on the germination of *Orobanche* seed. *Pest Articles and News Summaries* 23: 206-208.

Krishnamurthy, S. and U.M. Rao. (1976). Control of *Orobanche* through crop rotation. *Indian Farming* 25: 23.

LaBonte, D.R. and R.L. Darding. (1988). Noncompetitive effects of morning glory on the growth of soybeans. *Trans. Ill. Academie des Sciences* 81: 39-44.

Labrada, R. and R. Perez. (1988). Non-chemical control methods for *Orobanche ramosa* (in Spanish). *Agrotecnica de Cuba* 20: 35-40.

Lahiri, A.N. and B.C. Kharabanda. (1962). Germination studies of arid zone plants. II. Germination inhibitors in the spikelet glumes of *Lasiurus sindicus*, *Cenchrus ciliaris* and *Cenchrus setigerus*. *Annals of Arid Zone* 1: 114-126.

Leather, G.R. (1983). Weed control using allelopathic crop plants. *Journal of Chemical Ecology* 9: 983-990.

Leather, G.R. (1987). Weed control using allelopathic sunflowers and herbicides. *Plant and Soil* 8: 17-23.

Leela, D. (1984). Allelopathy in *Oxalis corniculata*. *Pesticides* 18: 16-19.

Le Tourneau, D. and H.G. Heggeness. (1957). Germination and growth inhibitors in leafy spurge and quackgrass rhizomes. *Weeds* 5: 12-19.

Le Tourneau, D., G.D. Failes and H.G. Heggeness. (1956). The effects of aqueous extracts of plant tissue on germination of seeds and growth of seedlings. *Weeds* 4: 363-368.

Levin, D.A. (1976). The chemical defenses of plants to pathogens and herbivores. *Annual Review of Ecology and Systematics* 7: 121-159.

Levitt, J., J.V. Lovett, and P.R. Garlick. (1984). Effect of allelochemicals of *Datura stramonium* L. (Thorn-apple) on root tip ultrastructure of *Helianthus annuus* L. *New Phytologist* 97: 217-218.

Li, H.H., H. Nishimura, K. Hasegawa, and J. Mizutani. (1992). Allelopathy of *Sasa cernua*. *Journal of Chemical Ecology* 18: 1785-1796.

Li, H.H., M. Inoue, H. Nishimura, J. Mizutani, and E. Tsuzuki. (1993). Interactions of trans-cinnamic acids, its related phenolic allelochemicals, and abscisic acid in seedling growth and seed germination of lettuce. *Journal of Chemical Ecology* 19: 1775-1787.

Linke, K.H., H. Schnell, and M.C. Saxena. (1991). Factors affecting the seed bank of *Orobanche crenata* in fields under lentil based cropping systems in northern Syria. In *Proceedings of the 5th International Symposium on Parasitic Weeds*, eds. J.K. Ransom, L.J. Musselman, A.D. Worsham, and C. Parker. Nairobi, 1991. Nairobi: CIMMYT, pp: 321-327.

Locken, L.J. and R.G. Kelsey. (1987). Cnicin concentrations in *Centaurea maculosa* spotted knapweed. *Biochemical Systematics and Ecology* 15: 313-320.

Lodhi, M.A.K. (1979). Germination and decreased growth of *Kochia scoparia* in relation to its autoallelopathy. *Canadian Journal of Botany* 57: 1083-1088.

Lovett, J.V. (1982). Allelopathy and self-defense in plants. *Australian Weeds* 2: 33-36.

Lovett, J.V. (1990). Chemicals in plant protection: is there a natural alternative? In *Alternative to the Chemical Control of Weeds*, eds. C. Bassett, L.J. Whitehouse, and J.A. Zabkiewicz. FRI Bulletin 155. Rotorua, New Zealand: Ministry of Forestry, pp. 57-65.

Lovett, J.V. and A.M. Duffield (1981). Allelochemicals of *Camelina sativa*. *Journal of Applied Ecology* 18: 283-290.

Lovett, J.V. and J. Levitt. (1981). Allelochemicals in a future agriculture. In *Biological Husbandry*, ed. B. Stonehouse. London: Butterworth, pp. 169-181.

Lovett, J.V., J. Levitt, A.M. Duffield, and N.G.Smith (1981). Allelopathic potential of *Datura stramonium* L. (Thorn-apple). *Weed Research* 21: 165-170.

Lovett, J.V. and J.A. Lynch. (1979). Studies of *Salvia reflexa* Hornem. 1. Possible competitive mechanisms. *Weed Research* 19: 351-357.

Lovett, J.V and M.D. Speak. (1979). Studies on *Salvia reflexa*. Examination of specialized leaf surface structures. *Weed Research* 19: 359-362.

Lovett, J.V. and W.C. Potts. (1987). Primary effects of allelochemicals of *Datura stramonium* L. *Plant and Soil* 98: 137-144.

Lydon, J., J.R. Teasdale, and P.K. Chen. (1997). Allelopathic activity of annual wormwood (*Artemisia annua*) and the role of artemisinin. *Weed Science* 45: 807-811.

Macias, F.A., J.C.G. Galindo, J.M.G. Molinillo, and H.G. Cutler, eds. (1999). *Recent Advances in Allelopathy: A Science for the Future*. Vol. 1. Universidad de Cadiz, Cadiz, Spain: Servicio Publicationes.

Macias, F.A., R.M. Varela, A. Torres, and J.M.G. Molinillo. (1994). Potential of sunflower (*Helianthus annuus* L.) as source of natural herbicides models. In *International Symposium on Allelopathy in Sustainable Agriculture, Forestry and Environment*. Hisar, India: Indian Society of Allelopathy, pp. 76 (Abstract).

Mallik, M.A.B., R. Puchala and F.A. Grosz. (1994). A growth-inhibitory factor from lambsquarters (*Chenopodium album*). *Journal of Chemical Ecology* 20: 957-967.

Mann, H.H. and T.W. Barnes. (1945). The competition between barley and certain weeds under controlled conditions. *Annals of Applied Biology* 32: 15-24.

Mann, H.H. and T.W. Barnes. (1947). *Annals of Applied Biology* 34: 252. Cited by A.R. Putnam and L.A. Weston. (1986) Adverse impacts of allelopathy in agricultural systems. In *The Science of Allelopathy*, eds. A.R. Putnam and C.S. Tang. New York: John Wiley and Sons, pp. 43-56.

Mann, H.H. and T.W. Barnes. (1950). The competition between barley and certain weeds under controlled conditions. IV. Competition with *Stellaria media*. *Annals of Applied Biology* 37: 139-148.

Manners, G.D. and D.S. Galitz. (1986). Allelopathy of small everlasting (*Antennaria microphylla*): Identification of constituents phytotoxic to leafy spurge (*Euphorbia esula*). *Weed Science* 34: 8-12.

Manners, G.D. (1987). The role of phytochemistry in attacking the leafy spurge (*Euphorbia esula*) problem. In *Allelochemicals: Role in Agriculture and Forestry*, ed. G.R. Waller. ACS Symposium Series 330. Washington, DC: American Chemical Society, pp. 228-237.

Mares, D. (1987). Antimicrobial activity of protoanemonin, a lacton from ranunculaceous plants. *Mycopathologia* 98: 133-140.

Martin, P. and B. Rademacher. (1960). Studies on the mutual influences of weeds and crops. *Symposium of the British Ecological Society* 1: 143-152.

Mason-Sedun, W. (1986). *Differential Phytotoxicity of Residues from the Genus Brassica*. Thesis, University of New England.

Mason-Sedun, W., R.S. Jessop, and J.V. Lovett. (1986). Differential phytotoxicity among species and cultivars of the genus *Brassica* to wheat. 1. Laboratory and field screening of species. *Plant and Soil* 93: 3-16.

Massantini, F., F. Caporalli, and G. Zellini. (1977). Evidence for allelopathic control of weeds in lines of soybean. *Proceedings of European Weed Research Society Symposium on Weed Control*, Uppsala, Vol. 1, pp. 23-28.

Mattner, S.W. and D.G. Parbery. (1999). Rust-enhanced allelopathy of perennial ryegrass against white clover. In *Second World Congress on Allelopathy: Critical Analysis and Future Prospects*. Lakehead University, Thunder Bay, Ontario, Canada. August 8-13. Abstract no. 134.

McNaughton, S.J. (1968). Autotoxic feedback in relation to germination and seedling growth in *Typha latifolia*. *Ecology* 49: 367-369.

Megharaj, M., A.P. Rao, K. VenKateswarlu, and A.S. Rao. (1987). Toxicity of *Parthenium hysterophorus* to *Chlorella vulgaris* and *Synechococcus elongatus*. *Plant and Soil* 103: 292-294.

Meissner, R., P.C. Nel, and E.A. Beyers. (1986). Allelopathic influence of *Tagetes* and *Bidens*-infested soils on seedling growth of certain crop species. *South African Plant and Soil* 3: 176-180.

Meissner, R., P.C. Nel, and E.A. Beyers. (1989). Allelopathic effects of *Cynodon dactylon*-infested soil on early growth of certain crop species. *Applied Plant Science* 3: 125-126.

Melkania, N.P., J.S. Singh, and K.K.S. Bisht. (1982). Allelopathic potential of *Artemisia vulgaris* L. and *Pinus roxburghii* Sargent: A bioassay study. *Proceedings of the Indian National Science Academy* B 48: 685-688.

Menges, R.M. (1987). Allelopathic effects of palmer amaranth (*Amaranthus palmeri*) on other plant residues in soil. *Weed Science* 35: 339-347.

Menges, R.M. (1988). Allelopathic effects of palmer amaranth (*Amaranthus palmeri*) on seedling growth. *Weed Science* 36: 325-328.

Menges, R.M. and S. Tamez. (1981). Response of onion (*Allium cepa*) to annual weeds and post-emergence herbicides. *Weed Science* 29: 74-79.

Mercer, K.L., D.S. Murray, and L.M. Verhalen. (1987). Interference of unicorn-plant (*Proboscida louisianica*) with cotton (*Gossypium hirsutum*). *Weed Science* 25: 807-812.

Merrill, G.B. (1989). Eupatoriochromene and encecalin, plant growth regulators from yellow starthistle (*Centaurea solstitialis*). *Journal of Chemical Ecology* 15: 2073-2087.

Mersie, W. and M. Singh. (1987a). Allelopathic effects of *Lantana* on some agronomic crops and weeds. *Plant and Soil* 98: 25-30.

Mersie, W. and M. Singh. (1987b). Allelopathic effects of parthenium (*Parthenium hysterophorus* L.) extracts and residues on some agronomic crops and weeds. *Journal of Chemical Ecology* 13: 1739-1747.

Mersie, W. and M. Singh. (1988). Effects of phenolic acids and ragweed parthenium (*Parthenium hysterophorus*) extract on tomato (*Lycopersicon esculentum*) growth and nutrient and chlorophyll content. *Weed Science* 36: 278-281.

Misra, S.B. and S.N. Dixit. (1978). Antifungal properties of leaf extract of *Ranunculus sceleratus*. *Experientia* 34: 1442-1443.

Misra, S.B., R.B.R. Misra, and S.N. Dixit. (1974). Screening of higher plant for antifungal activity. *Proceedings of the National Academy of Science (India)* 76: 203-205.

Mizutani, J. and A. Yamane. (1991). Chenodynamics of glucosinolates in *Kireha inugarashi*, *Rorippa sylvestris*. *Weed Research (Japan)* 36: 68-72.

Mohnot, K. and S. Soni. (1976). Observations on the allelopathic factor in air-dried leaves of *Salvadora oleoides*. *Comparative Physiology and Ecology* 1: 125-128.

Mohnot, K. and S. Soni. (1977). Ecophysiological studies on desert plants. II. Growth retarding factor in air-dried stem of *Solanum surrattense* Burm. F. *Comparative Physiology and Ecology* 2: 97-100.

Molisch, H. (1937). *Der Einfluss einer pflane auf die andere-Allelopathie*. Fischer, Jena.

Mori, A., N. Enoki, K. Shinozuka, C. Nishino, and M. Fukushima. (1987). Antifungal activity of fatty acids against *Pyricularia oryae* related to antifungal constituents of *Miscanthus sinensis*. *Agricultural and Biological Chemistry* 51: 3403-3405.

Mubarak, B. and F. Hussain. (1978). Biochemical inhibition exhibited by *Datura innoxia* seeds. *Pakistan Journal of Botany* 10: 149-156.

Muir, A.D. and W. Majak. (1983). Allelopathic potential of diffuse knapweed (*Centaurea diffusa*) extracts. *Canadian Journal of Botany* 63: 989-996.

Muller, C.H. (1965). Inhibitory terpenes volatilized from *Salvia* shrubs. *Bulletin of the Torrey Botanical Club* 92: 38-45.

Muller, C.H. (1969). Allelopathy as a factor in ecological process. *Vegetatio* 18: 348-357.

Muller, C.H., W.H. Muller, and B.L. Haines. (1964). Volatile growth inhibitors produced by shrubs. *Science* 143: 471-473.

Muller, W.H. (1986). Allelochemical mechanisms in the inhibition of herbs by chaparral shrubs. In *The Science of Allelopathy*, eds. A.R. Putnam and C.S. Tang. New York: John Wiley and Sons, pp. 189-199.

Muller, W.H. and C.H. Muller. (1964). Volatile growth inhibitors produced by *Salvia* species. *Bulletin of the Torrey Botanical Club* 91: 327-330.

Murthy, M.S. and T. Zakharia. (1980). Allelopathic potential of some common weeds on bajra crop. Indian *Journal of Experimental Biology* 18: 91-93.

Musselman, L.J. (1980). The biology of *Striga*, *Orobanche* and other root parasitic weeds. *Annual Review of Phytopathology* 18: 463-489.

Musselman, L.J., J.L. Riopel, and R.E. Eplee. (1978). Recent research on the biology of *Striga asiatica* (witchweed) in the United States. *Paper presented at International Striga Workshop*. Khartoum, 1978.

Naqvi, H.H. and C.H. Muller. (1975). Biochemical inhibition (Allelopathy) exhibited on Italian ryegrass (*Lolium multiflorum* L.). *Pakistan Journal of Botany* 7: 139-147.

Narwal, S.S., ed. (1994). *Allelopathy in Crop Production*. Jodhpur, India: Scientific Publisher. 288 p.

Neill, R.L. and E.L. Rice. (1971). Possible role of *Ambrosia psilostachya* on patterning and succession in old-fields. *American Midland Naturalist* 86: 344-357.

Newman, E.I. (1978). Allelopathy, adaptation or accident? In *Biochemical Aspects of*

Plant and Animal Coevolution, ed. J.B. Harborne. New York: Academic Press, pp. 328-342.

Newman, E.I. (1982). The possible relevance of allelopathy to agriculture. *Pesticide Science* 13: 575-582.

Newman, E.I. and A.D. Rovira. (1975). Allelopathy among British grassland species. *Journal of Ecology* 63: 727-737.

Newman, E.I. and M.H. Miller. (1977). Allelopathy among some British grassland species. II. Influence of root exudates on phosphorus uptake. *Journal of Ecology* 65: 399-411.

Norby, R.J. and T.T. Kozlowski. (1980). Allelopathic potential of ground cover species on *Pinus resinosa* seedlings. *Plant and Soil* 57: 363-374.

Norrington-Davies. (1980). Paper XI. Allelopathy. ADAS Conference, 1979. H.M.S.O., 127-146.

Numata, M., A. Kobayashi, and N. Ohga. (1973). Studies on the allelopathic substances concerning the formation of the urban flora. In *Fundemental Studies in the Characteristics of Urban Ecosystems*, ed. M. Numata, pp. 59-64.

Obied, K. (1996). *Alleopathic Effects of Certain Weed Species on Different Vegetable Crops*. M.Sc. Thesis. University of Jordan, Amman, Jordan, pp. 114.

Oleszek, W. (1987). Allelopathic effects of volatiles from cruciferae species on lettuce, barnyardgrass and wheat growth. *Plant and Soil* 102: 271-273.

Olofsdotter, M., D.C. Navarez, and K. Moody. (1995). Allelopathic potential in rice (*Oryza sativa* L.) germplasm. *Annals of Applied Biology* 127: 543-560.

Osvald, H. (1953). On antagonism between plants. In *Proceedings of the 7th International Botanical Congress*, Stockholm, pp. 167-170.

Oudhia, P. and R.S. Tripathi. (1999). Allelopathic effects of *parthenium hysterophorus*, *Lantana camara* and *Ageratum conyzoides* on germination and seedling vigour of wheat and selected rabi weeds. In *Second World Congress on Allelopathy: Critical Analysis and Future Prospects*. Lakehead University, Thunder Bay, Ontario, Canada. August 8-13. Abstract no. 142.

Panchuk, M.A. and N.L. Prutenskaya. (1973). On the problem of the presence of allelopathic properties in wheat-wheat grass hybrids and their initial forms. In *Physiological-Biochemical Basis of Plant Interactions in Phytocenosis*, ed. A.M. Grodzinsky. Kiev: Naukova Dumka, pp. 44-47 (In Russian with English summary).

Pandya, S.M. (1975). Effect of *Celosia argentea* extracts on root and shoot growth of bajra seedlings. *Geobios* 2: 175-178.

Parenti, R.L. and E.L. Rice. (1969). Inhibitional effects of *Digitaria sanguinalis* and possible role in old-field succession. *Bulletin of the Torrey Botanical Club* 96: 70-78.

Parker, C. and C.R. Riches. (1993). *Parasitic Weeds of the World: Biology and Control*. Wallingford, UK: CAB International.

Patterson, D.T. (1985). Comparative ecophysiology of weeds and crops. In *Weed Physiology Vol. 1. Reproduction and Ecophysiology*, ed. S.O. Duke. Boca Raton, FL: CRC Press, Inc., pp. 102-129.

Patterson, D.T. (1986). Allelopathy. In *Research Methods in Weed Science*, ed. N.D. Camper. 3rd edition. Champaign, IL: Southern Weed Science Society, pp. 111-134.

Pavlychenko, T.K. and J.T. Harrington. (1934). Competitive efficiency of weeds of cereal crops. *Canadian Journal of Research* 10: 77-94.

Penn, D.J. and J.M. Lynch. (1982). New Phytologist 90: 51-55. Cited by J.M. Lynch. Allelopathy involving microorganisms: Case histories from the United Kingdom. In *Allelochemicals: Role in Agriculture and Forestry*, ed. G.R. Waller, ACS Symposium Series 330. Washington, DC: American Chemical Society, pp. 44-52.

Perez, F.J. and J. Ormeno-Nunze. (1993). Weed growth interference from temperate cereals: The effect of a hydroxamic acids-exuding rye (*Secale cereale* L.) cultivar. *Weed Research* 33: 115-119.

Peters, E.J. and A.H.B. Mohamad Zam. (1981). Allelopathic effects of tall fescue (*Festuca arundinacea*) genotypes. *Agronomy Journal* 73: 56-58.

Porwal, M.K. and O.P. Gupta. (1986). Allelopathic influence of winter weeds on germination and growth of wheat. *International Journal of Tropical Agriculture* 4: 276-279.

Putnam, A.R. (1983). Allelopathic chemicals: Nature's herbicide in action. *Chemical and Engineering News* 61: 34-45.

Putnam, A.R. (1985). Weed allelopathy. In In *Weed Physiology Vol. 1. Reproduction and Ecophysiology*, ed. S.O. Duke. Boca Raton, FL: CRC Press, Inc., 131-155.

Putnam, A.R. (1988). Allelochemicals from plants as herbicides. *Weed Technology* 2: 510-518.

Putnam, A.R. and C.S. Tang. (1986). Allelopathy: State of the Science. In *The Science of Allelopathy*, eds. A.R. Putnam and C.S. Tang. New York: John Wiley and Sons, Inc., pp. 1-19.

Putnam, A.R. and W.B. Duke. (1974). Biological suppression of weeds: Evidence for allelopathy in accessions of cucumber. *Science* 185: 370-372.

Putnam, A.R. and W.B. Duke (1978). Allelopathy in agro-ecosystems. *Annual Review of Phytopathology* 16: 431-457.

Qasem, J.R. (1987). *Fat-hen (Chenopodium album L.) and Groundsel (Senecio vulgaris) Interference with Certain vegetable Crops, with Special reference to Nutrition*. Thesis, Wye College, University of London., pp. 340.

Qasem, J.R. (1990). Aqueous extract effect of nettle-leaved goosefoot (*Chenopodium murale* L.) on wheat and barley. *Research Journal of Aleppo Unversity (Agriculture Science Series)* 14: 37-53.

Qasem, J.R. (1993). Allelopathic effect of nettle-leaved goosefoot (*Chenopodium murale*) on wheat and barley. *Dirasat 20B*: 80-94.

Qasem, J.R. (1994a). Allelopathic effect of white top (*Lepidium draba*) on wheat and barley. *Allelopathy Journal* 1: 29-40.

Qasem, J.R. (1994b). Allelopathic effects of pigweed (*Amaranthus* spp.) on barley (*Hordeum vulgare*). *Dirasat 21B*: 101-120.

Qasem, J.R. (1995a). Allelopathic effect of some arable land weeds on wheat (*Triticum durum* L.): A survey. *Dirasat 22*: 81-97 (In Arabic with English summary).

Qasem, J.R. (1995b). The allelopathic effect of three *Amaranthus* spp. (Pigweeds) on wheat (*Triticum durum*). *Weed Research* 35: 41-49.

Qasem, J.R. (1995c). Allelopathic effect of *Amaranthus retroflexus* and *Chenopodium murale* on vegetable crops. *Allelopathy Journal* 2: 49-66.

Qasem, J.R. (1996a). Fungicidal activity of *Ranunculus asiaticus* and other weeds against *Fusarium oxysporium* f. sp. *lycopersici*. *Annals of Applied Biology* 128: 533-540.

Qasem, J.R. (1996b). Aqueous extract effects of some common weed species against certain plant pathogenic fungi. *ACTES, Institute of Agronomy and Veterinary, (Maroc)* 16: 11-19.

Qasem, J.R. (1996c). Fungitoxicity of weed extracts to tomato wilt pathogen (*Fusarium oxysporum* f. sp. *lycopersici*). *Emirates Journal of Agricultural Sciences* 8: 103-112.

Qasem, J.R. (1999a). Biological activity of corn buttercup (*Ranunculus arvensis* L.). In *Recent Advances in Allelopathy: A Science for the Future*. Vol. 1, eds. F.A. Macias, J.C.G. Galindo, J.M.G. Molinillo, and H.G. Cutler. Universidad de Cadiz, Cadiz, Spain: Servicio Publicationes, pp. 287-300.

Qasem, J.R. (1999b). Allelopathic potential of *Cardaria draba* and *Salvia syriaca* on vegetable crops. In *Second World Congress on Allelopathy: Critical Analysis and Future Prospects*. Lakehead University, Canada. August 8-13. Abstract no. 150.

Qasem, J.R. and B.E. Abu-Irmaileh. (1985). Allelopathic effect of *Salvia syriaca* L. (Syrian sage) in wheat. *Weed Research* 25: 47-52.

Qasem, J.R. and H.A. Abu-Blan. (1995). Antifungal activity of aqueous extracts from some common weed species. *Annals of Applied Biology* 127: 215-219.

Qasem, J.R. and H.A. Abu-Blan. (1996). Fungicidal activity of some common weed extract against different plant pathogenic fungi. *Journal of Phytopathology* 144: 157-161.

Qasem, J.R. and M.A. Kasrawi. (1995). Variation of resistance to broomrape (*Orobanche ramosa*) in tomatoes. *Euphytica* 81: 109-114.

Qasem, J.R. and T.A. Hill. (1989a). Possible role of allelopathy in the competition between tomato, *Senecio vulgaris* L. and *Chenopodium album* L. *Weed Research* 29: 349-356.

Qasem, J.R. and T.A. Hill. (1989b). On difficulties with allelopathy methodology. *Weed Research* 29: 345-347.

Qasem, J.R., A.S. Al-Abed, and H.A. Abu-Blan. (1995). Antifungal activity of clammy inula (*Inula viscosa*) on *Helminthosporium sativum* and *Fusarium oxysporum* f.sp. *lycopersici*. *Phytopathologia Medeteranea* 34: 7-14.

Quinn, J.A. (1974). *Convolvulus sepium* in old field succession on the New Jersey Piedmont. *Bulletin of the Torrey Botanical Club* 101: 89-95.

Qureshi, Z.M., F. Hussain, and S. Shaukat. (1987). Allelopathic effect of *Avena fatua* Linn. On some crops. In *Medern Trend of Plant Science Research in Pakistan*, eds. I. Ilahi and F. Hussain. Proceedings of the Third National Conference of Plant Scientists. Peshawar: Department of Botany, University of Peshawar, pp. 53-58.

Rafiq, M., M.A. Nasir, and M.A.R. Bhatti. (1984). Antifungal properties of certain common wild plants against fungi. *Pakistan Journal of Agricultural Research* 5: 236-238.

Rassmussen, J.A. and E.L. Rice. (1971). Allelopathic effects of *Sporobolus pyramidatus* on vegetational patterning. *American Midland Naturalist* 86: 309-326.

Rassmussen, J.A. and F.A. Einhellig. (1975). Non competitive effects of common milkweed, *Asclepias syriaca* L. on germination and growth of sorghum. *American Midland Naturalist* 94: 478-483.

Rassmussen, J.A. and F.A. Einhellig. (1979). Allelochemic effects of leaf extracts of *Ambrosia trifida* (Compositae). Southwest. *Nature* 24: 637-644.
Rice, E.L. (1964). Inhibition of nitrogen-fixing and nitrifying bacteria by seed plants. Ecology 45: 824-837.
Rice, E.L. (1965). Inhibition of nitrogen-fixing and nitrifying bacteria by seed plants. IV. The inhibitors produced by *Ambrosia elatior* L. and *Ambrosia psilostachya* D.C. *Southwestern Naturalist* 10: 248-255.
Rice, E.L. (1969). Inhibition of nitrogen-fixing and nitrifying bacteria by seed plants. VI. Inhibitors from *Euphorbia supina* Raf. *Physiologia Plantarum* 22: 1175-1183.
Rice, E.L. (1974). *Allelopathy*. New York: Academic Press, pp. 352.
Rice, E.L. (1983). *Pest Control with Nature's Chemicals*. Norman, OK: University of Oklahoma Press, pp. 224.
Rice, E.L. (1984). *Allelopathy*. 2nd edition. Orlando.FL: Academic Press, pp. 422.
Rice, E.L. (1995). *Biological Control of Weeds and Plant Diseases: Advances in Applied Allelopathy*. Norman, OK: University of Oklahoma Press, pp. 439.
Rice, E.L. and R.L. Parenti. (1967). Inhibition of nitrogen-fixing and nitrifying bacteria by seed plants. V. Inhibitors produced by *Bromus japonica* Thunb. *Southwestern Naturalist* 12: 97-103.
Rice, E.L., W.T. Penfound, and L.M. Rohrbaugh. (1960). Seed dispersal and mineral nutrition in succession in abandoned fields in central Oklahoma. *Ecology* 41: 224-228.
Riffle, M.S. (1988). *Biological and Biochemical Interactions Between Unicorn Plant (Proboscidea louisianica) and Cotton (Gossypium hirsutum)*. Dessertation. Oklahoma State University. Stillwater, OK, USA.
Riffle, M.S., G.R. Waller, D.S. Murray, and R.P. Sgaramello. (1990). Devil's claw (*Proboscidea louisianica*), essential oil and its components: Potential allelochemical agents on cotton and wheat. *Journal of Chemical Ecology* 16: 1927-1939.
Rizvi, S.J.H. and V. Rizvi, eds. (1992). *Allelopathy: Basic and Applied Aspects*. London: Chapman and Hall, pp. 480.
Ross, M.A. and C.A. Lembi. (1999). *Applied Weed Science*, 2nd edition. Upper Saddle River, NJ: Prentice-Hall Inc., pp. 452.
Sahid, I.B. and J.B. Sugau. (1993). Allelopathic effect of lantana (*Lantana camara*) and Siam weed (*Chromolaena odorata*) on selected crops. *Weed Science* 41: 303-308.
Sakashita, S.I., N. Hirai, T. Sano, C.U. Premasthira, Y. Asakawa, J. Harada, Y. Fujii and H. Ohigashl. (1999). Plant growth inhibitors of an allelopathy weed, *Sphenoclea zeylanica*. In *Second World Congress on Allelopathy: Critical Analysis and Future Prospects*. Lakehead University, Canada. August 8-13. Abstract no. 160.
Salas, M.C. and E. Vieitez. (1972). Activated de crecimiento de Ericaceae. *Anales de Edafologia y Agrobiologia* 31: 1001-1009.
Sankhla, N., D. Baxi, and U.N. Chatterji. (1965). Eco-physiological studies on arid zone plants. I. Phytotoxic effects of aqueous extracts of mosquite (*Prosopis juliflora* DC.). *Current Science* 34: 612-614.
Sarma, K.K.V. (1974). Allelopathic potential of *Digera arvensis* Forsk. On *Pennisetum typhoides* Stapf. et Hubb. *Geobios (Jodhpur)* 1: 137.
Sarma, K.K.V., G.S. Giri, and K. Subramanyam. (1976). Allelopathic potential of *Parthenium hysterophorus* Linn. on seed germination and dry matter production

in *Arachis hypogea* Willd., *Crotalaria juncea* Linn. and *Phaseolus mungo* Linn. *Tropical Ecology* 17: 76-77.

Saxena, M. (1992). Allelopathic potential of terrestrial plants against the growth of aquatic weeds. In *Proceeding First national Symposium on Allelopathy in Agroecosystem*, eds. P. Tauro and S.S. Narwal. Haryana Agricultural University, Hisar, India: Indian Society of Allelopathy, pp. 147-148.

Schreiber, M.M. and J.L. Williams, Jr. (1967). Toxicity of root residues of weed grass species. *Weeds* 15: 80-81.

Schumacher, W.J., D.C. Thill, and G.A. Lee. (1983). Allelopathic potential of wild oat (*Avena fatua*) on spring wheat (*Triticum aestivum*) growth. *Journal of Chemical Ecology* 9: 1235-1246.

Schutt, P., H.J. Schuck, A.V. Sydow, and H. Hatzelmann. (1975). Allelopathic effects of forest weeds. 1. Influence of water extracts on the development of root hairs of *Picea abies* seedlings. *Forstwissenschaftliches Centralblatt* 94: 43-53.

Seegal, B.C. and M. Holden. (1945). The antibiotic activity of extracts of Ranunculaceae. *Science* 101: 413-414.

Selleck, G.W. (1972). The antibiotic effects of plants in laboratory and field. *Weed Science* 20: 189-194.

Sen, D.N. (1976). Ecophysiological studies on weeds of cultivated fields with special references to bajra (*Pennisetum typhoideum* Rich.) and til (*Sesamum indicum* Linn.) crops. *Second Progress Report of Project, No. A7-CR-425*, Laboratory of Plant Ecology, University of Jodhpur, India.

Sen, D.N. and D.D. Chawan. (1970). Ecology of desert plants and observations on their seedlings. III. The influence of aqueous extracts of *Prosopis juliflora* DC on *Euphorbia caducifolia* Haines. *Vegetatio* 21: 277-298.

Sethi, A. and K. Mohnot. (1988). Allelopathic influence of leaf extracts of *Troanthema portulacastrum* on germination and growth of moth bean, *Vigna aconitifollus*. *Journal of Current Biosciences* 5: 61-63.

Sharma, K.D. and D.N. Sen. (1971). Growth regulators in the fruit pulp of *Solanum surattense* L. *Zeitschrift fuer Pflanzenphysiologie* 65: 458-460.

Sharma, K.D., K.L. Sidana, and N.R. Singhvi. (1982). Allelopathic effect of *Peganum harmala* Linn. on *Pennisetum typhoideum* L. (Bajra). *Indian Journal of Botany* 5: 115-119.

Sharma, V. and G.S. Nathawat. (1987). Allelopathic effecst of *Argemone mexicana* L. on species of *Triticum, Raphanus* and *Pennisetum*. *Current Science* 50: 427-428.

Simkins, G.S. and J.D. Doll. (1982). Effect of annual weed residues on the growth of yellow nutsedge (Abstract). In *Proceedings of the North Central Weed Control Conference*. Indianapolis, USA.

Singh, D. (1994). Scope of medicinal and aromatic plants in pest management. In *International Symposium, Allelopathy in Sustainable Agriculture, Forestry and Environment*. Hisar, India: Indian Society of Allelopathy, pp. 68 (Abstract).

Singh, H.P., D.R. Batish, and R.K. Kohli. (1999). Autotoxicity: concept, organisms and ecological significance. *Critical Reviews in Plant Sciences* 18: 757-772.

Singh, M., R.V. Tamma, and H.N. Nigg. (1989). Effect of ajoene, a compound derived from garlic (*Allium sativum*), on *Phytophthora drechsleri* f. sp. *cajana*. *Mycologia* 84: 105-108.

Singh, P.N. (1977). Effect of root exudates and extracts of *Solanum nigrum* and *Argemone mexicana* seedlings on rhizosphere mucoflora. *Acta Botanica Indica* 5: 123-127.

Singh, U.P., V.B. Pandey, R.N. Singh, and R.D.N. Singh. (1988). Antifungal activity of some new flavones and flavone glycosides of *Echinops echinatus*. *Canadian Journal of Botany* 66: 1901-1903.

Smith, A.E. (1989). The potential allelochemic characteristics of bitter sneezweed (*Helianthus amarum*). *Weed Science* 37: 665-669.

Solomon, M.J. and D.C. Bhandari. (1981). Allelopathic potential of *Gomphrena decumbens* on two rain-fed crops. *Geobios* 8: 9-12.

Song, Q., Y. Fu, J. Tang, Z. Feny, and C. Yang. (1999). Allelopathic potential of *Eupatorium adenophorum*. In *Second World Congress on Allelopathy: Critical Analysis and Future Prospects*. Lakehead University, Canada. August 8-13. Abstract no. 170.

Soni, S.R. and K. Mohnot. (1988). Presence of an autotoxic factor in fruit carp of *Echinops echinatus* Roxb. *Journal of Current Biosciences* 5: 101-104.

Spring, O. and A. Hager. (1982). Inhibition of elongation growth by two sesquiterpene lactones isolated from *Halianthus annuus* L. *Planta* 156: 433-440.

Spring, O., U. Rodon, and F.A. Macias. (1992). Sesquiterpenes from noncapitate glandular trichomes of *Helianthus annuus*. *Phytochemistry* 31: 1541-1544.

Srivastava, A.K. (1969). Effect of seed extract of *Heliotropium eichwaldi* Steud. on seed germination of *Phaseolus aureus* Roxb. *Current Science* 38: 440-441.

Srivastava, P.P. and L.L. Das. (1974). Effect of certain aqueous plant extracts on germination of *Cyperus rotundus* L. *Science and Culture* 40: 318-319.

Stachon, W.J. and R.L. Zimdahl. (1980). Allelopathic activity of Canada thistle (*Cirsium arvense*) in Colorado. *Weed Science* 28: 83-86.

Steenhagen, D.A. and R.L. Zimdahl. (1979). Allelopathy of leafy spurge (*Euphorbia esula*). *Weed Science* 27: 1-3.

Stevens, G.A., Jr. and C.S. Tang. (1985). Inhibition of seedling growth of crop species by recirculating root exudates of *Biden pilosa* L. *Journal of Chemical Ecology* 11: 1411-1425.

Stevens, K.L. and G.B. Merrill. (1985). Sesquiterpene lactones and allelochemicals from *Centaurea*. In *The Chemistry of Allelopathy*, ed. A.C. Thompson. Washington, DC.: American Chemical Society, pp. 83-98.

Stiven, G. (1952). Production of antibiotic substances by roots of a grass [*Trachypogon plumosus* (H.B.K.) Nees] and *Pentanisia variabilis* (E. Mey.) Harv. (Rubiaceae). *Nature (London)* 170: 712-713.

Sugha, S.K. (1978). Allelopathic potential of superb lily (*Glorisaa superba*). *Science and Culture* 44: 461-462.

Sugha, S.K. (1979). Effect of weed extracts on wheat germination. *Science and Culture* 45: 65-66.

Sun, W.H., Z.W. Yu, and S.W. Yu. (1988). Inhibitory effects of *Eichhorina crassipes* (Mart.) Solms on algae. *Acta Phytophysiologica Sinica* 14: 294-300.

Sundaramoorthy, S. and D.N. Sen. (1990). Allelopathic effects of *Tephrosia purpu-*

rea Pers. I. On germination and early seedling growth of some arid zone crops. *Journal of Indian Botanical Society* 69: 251-255.

Suseelamma, M. and R.R. VenkataRaju. (1992). Allelopathic effect of *Digera muricata* (L.). Mart. on the germination and early seedling growth of horsegram. In *Proceedings of the First National Symposium on Allelopathy in Agroecosystems*, eds. P. Tauro and S.S. Narwal. Hisar, India: Indian Society of Allelopathy, pp. 89-91,

Swain, T. (1977). Secondary compounds as protective agents. *Annual Review of Plant Physiology* 28: 479-501.

Szezepanska, W. (1971). Allelopathy among the aquatic plants. *Polskie Archiwum Hydrobiologii/ Polish Archives of Hydrobiology* 18: 17-30.

Szezpanski, A.J. (1977). Allelopathy as a means of biological control of water weeds. *Aquatic Botany* 3: 193-197.

Tames, R.S., M.D.V. Gesto, and E. Vieitez. (1973). Growth substances isolated from tubers of *Cyperus esculentus* var. *aureus*. *Physiologia Plantarum* 28: 195-200.

Tang, C.S. and C.C. Young. (1982). Collection and identification of allelopathic compounds from the undisturbed root system of blight limpograss (*Hemarthria altissima*). *Plant Physiology* 69: 155-160.

Tang, C.S., C.K. Wat, and G.H.N. Towers. (1987). Thiophenes and benzofurans in the undisturbed rhizosphere of *Tagetes patula* L. *Plant and Soil* 98: 93-97.

Tang, C.S., W.F. Cai, K. Kohl, and R.K. Nishimoto. (1995). Plant stress and allelopathy. In *Allelopathy: Organisms, Processes, and Application*, eds. Inderjit, K.M.M. Dakshini, and F.A. Einhellig. ACS Symposium Series No. 582. Washington, DC: American Chemical Society, pp. 142-157.

Theophrastus. (ca. 300 B.C.). *Enquiry into Plants and Minor Works on Odours and Weather Signs*. 2 Vols. Translated to English by A. Hort and W. Heinemann, London, 1916.

Thijs, H., J.D. Shann, and J.D. Weidenhamer. (1994). The effect of phytotoxin on competitive outcome in a model system. *Ecology* 75: 1959-1964.

Tinnin, R. and C.H. Muller. (1971). The allelopathic potential of *Avena fatua*: influence on herb distribution. *Bulletin of the Torrey Botanical Club* 98: 243-250.

Tiwari, S.P., S.S. Bhandauri, M.S. Voddaria, and D. Dayal. (1985). Allelopathic effects of weeds on soybean, groundnut and greengram. *Current Science* 54: 434-435.

Toma, M., M. Vinatoru, L. Ghinea, and M. Andrei. (1999). Identification of allelochemical in *Aristolochia clematitis* L. In *Second World Congress on Allelopathy: Critical Analysis and Future Prospects*. Lakehead University, Canada. August 8-13. Abstract no. 177.

Tomas-Barberan, F., E. Iniesta-Sanmartin, F. Tomas-Lorente, and A. Rumberto. (1990). Antimicrobial phenolic compounds from three Spanish *Helichrysum* species. *Phytochemistry* 29: 1093-1095.

Tomas-Barberan, F.A., A.D. Msonthi, and K. Hostettmann. (1988). Antifungal epicuticular methylated flavonoids from *Helichrysum nitens*. *Phytochemistry* 27: 753-755.

Toth, A., G. Benecs-Bardi, and I. Beres. (1999). Role of alleopathy in the mass propagation of *Ambrosia elatior* and *Cirsium arvense* in Hungary. In *Second*

World Congress on Allelopathy: Critical Analysis and Future Prospects. Lakehead University, Canada. August 8-13. Abstract no. 178.

Tripathi, N.N., A. Dikshit, R.D. Tripathi, and S.N. Dixit. (1983). Fungitoxic properties of *Iberis amara*. *International Journal of Crude Drug Research* 21: 67-71.

Tripathi, R.S., R.S. Singh, and J.P.N. Rai. (1981). Allelopathic potential of *Eupatorium adenophorum*, a dominant ruderal weed of Meghalaya. *Proceedings of the Indian National Science Academy, Part B, Biological Science* 47: 458-465.

Tsuzuki, E., J. Watanabe, and S. Shida. (1978). Studies on allelopathy among higher plants. I. Effects of residues and aqueous and various organic solvent extracts from the lawn weed pennywort (*Hydrocotyle sibthorpioides* Lam.) on the seed germination and growth of some vegetables and crop plants. *Bulletin of the Faculty of Agriculture of Miyazaki University* 25: 403-410 (Japanese with English summary).

Ueki, K. and M. Takahashi. (1984). Weed inhibition by *Lycoris radiata*. Herb. *Proceedings of Symposium on Allelopathy. (1984 International Chemical Congress of Pacific Basin Societies*, Dec. 16-21. Honolulu, Hawii) (Abstract).

VanderVeen, R. (1935). Arch. Koffiecult (Nederland India). 3: 65. Cited by A.R. Putnam and L.A. Weston. (1986). Adverse impacts of allelopathy in agricultural systems. In *The Science of Allelopathy*, eds. A.R. Putnam and C.S. Tang. New York: John Wiley and Sons, pp. 43-56.

Vezina, L. and D. Doyon. (1983). A note on the inhibitor effect of black knapweed (*Centaurea nigra*) residues. *PhytoProtection* 64: 77-81.

Vikherkova, M. (1970). Influence of active substances from rhizome of wheatgrass on growth and water balance of flax. In *Physiological-Biochemical Basis of Plant Interaction in Phytocenoses*, ed. A.M. Grodzinsky. Vol. 1. Keiv: Naukova Dumka, pp. 135-140 (In Russian, with English summary).

Wadhawani, C. and T.N. Bhardwaja. (1981). Effect of *Lantana camara* L. extract on fern spore germination. *Experientia* 37: 245-246.

Wali, M.K. and L.R. Iverson. (1978). Revegetation of coal mine spoils and autoallelopathy in *Kochia scoparia*. (Abstract). 144th National American Association of Advisory Scientific Meeting. Washington, DC pp. 121-122.

Waller, G.R., ed. (1987). *Allelochemicals: Role in Agriculture and Forestry*. ACS Symposium Series 330. Washington, DC: American Chemical Society.

Waller, G.R. (1989). Allelochemical action of some natural products. In *Phytochemical Ecology: Allelochemicals, Mycotoxins and Insect Phermones and Allomones*, eds. C.H. Chou and G.R. Waller. Academic Sinica Monograph Series No. 9. Taipei, Taiwan, ROC: Institute of Botany, Academia Sinica, pp. 129-154.

Waller, G.R. and H. Burstrom. (1969). Diterpenoid alkaloids as plant growth inhibitors. *Nature* 222: 576-578.

Wang, Y., M. Toyota, F. Krause, M. Hamburger, and K. Hostettmann. (1990). Polyacetylenes from *Artemisia borealis* and their biological activities. *Phytochemistry* 29: 3101-3105.

Wardle, D.A., K.S. Nicholson, and A. Rahman. (1993). Influence of plant age on the allelopathic potential of nodding thistle (*Carduus nutans* L.) against pasture grasses and legumes. *Weed Research* 16: 223-227.

Warrag, M.O.A. (1995). Autotoxic potential of foliage on seed germination and early growth of mesquite (*Prosopis juliflora*). *Journal of Arid Environment* 31: 415-421.

Weaver, T.W. and D. Klarich. (1977). Allelopathic effects of volatile substances from *Artemisia tridentata* Nutt. *American Midland Naturalist* 97: 508-512.

Wegmann, K. (1998). Progress in *Orobanche* research during the past decade. In *Current Problems of Orobanche Rresearch*, eds. K. Wegmann, L.J. Musselman, and D.M. Joel. Proceedings of the 4th International *Orobanche* Workshop, Sept. 23-26, 1998. Albena, Bulgaria: Institute for wheat and sunflower "Dobroudja," pp. 13-17.

Weidenhamer, J.D. and J.T. Romeo. (1989). Allelopathic properties of *Polygonella myriophylla*: Field evidence and bioassays. *Journal of Chemical Ecology* 15: 1957-1970.

Weidenhamer, J.D., D.C. Hartnett, and J.T. Romeo. (1989). Density dependent phytotoxicity: Distinguishing resource competition and allelopathy interference in plants. *Journal of Applied Ecology* 26: 613-624.

Weidenhamer, J.D., T.C. Morton, J.T. Romeo. (1987). Solution volume and number often overlooked factors in allelopathic bioassay. *Journal of Chemical Ecology* 13: 1481.

Weller, S.C., W.A. Skroch, and T.J. Monaco. (1985). Common bermudagrass (*Cynodon dactylon*) interference in newly planted peach (*Prunus persica*) trees. *Weed Science* 33: 50-56.

Weston, L.A. and A.R. Putnam. (1986). Inhibition of legume seedling growth residues and extracts of quackgrass (*Agropyron repens*). *Weed Science* 34: 366-372.

Whitehead, D.C., H. Dibb, and R.D. Hartley. (1982). Phenolic compounds in soil as influenced by the growth of different plant species. *Journal of Applied Ecology* 19: 579-588.

Whittaker, R.H. and P.P. Feeny. (1971). Allelochemicals: Chemical interactions between species. *Science* 171: 757-770.

Wickliffe, W.B., F.H. Yelverton, A.D. Worsham, and G.G. Nagabhushana. (1997). Increasing rye densities reduces need for herbicides in no-tillage crop production. *Weed Science Society of America. Abstract* 37: 82.

Willaim, R.D. and G.F. Warren. (1975). Competition between purple nutsedge and vegetables. *Weed Science* 23: 317-323.

Willis, R.J. (1985). The historical bases of the concept of allelopathy. *Journal of the History of Biology* 18: 71-102.

Wilson, R.E. and E.L. Rice. (1968). Allelopathy as expressed by *Helianthus annuus* and its role in old-field succession. *Bulletin of the Torrey Botanical Club* 95: 422-448.

Winter, A.G. (1961). New physiological and biological aspects in the interrelationships between higher plants. *Symposium of the Society for Experimental Biology* 15: 229-244.

Worsham, A.D., G.C. Klingman, and D.E. Moreland. (1962). Promotion of germination of *Striga asiatica* seed by coumarin derivatives and effect of seedling development. *Nature* 195: 199-200.

Wu, H., J. Pratley, D. Lemerle, and T. Haig. (1999). Crop cultivars with allelopathic capability. *Weed Research* 39: 171-180.

Wurzoburger, J. and Y. Leshem. (1969). Physiological action of the germination inhibitor in the husk of *Aegilops hotschyi* Bioss. *New Phytologist* 68: 337-341.

Yamane, A., J. Fujikura, H. Ogawa, and J. Mizutani. (1992). Isothiocyanates as allelopathic compounds from (*Rorippa indica*) Hiern (Cruciferae) roots. *Journal of Chemical Ecology* 18: 1941-1954.

Yenish, J.P., A.D. Worsham, and A.C. York. (1996). Cover crops for herbicide replacement in nontillage corn (*Zea mays*). *Weed Technology* 10: 815-821.

Young, C.C. and D.P. Bartholomew. (1981). Allelopathy in a grass-legume association: 1. Effects of *Hermatheria altissima* (Poir.) Stapf. and Hubb. root residues on the growth of *Desmodium intotum* (Mill.) Urb, and *Hemartheria* in a tropical soil. *Crop Science* 21: 770-774.

Zimdahl, R.L. (1999). *Fundamentals of Weed Science*. 2nd ed. San Diego: Academic Press, pp. 450.

Zuberi, M.I., A.R. Biswas, G.B. Ghosh, P.C. Roy, and A.M. Choudhury. (1989). Allelopathic effects of diffusates and extracts of *Striga densiflora* on germination and early growth of rapeseed. *Journa of the Indian Botanical Society* 68: 317-319.

Crop Allelopathy and Its Role in Ecological Agriculture

Daizy R. Batish
H. P. Singh
R. K. Kohli
Shalinder Kaur

SUMMARY. A number of crops exhibit allelopathic interactions that play a significant role in the complex environment of agroecosystems. Several studies have shown that allelopathic crops reduce growth, development and yield of other crops growing simultaneously or subsequently in the fields. Another aspect of interest regarding crop allelopathy is that allelochemicals may exhibit inhibitory effect on the same crop which is commonly called as *crop autotoxicity*. It is predominantly common in fields where sole cropping under reduced or no-tillage system is practiced. Though any crop part can be allelopathic, including even the pollens, but decomposing crop residues exhibit more influence on other plants. Furthermore, the extent of allelopathy by a crop plant varies with age, part and type of cultivar being used. Nowadays allelopathic crops are being used as an important tool in managing weeds and harmful pests under sustainable pest management programs. In this context several cover/smother and green manure crops with allelopathic

Daizy R. Batish is Reader (Associate Professor), H. P. Singh is Research Associate, R. K. Kohli is Professor, and Shalinder Kaur is PhD Student, Department of Botany, Panjab University, India.

Address correspondence to: Daizy R. Batish, Department of Botany, Panjab University, Chandigarh-160 014, India (E-mail: daizybatish@hotmail.com).

H. P. Singh is thankful to Council of Scientific and Industrial Research (CSIR), India for financial assistance.

[Haworth co-indexing entry note]: "Crop Allelopathy and Its Role in Ecological Agriculture." Batish, Daizy R. et al. Co-published simultaneously in *Journal of Crop Production* (Food Products Press, an imprint of The Haworth Press, Inc.) Vol. 4, No. 2 (#8), 2001, pp. 121-161; and: *Allelopathy in Agroecosystems* (ed: Ravinder K. Kohli, Harminder Pal Singh, and Daizy R. Batish) Food Products Press, an imprint of The Haworth Press, Inc., 2001, pp. 121-161. Single or multiple copies of this article are available for a fee from The Haworth Document Delivery Service [1-800-342-9678, 9:00 a.m. - 5:00 p.m. (EST). E-mail address: getinfo@haworthpressinc.com].

nature hold a good promise as well as challenge for the future as they have a potential to suppress noxious weeds. Likewise, the decomposing residues of such crops can also be used for managing harmful weeds and pests *vis-à-vis* maintaining the sustainability of the system. The traditional practice of crop rotation which declined with the discovery of synthetic herbicides can also be revived, if allelopathic potential of rotational crops is well understood. Selection of cultivars with high allelopathic potential/high competitiveness is another possible way by which crop allelopathy can be better utilized for weed control and yield enhancement. Crops with less allelopathic potential can be genetically improved by incorporating desired genes encoding the synthesis of allelochemicals. Pure allelochemicals extracted and identified from some crop plants can also be used as bioherbicides.

KEYWORDS. Crop allelopathy, crop residues, crop rotation, cover/smother crops, green manures, weed management

INTRODUCTION

Etymologically, crop is a cultivated plant that provides food, fodder, fiber and several other products. Van Raamsdonk (1995) defines crop plant as any set of plants cultivated by humans and used for a special purpose and it includes all cultigens, though not exclusively. Worldwide 25 crops, of which cereal crops like wheat, maize, rice are the leading ones are used (Wittwer, 1998). Any phenomenon linked with crops that may affect their productivity in any way, therefore, assumes great significance. Allelopathy is one such phenomenon that has been observed in the croplands since centuries. Theophrastus (300 B.C.) reported the *suppression of weeds* by chickpea (*Cicer arietinum*). Later, Plinius Secundus (1 A.D.) reported the *scorching up* of cornland by several crops like chickpea, barley (*Hordeum vulgare*), fenugreek (*Trigonella foenum-graecum*), and bitter vetch (*Vicia ervillia*). de Candolle (1832), a Swiss botanist and well-known taxonomist, observed that chemicals released from crop plants cause *soil sickness* and this problem can be controlled only if crop rotation is practiced. Earlier work on allelopathy was also demonstrated by Schreiner and Shorey (1907, 1908, 1909), Schreiner and Reed (1908), Schreiner and Sullivan (1909), and Pickering (1912, 1917). While studying the effect of ethylene, Molisch (1937) coined the term Allelopathy choosing two Greek words. The work in the field of allelopathy particularly in the agroecosystems continued at very slow pace even after post-Molisch period. It, however, gained momentum only in the last three decades or so. Such renewed interest in this field is due to several objectives including, the manipulations of allelopathy for crop improvement, conserving species diversity, weed management, environment protection through ecofriendly use of allelopathy/allelochemicals for the control of weeds, pests and plant diseases (Rizvi and Rizvi, 1992; Rice, 1995; Kohli, Batish, and Singh, 1998; Anaya, 1999; Chou, 1999; Rizvi et al., 1999). The horizons where the allelopathy can be used are unlimited and, therefore, phenomenal success can be expected.

Allelopathy in the croplands bears a great significance (Einhellig, 1985). The allelopathic effects of the crops can be summarized as follows:

- It affects the growth, productivity and yield of other crops.
- It may affect the same crop growing in monocultures or grown in succession.
- It causes soil sickness (at least partially, if not fully) and imbalance of nutrients and microbial populations.
- Crop allelopathy can be exploited to selectively suppress the weeds through various manipulations.

The first three of the above effects of crop allelopathy are negative in nature, and if the phenomenon is well understood, it can help solving problems of crop yield reduction. The fourth one, however, is catching the attention of the scientists worldwide who are trying to find the alternatives to the excessive use of synthetic chemicals.

ALLELOPATHIC EFFECTS OF CROPS ON OTHER CROP PLANTS

A number of crops have been reported to be allelopathic towards other crops grown either subsequently or simultaneously (Table 1). To enumerate, a few are *Avena sativa* (Guenzi and McCalla, 1966), *Brassica juncea* (Oleszek, 1987), *Cicer arietinum* (Lovett and Jessop, 1982), *Glycine max* (Tsuzuki and Kawagoe, 1984), *Medicago sativa* (Hegde and Miller, 1990), *Secale cereale* (Barnes et al., 1987) and *Triticum aestivum* (Lynch, 1977, 1978; Thorne et al., 1990; Blum et al., 1992). Various parts of the crop plants are known to release allelochemicals through different means. Most common being the leachation (Overland, 1966; Barnes and Putnam, 1983; Kim and Kil, 1987) and decomposing crop residues (Guenzi, McCalla, and Norstadt, 1967; Chou and Patrick, 1976; Hegde and Miller, 1990). Besides, several crops, particularly the crucifers, are known to release the allelochemicals through volatilization (Petrova, 1977; Oleszek, 1987) and/or through root exudates (Tang and Young, 1982). An interesting aspect of crop allelopathy is

TABLE 1. List of allelopathic crops affecting other crops.

Crop Species	Test Crop	Source of Allelochemical	Reference(s)
Allium cepa *A. sativum*	-	-	Evenari, 1949.
Arachis hypogaea	*Gosyypium hirsutum*	Residues	Megie, Pearson, and Hiltbold, 1967.
Asparagus officinalis	*Lycopersicon esculentum* *Lactuca sativa* *Hordeum vulgare* *Lactuca sativa*	Crude aqueous extracts, root tissue amended soil, soil extracts	Shafer and Garrison, 1986; Hazebroek, Garrison, and Gianfagna, 1989; Hartung, Nair, and Putnam, 1990; Pederson et al., 1991.
Avena sativa	*Triticum aestivum* *Oryza sativa* *Lactuca sativa* *Zea mays* and other crops	Phenolics, Straw and mulches, aqueous extracts, residues	Nielsen, Cuddy, and Woods, 1960; Guenzi and McCalla, 1966; Guenzi, McCalla, and Norstadt, 1967; Martin, McCoy, and Dick, 1990; Kato-Noguchi, Mizutani, and Hasegawa, 1994; Kato-Noguchi et al., 1994.
Beta vulgaris	*Gossypium hirsutum*	Residues	Kalburtji and Gagianas, 1997.
Brassica campestris	*Vigna radiata*	Residue extracts	Kuo, Chou, and Park, 1981.
B. juncea	*Lactuca sativa* *Triticum aestivum*	Volatiles	Oleszek, 1987.
B. napus	*Linum usitatissimum* *Triticum aestivum*	Residues	Martin and Rademacher, 1960; Horricks, 1969; Lovett and Jessop 1982; Vera, McGegor, and Downey, 1987; Gubels and Kenaaschuk, 1989.
B. nigra	*Lactuca sativa* *Triticum aestivum*	Volatiles, aqueous extracts, residues	Mason-Sedun, Jessop, and Lovett, 1986; Oleszek, 1987.
B. oleracea var. *Italica*	*B. oleracea* var. *capitata*	Aqueous extracts	Santos and Leskovar, 1997.
Cajanus cajan	*Glycine max* *Dolichos lablab* *Cajanus cajan*	Litter	Hepperly and Diaz, 1983.
Camelina sativa	*Linum usitatissimum*	Growing plants	Grümmer and Beyer, 1960.
Camellia sinensis	-	-	Tsuzuki, Ashihara, and Waller, 1992.
Carthamus tinctorius	*Triticum aestivum*	Fresh and partly decayed residues	Lovett and Jessop, 1982.
Cicer arietinum	*Triticum aestivum*	Dry, fresh and partly decayed residues	Lovett and Jessop, 1982.
Citrullus vulgaris	*Lactuca sativa* *Lycopersicon esculentum*	Seed exudates	Kushima et al., 1998.
Coffea arabica	*Lactuca sativa* *Lolium multiflorum* *Phaseolus mungo*	Aqueous extracts, allelochemicals	Chou and Waller, 1980; Rizvi et al., 1987.
Colocasia esculenta	*Vigna radiata* *Zea mays* *Hibiscus esculentus*	Residue extracts	Pardales and Dingal, 1988.

Crop Species	Test Crop	Source of Allelochemical	Reference(s)
Cruciferous Plants	*B. juncea* *B. nigra* *Lactuca sativa* *Triticum aestivum*	Volatiles	Oleszek, 1987.
Cucumis sativus	*Lycopersicon esculentus*	Phenolics	Politycka Wojcik-Wojtkowiak, and Pudelski, 1984.
Cucurbita moschata	*Zea mays* *Phaseolus vulgaris*	Leachates	Anaya et al., 1987.
Fagopyrum cymosum	*Oryza sativa* other crops	Aqueous extracts	Tsuzuki, Yamamoto, and Shimizu, 1987.
F. cymosum	*Brassica kaber* var. *Pinnatifida*	Germinating seeds	Prutenskaya, 1972.
Festuca elatior (Kentucky-31)	*Brassica napus* L.	Aqueous extracts	Peters, 1968.
Glycine max	*Zea mays* *Triticum aestivum* *Sorghum vulgare* *Brassica rapa* *Medicago sativa* *Raphanus sativus*	Residues, aqueous extracts	Guenzi and McCalla, 1962; Lovett and Jessop, 1982; Tsuzuki and Kawagoe, 1984; Martin, McCoy, and Dick, 1990.
Hedera helix	*Bromus tectorum*	Root exudates	Muller, 1966.
Helianthus annuus	*Triticum aestivum* *Glycine max* *Sorghum* spp. *Lycopersicon esculentum* *Hordeum vulgare* *Phaseolus mungo* *P. aureus* *Trifolium alexandrinum*	Residues, soil, leachates, extracts, stubbles, plant parts mixed in soil,	Wilson and Rice, 1968; Lovett and Jessop, 1982; Irons and Burnside, 1982; Schon and Einhellig, 1982; Purvis and Jones, 1990; Kohli, 1993; Macias et al., 1993, 1996, 1997.
H. tuberosus	*Daucus carota* *Raphanus sativus* *Inula heleium* *Pisum sativum*	Leaf extracts	Politycka, Wojcik-Wojtkowiak, and Pudelski, 1984.
Hemarthria altissima	*Desmodium intortum* cv. Green leaf	Crop residues	Young and Bartholomew, 1981.
Hordeum vulgare	*Medicago sativa* *Triticum aestivum* *Raphanus* sp. *Brassica kaber* var. *pinnatifida* *Stellaria media* *Linum usitatissimum*	Aqueous extracts, residues, growing plants, straw and mulches	Overland, 1966; Prutenskaya, 1974; Gubels and Kenaaschuk 1989; Read and Jansen, 1989.
Ipomoea batatas	Nearby growing crops	Decaying residues	Walker, Hubbell, and Sabberry, 1989
Lens culinaris	*Triticum aestivum* and legumes	Decaying residues	Cochran, Elliott, and Papandick, 1977.
Lycopersicon esculentum	*Brassica campestris* *Zea mays* *Lactuca sativa* *Sorghum* spp.	Aqueous extracts and leachates, root exudates	Kim and Kil, 1987.
Manihot esculenta	*Phaseolus vulgaris* *Zea mays* *Manihot esculenta*	Fresh leaves	Altieri and Doll, 1978.

TABLE 1 (continued)

Crop Species	Test Crop	Source of Allelochemical	Reference(s)
Medicago sativa	*Lycopersicon esculentum* *Raphanus sativus* *Gossypium hirsutum* *Lactuca sativa* *Zea mays* *Triticum aestivum* Grasses *Cucumis sativus* *Brassica oleracea* *Trifolium repens*	Residues, aqueous extracts, plant parts, root exudates, alfalfa meal, alcoholic extracts, roots, root saponins	Mishustin and Naumova, 1955; Nielsen, Cuddy and Woods, 1960; Grant and Sallans, 1964; Guenzi, Kehr, and McCalla, 1964; Pederson, 1965; Oleszek and Jurzysta, 1987; Miller et al., 1988; Hall and Henderlong, 1989; Hegde and Miller, 1990; Ells and McSay, 1991; Gorski, Miersch, and Polszynski, 1991; Wyman-Simpson et al., 1991; Miersch et al., 1992; Waller, Jurzysta, and Thorne, 1993; Nakahisa et al., 1994; Kawata, Harada, and Matsunaka, 1996.
Melilotus alba	*Zea mays* *Triticum aestivum* *Sorghum vulgare*	Aqueous extracts	McCalla and Duley, 1948; Guenzi and McCalla, 1962.
Nicotiana rustica	*Brassica rapa* *Lactuca sativa* *Setaria italica* *Zea mays*	Root exudates	Ul Haq and Hussain, 1979.
Oryza perennis	*Brassica oleracea*	Water extracts	Chou, Chang, and Oka, 1991.
O. sativa	*Vigna radiata* *Glycine max*	Rice straw, rice callus, phytotoxins	Chou and Lin, 1976; Rice, Lin, and Huang, 1980, 1981; Yang and Futsuhara, 1991.
Panicum antidotale	*Brassica campestris* *Pennisetum americanum*	Aqueous extracts, root exudates	Begum and Hussain, 1980.
P. miliaceum	*Brassica kaber* var. *pinnatifida*	Plants	Prutenskaya, 1974.
Phaseolus vulgaris	*Zea mays* *Manihot esculenta* *Cucurbita moschata*	Fresh leaves, leachates	Altieri and Doll, 1978; Anaya et al., 1987.
Pisum sativum	*Triticum aestivum* Legumes and other crops	Residues, Hot water extracts	Cochran, Elliott, and Papandick, 1977; Lovett and Jessop, 1982; Schenk and Werner, 1991.
Raphanus sativus	*Lactuca sativa*	Plant residues	Beckman and Noffke, 1978.
Secale cereale	*Zea mays* *Lactuca sativa* *Lycopersicon esculentum* *Stellaria media* *Lepidium sativum* *Triticum aestivum* *Brassica kaber* var. *pinnatifida* *Nicotiana tabacum*	Whole plants, residues, aqueous extracts, root leachates	Hill, 1926; Patrick and Koch, 1958; Patrick, Toussoun, and Snyder, 1963; Overland, 1966; Prutenskaya, 1974; Chou and Patrick, 1976; Barnes and Putnam, 1983; Barnes et al., 1987; Raimbault, Vyn, and Tollenaar, 1991; Mwaja, Masiunas, and Weston, 1995.
Solanum myriacanthum	*Brassica campestris*	Aqueous extracts	Chattopadhyoy, 1995.

Crop Species	Test Crop	Source of Allelochemical	Reference(s)
S. tuberosum	Crop plants	Aqueous extracts	Nielsen, Cuddy, and Woods, 1960
Sorghum bicolor	*Triticum aestivum* *Zea mays* *Glycine max* *Spinacea oleracea* *Senecio vulgaris* *Solanum tuberosum* *Pennisetum americanum* *Setaria italica*	Residues, aqueous extracts, root exudates, sorgoleone	Guenzi, McCalla, and Norstadt, 1967; Hussain and Gadoon, 1981; Lovett and Jessop, 1982; Ruiz-sifre and Ries, 1983; Netzley and Butler, 1986; Rasmussen et al., 1992; Ben-Hammouda, Kremer, and Minor, 1995; Nimbal et al., 1996a.
Sorghum sudanense or Sudex	*Lycopersicon esculentum* *Setaria italica* *Lepidium sativum* *Medicago sativa* *Stellaria media* *Gossypium hirsutum* and others	Aqueous extracts, root leachates, green manure, residues	Overland, 1966; Megie, Pearson, and Hiltbold, 1967; Forney and Foy, 1985; Forney, Foy, and Wolf, 1985; Weston, Harmon, and Mueller, 1989.
Trifolium alexandrinum	*Allium cepa* *Daucus carota* *Lycopersicon esculentum*	Volatiles from residues	Bradow and Connick, 1990.
T. incarnatum	*Zea mays* *Lolium multiflorum* *Gossypium* spp. *Ipomoea lacunosa*	Aqueous extracts	White, Worsham, and Blum, 1989.
T. repens	Legumes and grasses	Aqueous extracts	Macfarlane, Scott, and Jarvis, 1982.
Triticum aestivum	*Gossypium* species *Avena sativa* *Hordeum vulgare* *Sorghum* spp. *Pennisetum* spp. *Zea mays* *Vigna unguiculata* *Trifolium incarnatum* *Cyamopsis tetragonoloba* *Brassica kaber* var. *pinnatifida* and other crops	Straw and mulches, residue, aqueous extracts, phenolics	Kimber, 1967; Prutenskaya, 1974; Lynch, 1977, 1978; Chapman and Lynch, 1983; Hicks et al., 1989; Blum et al., 1992; Narwal, Sarmah, and Nandal, 1997; Hamilton-Kemp, and Anderson, 1986; Waller et al., 1987.
Vicia faba	*Triticum aestivum* *Zea mays* *Lolium multiflorum* *Gossypium* spp. *Ipomoea lacunosa*	Residues, aqueous extracts	Lovett and Jessop, 1982; White, Worsham, and Blum, 1989.
Vigna radiata	*Lactuca sativa* *Lycopersicon esculentum*	Plant parts, extracts, exudates	Tang and Young, 1982; Chou, 1995. Chou et al., 1995.
Zea mays	*Phaseolus vulgaris* *Manihot esculenta* *Cucurbita moschata* *Citrullus lanatus* *Triticum aestivum* *Nicotiana tabacum* Other crops	Fresh leaves, aqueous extracts, pollen, residues	Patrick and Koch, 1958; Nielsen, Cuddy, and Woods, 1960; Guenzi, McCalla, and Norstadt, 1967; Altieri and Doll, 1978; Anaya et al., 1987; Cruz-Ortega, Anaya, and Ramos, 1988.

the allelochemicals released by pollen of some crop plants. Jimenez et al. (1983) reported allelopathic effect of pollen of corn (*Zea mays*). Later, Cruz-Ortega, Anaya, and Ramos (1988) reported that the ethanolic extracts of corn pollen inhibit the growth of watermelon (*Citrullus lanatus*). Anaya et al. (1992) identified allelochemical phenylacetic acid from corn pollen that inhibited weeds like *Echinochloa crus-galli* and *Amaranthus leucocarpus*.

The crop allelopathy becomes more apparent under no- or reduced tillage conditions or under stubble mulch farming where residues are left in the fields after harvesting/thrashing in order to protect soil from wind and water erosion. Upon decomposition, these release allelochemicals which accumulate and adversely affect crop productivity (Putnam and Duke, 1974; Chou and Lin, 1976; Rice 1984; Putnam, 1994). Einhellig (1996) reported that the amount of allelochemicals is substantially elevated under biotic and abiotic stress conditions such as mineral deficiency, UV radiations, temperature, water and even herbicide treatment. It, therefore, becomes difficult to manage allelopathy for enhancing the crop production. In order to enhance the crop productivity in agroecosystems, the allelopathic effects of the crop plants or their residues can be minimized if,

- the traditional practice of crop rotation is revived with new approach;
- crop residues are managed effectively in such a way that their effect on the crops is minimized, however, noxious weeds are suppressed selectively;
- crops are protected against the stress conditions so that the amount and extent of impact of allelochemical(s) is reduced;
- efforts should be made to understand the mechanism of actions of allelochemicals on other plants and their fate in soil.

CROPS AFFECTING THE WEEDS

Several crops are known to inhibit weeds (Table 2). This tendency in the crops pose challenge as well as promise for the future for the sustainable management of the weeds in the already deteriorating agroecosystems. Some important crops possessing weed-suppressing ability are *Helianthus annuus* (Leather, 1983, 1987), *Cucumis sativus* (Putnam and Duke, 1974), *Secale cereale*, *Trifolium incarnatum* and *T. subterraneum* (Lehman and Blum, 1997) and *Sorghum bicolor* (Nimbal et al., 1996a). Among these rye (*Secale cereale*) is one of the most promising crops being grown in several cropping systems and also provides large amount of biomass. Its residues considerably reduce weed mass which has been attributed to the presence of DIBOA and its degradation product BOA in aqueous extracts of its residues (Barnes, Putnam, and Burke, 1986; Barnes et al., 1987). Shilling et al. (1986) identi-

TABLE 2. Allelopathic effect of crops on weeds.

Crop Species	Weed Species	Allelochemical Source	Reference(s)
Allium cepa	*Amaranthus spinosus* *Kochia scoparia*	Mature plants	Macharia and Peffey, 1995.
Avena sativa	*Erysimum cheiranthoides* *Chenopodium album* *Descurainia sophia* *Thlaspi arvense* *Bromus tectorum* *Brassica kaber*	Mature plants, aqueous extracts, root exudates	Markova, 1972; Neustruyeva and Dobretsova, 1972; Moyer and Huang, 1997; Baghestani et al., 1999.
Beta vulgaris	*Agrostemma githago*	Seeds Inhibitors	Funke, 1941.
Brassica napus	*Amaranthus retroflexus* *Bromus tectorum* *Chenopodium album* *Descurainia sophia* *Thlaspi arvense*	Residues, aqueous extracts	Boydston and Hang, 1995; Moyer and Huang, 1997
Cajanus cajan	Weeds	Litter	Hepperly and Diaz, 1983.
Capsicum annuum	Weeds	Shoot leachates, root exudates	Gonzalez, Souto, and Reigosa, 1997.
Coffea arabica	*Festuca* spp. *Lolium multiflorum* *Amaranthus spinosus* other weeds	Aqueous extracts from leaves, roots and seed, 1,3,7-trimethyl xanthine	Chou and Waller, 1980; Anaya et al., 1978; Rizvi, Mukerji, and Mathur, 1980; Rizvi et al., 1987.
Cucumis sativus *Cucumis* spp.	*Brassica hirta* *Panicum miliaceum*	Plants suppress growth	Putnam and Duke, 1974.
Cucurbita moschata	*Portulaca olearacea* *Echinochloa crus-galli* *Amaranthus leucocarpus*	Leachates	Anaya et al., 1987.
Fagopyrum esculentum *F. cymosum*	*Stellaria media* and other weeds *Chenopodium album* *Brassica kaber* var. *Pinnatifida*	Aqueous extracts, growing plants and germinating seeds	Overland, 1966; Neustruyeva and Dobretsova, 1972; Prutenskaya, 1972; Tsuzuki, amamoto, and Shimizu, 1987
Festuca elatior (Kentucky-31)	Weeds	Aqueous extracts of root and leaves	Peters, 1968.
F. longifolia (Rebel and Kentucky-31)	*Lolium multiflorum* *Festuca rubra* *Poa pratensis* *Lolium perenne* *Festuca longifolia*	Leaf extracts	Buta and Spaulding, 1989.
Glycine max	*Lemna minor* and other weeds	Root exudates and decaying roots	Massantini, Caporali, and Zellini, 1977; Nelson, 1985.
Helianthus annuus	*Amaranthus retroflexus* *Brassica kaber* *Ipomoea purpurea* *Datura stramonium* *Chenopodium album* *Cirsium arvense* *Setaria faberi* *Stellaria media* *Capsella bursa-pactoris* *Nicotiana tabacum* other weeds	Aqueous extracts and leachates, soil amendment of plant parts (root, stem, leaf and seed hulls).	Overland, 1966; Hall, Blum, and Fites, 1982; Leather, 1983, 1987.

TABLE 2 (continued)

Crop Species	Weed Species	Allelochemical Source	Reference(s)
Hemarthria altissima	*Desmodium intortum*	Plant residues	Young and Bartholomew, 1981.
Hordeum vulgare	*Brassica kaber* *Stellaria media* *Descurainia sophia* *Thlaspi arvense* *Bromus tectorum* and other weed species	Aqueous extracts, root exudates	Overland, 1966; Prutenskaya, 1974; Christensen, 1995; Moyer and Huang, 1997; Baghestani et al., 1999.
Ipomoea batatas	*Cyperus esculentus* *Medicago sativa* and other weed species	Root periderm extracts, Aqueous and methanolic extracts, soil	Harrison and Peterson, 1986; Peterson and Harrison, 1991.
Lens culinaris	*Descurainia sophia* *Thlaspi arvense* *Bromus tectorum*	Aqueous extracts	Moyer and Huang, 1997.
Lupinus albus	*Chenopodium album* *Amaranthus retroflexus*	Root secretions	Dzyubenko and Petrenko, 1971.
Manihot esculenta	Weeds.	Fresh leaves	Altieri and Doll, 1978.
Medicago sativa	*Chenopodium album* *Amaranthus* spp. *Echinochloa crus-galli* *Bromus secalinus* *Taraxacum officinale* *Daubentonia punicea*	Plant residues and root saponins	Waller, Jurzysta, and Thorne, 1993; Chung and Miller, 1995c; Miller, 1996.
Melilotus alba	*Stellaria media* *Capsella bursa-pastoris* *Nicotiana tabacum*	Aqueous extracts	Overland, 1966.
Nicotiana tabacum *N. rustica*	*Setaria italica* *Chamomilla recutita* and other weeds	Root exudates, aqueous extracts	Bondev, Panaiot, and Mariana, 1983; Ul Haq and Hussain, 1979.
Oryza sativa (cultivars IR29, IR38, Brown Gora, YH1)	*Echinochloa crus-galli* *Heteranthera limosa* *Ammannia coccinea* *Bacopa rotundifolia* *Cyperus iria* *Cyperus difformis*	Rice straw, field grown rice seedlings and plants, aqueous extracts	Olofsdotter, Navarez, and Moody, 1995; Olofsdotter and Navarez, 1996; Olofsdotter et al., 1999.
Panicum antidotale	Grasses, *Setaria italica* *Sorghum almum*	Aqueous extracts, root exudates	Begum and Hussain, 1980.
Phaseolus vulgaris	*Portulaca oleracea* *Echinochloa crus-galli* *Amaranthus leucocarpus*	Fresh Leaves, leachates	Altieri and Doll, 1978; Anaya et al., 1987
Pisum sativum	*Chenopodium album*	Mature plants	Neustruyeva and Dobretsova, 1972.
Secale cereale	*Stellaria media* *Avena fatua* *Amaranthus retroflexus* *Chenopodium album* *Ambrosia artemisiifolia* *Echinochloa crus-galli* *Tetradymia* spp. *Ipomoea purpurea* *Amaranthus albus* *Brassica kaber* var. *pinnatifida*	Aqueous extracts, root exudates, growing plants, residues, plant phenolics	Osvald, 1953; Overland, 1966; Prutenskaya, 1974; Shilling, Liebl, and Worsham, 1985; Barnes, Putnam, and Burke, 1986; Pérez and Orñéno-Nuñez, 1991; Prezepiorkowski and Gorski, 1994; Mwaja, Masiunas, and Weston, 1995; Lehman and Blum, 1997.

Crop Species	Weed Species	Allelochemical Source	Reference(s)
Sorghum bicolor	*Eragrostis tef* *Lemna minor* *Abutilon theophrasti* *Setaria viridis* *Amaranthus hybridus* *Striga asiatica*	Root exudates, plant residues, germinating seeds, sorgoleone	Panasiuk, Bills, and Leather, 1986; Putnam, 1986; Netzley et al., 1988; Einhellig and Rasmussen, 1989; Einhellig and Souza, 1992; Nimbal et al., 1996b.
Sorghum sudanense	*Stellaria media*	Aqueous extracts	Overland, 1966.
Trifolium incarnatum	*Ipomoea lacunosa* *Amaranthus retroflexus*	Residue	White, Worsham, and Blum, 1989; Lehman and Blum, 1997.
T. pratense	*Sinapis arvensis*	Residues	Ohno et al., 2000.
T. subterraneum	*Ipomoea lacunosa* *Amaranthus retroflexus*	Residue	Lehman and Blum, 1997.
Triticum aestivum	*Ipomoea purpurea* *Amaranthus albus* *Ipomoea hederacea* *Brassica kaber* var. *pinnatifida* *Avena fatua* *Chenopodium album*	Wheat straw and residues, root exudates, plant phenolics	Neustruyeva and Dobretsova, 1972; Prutenskaya, 1974; Pérez and Orméno-Nuñez, 1991; Blum et al., 1992; Lehman and Blum, 1997; Baghestani et al., 1999.
Vicia villosa	*Ipomoea lacunosa* 13 Weed species	Residues, seed excretions	Lazauskas and Balinevichinte, 1972; White, Worsham, and Blum, 1989.
Zea mays	*Chenopodium album* *Amaranthus retroflexus* *Amaranthus leucocarpus* *Echinochloa crus-galli* *Portulaca oleracea* and other weeds	Fresh foliage, root secretions, pollen allelochemicals	Dzyubenko and Petrenko, 1971; Altieri and Doll, 1978; Anaya et al., 1987, 1992.

fied β-phenyllactic acid (β-PLA) and β-hydroxybenzoic acid (β-HBA) which were water-soluble, leachable and also inhibitory to weeds. Leather (1983) reported that aqueous extracts prepared from two varieties of sunflower inhibited several noxious weeds like velvetleaf (*Abutilon theophrasti*), jimsonweed (*Datura stramonium*), morning glory (*Ipomoea purpurea*) and wild mustard (*Brassica kaber* var. *pinnatifida*). Later, he extended his green house studies to fields comparing the use of sunflower and herbicide S-ethyl dipropyl carbamothioate (EPTC). The results indicated that weed biomass was reduced equally in all plots whether or not herbicide was applied (Leather, 1987). Kohli (1993) demonstrated that aqueous leachates of sunflower have the potential to suppress germination and growth of *Parthenium hysterophorus*, a noxious weed. Their mixture with some herbicides like glyphosate, paraquat, atrazine and 2,4-D not only increases the herbicidal potential but also reduces their level. Einhellig and Rasmussen (1989) based on their observations on sorghum cultivars on broad leaf weeds concluded that allelopathic suppression of weeds can reduce reliance on the use of herbicides.

Increased use of crop residues and reduced level of herbicides because of their complimentary action can be an important option for sustainable weed management. Thus, allelopathy can better be capitalized if it is studied simultaneously with various environmental interactions.

CROP AUTOTOXICITY

The allelochemicals released from a crop plant affecting the same crop is commonly termed as *Autotoxicity* or *Autointoxication* which serves as a cause or consequence of *soil sickness* or *soil fatigue* (Singh, Batish, and Kohli, 1999). The phenomenon is prevalent in those fields where same crop is repeatedly grown as monoculture and the residues of the previous harvest are left in the fields for decomposition under no-till or reduced-till crop practices, or even under stubble mulch farming. The crops (Table 3) exhibiting this phenomenon include *Cucumis sativus* (Yu and Matsui, 1997), *Medicago sativa* (Miller, 1983), *Asparagus officinalis* (Shafer and Garrison, 1986), *Oryza sativa* (Chou, 1995), *Triticum aestivum* (Kimber, 1973) and *Zea mays* (Yakle and Cruse, 1983, 1984), and vegetable crops (Yu, 1999). In case of wheat the stubbles or the straw of the crop are left in the fields mainly with the objective of improving soil moisture and preventing erosion. However, phytotoxins released from the stubbles/plant residues adversely affect the germination and growth of wheat itself, if grown in sequence (Guenzi and McCalla, 1966; Kimber, 1973; Lodhi, Bilal, and Malik, 1987; Young, Thorne, and Waller, 1989). The allelochemicals like *p*-coumaric, *p*-hydroxybenzoic, vanillic, and syringic acids have been identified from the straw residues (Guenzi and McCalla, 1966). Likewise, in corn continuous sole cropping in monoculture results in decline in its yield due to release of allelochemicals from the decomposing residues (Yakle and Cruse, 1983, 1984; Turco et al., 1990; Crookston et al., 1991). Turco et al. (1990) have pointed out that decomposing residues may even alter the microbial composition of the soil which may not be beneficial to the crop. Similar observations have been made in case of rice (*Oryza sativa*) and sugarcane (*Saccharum officinarum*) fields (Chou, 1995). Chou and Lin (1976) and Chou (1995) pointed out that allelochemicals from the decomposing residues and altered microbial composition are responsible for the decline in the crop yield. In some cases the autotoxins (allelochemicals showing autotoxicity) show synergism with the plant pathogens and thus enhance the disease symptoms, e.g., in *Asparagus* (Peirce and Colby, 1987). In still other cases the uptake of the nutrients is also affected by the autoallelochemicals (Yu and Matsui, 1997).

TABLE 3. List of crop plants showing autotoxicity.

Crop Plant	Source of Allelochemical	Reference(s)
Asparagus officinalis	Aqueous extracts, root exudates, root-incorporated soil, ferulic acid	Tang and Young, 1982; Yang, 1982; Young, 1984; Young and Chou, 1985; Shafer and Garrison, 1986; Hazebroek, Garrison, and Gianfagna, 1989; Wacker, Safir, and Stephens, 1990.
Avena sativa	Aqueous extracts	Kato-Noguchi et al., 1994.
Brassica oleracea var. *Italica*	Aqueous extracts	Santos and Leskovar, 1997.
B. napus	Microspore cultures	Kott et al., 1988.
Bromus inermis	Aqueous extracts, roots mixed in soil	Bennedict, 1941; Grant and Sallans, 1964.
Cenchrus ciliaris	Root exudates, water extracts	Akhtar, Naqvi, and Hussain, 1978.
Chloris gayana	Root exudates, shoot extracts	Khanum, Hussain, and Naqvi, 1979.
Chrysopogon aucheri	Root exudates and water extracts	Akhtar, Naqvi, and Hussain, 1978.
Coffea arabica	Seeds	Friedman and Waller, 1983.
Colocasia esculenta	Residue extracts	Pardales and Dingal, 1988.
Cucumis sativus	Root exudates, phenolic acids	Yu and Matsui, 1997; Politycka, Wojcik-Wojtkowiak, and Pudelski, 1984.
Helianthus annuus	Extracts, root exudates, leachates, soil	Wilson and Rice, 1968.
Hordeum vulgare	Continuous cropping	Muller-Wilmes and Zoschke, 1980.
Ipomoea batatas	Root periderm extracts	Peterson and Harrison, 1991.
Linum usitatissimum	Residues	Gubels and Kenaaschuk, 1989.
Medicago sativa	Aqueous extracts, soil, soil extracts, dried roots and shoots, whole plants, residues	Grant and Sallans, 1964; Webster, Khan, and Moore, 1967; Jensen et al., 1981; Miller, 1983; Miller et al., 1988; Hall and Henderlong, 1989; Hegde and Miller, 1990; Chung and Miller, 1995a,b; Chung et al., 2000.
Nicotiana tabacum	Decomposing residues	Patrick and Koch, 1958.
Oryza sativa	Decomposing straw, plant parts, soil, phenolics, root exudates	Sadhu and Das, 1971; Chandramohan, Purushothaman, and Kothandaraman, 1973; Chou, Lin, and Kao, 1977; Sadhu, 1975; Chou and Chiou, 1979; Chou et al., 1987; Tapaswi et al., 1991; Chou, 1995.
Panicum antidotale	Root exudates	Khanum, Hussain, and Naqvi, 1979.
Parthenium argentatum	Mature plants	Bonner and Galston, 1944.
Phalaris arundinacea	Aqueous extracts	Grant and Sallans, 1964.
Phaseolus vulgaris	Fresh leaves	Altieri and Doll, 1978.
Phleum pratense	Aqueous extracts	Grant and Sallans, 1964.
Saccharum officinarum	Old fields	Wang, Yang, and Chuang, 1967.

TABLE 3 (continued)

Crop Plant	Source of Allelochemical	Reference(s)
Solanum melongena	Roots mixed into soil	Lee, Im, and Bark, 1967.
Spinacea oleracea	Roots	Litav and Isti, 1974.
Trifolium pratense	Mature plants, aqueous extracts	Grant and Sallans, 1964; Tamura et al., 1967, 1969; Katznelson, 1972.
Triticale spp.	Aqueous chaff extracts	Salmon et al., 1986.
Triticum aestivum	Aqueous extracts, decaying residues, straw, allelochemicals, rhizosphere soil	Guenzi and McCalla, 1962; Guenzi, McCalla, and Norstadt, 1967; Kimber, 1967, 1973; Cochran, Elliott, and Papendick, 1977; Lynch, 1977, 1978; Tang and Waiss, 1978; Parsad and Rao, 1981; Lovett and Jessop, 1982; Lynch and Elliot, 1983; Hall et al., 1986; Srivastava, Totey, and Prakash, 1986; Lodhi, Bilal, and Malik, 1987; Waller et al., 1987; Azm and Alam, 1989; Young, Thorne, and Waller, 1989; Thorne et al., 1990.
Vigna radiata	Aqueous extracts	Chou et al., 1995; Waller et al., 1995.
Zea mays	Aqueous extracts, residues, old fields	Altieri and Doll, 1978; Yakle and Cruse, 1983; Martin, McCoy, and Dick 1990; Turco et al., 1990.

ALLELOCHEMICALS FROM CROP PLANTS

Crop plants are known to release a diversity of allelochemicals into environment which includes phenolics (Guenzi and McCalla, 1966; Rice, 1984; Wojeik-Wojtkowiak et al., 1990; Chou, 1995) or terpenoids (Macias et al., 1993) (Table 4). They generally represent secondary metabolites and provide plants several defense strategies. However, some crops are less active allelopathically as compared to their wild relatives or accessions and co-existing weeds because they are highly selected for physiological specialization of enhancing yields rather than developing endogenous chemicals for defense purposes (Lovett, Fraser, and Duffield, 1982; Lovett, 1985). In this process genes encoding allelopathic substances are lost gradually. However, this lost trait can be re-introduced by traditional breeding methods and by employing molecular-genetic techniques (Foley, 1999). Upon release into crop environment the nature and concentration of allelochemicals may change because of complex environmental conditions and microbial action. However, their action is often synergistic and their detrimental effects are dependent upon several abiotic and biotic factors (Einhellig, 1996). Noteworthy among crop allelochemicals are the hydroxamic acids which are found in members of

TABLE 4. List of some potential allelochemicals (other than phenolic acids) identified from some crop plants.

Allelochemical	Crop Plant	Reference(s)
Scopoletin	*Avena sativa* *Hordeum vulgare* *Triticum aestivum*	Fay and Duke, 1977; Baghestani et al., 1999.
Glucosinolates (4-pentenyl, 2-hydroxybutenyl, 3-butenyl, and 2-hydroxypentenyl), AITC (Allyl isothiocyanate), ITC (Isothiocyanate), OZT (oxazolidinethione)	*Brassica* spp. *B. nigra* *B. juncea* *B. napus*	Eberlein et al., 1998; Vaughn and Boydston, 1997; Vaughn, 1999.
Caffeine, theobromine	*Coffea arabica*	Friedman and Waller, 1983; Suzuki, Ashihara, and Waller, 1992.
Gossypol	*Gossypium hirsutum*	Hedin and McCarty, 1994.
Annuolides A-E, tambulin, kukulcanin B, heliannone A-C, guaianolides 1-6, germacranolides 7 and 15, heliangolides 8-11 and 13, melampolide 16, cis,cis-germacradienolide 14	*Helianthus annuus*	Macias et al., 1993, 1996, 1997.
Gramine, Hordenine	*Hordeum vulgare*	Hanson et al., 1981; Liu and Lovett, 1993.
Lepidimoide	*Lepidium sativum*	Yamada, Anai, and Hasegawa, 1995.
Medicarpin and 4-Methoxy medicarpin, Medicagenic acid, canavanine, saponins, Chlorogenic acid	*Medicago sativa*	Oleszek and Jurzysta, 1987; Miller et al., 1988; Dornbos, Spencer, and Miller, 1990; Gorski, Miersch, and Ploszynski, 1991; Waller, Jurzysta, and Thorne, 1993; Miller, 1996; Chung et al., 2000.
L-3,4-Dihydroxyphenyl alanine (L-DOPA)	*Mucuna pruriens* var. *utilis*	Fujii, 1999a.
2,4-Dihydroxy-1,4(*2H*)-benzoxazin-3-one (DIBOA), 2(*3H*)-Benzoxazolinone (BOA), 2,4-dihydroxy-7-methoxy-1,4-benzoxazin-3-one (DIMBOA)	*Secale cereale* *Triticum aestivum* *Zea mays* *Hordeum* spp. (wild barley)	Tipton et al., 1967; Barnes, Putnam, and Burke, 1986; Barnes and Putnam, 1987; Perez, 1990; Barria, Copaja, and Niemeyer, 1992; Mwaja, Masiunas, and Weston, 1995; Cambier, Hance, and Hoffmann, 2000.
Sorgoleone	*Sorghum bicolor*	Netzley and Butler, 1986; Einhellig and Souza, 1992.
Isoflavonoids (Ononin, Genistein, Biochanin A, Biochanin A-7-glucoside, Formononetin)	*Trifolium pratense*	Tamura et al., 1967, 1969.
Sterols and Ketosteroids	*Triticum aestivum*	Gaspar, Pereira, and Neves, 1999.
C-glycosyl flavonoids	*Vigna radiata*	Chou, 1995.

family Poaceae (like wheat, rye and maize) as glucosides of benzoxazinones (Cambier, Hance, and Hoffmann, 2000), and sorgoleone–a long chain hydroquinone from *Sorghum* spp. (Netzley and Butler, 1986). Cruciferous plants produce characteristic volatile glucosinolates which provide them excellent pesticidal property to be used as green manure and biofumigants (Olivier et al., 1999; Vaughn, 1999). The concentration of allelochemicals varies with age, cultivar, and plant organ, and their amount is often enhanced upon various biotic and abiotic stress factors (Ben-Hammouda et al., 1995; Einhellig, 1996; Nimbal et al., 1996a; Burgos, Talbert, and Mattice, 1999; Kato-Noguchi, 1999; Olivier et al., 1999; Cambier, Hance, and Hoffmann, 2000). Mechanism of action of crop allelochemicals is least understood because of their complex nature and origin, and synergistic action. However, some work has been done with sorgoleone which is known to inhibit respiration and photosynthesis by acting as PSII inhibitor (Rasmussen et al., 1992; Einhellig et al., 1993; Nimbal et al., 1996b; Gonzalez et al., 1997).

COVER CROPS/SMOTHER CROPS/GREEN MANURES IN RELATION TO ALLELOPATHY

Cover crops are those crops that are grown between periods of regular cropping mainly for the purpose of conserving soil and moisture, improving nutrient recycling, biomass production, lowering of temperature, suppressing noxious weeds and supplying forage especially in emergency (Barnes and Putnam, 1983; Putnam and DeFrank, 1983; Teasdale, 1993; McLenaghen et al., 1996; Swanton and Murphy, 1996; Gallandt, Liebman, and Huggins, 1999). The use of cover crops has been an age-old practice and was grown extensively in traditional agriculture. However, it found a little use with the discovery of synthetic herbicides that revolutionized agriculture by enhancing the crop yields.

Some important cover crops, viz., barley (*Hordeum vulgare*), sorghum (*Sorghum* spp.), corn (*Zea mays*), wheat (*Triticum aestivum*), rye (*Secale cereale*), buckwheat (*Fagopyrum esculentum*), velvetbean (*Mucuna pruriens*), crimson clover (*Trifolium incarnatum*), subterranean clover (*T. subterraneum*), hairy vetch (*Vicia villosa*) and *Ipomoea batatas* and *I. tricolor* possess weed suppressing ability (Gliessman and Garcia, 1979; Smeda and Putnam, 1988; Einhellig and Rasmussen, 1989; Anaya et al., 1990; Fujii, Shibuya, and Yasuda, 1991, 1992; Worsham and Blum, 1992; Dyck, Liebman, and Erick, 1995; Fujii, 1999b). Generally, the cover crops either frozen or killed with desiccants (like herbicide glyphosate) are helpful in killing weeds (Putnam, DeFrank, and Barnes, 1983). Weed suppression by cover crops can be brought about by either physical barrier reducing light intensity and lowering temperature (Teasdale, 1993) or by allelochemicals released from the residues (Putnam, 1994), and from microbes stimulated by cover crops (McCalla and Haskins, 1964); however, for better management these must be separated (Yenish, Worsham, and Chilton, 1995).

Cover crops are also frequently known as *Smother Crops* which suppress weeds due to their over-shadowing effect (Overland, 1966). They shade out weeds by their thick and dense stands and fast growth (Foley, 1999). Some important smother crops are alfalfa (*Medicago sativa*), buckwheat (*Fagopyrum esculentum*), foxtail millet (*Setaria italica*), rye (*Secale cereale*), and sorghum (*Sorghum* spp.). Some cover crops like legumes (*Mucuna* spp., *Canavalia* spp., *Trifolium* spp.), cruciferous plants (*Brassica* spp.) and *Ipomoea* spp. improve soil through addition of organic matter and nitrogen (in case of legumes) besides controlling weeds and are thus known as *Green Manures*.

Rye is an excellent example of cover/smother/green manure crop and weed suppression by rye cover has been shown to last for 4 weeks (Barnes and Putnam, 1987; Wagner-Riddle et al., 1997). Smeda and Weller (1996) reported that rye as a cover crop provides an effective suppression of weed growth in transplant tomatoes. Both root and shoot residues of rye have been found to decrease growth of *Echinochloa crus-galli* though root residues provided better and effective means of control (Hoffman et al., 1996a,b). Eskelson and Crabtree (1995) reported that use of buckwheat as cover crop effectively controls the weed Canada thistle (*Cirsium arvense*). Similarly, wheat grown as cover crop reduces the growth of weeds like *Echinochloa crus-galli*, *Heternathera limosa*, *Ammania coccinea* and *Cyperus esculentus* (Jordan et al., 1999) and under some conditions also reduced the need for in-season herbicides (Worsham, 1991; Yenish, Worsham, and York, 1996; Jordan et al., 1999). However, integrated weed management tactics such as reduced post-emergence herbicides and inter-row cultivation are required (William, Mortensen, and Doran, 1998; Shilling et al., 1995).

Although cover crops are very useful but there are several problems linked with their use and hence they have to be managed. Cover crops may interfere with the crop production, the allelochemicals released from them or their residues cause stress, increase incidence of insects or pests or compete with water in some drier areas. Phytotoxic leachates from cover crop residues of rye and subterranean clover reduce growth of several vegetables (Stirzaker and Bunn, 1996). Thus, understanding the interactions of the cover crops with various biotic and abiotic factors of agroecosystems becomes essential, if their use is revived in a big way.

ALLELOPATHY AND CROP ROTATION

Crop rotation is one of the traditional practice where some crops particularly the legumes are grown in short rotation with the main crops and has also been suggested as a one of the possible means to overcome the problem of soil sickness (de Candolle, 1832). It also minimizes the incidence of weeds and other pests in the soil and hence, enhances crop productivity. This practice like other traditional agricultural practices declined with the onset of synthetic chemicals. However, in the recent years it has resurfaced in order to maintain the sustainability of the agroecosystems (Wilson et al., 1986; Edwards, Thurlow, and Eason, 1988; Schreiber, 1992). In case of corn, it has been reported that there is 7-10% more yield if corn is rotated with soybean (*Glycine max*) than in corn-corn continuous cropping system (Kalantari, 1981). Likewise, Horst and Haerdter (1994) reported that rotation of maize with cowpea increases the yield and nutrient use efficiency of maize compare to sole cropping of maize. Besides, rotation of corn-soybean also provides economical and environmental advantages over monoculture corn (Kessavalou and Walters, 1997). Under no-till system or in the plowed plots, similar observation on corn-soybean rotation were made by Turco et al. (1990) and Crookston et al. (1991). These workers attributed the decline in corn yield in continuous cropping to allelochemicals released by decomposing corn residues and change in microbial population in the soil that is not beneficial for the crop. In another study pertaining to corn under no-till or reduced till-system, there was significant decrease in the populations of giant foxtail (*Setaria faberi*) weed when soybean-wheat-corn rotation was practiced thus bringing a significant drop in the input of herbicides (Schreiber, 1992).

Crop rotation practice can also substantially decrease the severity of plant diseases on crops. The severity of root rot disease (*Thielaviopsis basicola*) of tobacco (*Nicotiana tabacum*) was very less if tobacco-ryegrass-corn rotation was followed (Patrick, 1971). The incidence of *Fusarium* wilt was also substantially low when cotton-peppermint rotation is practiced (Li, 1988). In this context the allelochemicals released during decomposition of residues play an important role in maintaining the populations of plant pathogens at a low level particularly those borne in soil. In some cases, however, crop productivity may also be affected by the allelochemicals particularly the phenolics (Putnam and Duke, 1974; Chou and Lin, 1976; Barnes and Putnam, 1987). Therefore, research is required to work out the levels of crop residues being rotated so as to achieve the maximum benefit.

CROP RESIDUES AND ALLELOPATHY

Plant residues have been defined as plant or their parts left in the field for decomposition after the crop has been harvested/thrashed (Kumar and Goh, 2000). These material were earlier regarded as waste products which needs disposal, however, now these are considered as natural resource rather than waste because of their usefulness to the fields. They are now deliberately left in the fields under reduced- or no-till agriculture. Their presence improves the physical, chemical and biological properties. They enhance mineral cycling, particularly nitrogen when the crop is legume, conserve soil and moisture, and suppress weeds and pests under some situations. Their decomposition is brought about by the microbial action thus releasing a number of carbon end products and several other nutrients to the soil system. Their decomposition depends on the kind of plant residue, edaphic factors of the area and management factors (Kumar and Goh, 2000). Toxicity of residues decreases with passage of time (Bradow and Connick, 1990; Purvis, 1990).

Lynch (1983) reported that the decomposition of residues of alfalfa, corn, and soybean brought about by anaerobic microorganisms may lead to the production of low molecular weight acids, viz., acetic, propionic and butyric acid, which effect the succeeding crops in the next growing season (Voroney et al., 1991). The extent of inhibition depends upon combination and type of acids, acidity and the hybrids; therefore, evaluation of multiple cultivars and hybrids has been suggested as the possible remedy (Janovicek et al., 1997).

The main concern regarding the crop residues is their allelopathic effect on the other or same crop plant (Patrick and Koch, 1958; Guenzi and McCalla, 1962; Megie, Pearson, and Hiltbold, 1967; Horricks, 1969; Patrick, 1971; Kimber, 1973; Lynch, 1977; Vera, McGegor, and Downey, 1987; Waller et al., 1987; Thorne et al., 1990). The decomposing crop residue release a variety of allelochemicals, particularly the phenolics, in the soil causing adverse effects on the other plants (Guenzi and McCalla, 1966; Vaughan, Sparling, and Ord, 1983; Nelson, 1996). They are also known to reduce nodulation of legumes and biological nitrogen fixation due to their allelopathic effects (Rice, 1971, 1984; Putnam and Weston, 1986; Heckman and Kluchinski, 1995). In some cases their excessive presence has been seen to decrease even herbicide efficiency (Rule, 1991).

The presence of crop residues on the soil surface as mulch can suppress weeds and thus they can be helpful in reducing reliance on herbicides (Worsham, 1991; White, Sully, and Melville, 1989; Facelli and Pickett, 1991; Teasdale, Beste, and Potts, 1991; Weston, 1996). However, much depends on the amount and type of crop residues and several interacting environmental factors. Crop residues can, therefore, be very useful in maintaining the sustainability of agroecosystems, provided they are efficiently managed.

CROP ALLELOPATHY AND WEED AND PEST MANAGEMENT

In spite of the best efforts of man pests pose a serious threat to agriculture worldwide. Nearly 10% loss of agricultural production is caused by weeds alone (Altieri and Liebman, 1988). Weeds are unwanted, fast growing plants that compete with crops for resources like nutrients, water and space. Besides, they also exhibit allelopathic phenomenon leading to decrease in crop yield (Kohli, Batish, and Singh, 1998). For managing weeds in agroecosystems, large quantities of synthetic chemicals are used which although help in enhancing the crop productivity, yet seriously affect the quality of environment and human health. Of late, the use of allelopathic crops for managing weeds is being realized fast because many of them have potential to manage weeds and other pests.

Use of allelopathic cover or smother or companion crops are currently being tested and genetically improved for the purpose of pest management (Kohli, Batish, and Singh, 1998; Foley, 1999; Kohli et al., 1999; Wu et al., 1999). Rye (*Secale cereale*), alfalfa (*Medicago sativa*), sorghum (*Sorghum* sp.), buckwheat (*Fagopyrum cymosum*), sunflower (*Helianthus annuus*) and several other cruciferous plants are used as smother/cover/companion crops for weed management (Putnam and DeFrank, 1983; Putnam, DeFrank, and Barnes, 1983; Leather, 1987; Oleszek, 1987; Kohli, 1993; Lehman and Blum, 1997; Foley, 1999). However, weed suppressing ability of cover crops depends upon several factors like type and age of cultivar, nature (living or dead), spacing, type of weed species and weed density, and interactions between allelochemicals and soil factors (Fay and Duke, 1977; Hanson et al., 1981; Peters and Zam, 1981; Purvis and Jones, 1990; Blum et al., 1992; Lehman and Blum, 1997); and for their widespread recommendation, studies especially genetic approaches for their improvement, are required so as to bring about effective pest suppression. Generally, the crop cultivars or genotypes with more allelochemicals are considered to be highly competitive and can suppress weeds more than those which are less competitive (Christensen, 1995; Lemerle, 1996). Cereal crops, viz., wheat, oat and barley exhibiting more competitiveness have been found to contain higher amounts of vanillic and *o*-coumaric acid, and scopoletin and these can be used as possible indicators for selecting potential genotypes through breeding to be used under integrated weed management (IWM) programs (Baghestani et al., 1999).

Subterranean clover (*Trifolium subterraneum*) has a potential to suppress noxious weeds like *Panicum dichotomiflorum* and *Ipomoea hederacea* when used as living mulch in no-till system (Enache and Ilnicki, 1990). Residues from cover crops like barley, rye, triticale, wheat and hairy vetch reduce seedling emergence of weed such as *Amaranthus* spp. and *Setaria* spp. (Williams, Mortensen, and Doran, 1998). Residues and debris from legume crops like *Trifolium* spp. have a great potential for weed management (Dyck and Liebman, 1994; Liebman and Ohno, 1997; Ohno et al., 2000). *Pteridium aquilinum*-fern, can also be used as green manure or the cover crop for suppressing weeds and also phytopathogenic fungi (Nava, Fernandez, and del Amo, 1987).

Several members of family brassicaceae, viz., *Brassica hirta*, *B. juncea*, *B. nigra*, *B. campestris*, *B. napus* and *Lepidium sativum* provide an excellent source of green manure and can effectively be used for weed and pest management. These possess strong odor due to volatile glucosinolates which are phytotoxic in nature and upon enzymatic hydrolysis produce isothiocyanates which are promising weed suppressants (Vaughn and Boydston, 1997) and can be used as bioherbicides (Brown and Morra, 1995). Winter rape (*Brassica napus*) used as a green manure has a potential for controlling weeds and other pests besides controlling soil erosion (Boydston and Hang, 1995; Eberlein et al., 1998; Krishnan, Holshouser, and Nissen, 1998). Glucosinolates are found to be major chemical constituents from its roots and shoots and their concentration varies with cultivars. Cultivars and varieties of winter rape with higher amount of glucosinolates have been found to be better biofumigants (Eberlein et al., 1998).

DIBOA–a type of hydroxamic acid and an important allelochemical of rye, corn, barley, wheat, etc., has a strong activity against a number of diseases, pests and herbivores (Niemeyer, 1988).

Several wild varieties of modern day plants, viz., *Avena* sp. (Fay and Duke, 1977), *Cucumis sativus* (Putnam and Duke, 1978), *Glycine max* (Massantini, Caporali, and Zellini, 1977), *Helianthus annuus* (Leather, 1987; Kohli, 1993), *Brassica* sp. (Sarmah, Narwal, and Yadava, 1992) and *Oryza sativa* (Dilday et al., 1998; Olofsdotter et al., 1999) possess allelopathic traits that impart them greater resistance against weeds and pests. However, during the process of cultivation, the genes responsible for encoding allelochemicals got eliminated because they were never preferred to the ones encoding quality and quantity of produce. Nowadays, it is being realized that if these genes are incorporated in modern day cultivars, they can have self-protection against weeds and pests. Several genetic and molecular level studies are being undertaken to transfer the genes encoding allelochemicals through recombinant DNA technology and several other modern techniques.

Crop plants possess a diversity of allelochemicals which can be utilized as promising pesticides. Their use is considered to be environmentally safe besides they also provide new molecular targets, unlike the synthetic herbicides (Nimbal et al., 1996a; Duke et al., 1997, 2000; Macias et al., 1999).

Allelopathy can also be exploited for controlling plant diseases (Rice, 1995). Allelopathic crops, particularly the crucifers, bear a great significance in this regard. Their use reduces the disease incidence effectively particularly those caused by soil borne pathogens. Severity of root rot of peas caused by *Aphanomyces euteiches* was reduced when crucifers were introduced as green manure (Muehlchen, Rand, and Parke, 1990; William-Woodward et al., 1997). Among the various cruciferous crops white mustard (*Sinapis alba*) proved to be most promising (Muehlchen, Rand, and Parke, 1990). Isothiocyanates derived from myrosinase catalyzed hydrolysis of glucosinolates commonly found in crucifers inhibit mycelial growth of *Botrytis cinerea*, *Rhizopus stolonifer*, *Monilinia laxa*, *Mucor piriformis* and *Penicillium expansum* all fungal pathogens of fruit and vegetable crops (Mari et al., 1993). Species of *Brassica* with high concentration of allyl-isothiocyanate (AITC)–a breakdown product of allylglucosinolate, suppress *Fusarium sambucinum*–a cause of potato tuber dry rot (Mayton et al., 1996). To exploit this vast potential of allelochemicals combined efforts of scientists from different fields are needed in order to make further breakthrough in their wide spread use as pesticides.

Biotechnological approaches employing tissue or cell culture, through DNA recombinant technology, microbial fermentation and several other upcoming molecular-genetic techniques like Polymer Chain Reaction (PCR) and Random Amplified Polymorphic DNA (RADP) can help a lot for manipulating and improving crop production. This will enable the crop plants to resist the harmful effects of weeds, nematodes, pests and insects. Another biotechnological approach could be the synthesis of herbicide and pesticide based on natural product chemistry. This has opened a new era in weed and pest management programs. Several compounds have been commercialized and many more are in the process of commercialization (Kohli, Batish, and Singh, 1998). The use of allelochemicals as growth regulators is also an important step for the sustainable agriculture. Though this aspect has been studied to a lesser extent yet some of the promising chemical compounds in this direction are the sesquiterpene lactones (Macias et al., 1996; Batish et al., 1997).

REFERENCES

Akhtar, N., H.H. Naqvi, and F. Hussain. (1978). Biochemical inhibition (allelopathy) exhibited by *Cenchrus ciliaris* Linn. and *Chrysopogon aucheri* (Bioss.) Stapf. *Pakistan Journal of Forestry* 28: 194-200.

Altieri, M.A. and M. Liebman. (1988). *Weed Management in Agroecosystems: Ecological Approaches*. Florida: CRC Press.

Altieri, M.A., and J.D. Doll. (1978). The potential of allelopathy as a tool for weed management in crops. *Proceedings of National Academy of Sciences, U.S.A.* 24: 495-502.

Anaya, A.L. (1999). Allelopathy as a tool in the management of biotic resource in agroecosystems. *Critical Reviews in Plant Sciences* 18: 697-739.

Anaya, A.L., B.E. Hernandez-Bautista, M. Jimenez-Estrada, and L. Velasco-Ibarra. (1992). Phenylacetic acid as a phytotoxic compound of corn pollen. *Journal of Chemical Ecology* 18: 897-905.

Anaya, A.L., L. Ramos, R. Cruz, J.G. Hernandez, and V. Nava. (1987). Prospectives on allelopathy in Mexican traditional agroecosystem: A case study in Tlaxcala. *Journal of Chemical Ecology* 13: 2083-2101.

Anaya, A.L., M.R. Calera, R. Mata, and R. Pereda Miranda. (1990). Allelopathic potential of compounds isolated from *Ipomoea tricolor* Cav. (Convolvulaceae). *Journal of Chemical Ecology* 16: 2145-2152.

Anaya, A.L., S. Del Amo, G. Ruy-Ocotla, and L.M. Ortiz. (1978). Allelopathic potential of a coffee plantation. In *2nd International Congress of Ecology*. Jerusalem, Israel. pp. 8 (Abstract).

Azm, A.R. and S.M. Alam. (1989). Effects of some wild plant residues and wheat straw on germination and growth of wheat cultivars. *Cereal Research Communications* 17: 59-62.

Baghestani, A., C. Lemieux, G.D. Leroux, R. Baziramakenga, and R.R. Simard. (1999). Determination of allelochemicals in spring cereal cultivars of different competitiveness. *Weed Science* 47: 498-504.

Barnes, J.P. and A.R. Putnam. (1983). Rye residues contribute weed suppression in no-tillage cropping systems. *Journal of Chemical Ecology* 9: 1045-1057.

Barnes, J.P. and A.R. Putnam. (1987). Role of benzoxazinones in allelopathy by rye (*Secale cereale* L.). *Journal of Chemical Ecology* 13: 889-906.

Barnes, J.P., A.R. Putnam, and B.A. Burke. (1986). Allelopathic activity of rye (*Secale cereale* L.) In *The Science of Allelopathy*, eds. A.R. Putnam and C.S. Tang. New York: John Wiley and Sons, pp. 271-286.

Barnes, J.P., A.R. Putnam, B.A. Burke, and A.J. Aasen. (1987). Isolation and characterization of allelochemicals in rye herbage. *Phytochemistry* 26: 1385-1390.

Barria, B.N., S.V. Copaja, and H.M. Niemeyer. (1992). Occurrence of DIBOA in wild *Hordeum* species and its relation to aphid resistance. *Phytochemistry* 31: 89-91.

Batish, D.R., R.K. Kohli, D.B. Saxena, and H.P. Singh. (1997). Growth regulatory response of parthenin and its derivatives. *Plant Growth Regulation* 21: 189-194.

Beckmann, E.O. and K. Noffke. (1978). The effect on lettuce of addition of radish plant residues to various substances. *Gartenbauwissen chaft* 43: 70-78.

Begum, I. and F. Hussain. (1980). Allelopathic effects of *Panicum antidotale* Retz. *Pakistan Journal of Scientific Industrial Research* 23: 182-188.

Ben-Hammouda, M., R.J. Kremer, and H.C. Minor. (1995). Phytotoxicity of extracts from *Sorghum* plant components on wheat seedlings. *Crop Science* 35: 1652-1656.

Ben-Hammouda, M., R.J. Kremer, H.C. Minor, and M. Sarwar. (1995). A chemical basis for differential allelopathic potential of *Sorghum* hybrids on wheat. *Journal of Chemical Ecology* 21: 775-785.

Bennedict, H.M. (1941). The inhibitory effect of dead roots on growth of bromegrass. *Journal of American Society of Agronomy* 33: 1108-1109.

Blum, U., T.M. Gerig, A.D. Worsham, L.D. Holappa, and L.D. King. (1992). Allelopathic activity in wheat conventional and wheat no-till soil: Development of soil extract bioassays. *Journal of Chemical Ecology* 18: 2119-2221.

Bondev, I.A., P.P. Panaiot, and I.L. Mariana. (1983). Allelopathic relations between germination of tobacco (*Nicotiana tabacum*) and chamomilla (*Chamomilla recutita*). *Ekologiya (Sofia)* 0: 14-25.

Bonner, J. and A.W. Galston. (1944). Toxic substances from the culture media of guayule which may inhibit growth. *Botanical Gazette (Chicago)* 106: 185-198.

Boydston, R.A. and A. Hang. (1995). Rapeseed (*Brassica napus*) green manure crop suppresses weeds in potato (*Solanum tuberosum*). *Weed Technology* 9: 669-675.

Bradow, J.M. and W.J. Connick, Jr. (1990). Volatile seed germination inhibitors from plant residues. *Journal of Chemical Ecology* 16: 645-666.

Brown, P.D. and M.J. Morra. (1995). Glucosinolate-containing plant tissue as bioherbicide. *Journal of Agricultural and Food Chemistry* 43: 3070-3074.

Burgos, N.R., R.E. Talbert, and J.D. Mattice. (1999). Cultivar and age differences in the production of allelochemicals by *Secale cereale*. *Weed Science* 47: 481-485.

Buta, J.G. and D.W. Spaulding. (1989). Allelochemicals in tall fescue–abscissic and phenolic acids. *Journal of Chemical Ecology* 15: 1629-1636.

Cambier, V., T. Hance, and E. de Hoffmann. (2000). Variation of DIMBOA and related compounds content in relation to the age and plant organ in maize. *Phytochemistry* 53: 223-229.

Chandramohan, D., D. Purushothaman, and R. Kothandaraman. (1973). Soil phenolics and plant growth inhibition. *Plant and Soil* 39: 303-308.

Chapman, S.J. and J.M. Lynch. (1983). The relative role of microorganisms and their metabolites in phytotoxicity of decomposing plant residues. *Plant and Soil* 74: 457-459.

Chattopadhyoy, S.P. (1995). Allelopathic potential of *Solanum myriacanthum* (Solanaceae) in relation to seed germination and seedling growth of mustard (*Brassica*). *Acta Botanica Indica* 23: 29-31.

Chou, C.H. (1995). Allelopathy and sustainable agriculture. In *Allelopathy: Organisms, Processes and Applications*, eds. Inderjit, K.M.M. Dakshini, and F.A. Einhellig. ACS Symposium Series 582. Washington, DC: American Chemical Society, pp. 211-223.

Chou, C.H. (1999). Roles of allelopathy in plant biodiversity and sustainable agriculture. *Critical Reviews in Plant Sciences* 18: 609-636.

Chou, C.H., F.J. Chang, and H.S. Oka. (1991). Allelopathic potentials of wild rice, *Oryza perennis*. *Taiwania* 36: 201-210.

Chou, C.H. and G.R. Waller. (1980). Possible allelopathic constituents of *Coffea arabica*. *Journal of Chemical Ecology* 6: 643-654.

Chou, C.H., G.R. Waller, C.S. Cheng, C.F. Yang, and D. Kim. (1995). Allelochemical activity of naturally occurring compounds from mungbean (*Vigna radiata* L.) plants and their surrounding soil. *Botanical Bulletin of Academia Sinica* 36: 9-18.

Chou, C.H. and H.J. Lin. (1976). Autointoxication mechanism of *Oryza sativa*. I. Phytotoxic effects of decomposing rice residues in soil. *Journal of Chemical Ecology* 2: 353-367.

Chou, C.H., S. Hwang, C. Peng, Y. Wang, F. Hsu, and N. Chung. (1987). The selective allelopathic interaction of a pasture forest intercropping in Taiwan. *Plant and Soil* 98: 31-41.

Chou, C.H. and S.J. Chiou. (1979). Autointoxication mechanism of *Oryza sativa*. II. Effect of culture treatments on the chemical nature of paddy soil and on rice productivity. *Journal of Chemical Ecology* 5: 839-859.

Chou, C.H., T.J. Lin, and C.I. Kao. (1977). Phytotoxins produced during decomposition of rice stubbles in paddy soil and their effect on leachable nitrogen. *Botanical Bulletin of Academia Sinica* 18: 45-60.

Chou, C.H., and Z.A. Patrick. (1976). Identification and phytotoxic activity of compounds produced during decomposition of corn and rye residues in soil. *Journal of Chemical Ecology* 2: 369-387.

Christensen, S. (1995). Weed suppression ability of spring barley varieties. *Weed Research* 35: 241-247.

Chung, I.M. and D.A. Miller. (1995a). Difference in autotoxicity among seven alfalfa cultivars. *Agronomy Journal* 87: 596-600.

Chung, I.M. and D.A. Miller. (1995b). Effect of alfalfa plant and soil extracts on germination and growth of alfalfa. *Agronomy Journal* 87: 762-767.

Chung, I.M., and D.A. Miller. (1995c). Natural herbicide potential of alfalfa residues on selected weed species. *Agronomy Journal* 87: 920-925.

Chung, M., D. Seigler, D.A. Miller, and S.H. Kyung. (2000). Autotoxic compounds from fresh Alfalfa leaf extracts: Identification and Biological activity. *Journal of Chemical Ecology* 26: 315-327.

Cochran, V.L., L.F. Elliott, and R.I. Papandick. (1977). The production of phytotoxins from surface crop residues. *Soil Science Society of America Journal* 41: 903-908.

Crookston, R.K., J.E. Kurle, P.J. Copeland, J.H. Ford, and W.E. Lueschen. (1991). Rotational cropping sequence affects yield of corn and Soybean. *Agronomy Journal* 83: 108-113.

Cruz-Ortega, R., A.L. Anaya, and L. Ramos. (1988). Effects of allelopathic compounds of corn pollen on respiration and cell division of watermelon. *Journal of Chemical Ecology* 14: 71-86.

de Candolle, M.A.P. (1832). *Physiologie Vegetate*. Tome III. Paris: Bechet Jeune, Lib. Fac. Med., pp. 1474-1475.

Dilday, R.H., W.G. Yan, K.A.K. Moldenhauer, and K.A. Gravois. (1998). Allelopathic activity in rice for controlling major aquatic weeds. In *Allelopathy in Rice*, ed. M. Olofsdotter. Manila, Philippines: International Rice Research Institute, pp. 7-26.

Dornbos, D.L., Jr., G.F. Spencer, and R.W. Miller. (1990). Medicarpin delays alfalfa seed germination and seedling growth. *Crop Science* 30: 162-166.

Duke, S.O., F.E. Dayan, A. Hernández, M.V. Duke, and H.K. Abbas. (1997). Natural products as leads for new herbicide modes of action. In *Proceedings of The 1997 Brighton Crop Protection Conference-Weeds. Vol. 2*, eds. K.E. Pallett. Farnham, UK: British Crop Protection Council, pp. 579-586.

Duke, S.O., F.E. Dayan, J.G. Romagni, and A.M. Rimando. (2000). Natural products as sources of herbicides: current status and future trends. *Weed Research* 40: 99-111.

Dyck, E. and M. Liebman. (1994). Soil fertility management as a factor in weed control: the effect of crimson clover residue, synthetic nitrogen fertilizer, and their interactions on emergence and early growth of lambsquarters and sweet corn. *Plant and Soil* 167: 227-237.

Dyck, E., M. Liebman, and M.S. Erich. (1995). Crop-weed interference as influenced by a leguminous or synthetic fertilizer nitrogen source: I. Double cropping experiments with crimson clover, sweet corn and lambsquarters. *Agriculture Ecosystems and Environment* 56: 93-108.

Dzyubenko, N.N. and N.I. Petrenko. (1971). On biochemical interaction of cultivated plants and weeds. In *Physiological-Biochemical Basis of Plant Interactions in Phytocenoses*, ed. A.M. Grodzinsky. Vol. 2, Kiev: Naukova Dumka, pp. 60-66 (In Russian, English summary).

Eberlein, C.V., M.J. Morra, M.J. Guttieri, P.D. Brown, and J. Brown. (1998). Glucosinolate production by five field-grown *Brassica napus* cultivars used as green manures. *Weed Technology* 12: 712-718.

Edwards, J.H., D.L. Thurlow, and J.T. Eason. (1988). Influence of tillage and crop rotation on yield of corn, soybean, and wheat. *Agronomy Journal* 80: 76-80.

Einhellig, F.A. (1985). Effect of allelopathic chemicals on crop productivity. In *Bioregulators for Pest Control*, ed. P.A. Hedin. ACS Symposium Series 276. Washington, DC: American Chemical Society, pp. 109-132.

Einhellig, F.A. (1996). Interactions involving allelopathy in cropping systems. *Agronomy Journal* 88: 886-893.

Einhellig, F.A. and I.F. Souza. (1992). Phytotoxicity of sorgoleone found in grain Sorghum root exudates. *Journal of Chemical Ecology* 18: 1-11.

Einhellig, F.A. and J.A. Rasmussen. (1989). Prior cropping with grain *Sorghum* inhibits weeds. *Journal of Chemical Ecology* 15: 951-960.

Einhellig, F.A., J.A. Rasmussen, A.M. Hejl, and I.F. Souza. (1993). Effects of root exudate Sorgoleone on photosynthesis. *Journal of Chemical Ecology* 19: 369-375.

Ells, J.E. and A.E. McSay. (1991). Allelopathic effects of alfalfa plant residues on emergence and growth of cucumber seedlings. *HortScience* 26: 368-370.

Enache, A.J. and R.D. Ilnicki. (1990). Weed control by subterranean clover (*Trifolium subterraneum*) used as a living mulch. *Weed Technology* 4: 534-538.

Eskelsen, S.R. and G.D. Crabtree. (1995). The role of allelopathy in buckwheat (*Fagopyrum sagittatum*) inhibition of canada thistle (*Cirsium arvense*). *Weed Science* 43: 70-74.

Evenari, M. (1949). Germination inhibitors. *Botanical Review* 15: 153-194.

Facelli, J.M. and S.T.A. Pickett. (1991). Plant litter: light interaction and effect on an old-field plant community. *Ecology* 72: 1024-1031.

Fay, P.K. and W.B. Duke. (1977). An assessment of allelopathic potential in *Avena* germplasm. *Weed Science* 25: 224-228.

Foley, M.E. (1999). Genetic approach to the development of cover crops for weed management. *Journal of Crop Production* 2: 77-93.

Forney, D.R. and C.L. Foy. (1985). Phytotoxicity of products from rhizospheres of a Sorghum-Sudangrass hybrid (*S. bicolor* × *S. sudanense*). *Weed Science* 33: 597-604.

Forney, D.R., C.L. Foy, and D.D. Wolf. (1985). Weed suppression in no-till alfalfa (*Medicago sativa*) by prior cropping of summer annual forage grasses. *Weed Science* 33: 490-497.

Friedman, J. and G.R. Waller. (1983). Caffeine hazards and their prevention in germinating seeds of Coffee (*Coffea arabica* L.). *Journal of Chemical Ecology* 9: 1099-1106.

Fujii, Y. (1999a). Allelopathy of velvetbean: Determination and identification of L-DOPA as a candidate of allelopathic substances. In *Biologically Active Natural Products*, eds. H.G. Cutler and S.J. Cutler. Boca Raton: CRC Press, pp. 33-48.

Fujii, Y. (1999b). Allelopathy of hairy vetch and *Mucuna*: their application for sustainable agriculture. In *Biodiversity and Allelopathy: From Organisms to Ecosystems in the Pacific*, eds. C.H. Chou, G.R. Waller, and C. Reinhardt. Taipei, Taiwan, ROC: Academia Sinica, pp. 289-300.

Fujii, Y., T. Shibuya, and T. Yasuda. (1991). L-3,4-Dihydroxyphenylalanine as an allelochemical candidate from *Mucuna pruriens* (L.) DC. var. *utilis*. *Agricultural and Biological Chemistry* 55: 617-618.

Fujii, Y., T. Shibuya, and T. Yasuda. (1992). Allelopathy of velvetbean: its discrimination and identification of L-DOPA as a candidate of allelopathic substances. *Japanese Agricultural Research Quarterly* 28: 238-247.

Funke, G.L. (1941). Essai de phytosociologie experimentale. *Bulletin of the Soc. Hist. Nat. Toulouse* 76: 19-21.

Gallandt, E.R., M. Liebman, and D.R. Huggins. (1999). Improving soil quality: Implications for weed management. *Journal of Crop Production* 2: 95-121.

Gaspar, E.M.M., M. Pereira, and H.J. Chaves das Neves. (1999). Potential allelopathic sterols and ketosteroids from wheat straw (*Triticum aestivum*). In *Recent Advances in Allelopathy*, eds. F.A. Macias, J.C.G. Galindo, J.M.G. Molinillo, and H.G. Cutler. Vol. 1. Cadiz, Spain: Servicio De Publicaciones–Universidad de Cadiz, pp. 69-80.

Gliesmann, S.R. and E.R. Garcia. (1979). The use of some tropical legumes in accelerating the recovery of productivity of soils in lowland humid tropics of Mexico. In *Tropical Legumes: Resources for the Future*. NAS Publication no. 27. Washington, DC: National Academy of Sciences, pp. 292-303.

Gonzalez, L., X.C. Souto, and M.J. Reigosa. (1997). Weed control by *Capsicum annuum*. *Allelopathy Journal* 4: 101-110.

Gonzalez, V.M., J. Kazimir, C. Nimbal, L.A. Weston, and G.M. Cheniae. (1997). Inhibition of a photosystem II electron transfer reaction by the natural product sorgoleone. *Journal of Agriculture and Food Chemistry* 45: 1415-1421.

Gorski, P.M., J. Miersch, and M. Ploszynski. (1991). Production and biological activity of Saponins and canavanine in alfalfa seedlings. *Journal of Chemical Ecology* 17: 1135-1143.

Grant, E.A. and W.G. Sallans. (1964). Influence of plant extracts on germination and growth of eight forage species. *Journal of British Grassland Society* 19: 191-197.

Grümmer, G. and H. Beyer. (1960). The influence exerted by species of *Camelina* on flax by means of toxic substances. *Symposium of British Ecological Society* 1: 153-157.

Gubels, G.H., and E.O. Kenaaschuk. (1989). Agronomic performance of flax grown on canola, barley and flax stubble with and without tillage prior to seeding. *Canadian Journal of Plant Science* 69: 31-38.

Guenzi, W.D. and T.M. McCalla. (1962). Inhibition of germination and seedling development by crop residues and their phytotoxicity. *Agronomy Journal* 58: 303-304.

Guenzi, W.D. and T.M. McCalla. (1966). Phenolic acids in oats, wheat, sorghum, and corn residues and their phytotoxicity. *Agronomy Journal* 58: 303-304.

Guenzi, W.D., T.M. McCalla, and F.A. Norstadt. (1967). Presence and persistence of phytotoxic substances in wheat, oat, corn and sorghum residues. *Agronomy Journal* 59: 163-165.

Guenzi, W.D., W.D. Kehr, and T.M. McCalla. (1964). Water soluble phytotoxic substances in alfalfa foliage; variation and variety, cutting, year and stage of growth. *Agronomy Journal* 56: 499-500.

Hall, A.B., U. Blum, and R.C. Fites. (1982). Stress modification of allelopathy of

Helianthus annuus L. debris on seed germination. *American Journal of Botany* 69: 776-783.

Hall, K.C., S.J. Chapman, D.G. Christean, and M.B. Jackson. (1986). Abscissic acid in straw residues from autumn-sown wheat. *Journal of Science Food and Agriculture* 37: 219-222.

Hall, M.H. and P.R. Henderlong. (1989). Alfalfa autotoxic fraction characterization and initial separation. *Crop Science* 29: 425-428.

Hamilton-Kemp, T.R. and R.A. Andersen. (1986). Volatile compounds from wheat plants: Isolation, identification, and origin. In *Biogeneration of Aromas*, eds. T.H. Parliment and R. Croteau, ACS Symposium Series 317. Washington, DC: American Chemical Society, pp. 193-200.

Hanson, A.D., P.L. Traynor, K.M. Ditz, and D.A. Reicosky. (1981). Gramine in barley forage-effects of genotype and environment. *Crop Science* 21: 726-730.

Harrison, H.F., Jr. and J.K. Peterson. (1986). Allelopathic effects of sweet Potatoes (*Ipomoea batatas*) on yellow nutsedge (*Cyperus esculentus*) and alfalfa (*Medicago sativa*). *Weed Science* 34: 623-627.

Hartung, AC., M.G. Nair, and A.R. Putnam. (1990). Isolation and characterization of phytotoxic compounds from asparagus (*Asparagus officinalis*) roots. *Journal of Chemical Ecology* 16: 1707-1718.

Hazebroek, J.P., S.A. Garrison, and T. Gianfagna. (1989). Allelopathic substances in asparagus roots: Extraction, characterization and biological activity. *Journal of American Society of Horticultural Science* 114: 152-158.

Heckman, J.R. and D. Kluchinski. (1995). Soybean nodulation and nitrogen fixation on soil amended with plant residues. *Biology and Fertility of Soils* 20: 284-288.

Hedin, P.A. and J.C. McCarty, Jr. (1994). Effects of several commercial plant growth regulator formulations on yield and allelochemicals of cotton (*Gossypium hirsutum*). *Journal of Agriculture and Food Chemistry* 42: 1355-1357.

Hegde, R.S. and D.A. Miller. (1990). Allelopathy and autotoxicity in alfalfa: characterization and effects of preceding crops and residue incorporation. *Crop Science* 30: 1255-1259.

Hepperly, P.R. and M. Diaz. (1983). The allelopathic potential of pigeon peas in Puerto Rico. *Journal of Agricultural University of Puerto Rico* 67: 453-463.

Hicks, S.K., C.W. Wendt, J.R. Gannaway, and R.B. Baker. (1989). Allelopathic effects of wheat straw on cotton germination, emergence and yield. *Crop Science* 29: 1057-1061.

Hill, H.H. (1926). Decomposition of organic matter in soil. *Agricultural Research* 33: 77-99.

Hoffman, M.L., L.A. Weston, J.C. Snyder, and E.E. Regnier. (1996a). Separating the effects of sorghum (*Sorghum bicolor*) and rye (*Secale cereale*) root and shoot residues on weed development. *Weed Science* 44: 402-407.

Hoffman, M.L., L.A. Weston, J.C. Snyder, and E.E. Regnier. (1996b). Allelopathic influence of germinating seeds and seedlings of cover crops on weed species. *Weed Science* 44: 579-584.

Horricks, J.S. (1969). Influence on rape residue on cereal production. *Canadian Journal of Plant Science* 49: 632-634.

Horst, W.J. and R. Haerdter. (1994). Rotation of maize with cowpea improves yield and nutrient use of maize compared to maize monocropping in an alfa soil in Northern Guinea Savanna of Ghana. *Plant and Soil* 160: 171-183.

Hussain, F. and M.A. Gadoon. (1981). Allelopathic effects of *Sorghum vulgare* Pers. *Oecologia* 51: 284-288.

Irons, S.M. and O.C. Burnside. (1982). Competitive and allelopathic effects of sunflower (*Helianthus annuus*). *Weed Science* 39: 372-377.

Janovicek, K.J., T.J. Vyn, R.P. Voroney, and O.B. Allen. (1997). Early corn seedling growth responses to acetic, propionic and butyric acids. *Canadian Journal of Plant Science* 77: 333-337.

Jensen, E.H., G.J. Hartman, F. Lundian, S. Knapp, and B. Brookerd. (1981). Autotoxicity in alfalfa. *Nevada Agricultural Experiment Station Report R144*.

Jimenez, J.J., K. Schultz, A.L. Anaya, J. Hernandez, and O. Espejo. (1983). Allelopathy potential of corn pollen. *Journal of Chemical Ecology* 9: 1011-1025.

Jordan, D.L., P.K. Bollich, M.P. Braverman, and D.E. Sanders. (1999). Influence of tillage and *Triticum aestivum* cover crop on herbicide efficacy in *Oryza sativa*. *Weed Science* 47: 332-337.

Kalantri, I. (1981). Stimulation of corn seedling growth by allelochemicals from soybean residue. *PhD dissertation, Iowa State University, Ames. (Dissertation Abstract Order No. DA8209137)*.

Kalburtji, K.L. and A. Gagianas. (1997). Effects of sugar beet as a proceeding crop on cotton. *Journal of Agronomy and Crop Science* 178: 59-63.

Kato-Noguchi, H. (1999). Effect of light-irradiation on allelopathic potential of germinating maize. *Phytochemistry* 52: 1023-1027.

Kato-Noguchi, H., J. Mizutani, and K. Hasegawa. (1994). Allelopathy of Oats. II. Allelochemical effect of L-tryptophan and its concentration in oat root exudates. *Journal of Chemical Ecology* 20: 315-319.

Kato-Noguchi, H., S. Kosemura, S. Yamamura, J. Mizutani, and K. Hasegawa. (1994). Allelopathy of oats. I. Assessment of allelopathic potential of extract of oat shoots and identification of an allelochemical. *Journal of Chemical Ecology* 20: 309-314.

Katznelson, J. (1972). Studies in clover soil sickness. I. The phenomenon of soil sickness in berseem and Persian clover. *Plant and Soil* 36: 379-393.

Kawata, Y., I. Harada, and T. Matsunaka. (1996). Allelopathic interaction of alfalfa (*Medicago sativa* L.) root exudates. *Journal of Rakuno Gakven University of Natural Sciences* 21: 79-86.

Kessavalou, A. and D.T. Walters. (1997). Winter rye as a cover crop following soybean under conservation tillage. *Agronomy Journal* 89: 68-74.

Khanum, S., F. Hussain, and H.H. Naqvi. (1979). Allelopathic potentiality of *Chloris gayana* Kunth. and *Panicum antidotale* Retz. *Pakistan Journal of Forestry* 29: 245-249.

Kim, Y.S. and B.S. Kil. (1987). Bioassay on susceptivity of selected species to phytotoxic substances from tomato plants. *Korean Journal of Botany* 30: 59-67.

Kimber, R.W.L. (1967). Phytotoxicity from plant residues. I. The influence of rotted wheat straw on seedling growth. *Australian Journal of Agricultural Research* 18: 361-374.

Kimber, R.W.L. (1973). Phytotoxicity from plant residues. III. The relative effect of toxins and nitrogen immobilisation on germination and growth of wheat. *Plant and Soil* 38: 543-555.

Kohli, R.K. (1993). *Allelopathic Impact of Sunflower on Weeds and Crops of Punjab.* India: PSCST Project Report.

Kohli, R.K., D. Batish, and H.P. Singh (1998). Allelopathy and its implications in Agroecosystems. *Journal of Crop Production* 1: 169-202.

Kohli, R.K., D.R. Batish, H.P. Singh, and V. Arora. (1999). Allelopathy as a tool for weed and pest management. *Journal of Punjab Academy of Sciences* 1: 127-131.

Kott, L.S., L. Polsoni, B. Ellis, and W.D. Beversdorf. (1988). Autotoxicity in isolated microspore cultures of *Brassica napus*. *Canadian Journal of Botany* 66: 1665-1670.

Krishnan, G., D.L. Holshouser, and S.J. Nissen. (1998). Weed control in soybean (*Glycine max*) with green manure crops. *Weed Technology* 12: 97-102.

Kumar, K. and K.M. Goh. (2000). Crop residues and management practices: effects on soil quality, soil nitrogen dynamics, crop yield, and nitrogen recovery. *Advances in Agronomy* 68: 197-319.

Kuo, C.G., M.H. Chou, and H.G. Park. (1981). Effect of chinese cabbage (*Brassica campestris* subsp. *Pekinensis*) residue on mung bean (*Vigna radiata*). *Plant and Soil* 61: 473-478.

Kushima, M., H. Kakuta, S. Kosemura, S. Yamamura, K. Yamada, K. Yakotani-Tomita, and K. Hasegawa. (1998). An allelopathic substance exuded from germinating watermelon seeds. *Plant Growth Regulation* 25: 1-4.

Lazauskas, P. and Z. Balinevichiute. (1972). Influence of excretion from *Vicia villosa* Roth. Seeds or germination and primary growth of some crops and weeds. In *Physiological-Biochemical Basis of Plant Interactions in Phytocenoses*, ed. A.M. Grodzinsky. Vol. 3. Kiev: Naukova Dumka, pp. 76-79 (In Russian, English Summary).

Leather, G.R. (1983). Sunflowers (*Helianthus annuus*) are allelopathic to weeds. *Weed Science* 31: 37-42.

Leather, G.R. (1987). Weed control using allelopathic sunflowers and herbicide. *Plant and Soil* 98: 17-23.

Lee, I.K., O.A. Im, and M.S. Bark. (1967). Studies on the sick soil phenomenon of *Setaria italica* and *Solanum melongena*. Nuc. Eng. Stud. Ser. 7 (No.1, Pt-2). pp. 39-44 (In Korean, English Summary).

Lehman, M.E. and U. Blum. (1997). Cover crop debris effects on weed emergence as modified by environmental factors. *Allelopathy Journal* 4: 69-88.

Lemerle, D. (1996). The potential for selecting wheat varieties strongly competitive against weeds. *Weed Research* 36: 505-513.

Li, Y. (1988). A preliminary study on use of the medical herb peppermint in the control of cotton *Fusarium* Wilt. *Scientia Agriculturae Sinica* 21: 65-69.

Liebman, M. and T. Ohno. (1997). Crop rotation and legume residue effects on weed emergence and growth: application for weed management. In *Integrated Weed and Soil Management*, eds. J.L. Hatfield, D.D. Buhler, and B.A. Stewart. Chelsea, MI: Ann Arbor Press, pp. 181-221.

Litav, M. and D. Isti. (1974). Root competition between two strains of *Spinacia oleracea*. II. Effects of nutrient supply and non-simultaneous emergence. *Journal of Applied Ecology* 11: 1017-1025.

Liu, D.L. and J.V. Lovett. (1993). Biologically active secondary metabolites of barley. I. Developing techniques and assessing allelopathy in barley. *Journal of Chemical Ecology* 19: 2217-2230.

Lodhi, M.A.K., R. Bilal, and K.A. Malik. (1987). Allelopathy in agroecosystems: wheat phytotoxicity and its possible roles in crop rotation. *Journal of Chemical Ecology* 13: 1881-1891.

Lovett, J.V. (1985). Defensive strategies of plants, with special reference to allelopathy. *Papers and Proceedings of the Royal Society of Tasmania* 119: 31-37.

Lovett, J.V. and R.S. Jessop. (1982). Effects of residues of crop plants on germination and early growth of wheat. *Australian Journal of Agricultural Research* 33: 909-916.

Lovett, J.V., S.A. Fraser, and A.M. Duffield. (1982). Allelopathic activity of cultivated sunflowers. In *Proceedings Tenth International Sunflower Conference.* Surfers Paradise, pp. 198-201.

Lynch, J.M. (1977). Phytotoxicity of acetic acid produced in the anaerobic decomposition of wheat straw. *Journal of Applied Bacteriology* 42: 81-87.

Lynch, J.M. (1978). Production and phytotoxicity of acetic acid in anaerobic soils containing plant residues. *Soil Biology and Biochemistry* 10: 131-135.

Lynch, J.M. (1983). *Soil Biotechnology: Microbiological Factors in Crop Productivity.* Boston, MA: Blackwell Scientific Publishers.

Lynch, J.M. and L.F. Elliott. (1983). Minimizing the potential phytotoxicity of wheat straw by microbial degradation. *Soil Biology and Biochemistry* 15: 221-222.

MacFarlane, M.J., D. Scott, and P. Jarvis. (1982). Allelopathic effects of white clover. I. Germination and chemical bioassay. *New Zealand Journal of Agricultural Research* 25: 502-510.

Macharia, C. and E.B. Peffey. (1995). Suppression of *Amaranthus spinosus* and *Kochia scoparia*: Evidence of competition or allelopathy in *Allium fistulosum.* *Crop Protection* 14: 155-158.

Macías, F.A., A. Torres, J.M.G. Molinillo, R.M. Varela, and D. Castellano. (1996). Potential allelopathic sesquiterpene lactones from sunflower leaves. *Phytochemistry* 43: 1205-1215.

Macías, F.A., J.M.G. Molinillo, A. Torres, R.M. Varela, and D. Castellano. (1997). Bioactive Flavonoids from *Helianthus annuus* cultivars. *Phytochemistry* 45: 683-687.

Macias, F.A., J.M.G. Molinillo, J.C.G. Galindo, R.M. Varela, A. Torres, and A.M. Simonet. (1999). Terpenoids as potential use as natural herbicide templates. In *Biologically Active Natural Products: Agrochemicals*, eds. H.G. Cutler and S.J. Cutler. Boca Raton, USA: CRC Press, pp. 15-31.

Macías, F.A., R.M. Varela, A. Torres, and J.M.G. Molinillo. (1993). Potential allelopathic guaianolides from cultivar sunflower leaves, var. SH-222. *Phytochemistry* 34: 669-674.

Mari, M., R. Iori, O. Leoni, and A. Marchi. (1993). *In vitro* activity of glucosinolate-derived isothicyanates against postharvest fruit pathogens. *Annals of Applied Biology* 123: 155-164.

Markova, S.A. (1972). Experimental investigations of the influence of oats on growth and development of *Erysimum cheiranthoides* L. In *Physiological-Biochemical*

Basis of Plant Interactions in Phytocenoses, ed. A.M. Grodzinsky. Vol. 3. Kiev: Naukova Dumka, pp. 66-68 (In Russian, English summary).

Martin, P. and B. Rademacher. (1960). Studies on the mutual influences of weeds and crops. *Symposium of British Ecological Society* 1: 143-152.

Martin, V.L., E.L. McCoy, and W.A. Dick. (1990). Allelopathy of crop residues influences corn seed germination and early growth. *Agronomy Journal* 82: 555-560.

Mason-Sedun, W., R.S. Jessop, and J.V. Lovett. (1986). Differential phytotoxicity among species and cultivars of genus *Brassica* to wheat. I. Laboratory and field screening of species. *Plant and Soil* 93: 3-16.

Massantini, F., F. Caporali, and G. Zellini. (1977). Evidence for allelopathic control of weeds in lines of soybean. In *Proceedings EWRS Symposium on Methods of Weed Control and Their Integration*. Vol. 1. pp. 23-28.

Mayton, H.S., C. Olivier, S.F. Vaughn, and R. Loria. (1996). Correlation of fungicidal activity of *Brassica* species with allyl isothiocyanate production in macerated leaf tissue. *Phytopathology* 86: 267-271.

McCalla, T.M. and F.A. Haskins. (1964). Phytotoxic substances from soil microorganisms and crop residues. *Bacteriological Reviews* 28: 181-207.

McCalla, T.M. and F.L. Duley. (1948). Stubble mulch studies: Effects of sweet clover extract on corn germination. *Science* 108: 163.

McLenaghen, R.D., K.C. Cameron, N.H. Lampkin, M.L. Daly, and B. Deo. (1996). Nitrate leaching from plowed pasture and the effectiveness of winter catch crops in reducing leaching losses. *New Zealand Journal of Agricultural Research* 39: 413-420.

Megie, C.A., R.W. Pearson, and A.E. Hiltbold. (1967). Toxicity of decomposing crop residues to cotton germination and seedling growth. *Agronomy Journal* 57: 516-517.

Miersch, J., C. Jühlke, G. Sternkopf, and G.J. Krauss. (1992). Metabolism and exudation of canavanine during development of alfalfa (*Medicago sativa* L. cv. Verbo). *Journal of Chemical Ecology* 18: 2117-2129.

Miller, D.A. (1983). Allelopathic effects of alfalfa. *Journal of Chemical Ecology* 9: 1059-1072.

Miller, D.A. (1996). Allelopathy in forage crop systems. *Agronomy Journal* 88: 854-859.

Miller, R.W., R. Kleiman, G. Powell, and A.R. Putnam. (1988). Germination and growth inhibitors of alfalfa. *Journal of Natural Products* 51: 328-330.

Mishustin, E.N. and A.N. Naumova. (1955). Secretion of toxic substances by alfalfa and their effect on cotton and soil microflora. *Akademiya Nauk SSSR Izvestiya Seriya Biologicheskaya* 6: 3-9 (In Russian with English summary).

Molisch, H. (1937). *Der Einflus einer Pflanze auf die andere-Allelopathie*. Jena: Fischer.

Moyer, J.R. and H.C. Huang. (1997). Effect of aqueous extracts of crop residues on germination and seedling growth of ten weed species. *Botanical Bulletin of Academia Sinica* 38: 131-139.

Muehlchen, A.M., R.E. Rand, and J.L. Parke. (1990). Evaluation of crucifer green manures for controlling aphanomyces root rot of peas. *Plant Disease* 74: 651-654.

Muller, C.H. (1966). The role of allelochemical inhibition (allelopathy) in vegetational composition. *Bulletin of the Torrey Botanical Club* 93: 332-351.

Muller-Wilmes, U. and M. Zoschke. (1980). Allelopathie-eine mögliche Ursache fur Vertraglichkeits-beziehungen der kulturpflanzen. *Angewandte Botanik* 54: 109-123.

Mwaja, V.Y., J.B. Masiunas, and L.A. Weston. (1995). Effects of fertility on biomass, phytotoxicity and allelochemical content of cereal rye. *Journal of Chemical Ecology* 21: 81-94.

Nakahisa, K., E. Tsuzuki, H. Terao, and S. Kosemura. (1994). Study on allelopathy of alfalfa (*Medicago sativa*): II. Isolation and identification of allelopathic substances in alfalfa. *Japanese Journal of Crop Science* 63: 278-284.

Narwal, S.S., M.K. Sarmah, and D.P.S. Nandal. (1997). Allelopathic effects of wheat residues on growth and yield of fodder crops. *Allelopathy Journal* 4: 111-120.

Nava, R.V., L.E. Fernandez, and R.S. del Amo. (1987). Allelopathic effects of green fronds of *Pteridium aquilinum* on cultivated plants, weeds, phytopathogenic fungi and bacteria. *Agriculture Ecosystems and Environment* 18: 357-379.

Nelson, C.J. (1996). Allelopathy in cropping systems. *Agronomy Journal* 88: 991-996.

Nelson, L.S. (1985). Isolating Potential Allelochemicals from Soybean-Soil Residues. *MS Thesis, Iowa State University, Ames.*

Netzley, D.H. and L.G. Butler. (1986). Roots of sorghum exudate hydrophobic droplets containing biologically active components. *Crop Science* 26: 776-778.

Netzley, D.H., J.L. Reopel, G. Ejeta, and L. Butler. (1988). Germination stimulants of witch-weed (*Striga asiatica*) from hydrophobic root exudate of Sorghum (*Sorghum bicolor*). *Weed Science* 36: 441-446.

Neustruyeva, S.N. and T.N. Dobretsova. (1972). Influence of some summer crops on white goosefoot. In *Physiological-Biochemical Basis of Plant Interactions in Phytocenoses*, ed. A.M. Grodzinsky. Vol. 3. Kiev: Naukova Dumka, pp. 68-73 (In Russian, English Summary).

Nielsen, K.F., T. Cuddy, and W. Woods. (1960). The influence of the extract of some crops and soil residues on germination and growth. *Canadian Journal of Plant Science* 40: 188-197.

Niemeyer, H.M. (1988). Hydroxamic acids (4-hydroxy-1,4-benzoxazine-3-ones), defense chemicals in the gramineae. *Phytochemistry* 27: 3349-3358.

Nimbal, C.I., J. Pedersen, C.N. Yerkes, L.A. Weston, and S.C. Weller. (1996a). Phytotoxicity and distribution of sorgoleone in grain sorghum germplasm. *Journal of Agricultural and. Food Chemistry* 44: 1343-1347.

Nimbal, C.I., C.N. Yerkes, L.A. Weston, and S.C. Weller. (1996b). Herbicidal activity and site of action of natural product sorgoleone. *Pesticide Biochemistry and Physiology* 54: 73-83.

Ohno, T., K. Doolan, L.M. Zibilske, M. Liebman, E.R. Gallandt, and C. Berube. (2000). Phytotoxic effects of red clover amended soils on wild mustard seedling growth. *Agriculture Ecosystems and Environment* 78: 187-192.

Oleszek, W. (1987). Allelopathic effects of volatiles from some cruciferae species on lettuce, barnyard grass and wheat growth. *Plant and Soil* 102: 271-273.

Oleszek, W. and M. Jurzysta. (1987). An allelopathic potential of alfalfa root medicagenic acid glycosides and their fate in soil environment. *Plant and Soil* 98: 67-80.

Olivier, C., S.F. Vaughn, E.S.G. Mizubuti, and R. Loria. (1999). Variation in allyl isothiocyanate production within *Brassica* species and correlation with fungicidal activity. *Journal of Chemical Ecology* 25: 2687-2701.

Olofsdotter, M. and D. Navarez. (1996). Allelopathic rice for *Echinochloa crus-galli* control. *Paper Presented at the Second International Weed Control Congress.* Copenhagen, Denmark. pp. 1175-1181.

Olofsdotter, M., D. Navarez, and K. Moody. (1995). Allelopathic potential in rice (*Oryza sativa* L.) germplasm. *Annals of Applied Biology* 127: 543-560.

Olofsdotter, M., D. Navarez, M. Rebulanan, and J.C. Streibig. (1999). Weed suppressing rice cultivars–does allelopathy play a role? *Weed Research* 39: 441-554.

Osvald, H. (1953). On antagonism between plants. In *Proceedings of 7th International Botanical Congress*, Stockholm, pp. 167-177.

Overland, L. (1966). The role of allelopathic substances in the "smother crop" barley. *American Journal of Botany* 53: 423-432.

Panasiuk, O., D.D. Bills, and G.R. Leather. (1986). Allelopathic influence of *Sorghum bicolor* on weeds during germination and early development of seedling. *Journal of Chemical Ecology* 12: 1533-1543.

Pardales, J.R., Jr. and A.G. Dingal. (1988). An allelopathy factor in taro residues. *Tropical Agriculture* 65: 21-24.

Patrick, Z.A. (1971). Phytotoxic substances associated with decomposition in soil of plant residues. *Soil Science* 111: 13-18.

Patrick, Z.A. and L.W. Koch. (1958). Inhibition of respiration, germination and growth by substances arising during decomposing of certain plant residues in soil. *Canadian Journal of Botany* 36: 621-647.

Patrick, Z.A., T.A. Toussoun, and W.C. Snyder. (1963). Phytotoxic substances in arable soils associated with decomposition of plant residues. *Phytopathology* 53: 152-161.

Pedersen, C.T., G.R. Safir, J.O. Siqueira, and S. Parent. (1991). Effect of phenolic compounds on asparagus mycorrhiza. *Soil Biology and Biochemistry* 23: 491-494.

Pederson, M.W. (1965). Effect of alfalfa saponin on cotton seed germination. *Agronomy Journal* 57: 516-517.

Peirce, L.C. and L.W. Colby. (1987). Interaction of asparagus root filtrate with *Fusarium oxysporum* f. sp. *asparagi*. *Journal of American Society for Horticultural Science* 112: 35-40.

Pérez, F.J. (1990). Allelopathic effects of hydroxamic acids from cereals on *Avena sativa* and *A. fatua*. *Phytochemistry* 29: 773-776.

Pérez, F.J. and J. Ormeño-Nuñez. (1991). Difference in hydroxamic acid content in roots and root exudates of wheat (*Triticum aestivum* L.) and rye (*Secale cereale* L.): Possible role in allelopathy. *Journal of Chemical Ecology* 17: 1037-1043.

Peters, E.J. (1968). Toxicity of tall fescue to rape and birdsfoot trefoil seeds and seedlings. *Crop Science* 8: 650-653.

Peters E.J. and A.H.B.M. Zam. (1981). Allelopathic effects of tall fescue genotypes. *Agronomy Journal* 73: 56-59.

Peterson, J.K. and H.F. Harrison, Jr. (1991). Differential inhibition of seed germination by sweet potato (*Ipomoea batatas*) root periderm extracts. *Weed Science* 38: 119-123.

Petrova, A.G. (1977). Effect of phytoncides from soybean, gram, chickpea and bean on uptake of phosphorus by maize. In *Interaction of Plants and Microorganisms in Phytocenoses*, A.M. Grodzinsky. Kiev: Naukova Dumka, pp. 91-97 (In Russian, English Summary).

Pickering, S.U. (1912). The effect of grass on plants. *Nature (London)* 89: 399.

Pickering, S.U. (1917). The effect of one plant on another. *Annals of Botany (London)* 31: 181-187.

Plinius Secundus, C. (I.A.D.). *Natural History.* 10 Vols. Translated to English by H. Rackam, W.H.S. Jones, and D.E. Eichholz. Cambridge, Massachusetts: Harvard University Press, pp. 1938-1963.

Politycka, B., D. Wójcik-Wojtkowiak, and T. Pudelski. (1984). Phenolic compounds as a cause of phytotoxicity in greenhouse substrates repeatedly used in cucumber growing. *Acta Horticulturae* 156: 89-94.

Prasad, R. and B.R.R. Rao. (1981). Allelopathy in spring wheat mixtures (cultivars). *Experientia* 37: 1078.

Prezepiorkowski, T. and S.F. Gorski. (1994). Influence of rye (*Secale cereale*) plant residue on germination and growth of three triazine-resistant and susceptible weeds. *Weed Technology* 8: 744-747.

Prutenskaya, N.I. (1972). Presence of inhibitors and stimulators of *Sinapis arvensis* L. in germinating seeds of cultivated plants. In *Physiological-Biochemical Basis of Plant Interactions in Phytocenoses*, ed. A.M. Grodzinsky. Vol. 3. Kiev: Naukova Dumka, pp. 73-75 (In Russian, English Summary).

Prutenskaya, N.I. (1974). Peculiarities of Interaction between *Sinapis arvensis*. L. and cultivated plants. In *Physiological-Biochemical Basis of Plant Interactions in Phytocenoses*, ed. A.M. Grodzinsky. Vol. 5. Kiev: Naukova Dumka, pp. 66-68 (In Russian, English Summary).

Purvis, C.E. (1990). Differential responses of wheat to retained crop stubbles. I. Stubble type and degree of decomposition. *Australian Journal of Agricultural Research* 41: 225-242.

Purvis, C.E. and G.P.D. Jones. (1990). Differential responses of wheat to retained crop stubbles. II. Other factors influencing allelopathic potential: intraspecific variation, soil type and stubble quantity. *Australian Journal of Agricultural Research* 41: 243-252.

Putnam, A.R. (1986). Allelopathy: can it be managed to benefit horticulture? *HortScience* 21: 411.

Putnam, A.R. (1994). Phytotoxicity of plant residues. In *Managing Agricultural Residues*, ed. P.W. Unger. New York: CRC Press, pp. 285-314.

Putnam, A.R. and J. DeFrank. (1983). Use of phytotoxic plant residues for selective weed control. *Crop Protection* 2: 173-181.

Putnam, A.R. and L.A. Weston. (1986). Adverse impacts of allelopathy in agricultural systems. In *The Science of Allelopathy*, eds. A.R. Putnam and C.S. Tang. New York: John Wiley and Sons, pp. 43-56.

Putnam, A.R. and W.B. Duke. (1974). Biological suppression of weeds: Evidence for allelopathy in accessions of cucumber. *Science* 185: 370-372.

Putnam, A.R. and W.B. Duke. (1978). Allelopathy in agroecosystems. *Annual Review of Phytopathology* 16: 431-451.

Putnam, A.R., J. DeFrank, and J.P. Barnes. (1983). Exploitation of allelopathy for weed control in annual and perennial cropping systems. *Journal of Chemical Ecology* 9: 1001-1010.

Raimbault, B.A., T.J. Vyn, and M. Tollenaar. (1991). Corn response to rye cover crop, tillage methods, and planter options. *Agronomy Journal* 83: 287-290.

Rasmussen, D.A., A.M. Hejl, F.A. Einhellig, and T.A. Thomas. (1992). Sorgoleone from root exudate inhibits mitochondrial functions. *Journal of Chemical Ecology* 18: 197-207.

Read, J.J. and E.H. Jansen. (1989). Phytotoxicity of water-soluble substances from alfalfa and barley soil extracts on four crop species. *Journal of Chemical Ecology* 15: 619-628.

Rice, E.L. (1971). Some possible roles of inhibitors in old-field succession. In *Biochemical Interactions Among Plants*. Washington, ed. US National Commission for IBP, DC: National Academy of Sciences, pp. 128-132.

Rice, E.L. (1984). *Allelopathy*. 2nd ed. Academic Press, Orlando, Florida.

Rice, E.L. (1995). *Biological Control of Weeds and Plant Diseases–Advances in Applied Allelopathy*. USA: University of Oklahoma Press.

Rice, E.L., C.Y. Lin, and C.Y. Huang. (1980). Effects of decaying rice straw on growth and nitrogen fixation of a bluegreen alga. *Botanical Bulletin of Academia Sinica* 21: 111-117.

Rice, E.L., C.Y. Lin, and C.Y. Huang. (1981). Effects of decomposing rice straw on growth and nitrogen fixation by *Rhizobium*. *Journal of Chemical Ecology* 7: 333-344.

Rizvi, S.J.H., D. Mukerji, and S.N. Mathur. (1980). A new report on a possible source of natural herbicide. *Indian Journal of Experimental Biology* 18: 777-778.

Rizvi, S.J.H., M. Tahir, V. Rizvi, R.K. Kohli, and A. Ansari. (1999). Allelopathic interactions in agroforestry systems. *Critical Reviews in Plant Sciences* 18: 773-796.

Rizvi, S.J.H. and V. Rizvi. (1992). Exploitation of allelochemicals in improving crop productivity. In *Allelopathy: Basic and Applied Aspects*, eds. S.J.H. Rizvi and V. Rizvi. London: Chapman and Hall, pp. 443-472.

Rizvi, S.J.H., V. Rizvi, D. Mukerji, and S.N. Mathur. (1987). 1,3,7-trimethylxanthine, an allelochemical from seeds of *Coffea arabica* some aspects of its mode of action as a natural herbicide. *Plant and Soil* 98: 81-91.

Ruiz-Sifre, G.V. and S.K. Ries. (1983). Response of crops to sorghum residues. *Journal of American Society for Horticultural Science* 108: 262-266.

Rule, J.S. (1991). The effect of straw disposal method on weed populations and efficiency of herbicides on *Alopecurus myosuroides*, *Bromus sterilis* and *Bromus commutatus* in winter wheat crops. In *Proceedings of the 1991 Brighton Crop Protection Conference–Weeds. Vol. 2*. Farnham, UK: British Crop Protection Council, pp. 798-806.

Sadhu, M.K. (1975). Nature of inhibitory substances in root exudates of rice seedlings. *Indian Journal of Experimental Biology* 13: 577-579.

Sadhu, M.K. and T.M. Das. (1971). Root exudates of rice seedlings: The influence of one variety on another. *Plant and Soil* 34: 541-546.

Salmon, D.F., J.H. Helm, T.R. Duggan, and D.M. Lakeman. (1986). The influence of chaff extracts on germination of spring triticale. *Agronomy Journal* 78: 863-867.

Santos, J.R.A. and D.I. Leskovar. (1997). Interference from broccoli residue on *Brassica* germination and seedling growth. *Journal of American Society for Horticultural Science* 122: 715-720.

Sarmah, M.K., S.S. Narwal, and J.S. Yadava. (1992). Smothering effect of *Brassica* species on weeds. In *Proceedings First National Symposium on Allelopathy in Agroecosystems*, eds. P. Tauro and S.S. Narwal. Hisar: Indian Society of Allelopathy, pp. 51-55.

Schenk, S.U., and D. Werner. (1991). β-(3-Isoxazolin-5-on-2-yl)-alanine from *Pisum*: Allelopathic properties and antimycotic bioassay. *Phytochemistry* 30: 467-470.

Schon, M.K. and F.A. Einhellig. (1982). Allelopathic effects of cultivated sunflower on grain sorghum. *Botanical Gazette* 143: 505-510.

Schreiber, M.M. (1992). Influence of tillage, crop rotation, and weed management on giant foxtail (*Setaria faberi*) population dynamics and corn yield. *Weed Science* 40: 645-653.

Schreiner, O. and E.C. Shorey. (1907). The presence of secondary decomposition products of proteids in soil. *Journal of Biological Chemistry* 5: 38-39.

Schreiner, O. and E.C. Shorey. (1908). The isolation of picoline carboxylic acid from soils and its relation to soil fertility. *Journal of American Chemical Society* 30: 1295-1307.

Schreiner, O. and E.C. Shorey. (1909). The isolation of harmful organic substances from soils. *United States Department of Agriculture Bureau of Soils Bulletin No. 53*.

Schreiner, O. and H.S. Reed. (1908). The toxic action of certain organic plant constituents. *Botanical Gazette* 45: 73-102.

Schreiner, O. and M.X. Sullivan. (1909). Soil fatigue caused by organic compounds. *Journal of Biological Chemistry* 6: 39-50.

Shafer, W.E. and S.A. Garrison. (1986). Allelopathic effects of soil incorporated asparagus roots on lettuce and asparagus seedling emergence. *HortScience* 21: 82-84.

Shilling, D.G., B.J. Brecke, C. Hiebsch, and J. MacDonald. (1995). Effect of soybean (*Glycine max*) cultivar, tillage, and rye (*Secale cereale*) mulch on sicklepod (*Senna obtusifolia*). *Weed Technology* 9: 339-344.

Shilling, D.G., L.A. Jones, A.D. Worsham, C.E. Parker and R.F. Wilson. (1986). Isolation and identification of some phytotoxic compounds from aqueous extracts of rye (*Secale cereale* L.). *Journal of Agricultural and Food Chemistry* 34: 633-638.

Shilling, D.G., R.A. Liebl, and A.D. Worsham. (1985). Rye and wheat mulch: The suppression of certain broad-leaved weeds and the isolation and identification of phytotoxins. In *The Chemistry of Allelopathy*, ed. A.C. Thompson. ACS Symposium Series No. 268. Washington, DC: American Chemical Society, pp. 243-271.

Singh, H.P., D.R. Batish, and R.K. Kohli. (1999). Autotoxicity: concept, organisms and ecological significance. *Critical Reviews in Plant Sciences* 18: 757-772.

Smeda, R.J. and A.R. Putnam. (1988). Cover crop suppression of weeds and influence on strawberry yields. *Horticultural Science* 23: 132-134.

Smeda, R.J. and S.C. Weller. (1996). Potential of rye (*Secale cereale*) for weed management in transplant tomatoes (*Lycopersicon esculentum*). *Weed Science* 44: 596-602.

Srivastava, P.C., N.G. Totey, and O. Prakash. (1986). Effect of straw extract on water absorption and germination of wheat (*Triticum aestivum* L. variety RR-21) seeds. *Plant and Soil* 91: 43-45.

Stirzaker, R.J. and D.G. Bunn. (1996). Phytotoxicity of ryegrass and clover cover crops, and a lucerne alley crop for no-till vegetable production. *Biological Agriculture and Horticulture* 13: 83-101.

Suzuki, T., H. Ashihara, and G.R. Waller. (1992). Purine and purine alkaloid metabolism in *Camellia* and *Coffea* plants. *Phytochemistry* 31: 2575-2584.

Swanton, C.J. and S.D. Murphy. (1996). Weed science beyond the weeds: the role of integrated weed management (IWM) in agroecosystem health. *Weed Science* 44: 437-445.

Tamura, S., C. Chang, A. Suzuki, and S. Kumai. (1967). Isolation and structure of a novel isoflavone derivative in red clover. *Agricultural and Biologial Chemistry* 31: 1108-1109.

Tamura, S., C. Chang, A. Suzuki, and S. Kumai. (1969). Chemical studies on "clover sickness." Part I. Isolation and structural elucidation of two new isoflavonoids in red clover. *Agricultural and Biologial Chemistry* 33: 391-397.

Tang, C.S. and A.C. Waiss, Jr. (1978). Short-chain fatty acids as growth inhibitors in decomposing wheat straw. *Journal of Chemical Ecology* 4: 225-232.

Tang, C.S. and C.C. Young. (1982). Collection and identification of allelopathic compounds from the undisturbed root system of bigalta limpograss (*Hemarthria altissima*). *Plant Physiology* 69: 155-160.

Tapaswi, P.K., R.N. Banerjee, D.K. Bagchi, S. Adhikary, and R.L. Brahmachary. (1991). Interaction of two varieties of rice–a study of intervarietal interaction. *Indian Biologist* 23: 8-9.

Teasdale, J.R. (1993). Interaction of light, soil moisture, and temperature with weed suppression by hairy vetch residues. *Weed Science* 41: 46-51.

Teasdale, J.R., C.E. Beste, and W.E. Potts. (1991). Response of weeds to tillage and cover crop residues. *Weed Science* 39: 195-199.

Theophrastus. (ca. 300BC). *Enquiry into Plants and Minor Works on Odours and Weather Signs*. 2 Vols. Translated to English by A. Hort. and W. Heinemann. London, 1916.

Thorne, R.L.Z., G.R. Waller, J.K. McPherson, E.G. Krenzer, Jr., and C.C. Young. (1990). Autotoxic effects of old and new wheat straw in conventional-tillage wheat soil. *Botanical Bulletin of Academia Sinica* 31: 35-39.

Tipton, C.L., J.A. Klun, R.R. Husted, and M.D. Pierson. (1967). Cyclic hydroxamic acids and related compounds from maize. Isolation and characterization. *Biochemistry* 9: 2866-2870.

Tsuzuki, E., H. Ashihara, and G.R. Waller. (1992). Purine and purine alkaloid metabolism in *Camellia* and *Coffea* plants. *Phytochemistry* 31: 2575-2584.

Tsuzuki, E. and H. Kawagoe. (1984). Studies of allelopathy among higher plants: IV Allelopathy in leguminous crops. *Bulletin of the Faculty of Agriculture of the Miyazaki University* 31: 189-196.

Tsuzuki, E., Y. Yamamoto, and T. Shimizu. (1987). Fatty acids in buckwheat are growth inhibitors. *Annals of Botany* 60: 69-70.

Turco, R.F., M. Bischoff, D.P. Breakwell, and D.R. Griffith. (1990). Contribution of soil-borne bacteria to the rotation effect in corn. *Plant and Soil* 122: 115-120.

Ul Haq, I. and F. Hussain. (1979). Effect of root exudates of tobacco (*Nicotiana rustica* L.) on some cultivated plants. *Pakistan Tobacco* 3: 7-19.

Van Raamsdonk, L.W.D. (1995). The cytological and genetical mechanisms of plant domestication exemplified by four crop models. *Botanical Review* 61: 367-399.

Vaughan, D., G.P. Sparling, and B.G. Ord. (1983). Amelioration of the phytotoxicity of phenolic acids by some soil microbes. *Soil Biology and Biochemistry* 5: 613-614.

Vaughn, S.F. and R.A. Boydston. (1997). Volatile allelochemicals released by crucifer green manures. *Journal of Chemical Ecology* 23: 2107-2116.

Vaughn, S.F. (1999). Glucosinolates as natural pesticides. In *Biologically Active Natural Products*, eds. H.G. Cutler and S.J. Cutler. Boca Raton: CRC Press, pp. 81-92.

Vera, C.L., D.I. McGegor, and R.K. Downey. (1987). Detrimental effects of volunteer *Brassica* on production of certain cereal and oilseed crops. *Canadian Journal of Plant Science* 67: 983-995.

Voroney, R.P., B.J. Farquarson, K.J. Janovicek, M. Fortin, T.J. Vyn, and E.G. Beauchamp. (1991). Yield reducing affects of crop residues in conservation tillage. *Final report to Agriculture Canada (SWEEP)*.

Wacker, T.L., G.R. Safir, and C.T. Stephens. (1990). Effects of ferulic acid on *Glomus fasciculatum* and associated effects on phosphorus uptake and growth of asparagus (*Asparagus officinalis* L.). *Journal of Chemical Ecology* 16: 901-909.

Wagner-Riddle, C., T.J. Gillespie, L.A. Hunt, and C.J. Swanton. (1997). Modelling a rye cover crop and subsequent soybean yield. *Agronomy Journal* 89: 208-218.

Walker, D., T.J. Hubbell, and J.E. Sedberry. (1989). Influence of decaying sweet potato crop residues on nutrient uptake of sweet potato plants. *Agriculture Ecosystems and Environment* 26: 45-52.

Waller, G.R., C.S. Cheng, C.H. Chou, D. Kim, C.F. Yang, S.C. Huang, and Y.F. Lin. (1995). Allelopathic activity of naturally occurring compounds from mung beans (*Vigna radiata*) and their surrounding soil. In *Allelopathy: Organisms, Processes and Applications*, eds. Inderjit, K.M.M. Dakshini, and F.A. Einhellig. ACS Symposium Series No. 582. Washington, DC.: American Chemical Society, pp. 242-259.

Waller, G.R., E.G. Krenzer, Jr., J.K. McPherson, and S.R. McGown. (1987). Allelopathic compounds in soil from no-tillage vs. conventional tillage in wheat production. *Plant and Soil* 98: 5-15.

Waller, G.R., M. Jurzysta, and R.L.Z. Thorne. (1993). Allelopathic activity of root Saponins from alfalfa (*Medicago sativa* L.) on weeds and wheat. *Botanical Bulletin of Academia Sinica* 34: 1-11.

Wang, T.S.C., T. Yang, and T. Chuang. (1967). Soil phenolic acids as plant growth inhibitors. *Soil Science* 103: 239-246.

Webster, G.R., S.U. Khan, and A.W. Moore. (1967). Poor growth of alfalfa (*Medicago sativa* L.) on some Alberta soils. *Agronomy Journal* 59: 37-41.

Weston, L.A. (1996). Utilization of allelopathy for weed management in agroecosystems. *Agronomy Journal* 88: 860-866.

Weston, L.A., R. Harmon, and S. Mueller. (1989). Allelopathic potential of Sorghum-sudangrass hybrid (Sudex). *Journal of Chemical Ecology* 15: 1855-1865.

White, I., M.J. Sully, and M.D. Melville. (1989). Use and hydrological robustness of time-to-incipient-ponding. *Soil Science Society of America Journal* 29: 1343-1346.

White, R.H., A.D. Worsham, and U. Blum. (1989). Allelopathic potential of legume debris and aqueous extracts. *Weed Science* 37: 674-679.

Williams, M.M., II, D.A. Mortensen, and J.W. Doran. (1998). Assessment of weed and crop fitness in cover crop residues for integrated weed management. *Weed Science* 46: 595-603.

Williams-Woodward, J.L., F.L. Pfleger, V.A. Fritz, and R.R. Allmaras. (1997). Green manures of oat, rape and sweet corn for reducing common root rot in pea (*Pisum sativum*) caused by *Aphanomyces euteiches*. *Plant and Soil* 188: 43-48.

Wilson, H.P., M.P. Mascianica, T.E. Hines, and R.F. Walden. (1986). Influence of tillage and herbicides on weed control in a wheat (*Triticum aestivum*)-soybean (*Glycine max*) rotation. *Weed Science* 34: 590-594.

Wilson, R.E. and E.L. Rice. (1968). Allelopathy as expressed by *Helianthus annuus* and its role in old-field succession. *Bulletin of the Torrey Botanical Club* 95: 432-448.

Wittwer, S.H. (1998). The changing global environment and world crop production. *Journal of Crop Production* 1: 291-299.

Wojcik-Wojtkowiak, D., B. Politycka, M. Schneider, and J. Perkowski. (1990). Phenolic substances as allelopathic agents arising during the degradation of rye (*Secale cereale*) tissues. *Plant and Soil* 124: 143-147.

Worsham, A.D. (1991). Allelopathic cover crops to reduce herbicide inputs. *Proceedings of Southern Weed Science Society* 44: 58-69.

Worsham, A.D. and U. Blum. (1992). Allelopathic cover crops to reduce herbicide inputs in cropping systems. In *Proceedings of First International Weed Control Congress*, ed. R.G. Richardson. Melbourne, Australia: Weed Science Society of Victoria, pp. 577-579.

Wu, H., J. Pratley, D. Lemerle, and T. Haig. (1999). Crop cultivars with allelopathic capability. *Weed Research* 39: 171-180.

Wyman-Simpson, C.L., G.R. Waller, M. Jurzysta, J.K. McPherson, and C.C. Young. (1991). Biological activity and chemical isolation of root saponins of six cultivars of alfalfa (*Medicago sativa* L.). *Plant and Soil* 135: 83-94.

Yakle, G.A. and R.M. Cruse. (1983). Corn plant residue age and placement effects on early corn growth. *Canadian Journal of Plant Science* 63: 871-877.

Yakle, G.A. and R.M. Cruse. (1984). Effects of fresh and decomposing corn plant residue extracts on corn seedling development. *Soil Science Society of America Journal* 48: 1143-1146.

Yamada, K., T. Anai, and K. Hasegawa. (1995). Lepidimoide, an allelopathic substance in the exudates from germinating seeds. *Phytochemistry* 39: 1031-1032.

Yang, H.J. (1982). Autotoxicity of *Asparagus officinalis* L. *Journal of American Society for Horticultural Science* 107: 860-862.

Yang, Y.S. and Y. Futsuhara. (1991). Inhibitory effects of volatile compounds released from rice callus on soybean callus growth: Allelopathic evidence observed *in vitro* cultures. *Plant Science* 77: 103-110.

Yenish, J.P., A.D. Worsham, and A.C. York. (1996). Cover crops for herbicide replacement on no-tillage corn (*Zea mays*). *Weed Technology* 10: 815-821.

Yenish, J.P., A.D. Worsham, and W.S. Chilton. (1995). Disappearance of DIBOA-glucoside, DIBOA and BOA from rye (*Secale cereale* L.) cover crop residues. *Weed Science* 43: 18-20.

Young, C.C. (1984). Autointoxication in root exudates of *Asparagus officinalis* L. *Plant and Soil* 82: 247-253.

Young, C.C. and D.P. Bartholomew. (1981). Allelopathy in a grass-legume association: I. Effects of *Hemarthria altissima* (Poir.) Stapf. and Hubb. root residues on growth of *Desmodium intortum* (Mill.) Urb, and *Hemarthria altissima* in a tropical soil. *Crop Science* 21: 770-774.

Young, C.C., R.L.Z. Thorne, and G.R. Waller. (1989). Phytotoxic potential of soils and wheat straw in rice rotation cropping systems of subtropical Taiwan. *Plant and Soil* 120: 95-100.

Young, C.C. and T.C. Chou. (1985). Autointoxication in residues of *Asparagus officinalis* L. *Plant and Soil* 85: 385-393.

Yu, J.Q. (1999). Autotoxic potential of vegetable crops. In *Allelopathy Update: Basic and Applied Aspects*, ed. S.S. Narwal, NY: Science Publishers, Inc., pp. 149-162.

Yu, J.Q. and Y.H. Matsui. (1997). Effects of root exudates of cucumber and allelochemicals on ion uptake by cucumber seedling. *Journal of Chemical Ecology* 23: 817-827.

Weed Control with Crop Residues in Vegetable Cropping Systems

Allen V. Barker
Prasanta C. Bhowmik

SUMMARY. Crop residues have many potential uses in cropping systems, among which are imparting weed control. The principal objective of this research was to evaluate residues from corn (*Zea mays* L.), soybean (*Glycine max* Merr.), and sunflower (*Helianthus annuus* L.) for their effects on weed control and on crop productivity of tomato (*Lycopersicon esculentum* Mill.) and summer squash (*Cucumis pepo* L. var. *melopepo* Alef.). In one experiment, residues were imported to the vegetable plots and applied as surface mulches or incorporated shallowly into the ground. In another experiment, the residues were grown on site as cover crops in the year preceding vegetable production and disked into the plots. Application of imported residues was more effective in weed control and yield enhancement than the cover crop residues. Imported residues were effective in control of early emerging weeds, whereas with the cover crops supplemental weed management was required for early weed control. Weed control did not differ substantially with species of residue, but control increased as amount of incorporated residues increased from 6 to 24 Mg/ha. Weed control with residue incorporation at 6 Mg/ha was as effective as a mulch of 24

Allen V. Barker and Prasanta C. Bhowmik are Professors, Department of Plant and Soil Sciences, Bowditch Hall, University of Massachusetts, Amherst, MA 01003.

Address correspondence to: Allen V. Barker, Plant and Soil Sciences, Bowditch Hall, University of Massachusetts, Amherst, MA 01003 (E-mail: barker@pssci.umass.edu).

Massachusetts Agricultural Experiment Station Paper No. 3224.

[Haworth co-indexing entry note]: "Weed Control with Crop Residues in Vegetable Cropping Systems." Barker, Allen V., and Prasanta C. Bhowmik. Co-published simultaneously in *Journal of Crop Production* (Food Products Press, an imprint of The Haworth Press, Inc.) Vol. 4, No. 2 (#8), 2001, pp. 163-183; and: *Allelopathy in Agroecosystems* (ed: Ravinder K. Kohli, Harminder Pal Singh, and Daizy R. Batish) Food Products Press, an imprint of The Haworth Press, Inc., 2001, pp. 163-183. Single or multiple copies of this article are available for a fee from The Haworth Document Delivery Service [1-800-342-9678, 9:00 a.m. - 5:00 p.m. (EST). E-mail address: getinfo@haworthpressinc.com].

Mg/ha. If weed control was imparted by the residues, crop yields with any residue treatment were equivalent to those from plots kept relatively weed-free by tillage. Yields did not vary with amounts of residues applied. Weed control by the residues was effective in growing seasons or months of growing seasons with below-normal or near-normal rainfall. High rainfall destroyed the weed-controlling potential of the mulched or incorporated imported residues or cover crop residues. This research demonstrated that crop residues impart weed control and that in years with adequate control, yields of crops will be equal to those in plots with rigorous control by tillage. In times of high rainfall during crop production or with use of residues from cover crops, supplemental weed management practices will be needed.

KEYWORDS. Allelopathy, *Lycopersicon esculentum*, *Cucumis pepo*, *Zea mays*, *Glycine max*, *Helianthus annuus*, phytotoxicity, mulch

INTRODUCTION

Employment of crop residues in crop management systems is associated commonly with conservation tillage, which has among its goals, those of reducing soil erosion and water loss (Follett, Steward, and Cole, 1987). In utilization of residues, consideration is also given as to whether the residues enrich or lower soil fertility through addition or removal of available nitrogen (Smith and Sharpley, 1990; Mahler and Hemamda, 1993; Tollenaar, Mihajlovic, and Vin, 1993). Organic residues also impart effects on crop productivity that are not attributed to factors of soil conservation or fertility. Weed or crop residues at or near the soil surface may release compounds that hinder growth of crop or weed seedlings through effects known as allelopathy (Putnam and Duke, 1978; Rice 1984, 1995; Rizvi and Rizvi, 1992; Kohli, Batish, and Singh, 1998). Allelopathy may be exerted by the vegetative residues or from seeds (Friedman and Waller, 1983; Putnam, 1986; Chung and Miller, 1995).

Allelopathy can cause great economic losses in crops; however, emphasis is now placed on employment of allelopathic crop residues for weed control (Fortin and Pierce, 1990; Putnam, 1986; Rice, 1995; Chung and Miller, 1995). Previous work has been largely with agronomic crops, in part because of the large amounts of residues from these crops and because of crop rotations. Increases in practices of conservation tillage and integrated pest management have heightened interest in employment of crop residues in vegetable cropping systems (Putnam, 1986; Masiunas, Weston, and Weller, 1995; Mwaja, Masiunas, and Weston, 1995). Since vegetable crops leave little surface residues, employment of organic residues in production of these crops will have to be through production of cover crops which precede vegetables or through importing of residues from other fields.

In the present study, three crops, viz., corn (*Zea mays* L.), soybean (*Glycine max* Merr.), and sunflower (*Helianthus annuus* L.) suggested to have potential allelopathic effects and also potential to increase soil fertility through nutrient enrichment were selected. The objective was to evaluate these residues for weed control and other yield-enhancing benefits in vegetable crops. Crop residues were imported from other fields or grown onsite and assessed for effects on weed control and zuccini summer squash (*Cucumis pepo* L. var. *melopepo* Alef.) and tomato (*Lycopersicon esculentum* Mill.) fruit production. This research was initiated in May 1992 and was conducted in the 1992, 1993, and 1994 growing seasons.

MATERIALS AND METHODS

Application of Residues Imported from Other Fields

In June 1992, corn (*Zea mays* L.) silage, cracked soybean (*Glycine max* Merr.) seeds, and cracked striped sunflower (*Helianthus annuus* L.) seeds were applied to Hadley very fine sandy loam (Typic Udifluvent, coarse-silty, mixed, nonacid, mesic) at South Deerfield, Massachusetts. Seeds were cracked with a hammer mill. The seeds were used in 1992 to facilitate initiation of the experiments so that protocols for evaluating residues could be developed. Sources of sunflower and soybean vegetative residues were not available in the area and had to be grown. In 1993 and 1994, the sunflower and soybean seeds were replaced by sunflower silage and by soybean hay, grown on the farm at South Deerfield. The corn silage used in 1992 was obtained from the University dairy farm and was from an unidentified cultivar. Corn silage used in 1993 and 1994 was made from sweet corn 'Silver Queen'. Sunflower silage was from a striped-seeded cultivar. After harvest with a forage chopper, corn and sunflower silages were stored in 6 m diameter silos made from 1.2 m lath-type snow fence, covered with plastic, and overwintered. Silages of corn and sunflower were chosen because these materials can be harvested and applied easily compared to the unchopped residues. Soybean hay was bales of cv. 'Williams' harvested at initiation of bloom and stored indoors.

Soon after June 1 of each year, after the danger of frost, treatments were applied for immediate planting of the squash and tomatoes. The residues were incorporated shallowly by raking into the soil at 6, 12, and 24 Mg/ha.

Residues were applied also as mulches at 24 Mg/ha in separate treatments. Amounts of application are expressed as dry tonnages, although the materials were applied without drying. The mulches were about 4 cm thick with the vegetative materials and about 2 cm thick with the seeds. In 1992, the land was fertilized with 135 kg N, 60 kg P, and 110 kg K per ha broadcasted and disked in before the residues were applied. In 1993 and 1994, fertilization used was 135 kg N, 30 kg P, and 90 kg K per ha (to apply less P than in the previous year in this P-rich soil according to soil tests). No irrigation was used in the experiments. In all years, 'Celebrity' tomato seedlings were transplanted, and 'Black Beauty' summer squash was direct-seeded following application of the residues. Treatments were replicated in four randomized complete blocks. In each block, one plot was uncultivated to give a weedy control. Also, in each block, one plot was hoed weekly to remove all weeds, giving a clean control.

In 1992, plots were 3.66 m × 3.66 m (12 ft × 12 ft) with 4 plants spaced at 0.9 m in 3 rows spaced at 1.2 m. In 1993 and 1994 experiments were conducted on an adjacent site to incorporate some changes in plot layout and to give fresh land for the application of the vegetative residues. The spacings in 1992 were too dense for convenient harvests. In 1993 and 1994, plants were spaced 1.2 m apart in 2 rows spaced at 1.8 m; therefore, the density in 1993 and 1994 was half of that in 1992 (6 plants per plot compared to 12 plants per plot).

Rainfall for the region was above normal in 1992 (Table 1), was 70% of normal in 1993, and except for June was 1.5 to 2 times normal in 1994. Daytime mean air temperatures of 22°C (72°F) during the growing seasons of 1992, 1993, and 1994 were normal temperatures for the region. General procedures were to count numbers of weeds emerging weekly or biweekly after application of treatments to get an assessment of the efficacy of the residues in weed control early in the development of the crops. Weed populations are reported from counts in a 0.25 m^2 quadrat (a 0.5 m square) in each plot. Counts

TABLE 1. Precipitation for 28-day periods ending on the last day of the last full week of June, July, or August during the three years of the study.

Year	Monthly rainfall, cm and % of normal					
	June		July		August	
	cm	%	cm	%	cm	%
1992	13.7	176	9.9	119	10.0	113
1993	6.2	79	5.0	59	6.1	70
1994	3.6	46	17.1	201	13.4	150

continued for 6 weeks in 1992 and for 4 weeks in 1993 and 1994. These counts are reported separately when significant and are pooled when insignificant by dates. Similarly, weed emergence was counted for individual plots of tomato or squash, and the data are reported separately when significant by crop and are pooled when insignificant. Generally, masses of the early emerging weeds were small and were not measured. At the end of August of each year, 12 weeks from planting, weed masses were measured by harvesting shoots from the 0.25-m^2 quadrate, when tomato productivity was waning.

Harvests of squash and tomato were made weekly as soon as harvestable produce was available and until production was ended by crop senescence or by a killing frost. The same plants from middles of the plots were harvested each week. Fruits were not graded. Rotten or malformed fruits were not included in the harvest, but oversized fruits (mainly squash) were included. In 1994, the wet season, rotting of fruits contributed to lower yields. No significant differences in yields by treatments occurred during the weekly harvests in any season; therefore, only total harvests for each season are reported.

Incorporation of Residues Grown on Site

Sweet corn 'Silver Queen', soybean 'Williams', and striped-seed sunflower were seeded in mid-June 1992 and mid-June 1993 in plots adjacent to the plots to which the imported residues were applied. Seeding of each crop was in 3.6 × 3.6 m^2 plots with 30 cm (1 ft) or 90 cm (3 ft) between-row spacing to give different densities and succulence of residues. In each block, one plot in which no crop was grown was kept weedfree by hoeing throughout the season, and in one plot, weeds were uncontrolled. Treatments were arranged in four randomized complete blocks. Weeds were controlled in each seeded plot by hoeing or pulling throughout the season to keep the cover crops essentially weed free. Very little hand labor was required to control weeds in the stands of sunflower, which had an observable suppressive effect on weed establishment and growth. Stands of soybeans had little observable suppressive effects on weeds in the crop. Weed growth in corn stands was between that in the sunflowers and soybeans. Shoot biomass in each treatment was determined in the last week of September, which was about one week before a killing frost. Corn at this time in the 30 cm rows was lodged and without grain, and corn in the 90 cm rows was standing with grain in the milk stage. Soybeans remained standing in each density of planting and had filled green pods. Sunflowers in the 30 cm rows was lodged and without grain, whereas sunflowers in the 90 cm rows were standing with apparently mature seeds. The amount of dry matter produced by the crops and weeds is reported in Table 2. Corn and sunflowers produced more dry matter and were more uniform in production between the years than the soybeans. Mean dry matter production for all crops was higher in 1992 than in 1993. In general, dry

TABLE 2. Dry matter added by crops or weeds grown on site for the 1993 and 1994 experiments.

Crop residue		Dry matter added (Mg/ha)	
Plants	Spacing (cm)	1993	1994
Corn	30	16.1	9.8
Corn	90	9.8	9.1
Soybean	30	8.6	6.3
Soybean	90	6.2	3.4
Sunflower	30	13.9	7.6
Sunflower	90	10.6	7.5
Weeds	N/A*	15.1	4.7

*N/A, weeds were at a volunteered spacing.

matter production for the on-site production was between the lowest and middle amounts applied by the imported residues. After the killing frost, the crops were cut and laid uniformly by hand across the plots, and then all plots were disked, incorporating residues into the top 8 cm of soil. Volunteer plants from the cover crops did not present weed problems in any plots. Weedy plots were covered with lambsquarters (*Chenopodium album* L.) and redroot pigweed (*Amaranthus retroflexus* L.), each of which produced seeds.

No fertilizers were applied to the plots on which the residues were grown. Nutrients absorbed or fixed by the residue crops were returned to the soil when the residues were incorporated. Nitrogen in the residues varied with the crop, with corn having 1 to 1.2% N (dry wt), soybeans having 2.3 to 3.2% N, sunflowers having 1.6 to 1.7% N, and weeds having 1.4 to 1.9% N in the aerial parts, with higher analyses occurring in 1992 than in 1993. Elemental composition did not differ with density of planting. The plots were established in 1992 and in 1993 for tomato and squash grown in June 1993 and June 1994. At the end of May 1993 and May 1994, the plots were disked lightly (about 7 cm deep) to kill weeds (mainly *Galinsoga parviflora* Cav.) that grew from the time of disking in the fall until the spring and to prepare the seedbed. Planting of tomato and squash and measurements of weed populations and recording of crop yields followed the protocols used for the plots with imported residues in 1993 and 1994.

Data Processing

All data were processed by analysis of variance with comparison of means against the weedy or clean controls by least significant difference (Steel and Torrie, 1980).

RESULTS

Application of Residues Imported from Other Fields

These experiments were conducted in 1992, 1993, and 1994. Residues of corn, soybean, or sunflower were grown off site and imported and applied on the plots by incorporation or by mulching.

Weed Control

A diverse weed population, dominated by redroot pigweed, galinsoga (*Galinsoga parviflora* Cav.), and lambsquarters, emerged in all treatments. Records were kept by weed species, but no trends by species were noted; hence, total emerging weed populations are reported. In 1992, weeds were counted weekly, and averages are reported for a 6-week period (Table 3). Plots with mulches or incorporated plant materials (corn silage, soybean seeds, sunflower seeds) had fewer emerging weeds than plots with no applied materials and hoed free of weeds weekly (clean plots). Control of emergence was equal among species of materials, but increasing the amounts of applied materials further reduced the weed population. The highest amount of incorporated materials gave better control of emergence than application of the same amount of materials as mulches (Table 3).

At 12 weeks after planting in 1992, weed biomass were estimated (Table 3). At this time, mulching of residues gave better control than incorporation, and except for the soybean residues, increasing the amounts of incorporated residue increased suppression of weed growth. In the soybean-treated plots, control was variable, ranging from complete to no control. On average, all residues in any mode of application suppressed weed growth relative to the weedy control. The residues generally were not as effective in weed suppression as the clean hoeing throughout the season. The weeds at the final harvest were mostly galinsoga, which made a large surge of growth about 2 weeks before the 12-week harvest (Table 3).

In 1993, weed counts at 2 weeks are presented as representative of populations from a series of counts of emergence. At 2 weeks, diminished populations of early emerging weeds reflect the immediate effects of the residues (Table 4). The 12 and 24 Mg/ha incorporations of residues were strongly inhibitory to early weed establishment, giving as good or better control as clean tillage (Table 1). In 1993, a harvest of weeds was made in late August about 12 weeks from planting (Table 4). The biomass of weeds in each plot receiving residues was much lower than that in the weedy control and with only two exceptions was not above that of the clean control. Weed suppression in 1993 was the best among the three seasons (Table 4). The effectiveness is

TABLE 3. Mean weekly emergence and biomass of weeds in plots with imported residues in 1992.

Residue	Amount and method of application of residues			
	Incorporated (Mg/ha)			Mulched (Mg/ha)
	6	12	24	24
	----- Number of emerging weeds/m^2 per week -----			
Corn	31	14	9	43
Soybean	33	18	14	46
Sunflower	33	26	10	39
Weedy, 116; Clean, 73				
	------ Biomass of weed shoots (g dry wt/m^2) ------			
Corn	107	78	50	74
Soybean	48	165	222	6
Sunflower	137	66	18	33
Weedy, 654; Clean, 48				

For weed emergence, treatment means, LSD ($P \leq 0.05$) = 19; effects of crop and interaction of crop × treatment were insignificant.
For biomass (12 weeks), treatment means, LSD ($P \leq 0.05$) = 177; tomato 160, squash 84, $P \leq 0.05$ by F-test; interaction of crops × treatments was insignificant.

attributed in part to the incapacity of weeds to grow during the dry weather of this season. Rainfall during June, July, and August 1993 was 70% of normal.

In 1994, counts of weed emergence were made at 1 week and 2 weeks after planting (Table 5). No significant differences in weed populations occurred on these dates, so data are pooled. In early and middle June, rainfall was below normal, and crop residues had a suppressing effect on weed emergence. Weed emergence in plots with residues was as low as or lower than that in the clean-tilled plots and was well below that of the weedy plots. Weed populations did not differ statistically with species of residues or methods of application in 1994 (Table 5).

Weed growth was promoted by rains in late June 1994, so that by 4 weeks after planting, masses of weeds in the plots could be measured. Residues imparted no significant weed control on this date relative to the weedy plots. Average dry weights of weed shoots from plots with residues was 199 g/m^2 compared to 267 g/m^2 from the weedy plots. No weeds were harvested from the clean-tilled plots. Weed masses did not differ between tomato and squash crops. In 1994, weed dry masses among the residue-treated plots did not differ, and no control was imparted with a weed harvest at 12 weeks after planting. Weed masses averaged 616 g/m^2 in the residue-treated plots and 599 g/m^2 in the weedy plots. No weeds were harvested from the clean-tilled

TABLE 4. Early emergence (2 weeks) and biomass (12 weeks) of weeds in plots with imported residues in 1993.

Residue	Amount and method of application of residues			
	Incorporated (Mg/ha)			Mulched (Mg/ha)
	6	12	24	24
	---------- Weed population, number/m^2 ----------			
Corn	188	67	5	14
Soybean	154	90	7	40
Sunflower	226	94	25	40
Weedy, 210; Clean, 168				
	------ Biomass of weed shoots (g dry wt/m^2) ------			
Corn	28	16	7	14
Soybean	20	18	17	41
Sunflower	13	25	11	75
Weedy, 560; Clean, 16				

For emerging weeds (2 weeks), LSD ($P \leq 0.05$) = 23. Effects of crop and interaction of crop × residue were insignificant.
For biomass (12 weeks), LSD ($P \leq 0.05$) = 29; Effects of crop and interaction of crop × residue were insignificant.

TABLE 5. Early emergence of weeds in plots with imported residues in 1994 (means of counts taken 1 and 2 weeks after planting).

Residue	Amount and method of application of residues			
	Incorporated (Mg/ha)			Mulched (Mg/ha)
	6	12	24	24
	---------- Weed population, number/m^2 ----------			
Corn	88	44	18	27
Soybean	82	59	7	9
Sunflower	90	46	29	16
Weedy, 202; Clean, 63				

Treatment mean, LSD ($P \leq 0.05$) = 32. Effects of crop and interaction of crop × treatment were insignificant. Biomass of weeds are reported in text.

plots. In the weedy plots and in the plots treated with corn or soybean residues, redroot pigweed was 75 to 90% and lambsquarters was 5 to 15% of the weed population by visual indexing. Velvetleaf (*Abutilon theophrasti* Medic.) was the major weed in the plots treated with sunflower residues. The percentage of velvetleaf in the population ranged from 30 to 80% and increased as the amount of sunflower residues increased. It was ascertained after the residues had been applied that the sunflower silage had come from a field in which the sunflower planting was infested with velvetleaf. Seeds of the velvetleaf apparently remained viable through the ensiling.

Crop Yields

Yields of crops were estimated by weekly harvest of fruits from initiation of production until productivity was ended by senescence or cold weather. Variability due to uneven short-term productivity by individual plants caused a lack of statistical significance of weekly yields by treatments; therefore, weekly yields were pooled to give total seasonal production. Total seasonal yields differed significantly among treatments. In 1992, total tomato or squash fruit yields in managed plots, that is, those receiving residues or weekly cultivation, were higher than those in the weedy plots receiving no management (Table 6). Tomato yields were equal between the residue-treated plots and clean-tilled plots. Tomato yields rose insignificantly as the amount of residues applied by incorporation rose and did not differ between mulched or incorporated applications. Squash yields did not change with amount of a given residue applied, but yields were higher with the application of soybean or sunflower residues than with corn residues or with clean tillage (Table 6).

For plots on which residues were applied, yields in 1993 were quite simi-

TABLE 6. Total fruit yields of tomato and squash for the entire 1993 harvest in plots with imported residues.

Residue	Amount and method of application of residues			
	Incorporated (Mg/ha)			Mulched (Mg/ha)
	6	12	24	24
	kg fruits/m^2			
	Tomato			
Corn	6.5	5.7	7.9	6.9
Soybean	5.3	5.7	7.3	6.8
Sunflower	5.5	6.6	7.1	7.5
Weedy, 1.1; Clean, 7.4				
	Squash			
Corn	8.6	9.7	7.3	7.2
Soybean	10.9	10.5	10.6	11.1
Sunflower	10.7	13.6	9.1	9.6
Weedy, 3.3; Clean, 7.1				

LSD ($P \leq 0.05$) = tomato, 2.3; squash, 2.2.

lar to those in 1992 in spite of the dry weather (Table 7). Crop appearance during growth indicated that the plants were not suffering from drought. In 1993, yields of both crops in all managed systems (residues applied or clean tilled) exceeded yields in the weedy plots. Yields were not affected significantly whether the residues were applied as mulches or incorporated. Increasing the amount of incorporated residue increased the yields of squash but not of tomato (Table 7). Tomato yields in the residue-treated plots markedly exceeded yields in the clean-tilled plots in 1993. Squash yields from the residue-treated plots exceeded yields from the clean-tilled plots but less consistently or less markedly than with tomato.

In 1994, yields were smaller than those in the previous two years, reflecting the poor weed control and high percentage of rotten fruits in the wet season (Table 8). Residue applications of 24 Mg/ha as mulch or by incorporation were needed to enhance yields above the weedy control. Yields from treated plots were usually below those of the clean control. Yields increased as amounts of corn or soybean residues increased but not with sunflower residue (Table 8).

Incorporation of Residues Grown on Site

In these experiments conducted in 1993 and 1994, residues were grown on site in 30-cm or 90-cm row spacings in the year previous to planting of tomato and squash and disked into the soil. Treatments referred to as *clean*

TABLE 7. Total fruit yields of tomato and squash for the entire 1993 harvest in plots with imported residues.

Residue	Amount and method of application of residues			
	Incorporated (Mg/ha)			Mulched (Mg/ha)
	6	12	24	24
	kg fruits/m^2			
	Tomato			
Corn	6.2	7.0	6.4	6.1
Soybean	6.6	6.3	6.6	7.4
Sunflower	5.2	5.6	5.2	6.8
Weedy, 1.4; Clean, 3.6				
	Squash			
Corn	8.8	8.9	10.8	8.7
Soybean	8.0	9.9	11.0	8.8
Sunflower	8.6	8.7	9.6	10.0
Weedy, 3.5; Clean, 7.4				

LSD ($P \leq 0.05$) = tomato, 1.8; squash, 1.8.

TABLE 8. Total fruit yields of tomato and squash for the entire 1994 season in plots with imported residues.

Residue	Amount and method of application of residues			
	Incorporated (Mg/ha)			Mulched (Mg/ha)
	6	12	24	24
	kg fruits/m^2			
	Tomato			
Corn	1.3	1.6	2.3	2.3
Soybean	0.8	1.2	2.7	2.1
Sunflower	1.6	1.6	1.4	1.5
Weedy, 1.5; Clean, 4.1				
	Squash			
Corn	2.6	2.9	3.2	2.9
Soybean	1.8	2.1	3.2	2.7
Sunflower	2.8	2.6	2.2	2.5
Weedy, 2.2; Clean, 4.2				

LSD ($P \leq 0.05$) = tomato, 0.3; squash, 1.2.

were plots tilled in the previous year to control weeds and kept weed-free by tillage during the current crop-growing season. *Weedy* plots were those in which no control was administered, either in the season of production of the residues or during the crop growing season.

Weed Control

Only plots with corn residues had visual signs of weed control from time of disking of residues in the fall until time for planting of main crops in the springs of 1993 and 1994. A spring disking and hoeing were needed to prepare all plots for planting. For 1993, populations of emerging weeds are reported for counts at 2 weeks and 4 weeks after seeding or transplanting (Table 9). At 2 weeks, weed infestations were severe in all plots, and no treatment was imparting control. The plots were hoed again to institute a new base for measuring control, and in the following 2 weeks, control with the residues relative to weedy plots was apparent. Except with soybean or sunflower residues in 30-cm rows, control of weed emergence by the residues was as good as in the clean-tilled plots, hoed weekly. At the end of August 1993, about 12 weeks from planting, residues imparted weed control in all cases relative to the weedy control and in all cases but one (corn in 90-cm rows) gave control equivalent to the clean-tilled control (Table 10).

In 1994, at 2 weeks after planting, residues gave no significant weed

TABLE 9. Densities of early emerging weeds in 1993, 2 weeks and 4 weeks after planting in plots in which residues were grown on-site in 30-cm or 90-cm rows.

Crop residue	Time of measurement after planting			
	2 weeks		4 weeks	
	30-cm rows	90-cm rows	30-cm rows	90-cm rows
	number of weeds/m^2			
Corn	44	200	20	22
Soybean	259	314	68	44
Sunflower	268	218	46	29
Weedy	925		275	
Clean	216		23	

LSD ($P \leq 0.05$) = 51, 2 weeks; 22, 4 weeks. On these dates, crop (tomato or squash) had no significant effect on weed populations and did not interact with treatments.

TABLE 10. Dry weights of weeds at 12 weeks after planting in 1993 in plots in which residues were grown on-site.

Crop residue	Row width for residue production	
	30-cm	90-cm
	Dry weights, g/m^2	
	Tomato	
Corn	150	317
Soybean	161	181
Sunflower	90	119
Weedy, 629; Clean, 214		
	Squash	
Corn	21	38
Soybean	53	93
Sunflower	30	32
Weedy, 818; Clean, 51		

LSD ($P \leq 0.05$) = 30 for all means. Results of ANOVA: residues, significant ($P \leq 0.01$); crops, ($P = 0.07$); interaction, ($P \leq 0.01$).

control compared to the weedy plots and, in general, were not as effective as the clean control, hoed weekly. By 4 weeks, the residues had imparted weed control relative to the weedy plots and were as effective as the clean-tilled control (Table 11). By the end of August after two months of wet weather, weed control by the residues was not apparent (Table 12).

TABLE 11. Densities of early emerging weeds at 2 and 4 weeks after planting in 1994 in plots in which residues were grown on-site.

Crop residue	Date and density of residue production			
	2-wk		4-wk	
	30-cm rows	90-cm rows	30-cm rows	90-cm rows
	number/m^2			
Corn	88	54	137	194
Soybean	130	170	384	210
Sunflower	99	99	244	169
Weedy	150		518	
Clean	62		266	

Treatments, LSD ($P \leq 0.05$) = 108. Means for dates (106 weeds/m^2 at 2 week and 265 weeds/m^2 at 4 week) are significantly different ($P \leq 0.05$). Weed emergence did not differ between tomato and squash plots on these dates.

TABLE 12. Late season (12 weeks) dry weights of weeds in 1994 in plots in which residues were grown on-site.

Crop residue	Crop and density of residue production			
	Tomato		Squash	
	30-cm rows	90-cm rows	30-cm rows	90-cm rows
	g/m^2			
Corn	626	793	525	312
Soybean	814	896	392	470
Sunflower	702	778	269	400
Weedy	752		580	
Clean	0		0	

Treatment LSD ($P \leq 0.05$) = 238. Means for tomatoes (670 g/m^2) significantly greater than means for squash (368 g/m^2) ($P \leq 0.05$); interaction of crop × treatment was insignificant.

Crop Yields

For the plots with residues grown onsite in 1992 and the main crops grown in 1993, best tomato yields were in plots in which corn was densely planted (30-cm rows) (Table 13). This treatment produced the most residue (Table 2) and gave the best weed control. Yields from this plot exceeded those from the clean-tilled plot. Other treatments produced yields lower than or about equal

TABLE 13. Total yields of tomato and squash in plots in which the crop residues were grown on site in 1992 and 1993 and left on the soil surface for the 1993 and 1994 main crops.

Crop residue	Year of harvest and density of residue production			
	1993		1994	
	90-cm rows	30-cm rows	90-cm rows	30-cm rows
	Crop yields (kg/m^2)			
	Tomato			
Corn	6.6	4.5	3.2	3.2
Soybean	3.8	4.0	2.5	1.5
Sunflower	3.6	5.4	2.7	2.5
	Weedy, 1.9; Clean, 5.6		Weedy, 1.9; Clean, 4.5	
	Squash			
Corn	7.4	7.2	3.1	3.0
Soybean	7.7	7.4	3.5	2.4
Sunflower	7.4	6.5	3.4	2.2
	Weedy, 2.1; Clean, 5.7		Weedy, 2.4; Clean, 3.2	

Treatments in 1993, LSD ($P \leq 0.05$) = 0.8 for tomato; 2.3 for squash.
Treatments in 1994, LSD ($P \leq 0.05$) = 1.4 for tomato; 0.8 for squash.

to those in the clean-tilled plots. All managed treatments (residues from 30-cm rows or from 90-cm rows or clean-tilled) gave higher tomato yields than the weedy plots. Squash yields from the plots with residues grown onsite were the same regardless of density of the residues. However, yields of squash from all of the plots with residues exceeded yields from the clean-tilled plots or from the weedy plots. In 1994, the residues did not enhance tomato yields relative to those harvested from the weedy plots, and in most cases yields were below those of the clean control (Table 13). On the other hand, relative to the weedy plots, squash yields were enhanced by the residues grown in 30-cm rows, and these yields were equivalent to those from the clean control.

DISCUSSION

Residues of corn, soybean, and sunflower were evaluated for their efficacy in weed control and yield enhancement of tomato and squash. In one experiment for 3 years, residues were imported, that is, grown off-site, for application. These residues were applied as a mulch (24 Mg/ha) or incorporated into the soil in amounts of 25%, 50%, and 100% of the mulch tonnage. The goal was to assess the mulch as a barrier or allelopathic inhibitor of weeds and to assess the incorporated residues as allelopathic inhibitors. Although growers may prefer to use vegetative residues, in the first year of the study, soybean and sunflower seeds were used because vegetative residues of these crops were not available locally. The first year allowed for assessment of soybean and sunflower seeds as mulches and allelopathic inhibitors and allowed for development of experimental design and protocols for assessment of the efficacy of the treatments. Corn silage was a common treatment in all years of the study. Corn silage was readily available in the area, and its use facilitated harvest and application of corn residues.

Another aspect of the study involved growing corn, soybean, and sunflower crops (*residue crops*) at two densities on sites, disking these crops into the ground after a frost, and assessing the effects of their residues on weed control and tomato and squash productivity (*main crops*). To help avoid volunteer plants in the main crops, the residue crops were planted late in the growing season to lessen seed production. Plant size in the sparse plantings (90-cm row spacing) was larger than that in the dense planting (30-cm row spacing), but total dry matter production in the dense planting exceeded that in the sparse planting. Dry matter production varied between two growing seasons of the residue crops and lay between the lowest and highest amounts of crop residues imported into the plots in the experiment mentioned above. In a dry season averaging 70% of normal rainfall, crop residue production was less than that in a season in which rainfall averaged slightly above normal.

Weed Control

Treatments employed in this study imparted benefits in weed control and subsequent crop yield. With the imported residues, weed control increased, with few anomalies, as the tonnages of residue application increased during years in which rainfall was below or slightly above normal. Mulches gave the same level of weed control as residues incorporated at the same tonnages. Control of emerging weeds and weed mass with the residues was equivalent to or better than that with frequent tillage. Generally, significant differences in weed control did not occur among species of residues. Release of allelopathic compounds from surface-applied or shallowly incorporated crop residues is suggested as a factor in weed control (Rice, 1984; Martin, McCoy, and Dick, 1990; Mwaja, Masiunas, and Weston, 1995). The residues used in this study are known to impart allelopathic effects (Leather, 1983; Martin, McCoy, and Dick, 1990; Macias et al., 1993). Decaying of residues is reported as an important factor in release of allelopathic compounds (Schreiber and Williams, 1967; Putnam and Duke, 1978). This effect was noted in the present study, which indicated that 6 Mg residues incorporated into soil per hectare were as effective as 24 Mg/ha applied by incorporation or as mulch. Efficacy of crop residues declines with aging in the field, but corn residues are phytotoxic for as long as 22 weeks (Guenzi, McCalla, and Norstadt, 1967). In this study, residues remained in the soil from imported applications at least through the duration of the harvest. Corn residues were the most persistent and soybean residues were the least persistent by visual observations. In the first season in plots treated with seed residues, a surge of weed growth was noted before the end of the season, indicating that the efficacy of these residues had declined. Seed residues, however, were very effective in controlling emerging weeds during high rainfall in June 1992.

In each season, air temperature averages were normal for the region, about 22°C daytime, which is a favorable temperature for microbial production of phytotoxic compounds (Jessop and Stewart, 1983). Crop residues from a previous crop have been noted to retard soil warming and crop emergence and development (Fortin and Pierce, 1990; Fortin and Hamill, 1994). By the time that residues were applied in the current study, soils had warmed to 22°C to 15 cm depths, and physical effects of the residues were perceived as being beneficial in moisture conservation with no effects on soil temperature. In a wet season with rainfall about 170% of normal in July and August, the residues gave no weed control, except that which occurred on early emerging weeds in June when rainfall was below normal in that season. Although the lack of effectiveness of crop residues in controlling weeds may be due to leaching of allelopathic materials into the soil or to their accelerated decomposition, the perceived result in the wet season was that weeds responded to the additional moisture and were less affected by allelopathic factors or by physical barriers (mulches) than in the dry season. Purvis (1990) reported an inhibitory effect of wheat (*Triticum aestivum* L.) residues on a current wheat crop during cool, dry weather but noted a stimulating effect during wet, mild weather. Putnam (1986) commented on the importance of relative dry conditions for effectiveness of crop residues in weed control. Growers cannot predict whether a season will be wet or dry but should be cautioned that crop residues may not control weeds in seasons with rainfall far in excess of normal. With irrigated crops to which residues are applied, precision irrigation is suggested to place water near the plants rather than between the rows where weed growth might be stimulated by irrigation and uncontrolled by residues.

Weed control was better in squash plots than in tomato plots as the season progressed because the leaf canopy of squash covered the ground in about 6 weeks, whereas 8 or more weeks were needed for the tomato canopy to cover the plots. In general, the residues were better in imparting control of early emerging weed than in control later in the season. Some of the effects observed here were due to the early part of the growing season being dry and later parts being wet. Decomposition of residues may be necessary for release of allelopathic agents, but fresh residues generally are more effective in weed control than aged residues (Guenzi et al., 1967; Yakle and Cruse, 1984). Aging of residues likely lessens their effectiveness in weed control in the field. Hence, it is suggested that residues would be more useful in crops that grow rapidly and develop leaf canopy coverage quickly than in crops that produce incomplete or slow leaf canopy coverage of the ground.

A severe infestation of velvetleaf (*Abutilon theophrasti* Medic.) weed occurred in the sunflower-amended plots in one experiment (1994). Seeds of the weed were imported into the plots with the residues. These results suggest that sunflower is not allelopathic to velvetleaf, for the weed infestation occurred in the field in which the sunflowers were grown for silage. In plots during on-site production of residues, weed control by apparent allelopathy in the sunflower planting was excellent, and once the sunflower crop was established, hand tillage was not required. Intensive attention to weed control by hand was needed throughout the growing season to prevent dominance of weeds in the soybean plots. Occasional hand-weeding and tillage were required with the corn crop. Growing of crop residues in a cover crop on site and disking them into the ground for tomato or squash production in the following year imparted weed control, but in each season, control of early emerging weeds was initially weak. Corn was the only on-site produced residue that imparted weed control before planting. With all plots, after preparation of the seed beds, another hoeing was needed to institute weed control. Without this tillage, weed control by cover crop residues alone likely would not have been adequate for crop production. Masiunas, Weston, and Weller (1995) reported that supplemental weed management was needed for commercial tomato production following rye (*Hordeum vulgare* L.) cover crops. Also, tillage seemed to increase the efficacy of residues, perhaps by promoting dispersal of the residues in the soil, promoting decay and release of allelopathic agents, or even drying the soil surface to create a mulch (Putnam and DeFrank, 1983; Scott et al., 1987). The perception that tillage improved efficacy of residues held with the imported residues as well. Biomass of residue crops is an important factor in weed control (Putnam and DeFrank, 1983), and apparently biomass of soybean residues produced on-site was insufficient to control weeds.

Crop Yields

Tomato yields rose slightly but seldom significantly with increased incorporations of residues, but squash yields did not. These trends suggest that a threshold level of weed control occurred with the lowest tonnages of incorporated residues and that nutritional benefits from the residues did not raise yields. Apparently enough nutrition was provided by the soil and by the fertilizers to sufficiently nourish the crops, even though the residues added substantial amounts of plant nutrients. Nitrogen concentrations were about 2.1% in the corn silage, 6.2% in soybean seeds, 3.1% in sunflower seeds, 1.60% in sunflower silage, and 3.15% in soybean hay. Some of the growth-inhibiting effects of residues on crops and weeds is attributed to immobilization of N and other nutrients (Mahler and Hemamda, 1993; Mwaja, Masiunas, and Weston, 1995; Smith and Sharpley, 1990). Tomato nutrition is hindered by allelopathy from weed residues (Qasem and Hill, 1989). Crop nutrition did not appear to be a factor in the current study, for none of the crops in any year showed any developmental signs of nutrient deficiencies, even in the presence of massive weed populations. In the dry year, 1993, when rainfall was 70% of normal, mulched or incorporated residues enhanced yields over those obtained with tillage. Moisture conservation through a mulching effect of the residues on or near the soil surface is suggested as enhancing yields. In the year with near-normal rainfall, 1993, only squash yields were enhanced by the residues relative to those obtained in tilled plots. In the wet year, 1994, no yield enhancement occurred with the residues. In 1994, many of the fruits rotted in the field, and yields of even the tilled plots were half of those in the previous years.

CONCLUSIONS

This research demonstrated that crop residues can be used to control weeds but that supplemental practices may be needed for commercially acceptable control in some circumstances. Control is achieved with mulched or incorporated residues. Varying thickness of mulches was not tested here, but incorporated residues imparted control even at relatively low application. Control may be fragile in wet years in which aggressive weed growth makes mulches or incorporated residues ineffective or in which the efficacy of the residues is dissipated by decay or leaching. Nutrients added with the mulches did not affect crop yields, likely because soil fertility was high in the fields of these experiments. Supplemental weed control was important in ensuring efficacy by residues, especially if control of emerging weeds was weak. Moisture conservation from the mulching effects of the residues in the soil or on the soil surface appeared to be important, for residue-treated plots substantially out-yielded tilled ones in a dry year.

REFERENCES

Chung, I.-M. and D.A. Miller. (1995). Natural herbicide potential of alfalfa residue on selected weed species. *Agronomy Journal* 87: 920-925.

Follett, R.F., J.W.B. Steward, and C.V. Cole. (ed.). (1987). *Soil fertility and organic matter as critical components of production systems.* Soil Science Society of America Special Publication 19. Madison, WI: Soil Science Society of America.

Fortin, M.-C. and A.S. Hamill. (1994). Rye residue geometry for faster corn development. *Agronomy Journal* 86: 238-243.

Fortin, M.-C. and F.J. Pierce. (1990). Development and growth effects of crop residues on corn. *Agronomy Journal* 82: 710-715.

Friedman, J. and G.R. Waller. (1983). Seeds as allelopathic agents. *Journal of Chemical Ecology* 9: 1107-1117.

Guenzi, W.D., T.M. McCalla, and F.A. Norstadt. (1967). Presence and persistence of phytotoxic substances in wheat, oat, corn, and sorghum residues. *Agronomy Journal* 59: 163-165.

Jessop, R.S., and L.W. Stewart. (1983). Effects of crop residues, soil type, and temperature on emergence and early growth of wheat. *Plant and Soil* 74: 101-109.

Kohli, R.K., D. Batish, and H.P. Singh. (1998). Allelopathy and its implications in agroecosystems. *Journal of Crop Production* 1: 169-202.

Leather, G.R. (1983). Sunflowers (*Helianthus annuus*) are allelopathic to weeds. *Weed Science* 31: 37-42.

Macias, F.A., R.M. Varela, A. Torres, and J.M.G. Molinillo. (1993). Potential allelopathic guaianolides from cultivar sunflower leaves, var. SH-222. *Phytochemistry* 34: 669-674.

Mahler, R.L. and H. Hemamda. (1993). Evaluation of the nitrogen fertilizer value of plant materials to spring wheat production. *Agronomy Journal* 85: 305-309.

Martin, V.L., E.L. McCoy, and W.A. Dick. (1990). Allelopathy of crop residues influences corn seed germination and early growth. *Agronomy Journal* 82: 555-560.

Masiunas, J.B., L.A. Weston, and S.C. Weller. (1995). The impact of rye cover crops on weed populations in a tomato cropping system. *Weed Science* 43: 318-323.

Mwaja, V.N., J.B. Masiunas, and L.A. Weston. (1995). Effects of fertility on biomass, phytotoxicity, and allelochemical content of cereal rye. *Journal of Chemical Ecology* 21: 81-96.

Purvis, C.E. (1990). Differential response of wheat to retained crop stubbles. I. Effect of stubble type and degree of decomposition. *Australian Journal of Agricultural Research* 41: 243-251.

Putnam, A.R. (1986). Allelopathy: Can it be managed to benefit horticulture? *HortScience* 21: 411-413.

Putnam, A.R. and J. DeFrank. (1983). Use of phytotoxic plant residues for selective weed control. *Crop Protection* 2: 173-181.

Putnam, A.R. and W.B. Duke. (1978). Allelopathy in agroecosystems. *Annual Reviews of Phytopathology* 16: 431-451.

Qasem, J.R. and T.A. Hill. (1989). Possible role of allelopathy in competition between tomato, *Senecio vulgaris* and *Chenopodium album* L. *Weed Research* 29: 349-356.

Rice, E.L. (1984). Manipulated ecosystems: Roles of allelopathy in agriculture. In *Allelopathy*, ed. E.L. Rice. 2nd ed. Orlando, FL: Academic Press, Inc., pp. 8-73.

Rice, E.L. (1995). *Biological Control of Weeds and Plant Diseases: Advances in Applied Allelopathy*. Norman: University of Oklahoma Press.

Rizvi, S.J.H. and V. Rizvi. (ed.). (1992). *Allelopathy: Basic and Applied Aspects*. London: Chapman and Hall.

Schreiber, M.M. and J.L. Williams. (1967). Toxicity of root residues of weed grass species. *Weeds* 15: 80-81.

Scott, T.W., J. Mt. Pleasant, R.F. Burt, and D.J. Otis. (1987). Contributions of ground cover, dry matter, and nitrogen from intercrops and cover crops in a corn polyculture system. *Agronomy Journal* 79: 792-798.

Smith, S.J. and A.N. Sharpley. (1990). Soil nitrogen mineralization in the presence of surface and incorporated crop residues. *Agronomy Journal* 82: 112-116.

Steel, R.G.D. and J.H. Torrie. (1980). *Principles and Procedures of Statistics*. 2nd Ed. New York: McGraw Hill.

Tollenaar, M., M. Mihajlovic, and T.J. Vin. (1993). Corn growth following cover crops: Influence of cereal cultivar, cereal removal, and nitrogen rate. *Agronomy Journal* 85: 251-255.

Yakle, G.A. and R.M. Cruse. (1984). Effects of fresh and decomposing corn plant residue extracts on corn seedling development. *Soil Science Society of America Journal* 48: 1143-1146.

Allelopathic Influence of Decomposing Wheat Residues in Agroecosystems

Ibrahim S. Alsaadawi

SUMMARY. The concept that some crop plants have allelopathic effects is receiving greater attention as a potential mean that can be exploited in various ways for controlling agricultural pests. Wheat residues have been reported to inhibit growth of certain weeds, crops that follow wheat crop in crop rotation, soil nitrification and biological nitrogen fixation. Thus, crop production can be enhanced by avoiding the inhibitory effects or by exploiting favorable interactions. The present article presents an overview of research on the allelopathic influence of wheat residues in agroecosystems. *[Article copies available for a fee from The Haworth Document Delivery Service: 1-800-342-9678. E-mail address: <getinfo@haworthpressinc.com> Website: <http://www.HaworthPress.com> © 2001 by The Haworth Press, Inc. All rights reserved.]*

KEYWORDS. Allelopathy, wheat residues, crops, weeds, nitrification, N_2 fixation

INTRODUCTION

Incorporation of crop residues in soil is a very potential tool of improving physical and chemical properties of soil and thereby enhancing plant growth

---

Ibrahim S. Alsaadawi is Chief Scientific Researcher, Agricultural Research Center, P.O. Box 765, Baghdad, Iraq.

The author wishes to thank his assistants Mr. Mohammed I. Al-Tikrity and Ziyad T. Balasem for their help in the preparation and typing of the manuscript.

[Haworth co-indexing entry note]: "Allelopathic Influence of Decomposing Wheat Residues in Agroecosystems." Alsaadawi, Ibrahim S. Co-published simultaneously in *Journal of Crop Production* (Food Products Press, an imprint of The Haworth Press, Inc.) Vol. 4, No. 2 (#8), 2001, pp. 185-196; and: *Allelopathy in Agroecosystems* (ed: Ravinder K. Kohli, Harminder Pal Singh, and Daizy R. Batish) Food Products Press, an imprint of The Haworth Press, Inc., 2001, pp. 185-196. Single or multiple copies of this article are available for a fee from The Haworth Document Delivery Service [1-800-342-9678, 9:00 a.m. - 5:00 p.m. (EST). E-mail address: getinfo@haworthpressinc.com].

and yield. However, plant growth and productivity are sometimes reduced (Waller et al., 1978; Chou, 1992). The reduction has been attributed in part to the release of phytotoxic compounds during organic matter decomposition or to the synthesis of these phytotoxins by microorganisms that utilize the residues as nutrient source (Chou, 1992; Alsaadawi, Zwain, and Shahata, 1998). This allelopathic phenomenon has been subjected to extensive studies since it shows a great possibility of using it to control weeds, eliminate the deleterious effects of some plant species, regulate the nitrogen budget in soil and exploit the beneficial interactions in crop rotation (Rice, 1984; Rizvi and Rizvi, 1992). Wheat (*Triticum aestivum* L.) is known to have allelopathic potential against crops and weeds (Putnam, DeFrank, and Barnes, 1983; Alsaadawi, Zwain, and Shahata, 1998). Most of the work on allelopathy of wheat has been carried out on the use of its residues to control weeds and test its effects on subsequent crop in crop rotation.

ALLELOPATHIC EFFECTS OF WHEAT RESIDUES ON CROPS

The early studies on allelopathy of wheat residues have been made by McCalla and Duley (1949) who found that mulching the soil with wheat straw at the rate of 2 to 4 tons per acre reduced seed germination of corn from 44 to 92%. They suggested that the detrimental effects of wheat residues might be due to a combination of toxins released from residues and microorganisms that grow more rapidly owing to nutrient supply from residues. Subsequent work by Guenzi and McCalla (1962) showed that most of crop residues tested including wheat residues, contain water-soluble phytotoxins that can inhibit growth of corn, wheat and sorghum. Additional research by Norstadt and McCalla (1963) revealed that a fungus, viz., *Penicillium urticae* is stimulated in stubble mulch plots and produces a potent toxin Patulin that inhibits the growth of wheat. This fungus constitutes 90% of the fungal population in Crofton silt loam soil where stubble-mulch wheat farming occurs in Lincoln, Nebraska, USA. Guenzi and McCalla (1966) showed that wheat residues contain water-soluble phytotoxins that can inhibit growth of wheat. Additional work by these two workers showed that *p*-coumaric, syringic, vanillic, ferulic and *p*-hydroxybenzoic acids were found in water and NaOH extracts of soil taken from no-till plots. The isolated phenolic acids were also detected in tilled plots but in low level. Maximum toxicity of wheat residues occurred near the harvest time and disappeared after 4 weeks.

Kimber (1973) conducted a series of experiments in Western Australia to determine why residues of wheat straw left in the field reduced yield of subsequent wheat crop. He concluded that both nitrogen immobilization and toxins released from wheat straw were responsible for the yield reduction of wheat grown in soil containing excess amount of wheat residues. Wheat residues were found to be the main cause of inhibition of wheat seed germination. Lodhi, Bilal, and Malik (1987) studied the allelopathic interactions in wheat-wheat and wheat-cotton rotation systems, which are commonly utilized in Pakistan. They found that seed germination and growth of wheat and cotton were significantly inhibited by the extracts of wheat mulch and associated soil. Five phenolic acids were isolated from wheat mulch and their possible role in rotation system was discussed.

Waller et al. (1978) found that the forage yield from wheat crop taken at early growth stage for the conventional-tillage system is twice that for no-tillage. After a series of tests they found that wheat soils worked under no tillage and conventional practices contain inhibitory substances and they believed that no tillage soil may have equaled or exceeded the conventional soil in allelopathic activity. Subsequent work on this problem, Cast et al. (1990) revealed that soil of no tillage system was more inhibitory to wheat growth than soil of conventional system and suggested that the phytotoxins responsible for inhibition are products of microbial activity.

On the southern high plains of Texas, many farmers are planting cotton *Gossypium hirsutum* into wheat straw. Hicks et al. (1989) conducted several experiments to test if wheat straw has allelopathic potential against several cotton cultivars. After a series of pot experiments, they found that cotton seedling development was inhibited by aqueous extracts of wheat, and they identified phytotoxin tolerant Paymaster 404 and non-tolerant Acala A246 cultivars. When the two cultivars were evaluated in the presence of wheat residues under field conditions, it was found that emergence was reduced by an average of 9% and 21% for tolerant and non-tolerant varieties, respectively.

Cox, Collins, and Caviness (1976) reported that the allelopathic effect of wheat straw is one of the major factors limiting yield of soybeans in a double cropping system. Collins and Caviness (1978) extended their work on the problem and screened several soybean cultivars for tolerance to phytotoxic effect of wheat residues. They found that the test cultivars differed significantly in their tolerance to phytotoxins present in the wheat residues.

Blum et al. (1991) determined the concentration of several phytotoxic compounds in soil samples taken from wheat no-till and conventional-till, and fallow conventional-till soybean cropping systems from July to October, 1989. They found that total phenolic content of soil sample from no till system was significantly higher than those from all other systems. Most of the phenolic acids appeared to be accumulated at the top 2.5 cm of the soil and occurred in bound forms in the soil samples.

In Iraq rice is cultivated shortly after wheat harvesting. The remaining straw and root residues of wheat are customarily mixed with the soil and submerged prior to planting. It has been observed that growth of rice is markedly decreased following wheat (Alsaadawi, Zwain, and Shahata, 1998).

The reduction in growth didn't appear after straw burning. A series of experiments were conducted to test the possible allelopathic effects of wheat residues on rice growth. Aqueous extracts at a rate of 2 and 4 g per 100 ml significantly reduced growth of root, shoot and whole plant of rice (Table 1). This indicates that the extracts contain water-soluble phytotoxins, which can be released easily during the decomposition of wheat residues in soil especially under waterlogged conditions. A subsequent greenhouse experiment demonstrated that the incorporation of relatively small amounts of wheat residues (2 and 4 g/kg soil) to soil caused a significant reduction in root, top and seedling growth of rice (Table 2), giving another strong evidence on allelopathic effects of wheat residues on rice growth.

To test the persistence of phytotoxicity of wheat residues in soil, wheat straw was mixed with soil at the rate of 0, 20 and 60 g/kg soil, and allowed to decompose for the time intervals of 2, 4, 6, 8, and 10 weeks under greenhouse conditions. The extracts of all wheat residues-soil proportions tested significantly inhibited growth of radicle, coleoptile and seedling growth of rice compared with growth in control soil. The phytotoxicity was obviously high

TABLE 1. Effect of wheat residue extracts on growth of rice seedlings (Alsaadawi, Zwain, and Shahata, 1998, reprinted with permission from the publisher).

Extract Concentration (g residue/100 ml distilled water)	Dry Weight (mg)* Root	Shoot	Whole Plant	Weight of Whole Plant as % of Control
Control (distilled water)	8.7	23.8	32.5	0
2	6.6	18.4	25.0	78
4	4.5	16.5	21.0	65
LSD = 0.05	1.2	4.5	5.2	-

*Each value is an average of 3 replicates and each replicate consists of 3 seedlings.

TABLE 2. Effect of wheat residues on growth of rice seedlings (Alsaadawi, Zwain, and Shahata, 1998, reprinted with permission from the publisher).

Wheat Residue Added ($g \cdot kg^{-1}$ soil)	Dry Weight (mg)* Root	Shoot	Whole Plant	Weight of Whole Plant as % of Control
Control (Peat moss)	24.9	53.4	78.3	0.0
2	14.1[a]	48.0[a]	58.6[a]	74.8
Control (Peat moss)	23.4	56.6	80.0	0.0
4	13.0[a]	53.8[a]	66.9[a]	83.6

Controls had similar rates of peat moss only.
*Each value is an average of 3 replicates and each replicate consists of 3 plants.
[a] Represents significant difference from respective control at 5% level.

after 2 weeks of decomposition, but gradually declined with increased decomposition period. The phytotoxicity became very low or disappeared in certain residues-soil rates after 8 weeks of decomposition. In Iraq rice fields are prepared immediately after wheat harvesting. The straw is mixed with soil and submerged prior to transplanting of rice seedlings. Therefore, it is possible that the phytotoxic substances, which are released during the decomposition periods, may be responsible for the reduction of rice seedling growth. In another experiment, four phytotoxins, viz., *p*-coumaric, *p*-hydroxybenzoic, vanillic and syringic acids were isolated from decomposing wheat residues. All identified phytotoxins appeared in high amount after 2 weeks of decomposition and then reduced to small amounts after 10 weeks. The trend of toxicity at different decomposition periods is apparently analogous with the trend of phytotoxin level, giving strong evidence on the allelopathic effect of wheat residues on rice growth.

ALLELOPATHIC EFFECTS OF WHEAT RESIDUES ON WEEDS

There is a strong evidence on the allelopathic effect of crop residues and their possible use to control weeds. One approach for utilization of crop residues is to plant allelopathic smother crops to suppress weed growth through competition and allelopathy after residue decomposition (Guenzi and McCalla, 1962; Collins and Caviness, 1978). Another approach is to desiccate 30-40 day old allelopathic plants by freezing or applying herbicides and then using their residues to suppress weeds (DeFrank and Putnam, 1978; Lehle and Putnam, 1982).

Wheat residues have been tested for allelopathic potential against several weed species. Putnam and DeFrank (1979) studied the allelopathic effects of residues of several smother crops including wheat on weed population. They found that wheat plants which were desiccated by the herbicide glyphosate and paraquate or by freezing in spring or fall reduced weed biomass by 76-88%. Additional work by these workers demonstrated that wheat straw reduced weed densities and biomass by an average of 90% compared with those plots without residues (Putnam and DeFrank, 1983). Most of the inhibitory effects were found to be due to phytotoxicity of wheat residues.

Liebl and Worsham (1983) revealed that the allelopathic influence of wheat mulch was partially responsible for reduced population of morning glory, *Ipomoea lacunosa*, and other weeds in no-till wheat cropping system. Subsequent experiments demonstrated the presence of ferulic acid and other compounds which are responsible for reducing growth of certain weeds in no-till cropping system.

Shilling et al. (1985) initiated a series of experiments to test the allelopathic potential of wheat and rye, *Secale cereale*, on certain weed species in different cropping systems. The results indicated that morning glory plant population and biomass were found to be greatly affected by wheat mulch. Planting corn no-till into desiccated green wheat cover crop reduced *Ipomoea* spp. biomass up to 97% compared to a non-mulched tilled treatment. However, there was no mulch effect in double-crop soybean system.

Narwal, Sarmah, and Tamak (1998) tested the allelopathic effects of residues of several smother crops including wheat on the population and biomass of weeds. They found that the phytotoxicity of wheat residues was greater in certain stages of growth and decreased in later stages. Wheat straw caused 16.8% suppression of broad leaf weed population but didn't reduce grassy weeds. In contrast, wheat straw significantly reduced biomass of both types of weeds. This reduction increased up to 120 days after sowing and decreased afterwards.

Lehman and Blum (1997) found that residue of wheat living tissue prior to glyphosate desiccation significantly inhibited emergence of ivy-leaf morning glory, *Ipomoea hederacea*, and red root pigweed, *Amaranthus retroflexus*. The toxicity of wheat debris continued even one month after glyphosate desiccation. Beyond this time, only pigweed emergence was inhibited by wheat debris. The allelopathic potential of wheat debris was found to be influenced by its age, water content, total carbon, total nitrogen, and pH of the soil.

Field observations in Iraq showed low population of *Echinochloa crus-galli* weed in rice fields previously cultivated with wheat. Zwain (1996) conducted several greenhouse experiments to test if wheat residues exhibit allelopathic potential against *E. crus-galli* and other important weed, *Amaranthus retroflexus*. Results indicated that aqueous extract of wheat residues at 2 and 4% concentration significantly reduced seed germination and growth of root, shoot and whole plant of the test species (Table 3). This inhibition

TABLE 3. Effects of wheat residue extracts on seed germination and growth of test weeds (Zwain, 1996).

Test Weeds	Extract Conc. (g residue/100 ml)	Dry Weight (mg)* Root	Shoot	Whole Plant	Germination (%)
A. retroflexus	0	50.7	60.8	111.5	100.0
	2	24.2	43.1	67.3	80.4
	4	17.0	32.6	49.6	70.6
	LSD (0.05)	12.1	6.5	10.4	3.6
E. crus-galli	0	52.9	91.6	144.5	100.0
	2	28.1	70.8	98.9	95.2
	4	28.9	62.7	91.6	87.6
	LSD (0.05)	4.3	5.2	8.4	3.3

* Average of 3 replicates and each replicate consists of 3 plants.

increased with increasing extract concentration. The aqueous extract was more inhibitory to *E. crus-galli* than to *A. retroflexus*. Subsequent experiment showed that soil with wheat residues suppressed growth of the test weed species, and this suppression effect increased with increasing amount of wheat residues (Table 4). The inhibitory effect was more pronounced in *A. retroflexus* than in *E. crus-galli*. When the phytotoxicity of wheat residues against *E. crus-galli* was evaluated over different times of decomposition using a method outlined previously (Alsaadawi, Zwain, and Shahata, 1998), it was found that the phytotoxicity persisted for 6 weeks and disappeared afterwards (Table 5). These results may explain the low population density of *E. crus-galli* in rice field.

ALLELOPATHIC EFFECTS OF WHEAT RESIDUES ON NITROGEN FIXATION AND NITRIFICATION

There is considerable evidence concerning the allelopathic effects on nitrogen cycle. However, most of the work has done on the two processes in nitrogen cycle namely nitrogen fixation and nitrification and little or even no information is available about the allelopathic effects on other processes of nitrogen transformation in nature such as ammonification and denitrification (Rice, 1992).

Nitrogen Fixation

Crop residues are reported to have potential allelopathic effects on nitrogen fixation, but information concerning the allelopathic effects of wheat

TABLE 4. Effect of different amounts of wheat residues on growth of test weeds (Zwain, 1996).

Test Weeds	Residue Added ($g \cdot kg^{-1}$ soil)	Dry Weight (mg)* Root	Shoot	Whole Plant	% Reduction of Whole Plant
A. retroflexus	Control	0.7	27.0	27.7	0.0
	2	0.5	5.0[a]	5.5[a]	80.0
	Control	0.5	30.8	31.3	0.0
	4	0.6[a]	5.3[a]	5.9[a]	81.0
E. crus-galli	Control	21.8	62.9	84.7	0.0
	2	4.6[a]	33.1[a]	37.7[a]	55.5
	Control	19.9	79.9	99.8	0.0
	4	5.8[a]	30.2[a]	36.0[a]	64.0

Controls had similar rate of peat moss only.
*Average of 3 replicates and each replicates consists of 3 plants.
[a] Represents significant difference from respective control at 5% level.

TABLE 5. Effect of different amounts of wheat residues decomposed at different periods in soil on growth of *Echinochloa crus-galli* (Zwain, 1996).

Decomposition Period (weeks)	Residues Added ($g \cdot kg^{-1}$ soil)	Seedling Length (cm)*	% of Control
2	0	13.1	0.0
	20	10.5	80.2*
	60	10.1	77.1*
4	0	10.7	0.0
	20	9.1	85.0*
	60	8.6	80.4*
6	0	12.8	0.0
	20	9.8	76.6*
	60	10.6	82.8*
8	0	9.3	0.0
	20	9.3	0.0
	60	9.2	99.0
10	0	9.8	0.0
	20	9.7	99.0
	60	9.4	96.0

* Means significantly different from respective control at 5% level.

residues on this process is not available. Previously, we found that the reduction in growth of rice following wheat was primarily due to phytotoxic compounds released from wheat straw after its decomposition in soil (Alsaadawi, Zwain, and Shahata, 1998). Additional investigation by Zwain, Alsaadawi, and Shahata (1998) indicated that incorporating wheat residues in soil significantly inhibited nitrogen fixation by *Nostoc musicorum* and *Anabaena cylindrica* which are the dominant blue green algae found in rice fields in Iraq (Table 6). In most cases, the inhibition continued at all decomposition periods and increased with increasing amount of wheat residues in soil. Thus, it appears that wheat residues could also exert indirect allelopathic effect on rice through reduction of nitrogen fixation of blue green algae. The work on the allelopathic effects of wheat residues was extended to study why poor growth of mungbean, *Vigna radiata*, is observed after wheat harvest. Zwain (1996) found that wheat residues at a rate of 4 and 8 g/kg soil significantly inhibited growth and nitrogen fixation in mungbeans.

Nitrification

Nitrification is an important process that causes nitrogen leaching in soil. Such leaching may severely reduce the productivity of plants, increase water pollution and disease incidence in plants (Huber et al., 1977). In order to

TABLE 6. Effect of rates of wheat residues decomposed in soil for different decomposition periods on N_2 fixation in *Anabaena cylindrica* and *Nostoc musicorum* (Zwain, Alsaadawi, and Shahata, 1998, reprinted with permission from the publisher).

Decomposition Period (weeks)	Residue rate ($g \cdot kg^{-1}$ soil)	*Anabaena cylindrica* C_2H_4 ($\mu mole \cdot g^{-1} \cdot h^{-1}$)	*Anabaena cylindrica* % of Control	*Nostoc musicorum* C_2H_4 ($\mu mole \cdot g^{-1} \cdot h^{-1}$)	*Nostoc musicorum* % of Control
2	0 (Control)	0.79a	100	0.74a	100
	20	0.69ab	87.34	0.62b	83.78
	80	0.65b	82.27	0.42c	56.75
4	0	1.06a	100	0.87a	100
	20	0.85b	80.18	0.86a	98.85
	80	0.84b	79.24	0.76b	87.35
6	0	0.83a	100	1.21a	100
	20	0.69b	83.13	1.14b	94.21
	80	0.68b	81.92	0.91c	75.20
8	0	0.72a	100	0.72a	100
	20	0.71a	98.61	0.60b	83.33
	80	0.55b	76.38	0.53c	73.61
10	0	1.17a	100	0.72a	100
	20	1.06b	90.59	0.71a	98.61
	80	1.00b	85.47	0.61b	84.72

*Numbers within each decomposition period for each species followed by the different letters are significantly different at 0.05 level according to Duncan's multiple range test.

enhance utilization of nitrogen fertilizers, nitrification inhibitors have been extensively used. Considerable work has been done on the allelopathic effects on nitrification in natural ecosystems and few have been conducted in agroecosystems.

Various crop species have been reported to inhibit nitrification in soil (Lodhi, 1981; Alsaadawi et al., 1986; Alsaadawi, 1988). The inhibition is caused by either crop residues (Lodhi, 1981; Alsaadawi et al., 1986; Alsaadawi, 1988) or by root exudation (Moore and Waid, 1971). Little work has been done on the allelopathic potential of wheat on nitrification. In a laboratory experiment, Moore and Waid (1971) found that root exudates of wheat caused a pronounced and persistent decrease in nitrification rate over a 80 days experimental period. Zwain (1996) evaluated the allelopathic effects of wheat residues on nitrification using soil incubation as explained by Alsaadawi (1988). He found that aqueous extract of wheat residues at a rate of 10% (w/v) and the incorporation of relatively small amount of wheat residues (g/kg soil) into the soil significantly inhibited nitrification rate over 35 days experimental period (Table 7). The aqueous extract caused high inhibition of nitrification at 10 and 15 days after incubation and started to decline afterwards. In contrast, wheat residues caused high reduction in nitrification rate

TABLE 7. Effect of aqueous extract and residues of wheat on nitrification (Zwain, 1996).

Treatments	% Nitrification*							
	Incubation Period (days)							
	0	5	10	15	20	25	30	35
Control (DW)	4.4	4.9	9.2	32.3	57.6	58.8	59.2	61.3
Wheat Extract (10%, w/v)	3.5 (20.5)	4.2 (14.3)	4.7 (48.9)	19.7 (39.0)	45.6 (20.8)	57.3 (2.6)	57.6 (2.7)	58.4 (4.7)
Peat Moss (Control) ($g \cdot kg^{-1}$ soil)	4.1	23.5	43.3	44.4	54.9	55.3	55.6	58.8
Wheat Residues ($g \cdot kg^{-1}$ soil)	3.7 (9.8)	4.3 (81.7)	4.9 (88.7)	5.4 (87.8)	16.4 (70.1)	20.9 (62.2)	39.7 (28.6)	40.9 (30.4)

Values in parenthesis represent % nitrification inhibition.

$$\% \text{ Nitrification} = \frac{\text{Amount of } NO_3}{\text{Amount of } (NO_3 + NH_4)} \times 100$$

$$\% \text{ Nitrification inhibition} = \frac{\text{Nitrification of control} - \text{Nitrification of treatment}}{\text{Nitrification of control}} \times 100$$

at 5, 10, 15, 20 and 25 days after incubation and decreased thereafter. Apparently the residue was more inhibitory than the aqueous extract.

CONCLUSION AND PROSPECTS

The data presented here demonstrate the potential allelopathic effects of wheat in agroecosystems. The strategies by which wheat and other crop residues can be used to control weeds and exploited the beneficial interactions in crop rotation have been established in several countries. Therefore, the time is appropriate to utilize allelopathic crop residues as one of the agricultural practices for the sustainable management of agroecosystems.

REFERENCES

Alsaadawi, I.S. (1988). Allelopathic suppression of nitrification by selected cultivars of *Helianthus annuus* L. *Journal of Chemical Ecology* 14: 722-742.

Alsaadawi, I.S., J.K. Al-Uqaili, A.J. Al-Rubeaa, and S.H. Al-Hadithy. (1986). Allelopathic suppression of weed and nitrification by selected cultivars of *Sorghum bicolor* (L.) Moench. *Journal of Chemical Ecology* 12: 21-29.

Alsaadawi, I.S., K.H.Y. Zwain, and H.A. Shahata. (1998). Allelopathic inhibition of growth of rice by wheat residues. *Allelopathy Journal* 5: 163-169.

Blum, V., T.R. Wentworth, K. Klein, A.D. Worsham, L.D. King, T.M. Gerig, and S.W. Lyu. (1991). Phenolic acid content of soils from wheat no-till, wheat-conventional till and follow conventional till soybeans cropping systems. *Journal of Chemical Ecology* 17: 1045-1068.

Cast, K.B., J.K. McPherson, A.J. Pollard, E.G. Krenzwer, and G.R. Waller. (1990). Allelochemical in soil from no-tillage *versus* conventional-tillage wheat. *Journal of Chemical Ecology* 16: 2277-2289.

Chou, C.H. 1992. Allelopathy in relation to agricultural productivity in Taiwan: problems and prospects. In *Allelopathy: Basic and Applied Aspects*, eds. S.J.H. Rizvi and V. Rizvi. London: Chapman and Hall Press, pp. 179-200.

Collins, F.C. and C.E. Caviness. (1978). Growth of Soybeans on wheat straw residue. Madison, WI: American Society of Agronomy, pp. 92-93 (abstract).

Cox, R.W., F.C. Collins, and C.E. Caviness. (1976). Effect of oat and wheat straw on soybeans in a double cropping system. Madison, WI: American Society of Agronomy, p. 83 (abstract).

DeFrank, J. and A.R. Putnam. (1978). Weed and crop response to allelopathic crop residues. *Proceedings of North Central Weed Control Conference* 33: 44.

Guenzi, W.D. and T.M. McCalla (1962). Inhibition of germination and seedling development by crop residues. *Proceedings of Soil Science Society of America* 26: 456-458.

Guenzi, W.D. and T.M. McCalla. (1966). Phytotoxic substances extracted from soil. *Proceedings of Soil Science Society of America* 30: 214-216.

Hicks, S.K., C.W. Wendt, J.R. Gannaway, and R.B. Baker. (1989). Allelopathic effects of wheat straw on cotton germination, emergence and yield. *Crop Science* 29: 1057-1061.

Huber, D.M., H.L. Warren, D.E. Nelson, and C.Y. Tsai. (1977). Nitrification inhibitors: New tools for food production. *Bioscience* 27: 523-400.

Kimber, R.W.L. (1973). Phytotoxicity from plant residues. III. The relative effect of toxins and nitrogen immobilization on the germination and growth of wheat. *Plant and Soil* 38: 543-555.

Lehle, F.R. and A.R. Putnam. (1982). Allelopathic potential of sorghum (*Sorghum bicolor*) herbage: Isolation of acidic inhibitors of seed germination. *Proceedings of Weed Science Society of America Meeting* 130: 67 (abstract).

Lehman, M.E. and U. Blum. (1997). Cover crop debris effects on weed emergence as modified by environmental factors. *Allelopathy Journal* 4: 69-88.

Liebl, R.A. and A.D. Worsham. (1983). Inhibition of morning glory (*Ipomoea lacunosa*) and certain other weed species by phytotoxic components of wheat (*Triticum aestivum* L.) straw. *Journal of Chemical Ecology* 9: 1027-1044.

Lodhi, M.A.K. (1981). Accelerated soil mineralization, nitrification, and revegetation of abandoned fields due to the removal of crop-soil phytotoxicity. *Journal of Chemical Ecology* 7: 685-694.

Lodhi, M.A.K., R. Bilal, and K.A. Malik. (1987). Allelopathy in agroecosystems: Wheat phytotoxicity and its possible role in crop rotation. *Journal of Chemical Ecology* 13: 1881-1891.

McCalla, T.M. and F.L. Duley. (1949). Stubble mulch studies. III. Influence of soil microorganisms and crop residues on the germination, growth and direction of root growth of corn seedling. *Soil Science Society of America Proceedings* 14: 196-199.

Moore, D.R. and J.S. Waid. (1971). The influence of washing of living roots on nitrification. *Soil Biology and Biochemistry* 3: 69-83.

Narwal, S.S., M.K. Sarmah, and J.C. Tamak. (1998). Allelopathic strategies for wheat management in rice-wheat rotation in northwestern India. In *Allelopathy in Rice*, ed. M. Olofsdotter. Proceeding of the workshop on Allelopathy in rice. Manila, Philippines: International Rice Research Institute, pp. 117-131.

Norstadt, F.A. and T.M. McCalla. (1963). Phytotoxic substances from a species of *Penicillium*. *Science* 140: 410-411.

Putnam, A.R. and J. DeFrank. (1979). Use of cover crops to inhibit weeds. *Proceedings of the IX International Congress of Plant Protection*, pp. 580-582.

Putnam, A.R. and J. DeFrank. (1983). Use of phytotoxic plant residues for selective weed control. *Crop Protection* 2: 173-181.

Putnam, A.R., J. DeFrank, and J.P. Barnes. (1983). Exploitation of allelopathy for weed control in annual and perennial cropping systems. *Journal of Chemical Ecology* 9: 1001-1010.

Rice, E.L. (1984). *Allelopathy*. 2nd edition. Orlando, Florida: Academic Press.

Rice, E.L. (1992). Allelopathic effects on nitrogen cycling. In *Allelopathy: Basic and Applied Aspects*, eds. S.J.H. Rizvi and V. Rizvi. London: Chapman and Hall Press, pp. 31-54.

Rizvi, S.J.H. and V. Rizvi. (1992). Exploitation of allelochemicals in improving crop productivity. In *Allelopathy: Basic and Applied Aspects*, eds. S.J.H. Rizvi and V. Rizvi. London: Chapman and Hall Press, pp. 443-467.

Shilling, D.G., R.A. Liebl, and A.D. Worsham. (1985). Rye (*Secale cereale*) and wheat (*Triticum aestivum*) mulch: The suppression of certain broad leaf weeds and the isolation and identification of phytotoxins. In *The Chemistry of Allelopathy: Biochemical Interaction Among Plants*, ed. A.C. Thompson. ACS Symposium Series 268. Washington, DC: American Chemical Society, pp. 243-271.

Waller, G.R., E.G. Krenzer, Jr., J.K. McPherson, and S.R. McGowan. (1978). Allelopathic compounds in soil from no tillage vs. conventional tillage in wheat production. *Plant and Soil* 98: 5-15.

Zwain, K.H.Y. (1996). *Allelopathic effects of wheat (Triticum aestivum* L.*) on some plant species and nitrogen cycle*. Ph.D. Dissertation, Al-Mustansriyah University, Baghdad, Iraq.

Zwain, K.H.Y., I.S. Alsaadawi, and H.A. Shahata. (1998). Effect of decomposing wheat residues on growth and biological nitrogen fixation of blue green algae. *Allelopathy Journal* 6: 13-20.

Significance of Allelopathy in Crop Rotation

A. P. Mamolos
K. L. Kalburtji

SUMMARY. Rotation systems and allelopathic interactions between plants-plants, plants-insects, plants-microorganisms would be important to exploit allelopathy in optimising the production of rotation systems. The objective of this study is to provide evidences for the importance of allelopathy in crop rotation for weed, insect and disease management with minimal application of external inputs. The relationships between allelochemicals and environmental factors are a key for the growth of plants under rotation. Examples from field crops, forage crops, horticultural species, weeds, and microbes provide evidences for the role allelopathy plays in crop rotation systems. In conclusion, the selection of certain plant sequences under standard environmental conditions may lead to suppression of weeds, insects and diseases and avoid yield decline. *[Article copies available for a fee from The Haworth Document Delivery Service: 1-800-342-9678. E-mail address: <getinfo@haworthpressinc.com> Website: <http://www.HaworthPress.com> © 2001 by The Haworth Press, Inc. All rights reserved.]*

KEYWORDS. Crop rotation, allelopathy, weed management, disease management, role of microbes

---

A. P. Mamolos is Senior Scientist, and K. L. Kalburtji is Associate Professor of Agricultural Ecology, Aristotle University of Thessaloniki, Faculty of Agriculture, Laboratory of Ecology and Environmental Protection, 540 06 Thessaloniki, Greece.

Address correspondence to: K. L. Kalburtji at the above address (E-mail: kalbourt@agro.auth.gr).

The authors would like to thank Professor P. A. Gerakis and A. Gagianas for their critical comments on this manuscript.

[Haworth co-indexing entry note]: "Significance of Allelopathy in Crop Rotation." Mamolos, A. P., and K. L. Kalburtji. Co-published simultaneously in *Journal of Crop Production* (Food Products Press, an imprint of The Haworth Press, Inc.) Vol. 4, No. 2 (#8), 2001, pp. 197-218; and: *Allelopathy in Agroecosystems* (ed: Ravinder K. Kohli, Harminder Pal Singh, and Daizy R. Batish) Food Products Press, an imprint of The Haworth Press, Inc., 2001, pp. 197-218. Single or multiple copies of this article are available for a fee from The Haworth Document Delivery Service [1-800-342-9678, 9:00 a.m. - 5:00 p.m. (EST). E-mail address: getinfo@haworthpressinc.com].

INTRODUCTION

The environmental and economic costs of cropping systems that are chemically intensive and contain little crop diversity have become increasingly apparent (Pimentel et al., 1987; Reganold, Papendick, and Parr, 1990). Crop diversity in time and space appears to be a critical element of sustainable agroecosystems that receive few external inputs (Wagstaff, 1987; Stiner and House, 1989; Bushnell, Francis, and King, 1991). Crop sequences, which include crops with different planting and maturation dates, competition and allelopathy characteristics, and management practices (tillage, mowing, and grazing), result in an unstable environment that prevents proliferation of weeds. Weed management has a key role in the design and function of agroecosystems. Akobundu (1987) estimated that weeds limit crop yields by 5% in the most developed countries, 10% in the less developed countries, and 25% in the least developed countries. Crop rotation is one of the most powerful cultural techniques available to farmers for reducing weed seed and seedling density (Liebman and Dyck, 1993).

Allelopathy is an important mechanism of plant interference mediated by the addition of plant-produced phytotoxins to the plant environment. Chemicals with allelopathic potential may be released into the environment mainly in the rhizosphere in sufficient quantities to affect neighboring plants (Rice, 1984). Allelopathy and competition are difficult, if not impossible, to separate in the field, but both have been well-documented in studies under controlled conditions (Inderjit and Dakshini, 1995). Much research has focused on the detrimental effects of living plants or their residues on growth of higher plants and crop yields. Problems of monoculture, autotoxicity and toxicity of mulch stubble have been attributed to allelochemicals. Putnam and Duke (1978) explored the possibility of utilizing allelopathic crops to control weed growth in farms. Also, Rizvi and Rizvi (1992), and Rizvi et al. (1999) discussed the important role of allelopathic compounds in relation to pest control. Allelopathy has direct and indirect effects on soil microorganisms (e.g., bacteria), which would determine the amount of available nitrogen for other organisms and plants (Rice, 1992). Many crops influence seed germination and plant growth of other crop species through allelochemicals (Choudhuri and Basu, 1989; Hicks et al., 1989; Kalburtzi, Gerakis, and Vokou, 1989; Kalburtzi, Veresoglou, and Gerakis, 1990; Kalburtji and Gagianas, 1997; Kohli, Batish, and Singh, 1998; Inderjit and Keating, 1999). This issue is of great importance for the selection of rotation sequence.

Understanding allelopathy, besides helping to decide which crops will be involved in a rotation system, may hold the key to new weed and pest management strategies. Allelopathic inhibition typically derives from the combined action of a group of allelochemicals. Ultimately, however, utilization of allelopathy in cropping systems will depend on better understanding of the chemical or chemicals involved and their behavior in agroecosystems. The objectives of this article are to summarize research that illustrates the action of allelochemicals and to provide evidence for the significance of allelopathy in relation to crop rotation for weed, insect and disease management with minimal application of external inputs, and also, for the sequence of the crops in a rotation system to avoid yield decline.

COMPETITION AND ALLELOPATHY

The growth and development of plants, under field conditions, are often modified by the presence of other plants of the same or different species. This interference arises from allelopathy and competition (Rice 1984). Thus, the relative importance of competition and allelopathy for ecosystems has been proved difficult to distinguish. Fuerst and Putnam (1983), in order to clarify the roles of competition and allelopathy, suggested a series of steps which included: (i) symptoms of competitive interference, (ii) demonstration that the plant, which is responsible for the effect, is correlated with reduced resources use by the plant that receives the effect, (iii) demonstration of which resources are limiting, and (iv) simulation of that interference, taking under consideration that the responsible plant for the effect is not present, by reduction of the resources at values similar to the ones found during interference. The above-mentioned researchers also stated that allelopathy included (i) identification of symptoms of interference, (ii) toxin production, (iii) simulation of the interference by supplying the toxin as it is supplied in nature, and (iv) quantification of the release, movement, and uptake of toxins. In practice, many of these steps are difficult to carry out. Dekker, Meggitt, and Putnam (1983) suggested that the use of replacement series (standard density with different proportion of plant species; Harper, 1977) makes available an experimental design for the study of allelopathy. Connolly (1988) and Thijs, Shann, and Weidenhamer (1994) found that experimental designs involving different plant densities are better for distinguishing between competition and allelopathy.

Muller (1966, 1969) highlighted the importance of allelopathy to the vegetation patterning. Allelopathic interactions may play a key role in influencing the distribution of vegetation in nature, the yield of various crop species, and weed interference (Del Moral and Muller, 1970; Putnam and Duke, 1978; Rice, 1984). Generally, allelochemicals negatively influence plant performance. As the concentration of allelochemicals increases, the negative effect increases until the plant dies. A large number of bioactive compounds may potentially be released in many communities, and a large number of organisms may be involved in any allelochemical response (Rice, 1984).

RELATIONSHIPS BETWEEN ALLELOCHEMICALS AND ENVIRONMENTAL FACTORS

Most of the natural products that cause allelopathy are secondary products, synthesized by plants and microorganisms, and most of the currently identified compounds are products of the shikimic acid and acetate pathways (Rice, 1984). Allelopathy is affected by many factors, e.g., light quality, light intensity, day length, nutrient limitation (N, P, K, B, Ca, Mg and S), soil moisture, temperature, age of plant organs, genotype, other allelochemicals, and plant density (Figure 1; Rice, 1974; Hall, Blum, and Fites, 1983; Einhellig and Eckrich, 1984; Field et al., 1992; Weidenhamer, 1996). These factors make the effects of allelochemicals very complicated. Furthermore, allelopathic inhibition is typically the result of the combined action of a group of allelochemicals (Einhellig, 1996).

Environmental stress on plant growth changes with time, with the same response happening with allelopathic effects (Einhellig, 1996). Sometimes, the release of allelochemicals depends on the growth stage. Schumacher, Thill, and Lee (1983) reported that wild oat (*Avena fatua* L.) becomes allelopathic to wheat (*Triticum aestivum* L.), when wheat reaches the four-leaf stage. Walnut trees (*Juglans nigra* L.) expressed their allelopathic effects after 15 to 25 years, when active concentrations of juglone have built up in the soil (Ponder and Tadros, 1985). Barnes and Putnam (1983) reported that green straw of rice (*Oryza sativa* L.) was more toxic than fully mature residues.

Allelochemicals are released into the atmosphere (essential oils; Vokou,

FIGURE 1. Environmental factors influencing release of allelopathic compounds from plants.

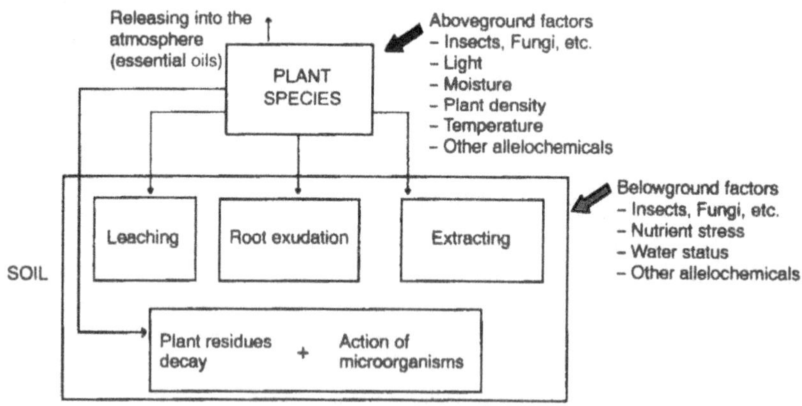

1992) or added to the soil through root exudations, litter decay, leaching, extracting and production by microorganisms (Figure 1; Rice, 1974). The amount of allelochemicals in the soil depends on the quantity of the plant biomass, the plant density, and the concentration and solubility of the specific allelochemicals (Weidenhamer, 1996). Losses of allelochemicals from soils occur in many ways, viz., leaching, chemical processes, microbial breakdown, and uptake by plants (Weidenhamer, 1996). Allelochemicals can be bound to soil organic matter or clay and become inactive (Daldon, Blum, and Weed, 1983, 1989). The toxicity of allelochemicals is a combination of concentration (availability in time) and flux rates (Williamson and Weidenhamer, 1990). These compounds affect soil microorganisms in ways that significantly alter the ecology of the field where the allelopathic plant and their residues are present. The activity and half-life of these allelochemicals derived from plants (crops and weeds) are important when minimum tillage practices are used for soil conservation.

THE BIOLOGICAL ROLE OF ALLELOCHEMICALS

Allelochemicals appear to have multiple functions (Saiki and Yoneda, 1981; Rice, 1984; Horsley, 1986; Lawrey, 1993). They may be antiherbivory, antiparasitic, antifungal, and antibacterial compounds, and phytotoxins. Further, the presence of synergistic interactions is established in several series of bioactive compounds (Feeny et al., 1988). Compounds implicated in autoallelopathy may play a positive role in nature by maintaining spacing of individuals to prevent build-up of predators or pathogenic fungi and bacteria, as well as, to promote outcrossing of individuals (Augspurger, 1983; Augspurger and Kelley, 1984; Singh, Batish, and Kohli, 1999).

The common allelochemicals include phenolics, terpenoids, alkaloids, coumarins, tannins, flavonoids, steroids, and quinones (Einhellig and Leather, 1988). Although most of the simple phenolic acids and flavonoids are allelochemicals, they seem to be weakly phytotoxic in soil and have little selectivity. Salicylic and *p*-hydroxybenzoic acids at high rates (56 to 112 kg/ha), are effective against weeds and are relatively non-selective (Duke and Lydon, 1993). Phenolic derivatives, such as the dihydroquinone sorgoleone, produced by sorghum (*Sorghum bicolor* L.), are extremely phytotoxic in hydroponic culture (Einhellig and Souza, 1992). Most phenolic compounds significantly affected weed growth in the field, while they influenced nutrient uptake (Booker, Blum, and Fiscus, 1992), nutrient cycling (Kalburtji, Mosjidis, and Mamolos, 1999), enzyme activity (Devi and Prasad, 1992), water relations (Barkosky and Einhellig, 1993), and photosynthesis and respiration (Hejl, Einhellig, and Rasmussen, 1993).

Terpenoids are important in crop-weed interactions. The monoterpene

1,8-cineole is strongly phytotoxic. Its derivative cinmethylin has been seriously considered for herbicide development (Grayson et al., 1987). Terpenoids such as taxol and alkaloids such as colchicine and vinblastine, which inhibit plant growth by interference with mitosis, have modes of action identical to certain synthetic herbicides (Vaughan and Vaughan, 1988). Alkaloids from lupine (*Lupinus albus* L.) are excreted to the rhizosphere and inhibit the germination of lettuce (*Lactuca sativa* L.) and grass (Wink, 1983). Rizvi, Mukerji, and Mathur (1981) proposed caffeine as a useful selective herbicide. Einhellig and Leather (1988) suggested that manipulation of timing, formulation, and application rate of an allelochemical as a herbicide can enhance phytotoxicity and improve weed control.

CROP ROTATION

Crop rotation involves growing different crops in systematic and recurring sequence on the same land, in contrast to monoculture in which a particular crop is planted repeatedly in the same field (Liebman and Dyck, 1993). One of the considerations when designing a rotation system is pest management. Especially one has to consider the biological and physical factors, which affect the plant species involved in the rotation and to find ways to influence these factors to achieve best pest management. If the farmer succeeds to minimize the input of pesticides through crop rotation, this will reflect to economical benefits.

The designing and managing of an effective and efficient crop rotation system are complex and in many cases site specific (Putnam, DeFrank, and Barnes, 1983; Liebman and Dyck, 1993). They depend on physical factors (e.g., soil characteristics, climate, etc.), economic factors, and management ability. The benefits to be gained from a well-designed rotation are considerable. Research has indicated that crop rotation can have a greater effect on weed species and densities than tillage practices (Lodhi, Bilal, and Malik, 1987; Weston, 1996). In designing a crop rotation for weed control, the overall key to success is the ecosystem diversity (Chou, 1993; Mongelli et al., 1997). Certain basic principles should be used as guidelines in selecting specific crops for a rotation system, which according to Lampkin (1994) could be the following: (a) alternating between autumn and spring germinating crops; (b) alternating between annual and perennial crops; (c) alternating between closed, dense crops which shade out weeds and open crops such as maize (*Zea mays* L.) which encourage weeds; (d) a variety of cultivation and cutting or topping operations (in particular the traditional cleaning crops, leys and green manures).

Allelopathy and Crop Rotation

Monoculture often leads to a decrease of crop productivity, presumably due to the imbalance of soil microorganisms, accumulation of phytotoxins, etc. Crop rotation is a good method to avoid this problem. Many crop rotation systems applied in many regions lead to higher productivity than monoculture (Liebman and Dyck, 1993; Lampkin, 1994). There is convincing evidence that allelopathic interactions play a crucial role in manipulated ecosystems (Putnam, DeFrank, and Barnes, 1983; Kohli, Batish, and Singh, 1998; Inderjit and Keating, 1999). It would be useful to see how allelopathy and crop rotation can confront problems of plant-plant, plant-microorganisms and plant-insect interactions. Much work has been done on the effect of weeds on crops, crops on weeds, and crops on crops (Kalburtji and Gagianas, 1997; Moyer and Huang, 1997; CruzOrtega et al., 1998; Inderjit and Dakshini, 1998; Krishnan, Holshouser, and Nissen, 1998; Mays, Sistani, and Bishnoi, 1998; Miro, Ferreira, and Aquila, 1998; Rawat et al., 1998). Less work has been done, but results are available to demonstrate the possibility of using allelochemicals as natural pesticides (e.g., Zúñiga, Salgado, and Corcuera, 1985; Chung and Miller, 1995b; Bjorkman, Blanchard, and Herman, 1998; Rizvi et al., 1999). Also, the role of allelopathy in providing protection of seeds before germination and the implication of allelopathy in nitrogen transformations in the soil are some aspects that have been identified (Rice, 1992; Zhang, 1997).

Often, crop rotation systems can reduce the toxic effects of allelochemicals. According to Chou (1992), pagnola grass (*Digitaria decumbens* Stent.) is a high productivity pasture and very stable in many fields in Taiwan. However, after several years of plantation the productivity declines due to autointoxication. Therefore, a crop rotation (pagnola grass-watermelon; *Citrullus lanatus* [Thunb.] Matsum. & Nakai.-pagnola grass) was suggested. After watermelon, the field was replanted with pagnola grass, and its yield was significantly increased up to 40%. The decay rate of allelochemicals is very important because a crop with greater resistance to allelochemicals can be selected to follow a crop producing phytotoxins in a crop rotation system. Some species that contain allelochemicals, e.g., sericea lespedeza (*Lespedeza cuneata* [Dum. de Cours] G. Don.), which contains tannins, had low decomposition rate and low nutrient release when tannins concentrations were high (Kalburtji, Mosjidis, and Mamolos, 1999). If such a species is included in a rotation system, it will probably influence the following crop and the weeds, not only through the release of allelochemicals but through the low availability of nutrients, as well.

Field Crops as Allelopathic Agents

The importance of allelopathy in agricultural practices has been increased with its exploitation in weed control (Rice, 1984). Many researchers reported that crop species and their residues such as wheat, barley (*Hordeum vulgare* L.), sorghum and sorghum-sudangrass hybrids have been effectively used to suppress a range of weed species (e.g., Putnam, DeFrank, and Barnes, 1983; Shilling, Liebl, and Worsham, 1985). Another approach to utilize allelopathy is to screen accessions of allelopathic crops for their ability to suppress weeds. A few crops have been evaluated in this aspect (Leather, 1983). Crop species, which influenced weed growth, are shown in Table 1. Sorghum was reported to provide a good weed killing capacity by Putnam, DeFrank, and Barnes (1983) and Leon (1976) showed that it is autotoxic and should be rotated with other crops for maximum yield. Alsaadawi et al. (1985) conducted a screening experiment to examine the influence of sorghum root

TABLE 1. Common crops utilized for weed interference, with their allelochemicals.

Scientific name	Allelochemicals	References
Avena sativa L.	Phenolic acids, scopoletin	Rice, 1984
Brassica nigra (L.) Koch.	Allyl isothiocyanate	Muller, 1969; Bell and Muller, 1973
Hordeum vulgare L.	Gramine	Hanson et al., 1983; Zúñiga, Salgado, and Corcuera, 1985
Lespedeza cuneata (Dum. Cours) G. Don.	Phenolic acids	Kalburtji and Mosjidis, 1992, 1993a,b
Melilotus spp.	Isoflavonoids, Phenolics	Rice, 1984
Secale cereale L.	Phenolic acids, benzoxazinones	Guenzi and McCalla 1966; Rice,1984; Barnes and Putnam, 1987; Nair, Whitenack, and Putnam, 1990; Mwaja, Masiunas, and Weston, 1995
Sorghum spp.	Phenolic acids, dhurrin, sorgoleone, *p*-hydroxy benzaldehyde, *p*-hydroxy benzoic acid, *p*-coumaric acid, chlorogenic acid, cyanogenetic glycosides	Nicollier, Pope, and Thompson,1983; Forney and Foy, 1985; Netzley and Butler, 1986; Weston, Harmon, and Mueller, 1989; Einhellig et al., 1993; Gubbiga, Worsham, and Corbin, 1996; Hoffman et al., 1996a,b
Trifolium spp.	Isoflavonoids, Phenolics	Rice, 1984
Triticum aestivum L.	Phenolic acids, simple acids	Guenzi and McCalla, 1966; Opoku, Vyn, and Voroney, 1997

exudates of 100 cultivars on inhibition of germination and seedling development of redroot pigweed (*Amaranthus retroflexus* L.) grown in sand. A high variability was observed in the ability of the root exudates of the test cultivars to influence seed germination or seedling growth of the weed. The inhibitory compounds occurring in sorghum plants are mostly phenols (Guenzi and McCalla, 1966).

Putnam (1990) found that living rye (*Secale cereale* L.) has strong interference ability against weeds and provides excellent weed control prior to annual crop establishment. Barnes et al. (1987) isolated and characterized the allelochemicals in rye. These were found to have strong inhibitory action on germinating dicotyledonous weed seedlings and some monocotyledonous weeds (Barnes and Putnam, 1983, 1986). Alborn, Stenhagen, and Leuschner (1992) reported that sorghum varieties were identified as highly and as less susceptible to shoot fly (*Atherigona soccata* [Rond.]) and spotted stem borers (*Chilo partellus* [Swinhoe.]) at the ICRISAT center, Patancheru, India. The same researchers stated that the chemical composition of sorghum is the major cause for this. The less susceptive varieties could be used in crop rotation systems in areas with high populations of the above mentioned insects. Gramine, an indolalkylamine (the side chain at position 3 is $-CH_2-N(Me)_2$) accumulated in barley seedlings (Smith, 1977; Hanson et al., 1983), increased the resistance of barley to aphid *Shizaphis graminum* (Rond.) (Zúñiga, Salgado, and Corcuera, 1985). Similar results to aphids have been observed from crops as wheat, rye, triticale (*Triticum aestivum* L. × *Secale cereale* L.) and maize, which contain hydroxamic acids (Arganoña et al., 1980). Cultivars of wheat, barley, rye, maize, triticale with high concentrations of gramine or hydroxamic acids are very important for rotation systems in areas with high aphid populations. This information is very useful for the production of natural pesticides and resistant varieties (Rizvi and Rizvi, 1992).

Several cereals have allelopathic effects on other crops or themselves. In Taiwan, rice planted twice a year in a monoculture system reduced the second crop yield by about 25% in areas of poor water drainage (Chou, 1990). Also, Chou (1993) found rice seedlings growing poorly in a decomposed rice straw and soil mixture. Compounds, which were autotoxic to rice, were *p*-hydroxybenzoic, syringic, vanillic, ferulic, acetic, *o*-hydroxyphenylacetic, propionic, and butyric acids. The benefits of rotating, cereals and legumes are well established (Gantzer et al., 1991; Mitchell et al., 1991; Lampkin, 1994). Mulvaney and Paul (1984) reported that a crop rotation system, which has been employed worldwide, involved maize and soybean (*Glycine max* [L.] Merr.). Maize yield was enhanced when sown after soybean, but it was reduced if the preceding crop was maize. They related their findings to the autotoxicity of maize. On the contrary, Crookston (1984) found that soybean was not beneficial as a preceding crop on maize. Mays, Sistani, and Bishnoi (1998) suggest that soybean root exudates are responsible for the decrease in wheat and triticale growth and yield. Sarabol and Anderson (1992) reported yield suppression of maize planted after sunflower. Kalburtji and Gagianas (1997) found that the residues of sugarbeet as a preceding crop suppressed cotton (*Gossypium hirsutum* L.) yield by 23% when compared to control (fallow in the preceding period). They also, observed a delay in the formation of cotton flowers or bolls. Early maturity is crucial for obtaining a profitable crop in marginal areas of cotton cultivation such as northern Greece (Kalburtji and Gagianas, 1997).

Forage Crops as Allelopathic Agents

Several pasture species such as bermudagrass (*Cynodon dactylon* [L.] Pers.), johnsongrass (*Sorghum halepense* [L.] Pers.), orchardgrass (*Dactylis glomerata* L.), pagnolagrass, redtop (*Agrostis gigantea* Roth.), rhodesgrass (*Chloris gayana* Kunth.), ryegrass (*Lolium multiflorum* Lam.), smooth bromegrass (*Bromus inermis* Leyss.), tall fescue (*Festuca arundinacea* Schreb.), timothy (*Phleum pratense* L.), western wheatgrass (*Pascopyrum smithii* [Rydb.] A. Löve), ladino clover (*Trifolium repens* L.), sericea lespedeza and alfalfa (*Medicago sativa* L.) were found to produce substances from living plants or release substances from residues that are toxic to the growth of other species (Chou and Young, 1975; Kochhar, Blum, and Reinert, 1980; Larson and Schwarz, 1980; Lolas and Coble, 1982; Luu, Matches, and Peters, 1982; Kalburtji and Mosjidis, 1992, 1993a,b; Ellis and McSay, 1984). In many pastures, the growth of newly planted grass is very high but productivity generally decreases with time (Chou and Young, 1975; Chou, 1989; Miller, 1992, 1996).

Planting alfalfa after alfalfa or other pasture species results in poor plant establishment and production (Rice, 1984; Chung and Miller, 1995a). Hegde and Miller (1990) studied the effect of alfalfa as a preceding crop on alfalfa or sorghum. They verified the existence of autoallelopathy (alfalfa on alfalfa) and allelopathy (alfalfa on sorghum). Both effects were implicated in the growth inhibition of the two crops. Apart from its inhibitory effects alfalfa showed some stimulatory effects. Chopped alfalfa applied to soil was reported to stimulate the growth of tomato (*Lycopersicon esculentum* Mill.), cucumber (*Cucumis sativus* L.), lettuce and several other crops (Einhellig, 1985). Grodzinsky (1992) reported black mustard (*Brassica nigra* [L.] Koch.) contributes to the accumulation of allelochemicals in soil and before they are recommended for any rotation system attention must be paid to this factor. Kalburtji and Mosjidis (1992, 1993a,b) reported that sericea lespedeza residues inhibited the growth of bermuda grass, tall fescue, italian ryegrass and bahiagrass (*Paspalum notatum* Flugge.), but not their germination and emergence.

Horticultural Species as Allelopathic Agents

Several vegetables suppress weeds but on the other hand they enhance the development of soil microflora and assist in the survival of useful insects (Grodzinsky, 1992). Winter rape (*Brassica napus* var. *oleifera* DC.) is sometimes proposed as an effective means of biological control against couch grass (*Agropyron repens* [L.]; Grodzinsky 1992). Bell and Muller (1973) reported a high allelopathic activity of black mustard, which leads to a reduction of weeds (90-96%). Santos and Leskovar (1997) in a greenhouse study found allelopathic effects from broccoli (*Brassica oleracea* L. subsp. *italica* L.) residue on cauliflower (*Brassica oleracea* L. var. *botrytis* L.) germination and seedling growth. Putnam and Duke (1974) found differences in toxicity among 500 cucumber accessions. One allelopathic accession reduced total fresh weight of smooth finger grass (*Panicum miliaceum* L.) and white mustard (*Brassica hirta* Moench.) by about 1/3rd of its biomass when compared to control (indicator plants grown in the absence of cucumbers). Hazebroek, Garrison, and Gainfagna (1989) reported that aqueous extracts from asparagus (*Asparagus officinalis* L.) roots inhibited tomato and lettuce seed germination. Many researchers have reported what is called the replant problem for apple, peach, almond and citrus trees (e.g., Martin et al., 1956; Börner, 1959; Patrick, Toussoun, and Koch, 1964). Patrick, Toussoun, and Koch (1964) reported that allelochemicals from peach roots are the main factor in the problem of replanting peach trees following the removal of an old peach orchard. Jose and Gillespie (1998) observed growth suppressions in maize and soybean growing in a black walnut alley and related them to juglone phytotoxicity.

Weeds as Allelopathic Agents

The phytotoxic effects of knotgrass (*Polygonum avicularia* L.) on bermuda grass (Alsaadawi, Rice, and Karns, 1983), johnsongrass on redroot pigweed (Mikulas, 1984), common buckwheat (*Fagopyrum sagittatum* Gilib.) on canada thistle (*Cirsium arvense* [L.] Scop.) (Eskelsen and Crabtree, 1995) are examples of weeds that negatively influence weeds. Abdul-Rahman and Habib (1986) investigated phytotoxicity of extracts of bermuda grass, johnsongrass and white amaranth (*Amaranthus albus* L.) against dodder (*Cuscuta campestris* L.). Habib and Abdul-Rahman (1988) tested the phytotoxic potential of the above weeds plus nettle-leaved goosefoot (*Chenopodium murale* L.) in controlling the dodder parasite under lathhouse and field conditions. In another experiment, Abdul-Rahman and Al-Naib (1986) found that aqueous extracts and root exudates of bermudagrass significantly inhibited germination and growth of johnsongrass and rough cocklebur (*Xanthium strumarium* L.). These aqueous extracts and root exudates were found to contain phenolic compounds. Einhellig and Leather (1988) used sorghum and soybean seedling bioassay to demonstrate inhibition from kochia (*Kochia scoparia* L.), jerusalem artichoke (*Helianthus tuberosus* L.), rough cocklebur, giant ragweed (*Ambrosia trifida* L.), and dock (*Rumex crispus* L.).

MICROBES AND THEIR ROLE IN CROP ROTATION SYSTEMS

Plant associated microbes and especially soil microbes play an important role in qualitative and quantitative availability of allelochemicals (Blum, Weed, and Daldon, 1987; Nair, Whitenack, and Putnam, 1990; Chase, Nair, and Mishra, 1991; Chase, Nair, and Putnam, 1991a,b). Chemicals in maize may be important in plant-microbes interactions. Hydroxamic acids appear to determine the resistance of maize to European corn borer *Ostrinia nubialis* Huebner (Klun, Tripton, and Brindley, 1967). Microbes utilize phenolic acids and flavonoids as a carbon source (Vaughan, Sparling, and Ord, 1983; Blum et al., 1987). Chapman and Lynch (1983) describe the role of microbes and their metabolites in phytotoxicity of decomposing plant residues. Zentmeyer (1963) reported the antifungal activity of alfalfa on *Phytophthora cinnamoni* Rands. and Levy et al. (1986) on important plant pathogens (*Aspergillus niger* Van. Tiegh., *Fusarium oxysporum* Schlecht., *Phytophthora cinnamoni* Rands., and *Sclerotium rolfsii* Sacc.). The antifungal activity of alfalfa might be useful in agriculture for plant protection. Qasem (1996) found that the aqueous extracts of falcaria (*Falcaria vulgaris* Beruh.), buttercup (*Ranunculus asiaticus* L.), cater pillar plant (*Scorpiurus muricatus* L.) and black nightshade (*Solanum nigrum* L.) were toxic to *Fusarium oxysporum* Schlecht f.sp. *lycopersici*. The growth of pathogens on resistant or susceptible hosts indicates the differential response of host cultivars to a pathogen. This reaction is very important for using resistant cultivars in rotation systems in areas with high pathogenic microbial population. The substances with antimicrobial action are defined as phytoalexins (Putnam and Tang, 1986).

For rice basic cropping systems sequences, such as maize-cabbage (*Brassica oleracea* L. var. *capitata*)-rice, rice-tobacco (*Nicotiana tabacum* L.)-rice, rice-cotton-rice, and rice-watermelon-rice are quite effective in combating root-knot nematode menace (*Meloidogyne* spp.) (Davide and Zorilla, 1983). Johnson (1985) reported that the following rotation systems were quite encouraging in controlling bitter gourd (*Momordica charantia* Linn.) infestation: sweet corn (*Zea mays* L.)-soybean-wheat-soybean-spinach (*Spinacia *oleracea* L.), turnip (*Brassica rapa* L.)-peanut (*Arachis hypogaea* L.)-snap bean (*Phaseolus vulgaris* L.), turnip-peanut-turnip, turnip-peanut-cucumber (*Cucumis sativus* L.)-turnip-soybean and turnip-field corn (*Zea mays* L.)-southern pea (*Pisum* spp.). Patrick and Koch (1963) and Chou and Patrick (1976) found that the crop rotation system, tobacco-ryegrass-maize reduced the problem from pathogen *Thielaviopsis basicola* (Berk. & Br.) Ferraris. and this was related to fungitoxins produced by ryegrass.

CONCLUSION AND PROSPECTS

Recently, a growing number of farmers and researchers, in developed and developing countries, have begun to question the sustainability of many conventional farming methods, and to suggest that agricultural sustainability depends upon the correct management of ecological processes related to nutrient cycling and regulation of pest populations. A crop rotation system for weed, insect and disease management with minimal application of external inputs is important in sustainable agriculture. The present study demonstrates the role of allelochemicals in crop rotation systems. Plant production could be enhanced simply by avoiding or exploiting inhibitory effects. Both actions can be achieved through a selection of crop plants based on their compatibility. Residue management, timing of operations, and proper agronomic practices need to be identified for specific areas of production to make use of allelopathic conditions. Allelochemicals may provide clues to new and environmentally safe herbicides and pesticides. Selecting allelopathic crops and including them in a rotation has shown promising results. New target sites of action can be exploited for phytotoxins. Thus, several allelochemicals have potential as herbicides and pesticides. The ability to understand the physiological basis for allelopathy in a plant species may allow the agronomist or ecologist to work closely with the molecular biologist or plant breeder to selectively enhance the traits responsible for weed or pest suppression.

The role of allelopathic plant species in crop rotation systems is important. Some of the issues for future research are: (a) the role of allelopathic plant species in relation to pest management under rotation systems in different environments, (b) the development of tillage techniques, fertilizer application, and irrigation to increase or decrease allelochemicals in relation to the crop rotation systems, (c) the transfer of appropriate genes to cultivars at aiming pest control, using either traditional plant breeding or molecular biological techniques.

REFERENCES

Abdul-Rahman, A.A.S. and F.A.G. Al-Naib. (1986). The effects of bermudagrass [*Cynodon dactylon* (L.) Pers.] on the germination and seedling growth of cotton and three weed species. *Journal of Agricultural Water Resource Research* 5: 115-128.

Abdul-Rahman, A.A.S. and S.A. Habib. (1986). Effectiveness of herbicides and some plant extracts in controlling dodder on alfalfa. *Journal of Agricultural Water Resource Research* 5: 53-64.

Akobundu, I.O. (1987). *Weed Science in the Tropics: Principles and Practice*. Chichester, UK: Wiley and Sons.

Alborn, H., G. Stenhagen, and K. Leuschner. (1992). Biochemical selection of sorghum crop varieties resistant to sorghum shoot fly (*Atherigona soccata*) and stem borer (*Chilo partellus*): role of allelochemicals. In *Allelopathy: Basic and Applied Aspects*, eds. S.J.H. Rizvi and V. Rizvi. London: Chapman & Hall, pp. 101-117.

Alsaadawi, I.S., J.K. Al-Uqaili, A.J. Al-Rubeaa, and S.M. Al-Hadithy. (1985). Effect of gamma irradiation on allelopathic potential of *Sorghum bicolor* against weeds and nitrification. *Journal of Chemical Ecology* 12: 1737-1745.

Alsaadawi, I.S., E.L. Rice, and T.K.B. Karns. (1983). Allelopathic effects of *Polygonum avicularia* L. III. Isolation characterization and biological activities of phytotoxins other than phenols. *Journal of Chemical Ecology* 9: 761-764.

Arganoña, V.H., J.G. Luza, H.M. Niemeyer, and L.J. Corcuera. (1980). Role of hydroxamic acids in the resistance of cereals to aphids. *Phytochemistry* 19: 1665-1668.

Augspurger, C.K. (1983). Seed dispersal of the tropical tree *Platypodium elegans* and the escape of its seedlings from fungal pathogens. *Journal of Ecology* 71: 759-771.

Augspurger, C.K. and C.K. Kelley. (1984). Pathogen mortality of tropical tree seedlings: Experimental studies of the effects of dispersal distance seedling density and light conditions. *Oecologia* 61: 211-217.

Barkosky, R.R. and F.A. Einhellig. (1993). Effect of salicylic acid on plant-water relationship. *Journal of Chemical Ecology* 19: 237-247.

Barnes, J.P. and A.R. Putnam. (1983). Rye residues contribute weed suppression in no-tillage cropping systems. *Journal of Chemical Ecology* 9: 1045-1057.

Barnes, J.P. and A.R. Putnam. (1986). Allelopathic activity of rye (*Secale cereale* L.). In *The Science of Allelopathy*, eds. A.R. Putnam and C.S. Tang. New York: John Wiley & Sons, pp. 271-286.

Barnes, J.P. and A.R. Putnam. (1987). Role of benzoxazinones in allelopathy by rye (*Secale cereale* L). *Journal of Chemical Ecology* 13: 889-906

Barnes, J.P., A.R. Putnam, B.A. Burke, and A.J. Aasen. (1987). Isolation and characterization of allelochemicals in rye herbage. *Phytochemistry* 26: 1385-1390.

Bell, D.T. and C.H. Muller. (1973) Dominance of California annual grasslands by *Brassica nigra*. *American Midland Naturalist* 90: 277-299.

Blum, U., S.B. Weed, and B.R. Daldon. (1987). Influence of various soil factors on the effects of ferulic acid on leaf expression of cucumber seedling. *Plant and Soil* 98: 111-130.

Bjorkman, T., L.M. Blanchard, and G.E. Harman. (1998). Growth enhancement of shrunken-2 (sh2) sweet corn by *Trichoderma harzianum* 1295-22: Effect of environmental stress. *Journal of the American Society for Horticultural Science* 123: 35-40.

Booker, F.L., U. Blum, and E.L. Fiscus. (1992). Short-term effects of ferulic acid on ion uptake and water relations in cucumber seedling. *Journal of Experimental Botany* 93: 649-655.

Börner, H. (1959). The apple replant problem. I. The excretion of phlorizin from apple root residues. *Contributions of the Boyce Thompson Institute* 20: 39-56.

Bushnell, J., C.A. Francis, and J. King. (1991). Design of resource efficient environmentally sound cropping systems. *Journal of Sustainable Agriculture* 1: 49-65.

Chapman, S.J. and J.M. Lynch. (1983). The relative roles of microorganisms and their metabolites in the phytotoxicity of decomposing plant residues. *Plant and Soil* 74: 457-459.

Chase, W.R., M.G. Nair, and A.R. Putnam. (1991). 22′-oxo-11′-azobenzene selective toxicity of rye (*Secale cereale* L.) allelochemicals to weed and crop species II. *Journal of Chemical Ecology* 17: 9-19.

Chase, W.R., M.G. Nair, and S.K. Mishra. (1991). 22′-oxo-11′-azobenzene microbial transformation of rye (*Secale cereale* L.) allelochemical in field soils by *Acinetobacter calcoaceticus*: III. *Journal of Chemical Ecology* 17: 1575-1584.

Chou, C.H. (1989). The role of allelopathy in biochemical ecology experience from Taiwan. *Biologia Plantarum* 31: 458-470.

Chou, C.H. (1990). The role of allelopathy in agroecosystems: from tropical Taiwan. In *Agroecology: Researching the Ecological Basis for Sustainable Agriculture. Ecological Studies 78*, ed. S.R. Gliessman. Berlin: Springer-Verlag, pp. 105-121.

Chou, C.H. (1992). Allelopathy in relation to agricultural productivity in Taiwan: problems and prospects. In *Allelopathy: Basic and Applied Aspects*, eds. S.J.H. Rizvi and V. Rizvi. London: Chapman & Hall, pp. 179-203.

Chou, C.H. (1993). The role of allelopathy in the diversity of plant-communities in Taiwan. *Botanical Bulletin of Academia Sinica* 34: 211-221.

Chou, C.H. and Z.A. Patrick. (1976). Identification and phytotoxicity of compounds produced during the decomposition of corn and rye residue in soil. *Journal of Chemical Ecology* 2: 369-387.

Chou, C.H. and C.C. Young. (1975). Phytotoxic substances in twelve subtropical grasses. *Journal of Chemical Ecology* 1: 183-193.

Choudhuri, N. and R.N. Basu. (1989). Effect of sugarbeet fruit inhibitors on seed germinability. *Seed Research* 17: 16-22.

Chung, I.M. and D.A. Miller. (1995a). Effect of alfalfa plant and soil extracts on germination and growth of alfalfa. *Agronomy Journal* 87: 762-767.

Chung, I.M. and D.A. Miller. (1995b). Natural herbicide potential of alfalfa residue on selected weed species. *Agronomy Journal* 87: 920-925.

Connolly, J. (1988). What is wrong with replacement series? *Tree* 3: 24-26.

Crookston, R.K. (1984). The rotation effect: What causes it to boost yields? *Crops and Soils* 2: 12-14.

CruzOrtega, R., A.L. Anaya, B.E. HernandezBautista, and G. LagunaHernandez. (1998). Effects of allelochemical stress produced by *Sycyos deppei* on seedling root ultrastructure of *Phaseolus vulgaris* and *Cucurbita ficifolia*. *Journal of Chemical Ecology* 24: 2039-2057.

Daldon, B.R., U. Blum, and S.B. Weed. (1983). Allelopathic substances in ecosystems: Effectiveness of sterile soil components in altering recovery of ferulic acid. *Journal of Chemical Ecology* 9: 1185-1201.

Daldon, B.R., U. Blum, and S.B. Weed. (1989). Differential sorption of exogenously applied ferulic *p*-coumaric, *p*-hydrohybenzoic and vanillic acids in soil. *Soil Science Society of American Journal* 53: 757-762.

Davide, R.G. and A. Zorilla. (1983). National Crop Protection Center (NCPC). *Leaflet No. 2.* Los Banos, Laguna, Philippines.

Dekker, J.H., W.F. Meggitt, and A.R. Putnam. (1983). Experimental methodologies to evaluate allelopathic plant interactions: The *Abutilon theophrasti-Glycine max* model. *Journal of Chemical Ecology* 9: 945-981.

Del Moral, R. and C.H. Muller (1970). Allelopathic effects of *Eucalyptus camaldulensis*. *American Midland Naturalist* 83: 254-282.

Devi, S.R. and M.N.V. Prasad. (1992). Effect of ferulic acid on growth and hydrolytic enzyme activities of germinating maize seeds. *Journal of Chemical Ecology* 18 1981-1990.

Duke, S.O. and J. Lydon. (1993). Natural phytotoxins as herbicides. In *Pest control with Enhanced Environmental Safety*, eds. S.O. Duke, J.J. Menn, and J.R. Plimmer. ACS Symposium Series 524. Washington, DC: American Chemical Society, pp. 110-111.

Einhellig, F.A. (1985). Effects of allelopathic chemicals on crop productivity. In *Bioregulators for Pest Control*, eds. ACS Symposium Series 276. Washington, DC: American Chemical Society, pp. 109-130.

Einhellig, F.A. (1996). Interactions involving allelopathy in cropping systems. *Agronomy Journal* 88: 886-893.

Einhellig, F.A. and P. Eckrich. (1984). Interactions of temperatures and ferulic acid stress on grain sorghum and soybeans. *Journal of Chemical Ecology* 10 161-170.

Einhellig, F.A. and G.R. Leather. (1988) Potentials for exploiting allelopathy to enhance crop production. *Journal of Chemical Ecology* 14: 1829-1844.

Einhellig, F.A. and I.F. Souza. (1992). Phytotoxicity of sorgoleone found in grain sorghum root exudates. *Journal of Chemical Ecology* 18: 1-11.

Einhellig, F.A., J.A. Rasmussen, A.M. Hejl, and I.F. Souza. (1993). Effects of root exudate sorgoleone on photosynthesis. *Journal of Chemical Ecology* 19: 369-375.

Ellis, J.E. and A.E. McSay. (1984). Allelopathic effect of plant residue on cucumber (*Cucumis sativus* L.) germination and seedling growth. *HortScience* 19: 534.

Eskelsen, S.R. and G.D. Crabtree. (1995). The role of allelopathy in buckwheat (*Fagopyrum sagittatum*) inhibition of canada thistle (*Cirsium arvense*). *Weed Science* 43: 70-74.

Feeny, P., K. Sachdev, L. Rosenberry, and M. Carter. (1988). Luteolin 7-*O*-(6′-*O*-malonyl)-*D*-glucoside and *trans*-chlorogenic acid: Oviposition stimulants for the black swallowtail butterfly. *Phytochemistry* 27: 3439-3448.

Field, C.B., F.S. Chapin, III, P.A. Matson, and H.A. Mooney. (1992). Responses of terrestrial ecosystems to the changing atmosphere: A resource-based approach. *Annual Review Ecology and Systematics* 23: 201-235.

Forney, D.R. and C.L. Foy. (1985). Phytotoxicity of products from rhizospheres of a sorghum-sudangrass hybrid (*S. bicolor* × *S. sudanense*). *Weed Science* 33: 597-604.

Fuerst, E.P. and A.R. Putnam. (1983). Separating the competitive and allelopathic components of interference: Theoretical principles. *Journal of Chemical Ecology* 9: 937-944.

Gantzer, C.J., S.H. Anderson, A.L. Thompson, and J.R. Brown. (1991). Evaluation of soil loss after 100 years of soil and crop management. *Agronomy Journal* 83: 74-77.

Grayson, B.T., K.S. Williams, P.A. Freehauf, R.R. Pease, W.T. Ziesel, R.L. Sereno, and R.E. Reinsfelder. (1987). The physical and chemical-properties of the herbicide cynmethylin (SD-95481). *Pesticide Science* 21: 143-153.

Grodzinsky, M. (1992). Allelopathic effects of cruciferous plants in crop rotation. In *Allelopathy: Basic and Applied Aspects*, eds. S.J.H. Rizvi and V. Rizvi. New York: Chapman and Hall, pp. 77-85.

Gubbiga, N.G., A.D. Worsham, and F.T. Corbin. (1996). Investigations into the growth suppressing effect of nicosulfuron treated johnsongrass (*Sorghum halepense*) on corn (*Zea mays*). *Weed Science* 44: 640-644.

Guenzi, W.D. and T.M. McCalla. (1966). Phenolic acids in oats wheat sorghum and corn residues and their phytotoxicity. *Agronomy Journal* 58: 303-304.

Habib, S.A. and A.A. Abdul-Rahman. (1988). Evaluation of some weed extracts against field dodder on alfalfa (*Medicago sativa*). *Journal of Chemical Ecology* 14: 443-452.

Hall, A.B., U. Blum, and R.C. Fites. (1983). Stress modification of allelopathy of *Helianthus annuus* (L.) debris on seedling biomass production of *Amaranthus retroflexus* L. *Journal of Chemical Ecology* 9: 1213-1222.

Hanson, A.D., K.M. Ditz, G.W. Singletary, and T.J. Leland. (1983). Gramine accumulation in leaves of barley grown under high temperature stress. *Plant Physiology* 71: 896-904.

Harper, J.L. (1977). *The Population Biology of Plants*. London, UK: Academic Press.

Hazebroek, J.P., S.A. Garrison, and T. Gianfagna. (1989). Allelopathic substances in asparagus roots: Extraction characterization and biological activity. *Journal of American Society of Horticultural Sciences* 114: 152-158.

Hegde, R.S. and D.A. Miller. (1990). Allelopathy and autotoxicity in alfalfa-characterization and effects of preceding crop and residue incorporation. *Crop Science* 30: 1255-1259.

Hejl, A.M., F.A. Einhellig, and J.A. Rasmussen. (1993) Effect of juglone on growth photosynthesis and respiration. *Journal of Chemical Ecology* 19: 559-568.

Hicks, S.K., C.W. Wendt, J.R. Gannaway, and R.B. Baker. (1989). Allelopathic effects of wheat straw on cotton germination emergence and yield. *Crop Science* 29: 1057-1061.

Hoffman, M.L., L.A. Weston, J.C. Snyder, and E.E. Regnier. (1996a). Separating the effects of sorghum (*Sorghum bicolor*) and rye (*Secale cereale*) root and shoot residues on weed development. *Weed Science* 44: 402-407.

Hoffman, M.L., L.A. Weston, J.C. Snyder, and E.E. Regnier. (1996b). Allelopathic influence of germinating seeds and seedlings of cover crops on weed species. *Weed Science* 44: 579-584.

Horsley, S.B. (1986). Evaluation of hayscented fern interference with black cherry. *American Journal of Botany* 73: 668-669.

Inderjit and K.M.M. Dakshini. (1995). On laboratory bioassays in allelopathy. *The Botanical Review* 61: 29-44.

Inderjit and K.M.M. Dakshini. (1998). Allelopathic interference of chickweed *Stellaria media* with seedling growth of wheat (*Triticum aestivum* L.). *Canadian Journal of Botany* 76: 1317-1321.

Inderjit and K.I. Keating. (1999). Allelopathy: Principles, procedures, processes and promises for biological control. *Advances in Agronomy* 67: 141-231.

Johnson, A.W. (1985). Specific crop rotation effects combined with cultural practices and nematicides. In *An Advanced Treatise on Meloidogyne. Vol. I. Biology and Control*, eds. J.N. Sasser and C.C. Carter. North Carolina, USA: Department of Plant Pathology and the US Agency for International Development of North Carolina State University Press, pp. 283-301.

Jose, S. and A.R. Gillespie. (1998). Allelopathy in black walnut (*Juglans nigra* L.) alley cropping. II. Effects of juglone on hydroponically grown corn (*Zea mays* L.) and soybean (*Glycine max* (L.) Merr.) growth and physiology. *Plant and Soil* 203: 199-205.

Kalburtji, K.L. and A. Gagianas. (1997). Effects of sugarbeet as a preceding crop on cotton. *Journal of Agronomy and Crop Science* 178: 59-63.

Kalburtji, K.L. and J.A. Mosjidis. (1992). Effects of sericea lespedeza residues on warm season grasses. *Journal of Range Management* 45: 441-444.

Kalburtji, K.L. and J.A. Mosjidis. (1993a). Effects of sericea lespedeza residues on cool season grasses. *Journal of Range Management* 46: 315-319.

Kalburtji, K.L. and J.A. Mosjidis. (1993b) Effects of sericea lespedeza root exudates on some perennial grasses. *Journal of Range Management* 46: 312-315.

Kalburtji, K.L., J.A. Mosjidis, and A.P. Mamolos. (1999). Litter dynamics of low and high tannin sericea lespedeza plants under field conditions. *Plant and Soil* 208: 271-281.

Kalburtzi, K.L., P.A. Gerakis, and D. Vokou. (1989). Allelopathy between wheat and faba beans as expressed by inhibitory effects on seed germination and growth. *Agricultura Meditterannea* 119: 372-382.

Kalburtzi, K.L., D.S. Veresoglou, and P.A. Gerakis. (1990). Effects of burnt or unburnt straw on wheat and faba beans as influenced by N fertilization. *Agriculture Ecosystems and Environment* 31: 173-185.

Klun, J.A., C.L. Tipton, and T.A. Brindley. (1967). 2,4-Dihydroxy-7-methoxy-14-benzoxazin-3-one (DIMBOA) an active agent in the resistance of maize to European corn borer. *Journal of Economic Entomology* 6: 1529-1533.

Kochhar, M., C.L. Blum, and R.A. Reinert. (1980). Effects of O_3 and (or) fescue on ladino clover: Interactions. *Canadian Journal of Botany* 58: 241-249.

Kohli, R.K., D. Batish, and H.P. Singh. (1998). Allelopathy and its implications in agroecosystems. *Journal of Crop Production* 1: 169-202.

Krishnan, G., D.L. Holshouser, and S.J. Nissen. (1998). Weed control in soybean (*Glycine max*) with green manure crops. *Weed Technology* 12: 97-102.

Lampkin, N. (1994). *Organic Farming*. UK: Farming Press Ltd.

Larson, M.M. and E.L. Schwarz. (1980). Allelopathic inhibition of black locust red clover and black alder by six common herbaceous species. *Forest Science* 26 511-520.

Lawrey, J.D. (1993). Chemical ecology of *Hobsonis christiansenii* a lichencolous hypomycetes. *American Journal of Botany* 80: 1109-1113.

Leather, G.R. (1983). Sunflowers (*Helianthus annuus*) are allelopathic to weeds. Weed *Science* 31: 37-42.

Leon, W.B. (1976). Phytotoxicité induité par les résidus de recolte de *Sorghum vulgare* dans les sols sableux d'ouset African. *Botanical Review* 45: 15-109.

Levy, M., U. Zehavi, M. Naim, and J. Polachek. (1986). An improved procedure for the isolation of medicagenic acid 3-*O*-*D*-glucopyranoside from alfalfa roots and its antifungal activity on plant pathogen. *Journal of Agriculture and Food Chemistry* 34: 960-963.

Liebman, M. and E. Dyck. (1993). Crop rotation and intercropping strategies for weed management. *Ecological Applications* 3: 92-122.

Lodhi, M.A.K., R. Bilal, and K.A. Malik. (1987). Allelopathy in agroecosystems: Wheat phytotoxicity and its possible roles in crop rotation. *Journal of Chemical Ecology* 13: 1881-1891.

Lolas, P.C. and H.D. Coble. (1982). Noncompetitive effects of johnsongrass (*Sorghum halepense*) on soybean (*Glycine max*). *Weed Science* 30: 589-593.

Luu, K.T., A.G. Matches, and E.J. Peters. (1982). Allelopathic effects of tall fescue on birdsfoot trefoil as influenced by N fertilization and seasonal changes. *Agronomy Journal* 74: 805-808.

Martin, J.P., L.J. Klotz, T.A. DeWolfe, and J.O. Ervin. (1956). Influence of some common soil fungi on growth of citrus seedling. *Soil Science* 81: 259-267.

Mays, D.A., K.R. Sistani, and U.R. Bishnoi. (1998) Adverse effects of soybeans on subsequent wheat and triticale crops. *Journal of Sustainable Agriculture* 12: 79-87.

Mitchell, C.C., R.L. Westerman, J.R. Brown, and T.R. Peck. (1991). Overview of long-term agronomic research. *Agronomy Journal* 83: 24-29.

Mikulas, J. (1984). Allelopathy of *Sorghum halepense* (L.) Pers. on weeds and crops. *Acta Agronomica Academiae Scientiarum Hungaricae* 33: 423-427.

Miller, D.A. (1992). Allelopathy in alfalfa and other forage crops in the United States. In *Allelopathy: Basic and Applied Aspects*, eds. S.J.H. Rizvi and V. Rizvi. New York: Chapman and Hall, pp. 169-177.

Miller, D.A. (1996). Allelopathy in forage crop systems. *Agronomy Journal* 88: 854-859.

Miro, C.P., A.G. Ferreira, and M.E.A. Aquila. (1998). Allelopathy of ripe fruits of *Ilex paraguariensis* ('mate') on corn cultivation. *Pesquisa Agropecuaria Brasileira* 33: 1261-1270.

Mitchell, C.C., R.L. Westerman, J.R. Brown, and T.R. Peck. (1991). Overview of long-term agronomic research. *Agronomy Journal* 83: 24-29.

Mongelli, E., C. Desmarchelier, J. Coussio, and G. Ciccia. (1997). The potential effects of allelopathic mechanisms on plant species diversity and distribution determined by the wheat rootlet growth inhibition bioassay in South American plants. *Revista Chilena de Historia Natural* 70: 83-89.

Moyer, J.R. and H.C. Huang. (1997). Effect of aqueous extracts of crop residue on germination and seedling growth of ten weed species. *Botanical Bulletin of the Academia Sinica* 38: 131-139.

Muller, C.H. (1966). The role of chemical inhibition (allelopathy) in vegetation composition. *Bulletin of the Torrey Botanical Club* 93: 332-351.

Muller, C.H. (1969). Allelopathy as a factor in ecological process. *Vegetatio* 18: 348-357.

Mulvaney, D.L. and L. Paul. (1984). Rotating crops and tillage. *Crops and Soils* 36: 18-19.

Mwaja, V.N., J.B. Masiunas, and L.A. Weston. (1995). Effects of fertility on biomass phytotoxicity and allelochemical content of cereal-rye. *Journal of Chemical Ecology* 21: 81-96.

Nair, M., C.J. Whitenack, and A.R. Putnam. (1990). 22′-Oxo-11′-azobenzene: A microbially-transformed allelochemical from 23-benzoxazolinone. *Journal of Chemical Ecology* 16: 353-364.

Netzley, D.H. and L.G. Butler. (1986) Roots of sorghum exude hydrophobic droplets containing biologically active components. *Crop Science* 26: 775-780.

Nicollier, J.F., D.F. Pope, and A.C. Thompson. (1983). Biological activity of dhurrin and other compounds from johnsongrass (*Sorghum halepense*). *Journal of Food Chemistry* 31: 744-748.

Opoku, G., T.J. Vyn, and R.P. Voroney. (1997). Wheat straw placement effects on total phenolic compounds in soil and corn seedling growth. *Canadian Journal of Plant Science* 77: 301-305.

Patrick, Z.A. and L.W. Koch. (1963). The adverse influence of phytotoxic substances from decomposing plant residues on resistance of tobacco to black root rot. *Canadian Journal of Botany* 41: 747-758.

Patrick, Z.A., T.A. Toussoun, and L.W. Koch. (1964). Effect of crop residue decomposition products on plant roots. *Annual Review of Phytopathology* 2: 267-292.

Pimentel, D.J., J. Allen, L. Guinand, R. Linder, P. McLaughlin, B. Meer, D. Musonda, D. Perdue, S. Poisson, S. Siebert, K. Stoner, R. Salazar, and A. Hawkins. (1987). World agriculture and soil erosion. *Bioscience* 37: 277-283.

Ponder, F., Jr. and S.H. Tadros. (1985). Juglone concentration in soil beneath black walnut interpreted with nitrogen-fixing species. *Journal of Chemical Ecology* 11: 937-942.

Putnam, A.R. (1990). Vegetable weed control with minimal herbicide inputs. *HortScience* 25: 155-158.

Putnam, A.R. and W.B. Duke (1974). Biological suppression of weeds: Evidence for allelopathy in accessions of cucumber. *Science* 185: 370-372.

Putnam, A.R. and W.B. Duke. (1978). Allelopathy in agroecosystems. *Annual Review of Phytopathology* 16: 431-451.

Putnam, A.R., J. DeFrank, and J.P. Barnes. (1983). Exploitation of allelopathy for weed control in annual and perennial cropping systems. *Journal of Chemical Ecology* 9: 1001-1010.

Putnam, A.R. and C.S. Tang. (1986). *The Science of Allelopathy*. New York: John Wiley & Sons.

Qasem, J.R. (1996). Fungicidal activity of *Ranunculus asiaticus* and other weeds against *Fusarium oxysporum* f. sp. *lycopersici*. *Annals of Applied Biology* 128: 533-540.

Rawat, M.S.M., G. Pant, D. Prasad, R.K. Joshi, and C.B. Pande. (1998). Plant growth inhibitors (Proanthocyanidins) from *Prunus armeniaca*. *Biochemical Systematics and Ecology* 26: 13-23.

Reganold, J.P., R.I. Papendick, and J.F. Parr. (1990). Sustainable agriculture. *Scientific American* 262: 112-120.

Rice, E.L. (1974). *Allelopathy*. New York: Academic Press.

Rice, E.L. (1984). *Allelopathy* 2nd edition. New York: Academic Press.

Rice, E.L. (1992). Allelopathic effects on nitrogen cycling. In *Allelopathy: Basic and Applied Aspects*, eds. S.J.H. Rizvi and V. Rizvi. New York: Chapman and Hall, pp. 31-58.

Rizvi, S.J.H., D. Mukerrji, and S.N. Mathur. (1981). 1,3,7-Trimethylxanthine–a new natural herbicide. *Agricultural and Biological Chemistry* 54: 1255-1256

Rizvi, S.J.H., M. Tahir, V. Rizvi, R.K. Kohli, and A. Ansari. (1999). Allelopathic interactions in agroforestry systems. *Critical Reviews in Plant Sciences* 18: 773-796.

Rizvi, S.J.H. and V. Rizvi. (1992). Exploitation of allelochemicals in improving crop productivity. In *Allelopathy: Basic and Applied Aspects*, eds. S.J.H. Rizvi and V. Rizvi. New York: Chapman and Hall, pp. 443-472.

Saiki, H. and K. Yoneda. (1981). Possible dual roles of an allelopathic compound *cis*-dehydromatricaria ester. *Journal of Chemical Ecology* 8: 185-193.

Santos, J.R.A. and D.I. Leskovar. (1997). Interference from broccoli residue on *Brassica* germination and seedling growth. *Journal of the American Society for Horticultural Science* 122: 715-720.

Sarabol, E. and I.C. Anderson. (1992). Improving yield of corn-soybean rotation: Role of allelopathy. In *Allelopathy: Basic and Applied Aspects*, eds. S.J.H. Rizvi and V. Rizvi. New York: Chapman and Hall, pp. 87-100.

Schumacher, W.J.D., D.C. Thill, and G.A. Lee. (1983). Allelopathic potential of wild oat (*Avena fatua*) on spring wheat (*Triticum aestivum*) growth. *Journal of Chemical Ecology* 9: 1235-1245.

Shilling, D.G., R.A. Liebl, and A.D. Worsham. (1985). Rye (*Secale cereale* L.) and wheat (*Triticum aestivum* L.) mulch: The suppression of certain broad-leaved weeds and the isolation and identification of phytotoxins. In *The Chemistry of Allelopathy*, eds. A.C. Thompson. ACS Symposium Series 268. Washington, DC: American Chemical Society, pp. 243-271.

Singh, H.P., D.R. Batish, and R.K. Kohli. (1999). Autotoxicity: Concept, organisms and ecological significance. *Critical Reviews in Plant Sciences* 18: 757-772.

Smith, T.A. (1977). Tryptamine and related compounds in plants. *Phytochemistry* 16: 171-175.

Stiner B.R. and House G.J. (1989) The search of sustainable agroecosystems. *Journal of Soil and Water Conservation* 44: 111-116.

Thijs, H., J.R. Shann, and J.D. Weidenhamer. (1994). The effect of phytotoxins on competitive outcome in a model system. *Ecology* 75: 1959-1964.

Vaughan, A.D. and K.C. Vaughan. (1988). Mitotic disrupters from higher plants and their potential uses as herbicides. *Weed Technology* 2: 533-539.

Vaughan, A.D., G.P. Sparling, and B.G. Ord. (1983). Amelioration of phytotoxicity of phenolic acids by some soil microbes. *Soil Biology and Biochemistry* 15: 613-614.

Vokou, D. (1992). The allelopathic potential of aromatic shrubs in phryganic (east Mediterranean) ecosystems. In *Allelopathy: Basic and Applied Aspects*, eds. S.J.H. Rizvi and V. Rizvi. London: Chapman & Hall, pp. 303-320.

Wagstaff, H. (1987). Husbandry methods and farm systems in industrialized countries which use lower levels of external inputs: A review. *Agriculture Ecosystems and Environment* 19: 1-27.

Weidenhamer, J.D. (1996). Distinguishing resource competition and chemical interference overcoming the methodological impasse. *Agronomy Journal* 88: 866-875.

Weston, L.A. (1996). Utilization of allelopathy for weed management in agroecosystems. *Agronomy Journal* 88: 860-866.

Weston, L.A., R. Harmon, and S. Mueller. (1989). Allelopathic potential of sorghum-sudangrass hybrid (sudex). *Journal of Chemical Ecology* 15: 1855-1865.

Williamson, G.B. and J.D. Weidenhamer. (1990). Bacterial degradation of juglone: Evidence against allelopathy? *Journal of Chemical Ecology* 16: 1739-1742.

Wink, M. (1983). Inhibition of seed germination by quinolizidine alkaloids. Aspects of allelopathy in *Lupinus albus* L. *Planta* 158: 365-368.

Zúñiga, G.E., M.S. Salgado, and L.J. Corcuera. (1985). Role of an indole alkaloid in the resistance of barley seedlings to aphids. *Phytochemistry* 24: 945-947.

Zentmeyer, G.A. (1963). Biological control of Phytophthora root rot of avocado with alfalfa meal. *Phytopathology* 53: 1383-1386.

Zhang, Q.S. (1997). Effects of soil extracts from repeated plantation woodland of Chinese fir on microbial activities and soil nitrogen mineralization dynamics. *Plant and Soil* 191: 205-212.

Selective Isolation and Screening of Soil Microorganisms for Metabolites with Herbicidal Potential

M. A. B. Mallik

SUMMARY. Crop losses from weeds worldwide average 10% annually. Herbicide application is almost essential in modern agriculture to maintain high productivity. Increased awareness of the environmental damage caused by continued use of conventional synthetic herbicides has aroused great interest in biodegradable, selective and environmentally friendly herbicides. The increasing incidence of herbicide resistance in weeds has stimulated research for herbicidal compounds with new or novel sites and modes of action. Microbial metabolites have become the focus of attention of researchers searching for natural product alternatives to conventional herbicides. Two well-known commercial herbicides of microbial origin are anisomycin (commercial analogue, methoxyphenon) and bialaphos. Both were isolated from members of actinomycetes. Several phytotoxins have been isolated from microbial metabolites, e.g., herbicidins A and B, herbimycins A and B, nigericin, tentoxin and others. The list of microbial phytotoxins continues to

---

M. A. B. Mallik is Emeritus Research Professor, Agricultural Research, Education and Extension Program, Langston University, Langston, Oklahoma, USA.

Address correspondence to: M. A. B. Mallik, Agricultural Research Program, P.O. Box 730, Langston University, Langston, OK 73050 USA (E-mail: mmallik @luresext.edu).

Critical review of the manuscript by Dr. Robert D. Williams, Agricultural Research Service, U.S. Department of Agriculture, and Dr. S. Hart, Agricultural Research and Extension Programs, Langston University is deeply appreciated. The author thanks Mrs. Nancy Smith for her assistance in preparing the manuscript.

[Haworth co-indexing entry note]: "Selective Isolation and Screening of Soil Microorganisms for Metabolites with Herbicidal Potential." Mallik, M. A. B. Co-published simultaneously in *Journal of Crop Production* (Food Products Press, an imprint of The Haworth Press, Inc.) Vol. 4, No. 2 (#8), 2001, pp. 219-236; and: *Allelopathy in Agroecosystems* (ed: Ravinder K. Kohli, Harminder Pal Singh, and Daizy R. Batish) Food Products Press, an imprint of The Haworth Press, Inc., 2001, pp. 219-236. Single or multiple copies of this article are available for a fee from The Haworth Document Delivery Service [1-800-342-9678, 9:00 a.m. - 5:00 p.m. (EST). E-mail address: getinfo@haworthpressinc.com].

increase. Actinomycetes are the major source of several herbicides and other bioactive compounds. The potential of discovering new phytotoxins from metabolites of microorganisms is still unlimited. Methods of isolation of actinomycetes, their screening for phytotoxins, separation of phytotoxins from metabolites and partial purification of the toxins are described. Included are screening of soil fungi for phytotoxin, mass culture of fungi, and bioassays.

KEYWORDS. Actinomycetes, agar plate screening, bioassay, fungi, herbicides, isolation of microbes, partial purification of phytotoxin, separation of phytotoxin

INTRODUCTION

Weeds are one of the largest causes of economic losses in crop production and cause more economic losses than any other pests. On a global scale herbicide accounted for 44% of the total pesticide sales in 1985 and total market value of this sale amounted to $16 billion. Less than 1% of this market is shared by pesticide of biological origin (Jutsum, 1988). In spite of advances made in weed control measure, weeds continue to cause serious crop losses. Herbicide application is likely to increase as a consequence of greater emphasis on reduced tillage and low input agriculture. Growers are apt to use herbicides only when infestation becomes visible, therefore post-emergence herbicides are likely to gain an increasing market share. Herbicides currently in use are generally broad spectrum affecting both target and non-target species. There is a need for more selective herbicides.

Although pesticide application is crucial for maintenance of high crop productivity, continued use or misuse of agricultural chemicals has resulted in serious environmental pollution. Increased public concern about the safety of pesticide application has resulted in a shift in our strategy of weed control. There is a tremendous interest in developing alternative weed control methods which should include more selective, biodegradable herbicides leaving very little residual effect, are less persistent, more target specific and environment friendly.

The increasing incidence of herbicide resistance in weed, reviewed by Altman (1993) suggests that there is a need for weed killing agent with different sites of action. The number of chemicals that must be screened to discover new herbicides with different sites and modes of action is becoming so large that the traditional approach is not economically feasible. An alternative approach is to search for naturally occurring phytotoxins which are more likely to have sites of action different from traditional ones. Microbial metabolites which offer exotic new chemistry are unlikely to be synthesized in the laboratories.

Lynch (1976) extensively reviewed microbial products in relation to plant growth. Soil microorganisms produce a wide range of bioactive compounds that are potentially phytotoxic to higher plants (Lynch, 1990). Research on isolation and screening of microbial metabolites to discover new herbicidal compounds has witnessed an upsurge, as evidenced by the number of reviews published during the last two decades (Fisher and Bellus, 1983; Misato and Yamaguchi, 1984; Omura, 1986; Cutler, 1988; Heisey et al., 1988; Hoagland, 1990; Duke and Abbas, 1995). Several microbial metabolites with herbicidal potential have been isolated (Misato, 1983; Sekizawa and Takematsu, 1983; Howell and Stipanovic, 1984; Misato and Yamaguchi, 1984).

Microbial phytotoxins for weed control have been in use for nearly four decades. Anisomycin, discovered from the culture filtrate of *Streptomyces toyocaensis* was the first microbial phytotoxin to be used as a commercial herbicide (Yamada et al., 1972). The phytotoxin is highly effective against barnyardgrass and crabgrass, but has no activity against turnip, tomato and other broadleaf horticultural species. Although anisomycin was not commercialized, its synthetic analogue methexyphenon or NK 049 (3′-3′-dimethyl-4-methoxy-benzophenon) was developed as a commercial herbicide for use in paddy fields (Yamada et al., 1974). Methoxyphenon is readily degraded in soil and its metabolites are not toxic to rice plants. Thus, if a bioactive compound has been detected in a metabolite, the compound may be chemically synthesized as has been done in case of anisomycin. The compound may also serve as a template from which other compound may be developed for a desired function.

Bialaphos was discovered in the metabolites of *Streptomyces viridochromogenes* as an antibacterial agent. Afterwards it was found to possess highly potent weed killing properties effective against both dicots and monocots. It can control regrowth of perennials longer than paraquat and has a long lasting effect similar to glyphosate (Sekizawa and Takematsu, 1983). It is currently being marketed in Japan (Duke, 1986). Bialaphos, an amino acid derivative, is readily metabolized to phospinothricin which is its herbicidal moiety. A synthetic version of phosphinothricin has been patented by a German Company as a herbicide (Fisher and Bellus, 1983). Bialaphos is highly biodegradable having a half-life of 2 to 3 days in soil. It has been noted that the application of bialaphos in combination with nitrogen fertilizer (e.g., urea, ammonium nitrate) enhances the herbicidal activity (Misato and Yamaguchi, 1984).

Herbicidins A and B from *Streptomyces saganonensis* are highly effective against barnyard grass, goosegrass, green panicum but does not affect paddy (Arai et al., 1976; Haneishi et al., 1976). Both are toxic to radish, tomato, smartweed (*Polygonum* sp.) and wild amaranth (*Amaranthus* sp.) (Cutler, 1986). Examples of other phytotoxins discovered from the metabolites of actinomycetes are listed in Table 1.

Gliocladium virens, a soil-borne non-phytopathogenic fungus, cultured on peat moss amended with ammonium nitrate, upon application to the soil controls a broad range of weeds by at least 90%. Herbicidal activity of the fungus is due to the production of phytotoxin viridiol (Jones and Hancock, 1990).

Two soil-borne plant pathogenic fungi *Phytophthora palmivora* and *Colle-*

TABLE 1. Examples of phytotoxin producing actinomycetes.

Actinomycetes	Phytotoxin	Effect	Reference
Unidentified	Uncharacterized	Inhibition	DeFrank and Putnam, 1985
Unidentified	Uncharacterized	Inhibition	Mishra et al., 1987
Unidentified	Uncharacterized	Inhibition	Mallik, 1997
Kitasatosporia phosalacinea	Phosalacine	Inhibition	Omura et al., 1984
Saccharothrix sp. *St-888*	Phosphonothrixin	Chlorosis, inhibition	Takahashi et al., 1995
Streptomyces hygroscopicus	Nigericin	Inhibition	Heisey and Putnam, 1990
S. hygroscopicus	Herbimycin Herbimycin B	Inhibition	Omura et al., 1979 Iwai et al., 1980
S. hygroscopicus	Geldanamycin	Necrosis	Heisey and Putnam, 1990
S. hygroscopicus	Hydantocidin	Inhibition	Takahashi et al., 1994
S. saganonensis	Herbicidins A & B	Pre-emergent inhibition	Haneishi et al., 1976 Arai et al., 1976
S. viridochromogenes	Bialaphos*	Inhibition	Sekizawa and Takematsu, 1983
Streptomyces sp. No. 638	Toyocamycin Anisomycin**	Inhibition	Yamada et al., 1972
Streptomyces sp.	Blasticidins	Inhibition	Scacchi et al., 1992
Streptomyces sp.	Gougerotin	Inhibition	Murao and Hyashi, 1983
Streptomyces sp.	Vulgamycin	Post-emergent inhibition	Babczinski et al., 1991

* This compound and its synthetic analogue are marketed as herbicide.

** A synthetic analogue of this compound is marketed as herbicide.

totrichum gloeosporoides (Peens) Sacc. f. sp. *Aeschynomene* are being actively marketed as herbicides (Diagle and Connick, Jr., 1990). The use of phytotoxins produced by microbes offer several advantages over using living microorganisms as herbicides. The logistics of preparation of inoculum, preservation, storing and application of living microorganisms are much more complicated. This subject is not under discussion here.

Biological control of weeds with plant pathogens is an attractive option (Charudattanan and Walker, 1982). Phytotoxins from plant pathogens are a good candidate for exploitation as a weed-killing agent. Durbin (1981) provided an excellent review on this subject. Several phytotoxins have been isolated from non-plant pathogenic soil fungi (Table 2). The list of new phytotoxins from soil fungi continues to grow.

TABLE 2. Examples of fungi producing phytotoxin.

Fungus	Phytotoxin	Effect	Reference
Alternaria alternata	AAL-toxin	Chlorosis	Abbas et al., 1995
A. alternaria (*A. tenuis*)	Tentoxin	Tentoxin	Lax and Shepherd, 1988
Aspergillus candidus	Terphenyllin Hydroxyterphenyllin	Inhibition	Cutler et al., 1978
A. niger	Orlandin	Inhibition	Cutler et al., 1979
A. repens and *A. flavus*	Cladosporin	Inhibition	Springer et al., 1981
A. terreus	Acetylaraton	Inhibition	Kamata, Saki, and Hirota, 1983
A. ustus	Pergillin Dihydroxypergillin	Inhibition	Cutler et al., 1980 Cutler et al., 1981
Cladosporium cladosporiodes	Cladosporin	Inhibition	Springer et al., 1981
Gliocladium virens	Viridiol	Necrosis	Howell and Stipanovic, 1984
Irpex pachydon	Irpexil	Anthocyanin inhibition	Fisher and Bellus, 1983
Nodulosporium hinnuleum	Desmethoxyviridiol	Inhibition	Cole et al., 1975
Paecilomyces varioti	Cornexistin	Inhibition	Takahashi et al., 1994
Penicillum cyclopium	Cyclopenol Cyclopenin	Inhibition	Cutler et al., 1984
P. urticae	Patulin	Inhibition	Norstadt and McCalla, 1963

The potential to discover new antibiotics is still unlimited, and "with regard to the application of microbial secondary metabolites for the benefit or mankind, the surface has only been scratched" (Demain, 1983). The vast majority of the antibiotics that have so far been discovered originate from actinomycetes. This chapter will be concerned with isolation and screening of soil microorganisms for phytotoxin, with special reference to actinomycetes.

ISOLATION

Generally, large number of isolates are screened with the hope that some useful microbes will be found. A degree of serendipity is not ruled out. The capacity of different microbial taxa for production of antibiotics and phytotoxin varies greatly. The vast majority of antibiotics are produced by actinomycetes and in particular by the genus *Streptomyces*.

Selection of material from which isolation of microbes is desired may be a matter of choice. Choice of natural materials such as soil, manure, compost or waste is based on the assumption that samples from diverse location and habitat will yield novel isolates and, therefore, novel metabolites. Very few attempts have been made for isolation and subsequent screening of acidophiles, basophiles and thermophiles. Uncommon habitats like swamps, acid and alkaline soils, and excreta of small herbivores are candidates for exploitation. Our discussion will be centered around isolation and screening of neutrophiles.

Pretreatment of Substrate

Soil samples are aseptically collected, after bringing to the laboratory are air-dried, and pulverized. Samples may be treated in various ways to enhance isolation of desired microbes. Methods of pretreatment of soil for isolation of different genera of actinomycetes have been described by Williams and Wellington (1982). Pretreatment of soil and roots at 40°C for 5 to 10 min often enhances the ratio of *Streptomyces* to other microbes. Enrichment of soil samples with chitin (1% w/w) and incubation for 2 to 3 weeks enhance the chance of isolation of more *Streptomyces*. Following the recommendation of Tsao, Leben, and Keitt (1960), we routinely added $CaCO_3$ to soil samples for raising pH to 7 or 8 and incubated for 2 to 3 weeks at 50% water holding capacity under humid condition. This pretreatment of soil samples encourages proliferation of *Streptomyces*.

Media

A wide range of media formulations are available (Shirling and Gottlieb, 1966; Waksman, 1967; Aaronson, 1970). Since most actinomycetes are neutrophiles, the pH of the media should be around 6.7 to 7.5. More commonly used media are: starch-casein agar, glucose-asparagine agar, casamino acids-yeast extract-glucose agar, glycerol-arginine agar, and AGS (Appendix). Inclusion of antifungal antibiotics in isolation media is essential when isolating from soil and similar substrates. Suggested antibiotics are actidione (cyclohexamide) 50-100 μg/ml, nystatin (mycostatin, dissolved in dimethyl sulfoxide) 50 μg/ml. Polymyxin B sulfate 5.0 μg/ml and sodium penicillin 1.0 μg/ml may be used, if necessary, to suppress the development of many true bacteria. These antibiotics are not effective against actinomycetes.

Procedure

One milliliter of soil of 10^{-4} to 10^{-6} dilution (in most situations) is plated out and thoroughly mixed with about 25-30 ml of desired agar medium. We routinely added 1 ml soil suspension of 10^{-3} to 10^{-5} dilution to 9 ml of 0.8% water agar, mixed thoroughly and overlaid on selected agar medium poured a day before (Mallik, 1997). This procedure allows the microbial colonies to appear on the surface of the agar plate. Incubation at 28 to 30°C for two weeks is usually sufficient. However, longer incubation periods may allow the isolation of slow growing actinomycetes.

Selection of Isolates

Colony selection is the most frustrating and time consuming process in isolation procedure. Colonies are selected on the basis of colony morphology, and color of surface and substratum. In spite of all precautions, it is difficult to avoid duplication. At the same time existence of biochemical difference between strains of the same species can not be ignored. Therefore, it is advisable to include all isolates in the screening process even if they have similar morphology.

Actinomycetes are aerobic. Colonies of actinomycetes on the isolation plates are easily distinguishable from those of fungi and bacteria. They are compact and often leathery. The well-developed colony produces abundant aerial hyphae which produce dry powdery spores. These spores can be easily picked up with a transfer needle wetted by touching on agar plate. If the aerial mycelium is not well developed, a flattened loop that can cut through tough mycelial mat may be necessary to transfer a colony. Colonies are picked and transferred to fresh agar plate of isolation medium. After checking for contamination, the colonies free of contamination are transferred to agar slants for maintenance and preservation. It is advisable to maintain cultures in more than one media including isolation medium. The suggested media for maintenance and preservation are given in the appendix.

Maintenance and Preservation

Agar slants in screw cap tube (cap coated with paraffin) with longer butt and shorter slants can hold cultures for at least four years held at room temperature. Lyophilization is the choice for long term preservation. It is suggested that an isolate is lyophilized once bioactivity has been established in it.

SCREENING

Actinomycetes

An inoculum from an agar slant is transferred onto an agar plate of suitable medium and spread on the surface. After incubation of 1 to 2 weeks at 28 to 30°C, growth from the agar plate is scraped and transferred into a tube containing 2 to 3 ml of sterilized distilled water. Tissues are fragmented by a glass rod and mixed thoroughly by repeated aspiration with a Pasteur pipet. A few drops of this suspension are placed in the center of an agar plate. Agar medium should be checked before hand to verify whether it supports growth of plant species used in the bioassay (glucose-asparagine agar is not inhibitory to germination of plant seeds and supports growth of most actinomycetes). The inoculum is spread by a bent glass rod over a 2 cm wide band across the center of the agar plate (Figure 1). Alternatively, the inoculum may be spread on one-half of the plate. The plates are incubated at 28 to 30°C under humid condition for 2 to 3 weeks. This incubation period is considered adequate for growth of microbial isolate and diffusion of metabolites into agar medium. At the end of the incubation period surface sterilized plant seeds are implanted close to the band on one or both sides of the band, as the case may be (Mallik, 1997). Germination and growth of seedlings are evaluated after incubation at 28 to 30°C for suitable length of period (Figure 2).

Soil Fungi

Few studies have involved large scale screening of soil saprophytic fungi for isolation of phytotoxin. Method of isolation and preliminary screening of soil fungi for the detection of bio-activity is similar to that described for actinomycetes. It involves placing a piece of fungal inoculum and spreading over one half of the plate. Following an incubation period of 5 to 10 days plant seeds are implanted. This procedure is suitable for fungi with restricted growth (e.g., *Penicillium*) and not for those which spread rapidly (e.g., *Trichoderma*). This method may limit detection of hydrophobic compound which diffuse less readily through agar.

FIGURE 1. Bioassay of actinomycetous isolates on agar medium. Note inhibition of cress and ryegrass by isolate CB31 and of cress by isolate CB6.

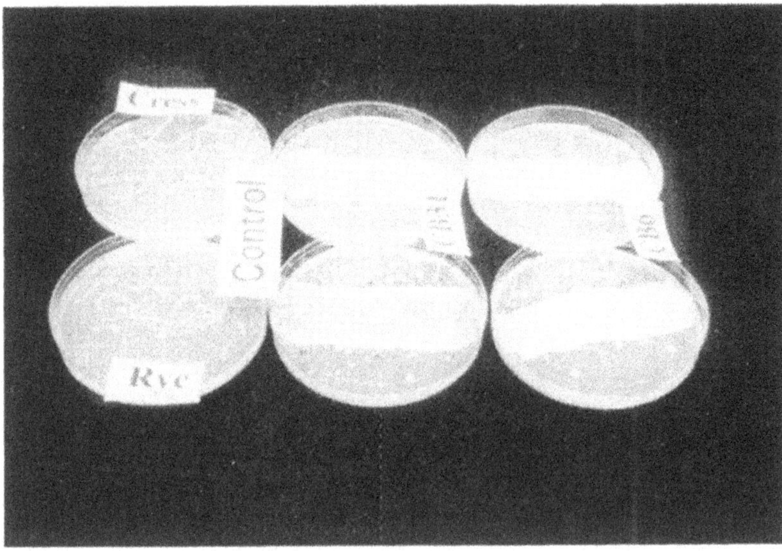

An alternative method involves screening directly the metabolite produced in broth by the fungi. The fungi are grown in broth as still culture for 5 to 10 days. The cell-free broth is placed in wells cut into water agar plates and the seeds are implanted next to the wells. The metabolite may also be added to the water agar and plated out; the plant seeds are implanted (Walton and Earle, 1984).

BIOASSAY

Bioassay is the most critical step in the isolation of bioactive compound and it should be done with utmost care. Plant seeds are most commonly used for convenience. However, not all plant seeds are suitable. Among dicots, cress, lettuce, and cucumber are a popular choice due to their sensitivity and fast growth rate. Among monocots, ryegrass, barnyardgrass and green foxtail are suitable for bioassay. Growth of radicle length and germination are more sensitive to toxin than hypocotyl/coleoptile growth. Pregerminated seeds with radicle protrusion of 1 to 2 mm length may be used to monitor root growth. This bioassay provides information on the potential of tissue damage, but not of chlorophyll damage or disruption of photosynthesis (as in tentoxin).

FIGURE 2. Procedure for screening of microbial isolates for bioactivity.

Inoculum on agar plate

↓

Transfer inoculum to a test tube with
2 to 3 ml sterilized water

↓

Fragment inoculum with a glass rod

↓

Suspension worked-up by a
Pasteur pipet

↓

Transfer 2 to 3 drops of inoculum
on a agar plate of desired medium

↓

Spread maximum in a 2-cm wide strip
or across half of the agar plate

↓

Incubate the inoculated agar plates
at 28 to 30°C for 15 to 20 days

↓

Implant surface sterilized seeds

↓

Examine the plates for germination
and growth of seedlings after 48 to 72 h
incubated at 28 to 30°C.

In addition to seeds, other plant materials have been used. *Chlorella pyrenoidosa* has been used to detect photosynthetic inhibitors (Kratky and Warren, 1971). *Chlamydomonas* is sensitive to inhibitors of cell division and enlargement, so it can provide indication of growth inhibition. *Lemna minor* bioassay has been used to determine growth inhibition (Einhellig, Leather, and Hobbs, 1985). The wheat coleoptile bioassay is very sensitive, capable of detecting a wide spectrum of bioactivity, and is highly reproducible (Cutler, 1986). Usually a bioactive compound has more than one bioactivity. A phytotoxin may also function as fungicide, bactericide and plant growth regulator. A new bioactive metabolite can become a template from which useful molecules for further application can be designed.

ISOLATION OF PHYTOTOXIN

Selected isolates are grown in broth as submerged culture with shaking. The majority of the actinomycetes, particularly *Micromonospora* and *Strep-*

tomyces grow as flakes or small globules leaving the broth clear. The mass of growth can be removed by centrifugation or filtration through coarse filter paper. Most phytotoxins are hydrophilic exotoxins, therefore, the broth should contain the active principle. Nutritional and environmental factors greatly affect the production of toxin. Screening for phytotoxin production should be carried out on several broth formulations because of great variation in the level of bioactive metabolite synthesis in different broth. Isolates producing toxin in agar medium may or may not produce the toxin in submerged culture. Warren, Prokop, and Grundy (1955) found that glucose-peptone-molasses broth yielded more toxin than other formulations (see appendix).

Mass Culture

Most phytotoxins are present at very low concentration in the broth; therefore, a large volume of starting material are needed for isolation of reasonable amount of toxin. The first step is to grow a primary inoculum to inoculate large volume of broth. To prepare a primary inoculum for broth culture, the growth of an isolate on agar plate is scraped into a tube containing about 2 to 3 ml of sterilized distilled water, the tissues are fragmented by a glass rod and thoroughly mixed by repeated aspiration with a Pasteur pipet. This inoculum is used to inoculate a flask containing 25 to 30 ml of broth selected for mass culture. The culture is grown for 5 to 10 days as shake culture. This inoculum is used to inoculate larger volume of broth; the ratio of the inoculum to broth is generally maintained at 1:20. The culture is grown for 5 to 15 days; the broth is sampled periodically to monitor toxin production by bioassay. A toxicity vs. incubation period curve is developed to determine the optimal incubation period for toxin production. Most broth formulations at full strength do not support plant growth; therefore, it is advisable to use a diluted broth. The cell-free diluted broth (1:1) is bioassayed on filter paper with plant seeds or any other bioassay method mentioned previously.

Reduction of broth volume of toxin containing metabolite can be conveniently done by evaporation under vacuum at about 50°C. Most low molecular weight phytotoxins are alcohol soluble. Cold methanol or ethanol at 50-80% (v/v) precipitates high molecular compounds which can be removed by centrifugation. Removal of high molecular weight compounds before solvent extraction minimizes emulsion formation. Methanol or ethanol should be removed by vacuum evaporation before extraction.

Fungi are grown for phytotoxin production on shredded wheat medium in Fernback flasks. Composition of the medium is given in the appendix. The culture is allowed to grow for 10 to 15 days at about 26 to 28°C. Alternatively, fungi may be grown in broth as still culture.

SOLVENT EXTRACTION

Most phytotoxins are soluble in water-immiscible organic solvent, e.g., dicloromethane, ether, ethyl acetate or chloroform. Extraction of cell-free broth with solvent can remove most of the phytotoxins in the broth. Concentration of the solvent should precipitate the phytotoxin. It can be dried over anhydrous magnesium sulfate. The extract is checked for phytotoxicity using an appropriate bioassay.

Not all microbial isolate that indicate bioactivity on agar plate are capable of releasing such compounds into broth culture. An inoculum of an isolate of this kind is spread evenly over the agar plates of the chosen medium. The culture is grown for 7 to 10 days. The whole culture including agar are cut up in small pieces and extracted with organic solvent (methanol or acetone). The organic solvent is reduced or evaporated, the residue is taken into water and extracted with organic solvent as described previously (Shomura et al., 1970; Maruyama et al., 1975).

The mass fungal culture is extracted with acetone (Walton and Earle, 1984). Acetone dissolves wide range of polar and non-polar metabolites. The fungal growth is macerated with a homogenizer to produce a pulp and is filtered. The clear filtrate is evaporated under reduced pressure to remove acetone. The aqueous phase is extracted three times with ethyl acetate or dicloromethane. The volume of extractant each time is equal to the aqueous phase. The solvent is evaporated under reduced pressure to near dryness and finally dried over nitrogen gas.

PURIFICATION

Purification of the phytotoxins can be achieved by ultra filtration (separation of different fractions based on molecular weights), adsorption chromatography, ion-exchange chromatography, thin layer chromatography (TLC) and HPLC. For preliminary purification TLC is more reproducible, convenient and fast. Different bands from preparatory TLC plates are scraped and extracted with organic solvent. Each fraction should be checked for phytotoxicity.

CONCLUSION

The increasing need for new herbicides with greater efficacy, higher target specificity and benign environmental properties continues to fuel interest in microbial metabolites as sources of new bioactive molecules. The increasing incidence of weeds with evolving resistance to available herbicides has focused attention of the researchers to search for molecules with new modes and new sites of action. Natural products from microbes offer the best opportunity for isolation of these chemicals. Microbial metabolites are unique sources of potential herbicides and as template for synthesis of new herbicidal compounds. Although phytotoxicity of many microbial metabolites has been established, their mechanisms and molecular site of action studies has infrequently been undertaken. Studies correlating chemical structure of the isolated metabolites with molecular site of action will be necessary to fully utilize a rational herbicide design approach. Multidisciplinary efforts of microbiologists, biochemists, plant physiologists and weed scientists will be needed to fully exploit microbial metabolites as commercial herbicides.

REFERENCES

Aaronson, S. (1970). *Experimental Microbial Ecology*. New York: Academic Press.

Abbas, H.K., T. Tanaka, S.O. Duke, and C.D. Boyette. (1995). Susceptibility of various crop and weed species to AALtoxin, a natural herbicide. *Weed Technology* 9: 125130.

Altman, J. (1993). Pesticide interaction in crop production. In *Pesticide Interaction in Crop Production, Beneficial and Deleterious Effects*, ed. J. Altman. Boca Raton, Florida: CRC, pp. 39.

Arai, M., T. Haneishi, N. Kitahara, R. Enokita, K. Kawakubo, and Y. Kondo. (1976). Herbicidins A and B, two new antibiotics with herbicidal activity, I. Producing organism and biological activities. *The Journal of Antibiotics* 29: 863-869.

Babczinski, P., M. Dorgerloh, A. Lobberding, H.J. Santel, R.R. Schmidt, P. Scmitt, and C. Wunsche. (1991). Herbicidal activity and mode of action of vulgamycin. *Pesticide Science* 33: 439-446.

Charudattan, R. and H.L. Walker. (eds.). (1982). *Biological Control of Weeds with Plant Pathogens*. New York: Wiley Interscience.

Cole, R.J., J.W. Kirksey, J.P. Springer, J. Clardy, H.G. Cutler, and K.H. Garren. (1975). Desmethoxyviridiol, a new toxin from *Nodulisporium hinnuleum*. *Phytochemistry* 14: 1429-1432.

Crawford, D.L., J.M. Lynch, J.M. Whipps, and M.A. Ousley. (1993). Isolation and characterization of actinomycete antagonists of a fungal root pathogen. *Applied and Environmental Microbiology* 59: 3899-3905.

Cutler, H.G. (ed.) (1988). *Biologically Active Natural Products, Potential Use in Agriculture*. ACS Symposium No. 380. Washington, DC: American Chemical Society.

Cutler, H.G. (1986). Isolating, characterizing and screening mycotoxin for herbicidal activity. In *The Science of Allelopathy*, eds. A.R. Putnam and C.S. Tang. New York: Wiley Interscience, pp. 147-170.

Cutler, H.G., F.G. Crumley, J.P. Springer, R.H. Cox, R.J. Cole, J.W. Dorner, and J.E. Thean. (1980). Pergillin: a non-toxic fungal metabolite with moderate plant growth inhibiting properties from *Aspergillus ustus*. *Journal of Agricultural and Food Chemistry* 28: 989-991.

Cutler, H.G., F.G. Crumley, J.P. Springer, and R.H. Cox. (1981). Dihydroxypergillin: a fungal metabolite with moderate plant growth inhibitory properties from *Aspergillus ustus*. *Journal of Agricultural and Food Chemistry* 29: 981-983.

Cutler, H.G., F.G. Crumley, R.H. Cox, J.H. Wells, and R.J. Cole. (1984). The biological properties of cyclopenin and cyclopenol. *Plant and Cell Physiology* 25: 257-263.

Cutler, H.G., F.G. Crumley, R.H. Cox, O. Hernandez, R.J. Cole, and J.W. Dorner. (1979). Orlandin: a nontoxic fungal metabolite with plant growth inhibiting properties. *Journal of Agricultural and Food Chemistry* 27: 592-595.

Cutler, H.G., J.H. LeFiles, F.G. Crumley, and R.H. Cox. (1978). Hydroxyterphenyllin: a novel fungal metabolite with plant growth inhibiting properties. *Journal of Agricultural and Food Chemistry* 26: 632-635.

DeFrank, J. and A.R. Putnam. (1985). Screening procedures to identify soil borne actinomycetes that can produce herbicidal compounds. *Weed Science* 33: 271-274.

Demain, A.L. (1983). New applications of microbial products. *Science* 219: 709-714.

Diagle, D.J. and W.J. Connick, Jr. (1990). Formulations and application technology for microbial weed control. In *Microbes and Microbial Products as Herbicides*, ed. R.E. Hoagland, ACS Symposium No. 439. Washington, DC: American Chemical Society, pp. 288-304.

Duke, S.O. 1986. Microbially produced phytotoxins as herbicides–a perspective. In *The Science of Allelopathy*, eds. A.R. Putnam and C.S. Tang. New York: Wiley Interscience, pp. 287-304.

Duke, S.O. and H.K. Abbas. (1995). Natural products with potential use as herbicides. In *Allelopathy: Organisms, Processes and Applications*, eds. Inderjit, K.M.M. Dakshini, and F.A. Einhellig. ACS Symposium No. 582. Washington, DC: American Chemical Society, pp. 348-362.

Durbin, R.D. (ed.). (1981). *Toxins in Plant Disease*. New York: Academic Press.

Einhellig, F.A., G.R. Leather, and L.L. Hobbs. (1985). Use of *Lemna minor* L. as a bioassay in allelopathy. *Journal of Chemical Ecology* 11: 65-72.

El-Nakeeb, M.A. and H.A. Lechevalier. (1963). Selective isolation of aerobic actinomycetes. *Applied Microbiology* 11: 75-77.

Fisher, H.P. and D. Bellus. (1983). Phytotoxin from microorganisms and related compounds. *Pesticide Science* 14: 334-346.

Haneishi, T., A. Terahara, H. Kayamori, J. Yabe, and M. Arai. (1976). Herbicidins A and B, two new antibiotics with herbicidal activity II. Fermentation, isolation and physico-chemical characterization. *The Journal of Antibiotics* 29: 887-875.

Heisey, R.M., S.K. Mishra, A.R. Putnam, J.R. Miller, C.J. Whitenack, J.E. Keller, and H. Huang. (1988). Production of herbicidal and insecticidal metabolites by soil microorganisms. In *Biologically Active Natural Products, Potential Use in Agriculture*, ed. H.G. Cutler, ACS Symposium No. 380. Washington, DC: *American Chemical Society*, pp. 65-78.

Heisey, R.M. and A.R. Putnam. (1990). Herbicidal activity of the antibiotics geldanamycin and nigericin. *Journal of Plant Growth Regulation* 9: 19-25.

Hoagland, R.E. (1990). Microbes and microbial products as herbicides, an overview.

In *Microbes and Microbial Product as Herbicides*, ed. R.E. Hoagland. ACS Symposium No 439. Washington, DC: American Chemical Society, pp. 2-52.

Howell, C.R. and R.D. Stipanovic. 1984. Phytotoxicity to crop plants and herbicidal effects on weeds of viridiol produced by *Gliocladium virens*. *Phytopathology* 74: 1346-1349.

Iwai, Y., A. Nakagawa, N. Sadakane, S. Omura, H. Oiwa, S. Matsumoto, M. Takahashi, T. Ikai, and Y. Ochiai. (1980). Herbimycin B, a new benzoquinonoid ansamycin with anti TBV and herbicidal activities. *The Journal of Antibiotics* 33: 114-119.

Jutsum, A.R. (1988). Commercial application of biological control: Status and prospects. *Philosophical Transaction of Royal Society of London B* 318: 357-373.

Jones, R.W. and J. Ge. Hancock. (1990). Soil borne fungi for biological control of weeds. In *Microbes and Microbial Products as Herbicides*, ed. R.E. Hoagland. ACS Symposium No. 439. Washington, DC. American Chemical Society, pp. 276-286.

Kamata, S., H. Saki, and A. Hirota. (1983). Isolation of acetylaranotin, bisdethiodi (Methlythio)-acetylaranotin and terrein as plant growth inhibitors from a strain of *Aspergillus terreus*. *Agricultural and Biological Chemistry* 47: 2637-2638.

Kratky, B.A. and G.F. Warren. (1971). The use of three simple, rapid bioassays on forty-two herbicides. *Weed Research* 11: 257-262.

Lax, A.R. and H.S. Shepherd. (1988). Tentoxin: a cyclic tetrapeptide having potential herbicidal usage. In *Biologically Active Natural Products, Potential Use in Agriculture*, ed. H.G. Cutler. ACS Symposium No. 380. Washington, DC: American Chemical Society, pp. 24-34.

Lynch, J.M. (1976). Products of soil microorganisms in relation to plant growth. *Critical Review of Microbiology* 5: 67-107.

Lynch, J.M. (1990). Microbial metabolites. In *The Rhizosphere*, ed. J.M. Lynch. West Sussex, England: John Wiley & Sons, pp. 177-206.

Mallik, M.A.B. (1997). Isolates of soil actinomycetes with potential for phytotoxin production. *Journal of Chemical Ecology* 23: 2683-2693.

Maruyama, H.B., Y. Suhara, J. Suzuki-Watanabe, Y. Maeshima, N. Shimizu, M. Ogura-Hamada, H. Fujimoto, and K. Takano. (1975). A new antibiotic, fumaramidmycin. I. Production, biological properties and characterization of producer strain. *The Journal of Antibiotics* 28: 636-547.

Misato, T. (1983). Recent status and future aspects of Agricultural antibiotics. In *Pesticide Chemistry: Human Welfare and the Environment, Vol. 2. Natural Products*, eds. N. Takahashi, H. Yoshioka T. Misato, and S. Matsunaka. Oxford, UK: Pergamon Press, pp. 241-246.

Misato, T. and I. Yamaguchi. (1984). Pesticides of microbial origin. *Outlook on Agriculture* 13: 117-120.

Mishra, S.K., W. H. Taft, A.R. Putnam, and S.K. Ries. (1987). Plant growth regulatory metabolites from novel actinomycetes. *Journal of Plant Growth Regulation* 6: 75-84.

Murao, S. and H. Hyashi. (1983). Gourgerotin, as a plant growth inhibitor, from *Streptomyces* sp. No. 179. *Agricultural and Biological Chemistry* 47: 1135-1136.

Norstadt, F.A. and T.M. McCalla. (1963). Phytotoxic substance from a species of *Penicillium*. *Science* 140: 410-411.

Omura, S. (1986). Philosophy of new drug discovery. *Microbiological Review* 50: 259-279.

Omura, S., Y. Iwai, Y. Takahashi, N. Sadakane, A. Nakagawa, H. Oiwa, Y. Hasegawa, and T. Ikai. (1979). Herbimycin, a new antibiotic produced by a strain of *Streptomyces*. *The Journal of Antibiotics* 32: 255-261.

Omura, S., M. Murata, H. Hanaki, K. Hinotozawa, R. Oiwa, and H. Tanaka. (1984). Phosalacine, a new herbicidal antibiotic containing phosphonothricin: fermentation, isolation biological activity and mechanism of action. *The Journal of Antibiotics* 37: 829-835.

Porter, J.N., J.J. Wilhelm, and H.D. Tresner. (1960). Method for the preferential isolation of actinomycetes from soils. *Applied Microbiology* 8: 174-178.

Scacchi, A., R. Bortolo, G. Cassani, G. Pirali, and E. Nielsen. (1992). Detection, characterization and phytotoxic activity of the nucleoside antibiotics, blasticidins and 5-hydroxy methyl-blasticidins. *Journal of Plant Growth Regulation* 11: 39-46.

Sekizawa, Y. and T. Takematsu. (1983). How to discover new antibiotics for herbicidal use. In *Pesticide* Chemistry: *Human Welfare and the Environment, Vol. 2. Natural Products,* eds. N. Takahashi, H. Yoshioka, T. Misato, and S. Matsunaka. Oxford, U.K.: Pergamon Press. pp. 261-268.

Shirling, E.B. and D. Gottlieb. (1966). Methods for characterization of *Streptomyces* species. *International Journal of Systematic Bacteriology* 16: 313-340.

Shomura, T., J. Yoshida, S. Amano, M. Kojima, S. Inouye, and T. Nida. (1970). Studies on actinomycetales producing antibiotics only on agar culture. I. Screening, taxonomy and morphology-productivity relationship of *Streptomyces halstedii,* Strain SF-1993. *The Journal of Antibiotics* 32: 427-435.

Springer, J.P., H.G. Cutler, F.G. Crumley, R.H. Cox, E.E. Davis, and J.E. Thean. (1981). Plant growth regulatory effects and stereochemistry of cladosporin. *Journal of Agricultural and Food Chemistry* 29: 853-855.

Takahashi, E., T. Kimura, K. Nakamura, M. Arashira, and M. Iida. (1995). Phosphonothrixin, a novel herbicidal antibiotic produced by *Saccharathrix* sp. ST 888, I. Taxonomy, fermentation, isolation and biological properties. *The Journal of Antibiotics* 48: 1124-1129.

Takahashi, S., M. Nakajima, T. Kinoshista, H. Haruyama, S. Sugai, T. Homma, S. Sato, and T. Haneishi. (1994). Hydantocidin and cornexistin, new herbicidal antibiotics. In *Natural and Engineered Pest Management Agents,* eds. P.A. Hedin, J.J. Menn, and R.M. Hollingworth. ACS Symposium No. 551. Washington, DC.: American Chemical Society, pp. 74-84.

Tsao, P.H., C. Leben, and G.W. Keitt. (1960). An enrichment method for isolating actinomycetes that produce diffusible antifungal antibiotics. *Phytopathology* 50: 88-89.

Waksman, S.A. (1967). *The Actinomycetes: A Summary of Current Knowledge*. New York: The Ronald Press Co.

Walton, J.D. and E.D. Earle. (1984). Isolation and bioassay of fungal phytotoxins. In *Cell Cultures and Somatic Cell Genetics of Plants, Vol. I. Laboratory Procedures and Their Applications,* ed. I.K. Vasil. New York: Academic Press, pp. 598-605.

Warren, H.B., Jr., J.F. Prokop, and W.E. Grundy. (1955). Nonsynthetic media for antibiotic producing actinomycetes. *Antibiotics and Chemotherapy* 5: 6-12.
Williams, S.T. and E.M.H. Wellington. (1982). Principles and problems of selective isolation of microbes. In *Bioactive Microbial Products: Search and Discovery*, eds. J.D. Bullock, L.J. Nisbet, and D.J. Winstanley. New York: Academic Press (for Society of Genetal Microbiology), pp. 9-26.
Yamada, O., Y. Kaise, F. Futatsuya, S. Ishida, K. Ito, H. Yamamoto, and K. Munakata. (1972). Studies on plant growth regulating activities of anisomycin and toycamycin. *Agricultural and Biological Chemistry* 36: 2013-2015.
Yamada, O., S. Ishida, F. Futatsuya, K. Ito, H. Yamamoto, and K. Munakata. (1974). Plant growth-regulating activities of 4-methoxybenzophenones. *Agricultural and Biological Chemistry* 38: 2017-2020.

APPENDIX

Formulations of a few selected media for growth of actinomycetes. All weights are in gram per liter, and agar at 15-18 g per liter.

A. Isolation Media

1. Glucose-Asparagine (Waksman, 1967)

Glucose 10 g, DL-asparagine 0.5 g, K_2HPO_4 0.5 g, tap/distilled water. pH 6.8 to 7.0.

2. Water-Yeast Extract (Crawford et al., 1993)

Yeast extract 0.25 g, K_2HPO_4 0.5 g, tap/distilled water or soil/compost extract. pH 7.0 to 7.2 (soil:water ratio may range from 1:9 to 1:1 depending on the richness of soil/compost).

3. Starch-Casein (Aaronson, 1970)

Soluble starch or glycerin 10 g, casein (dissolved in NaOH solution) 0.3 g, KNO_3 2.0 g, NaCl 2.0 g, K_2HPO_4 2.0 g, $MgSO_4.7H_2O$ 0.05 g, $CaCO_3$ 0.02 g, $FeSO_4.7H_2O$ 0.01 g. pH 7.0 to 7.5.

Cyclohexamide 50 μg/ml and mycostatin (dissolved in dimethyl sulphoxide) 50 μg/ml are added after autoclaving.

4. AGS Medium (El-Nakeeb and Lechevalier, 1963)

Arginine monohydrochloride 1.0 g, glycerol 12.5 g, K_2HPO_4 1.0 g, NaCl 1.0 g, $MgSO_4.7H_2O$ 0.50 g, $Fe_2(SO_4)_3.6H_2O$ 0.01 g,$CuSO_4.5H_2O$ 0.001 g, $ZnSO_4.7H_2O$ 0 001 g, $MnSO_4.H_2O$ 0.001 g. pH 7.0.

5. Glycerol-Arginine (Porter et al., 1960)

Glycerol 20.0 g, L-arginine 2.5 g, NaCl 1.0 g, $CaCO_3$ 0.01 g, $FeSO_4.7H_2O$ 0.10 g $MgSO_4.7H_2O$ 0.10 g, pimaricin 50 μg/ml, cyclohexamide 50 μg/ml and nystatin 50 μg/ml.

6. Casamino Acids-Yeast Extract-Glucose (Crawford et al., 1993)

Casamino acids (Difco) 0.03 g, yeast extract 0.3 g, glucose 0.3 g, K_2HPO_4 2.0 g.

B. Culture Maintenance in Agar Slant

1. Casamino Acids-Glucose (Crawford et al., 1993)

Casamino acids (Difco) 0.5 g, glucose 0.4 g, yeast extract 0.8 g, K_2HPO_4 2.0 g. pH 7.2

2. Oatmeal Agar (Shirling and Gottlieb, 1966)

Cook 20 g oatmeal in 1000 ml distilled water for 20-30 min, and strain through double layer of cheese cloth while hot. Add 1.0 ml of trace solution composed of $FeSO_4.7H_2O$ 0.10 g, $MnCl_2.4H_2O$ 0.10 g, and $ZnSO_4.7H_2O$ 0.10 g per 100 ml of distilled water. Make up the volume with distilled water to 1000 ml. Adjust to pH 7.2 with NaOH.

3. Yeast Extract-Malt Extract-Dextrose (Shirling and Gottlieb, 1966)

Bacto-yeast extract 4.0 g, Bacto-malt extract 10.0 g, Bacto-dextrose 4.0 g. Adjust to pH 7.2.

Media Nos. B2 and B3 are recommended for the study of morphology of actinomycetes.

C. Toxin Producing Broth

1. Peptone-Molasses-Corn Steep (DeFrank and Putnam, 1985)

Glucose 5.0 g, Bacto-peptone (Difco) 2.5 g, molasses (sulfur free) 5.0 g, corn steep solids 1.0 g. Corn steep is separately boiled in about 100 ml distilled water, filtered and the filtrate is added to the broth. Adjust to pH 7.0

2. Glucose-Peptone-Molasses (Warren et al., 1955)

Glucose 10.0 g, peptone 5.0 g, molasses 20.0 g. Adjust to pH 7.0

3. Soybean Meal-Corn Steep-Dextrin (Warren et al., 1955)

Soybean meal 10.0 g, corn steep liquid 20.0 g, dextrin 10.0 g, NaCl 5.0 g, $CaCO_3$ 2.0 g, K_2HPO_4 2.0 g. Adjust to pH 7.0

4. Shredded Wheat Medium for Fungi (Cutler et al., 1981)

Shredded wheat 100 g, Difco mycological broth (pH 4.8) 200 ml, yeast extract 2.0 g, sucrose 20.0 g.

The Use of Allelopathic Studies in the Search for Natural Herbicides

Francisco A. Macías
José M. G. Molinillo
Juan C. G. Galindo
Rosa M. Varela
Ana M. Simonet
Diego Castellano

SUMMARY. The use of herbicides is claimed to negatively affect the environment and actually it does not represent an appropriate tool for the control of some weeds developing resistance. Allelopathic studies offer a challenge for discovering new lead compounds with new target sites. Then, they might be able to control the actual resistant weeds. A Standard Bioassay with selected Standard Target Species (belonging to both monocotyledonous and dicotyledonous species) is proposed to overcome the low reproducibility and the difficulties in comparing results from different research groups. This bioassay includes a commercial herbicide of known levels of activity. Herein, we illustrate this procedure with the results of studies performed with different plants

---

Francisco A. Macías, Professor, José M. G. Molinillo, Associate Professor, Juan C.G. Galindo, Assistant Professor, Rosa M. Varela, Assistant Professor, Ana M. Simonet, Assistant Professor, Diego Castellano, Assistant Professor, Departamento de Química Orgánica, Facultad de Ciencias, Universidad de Cádiz, c/República Saharaui s/n. Apdo. 40, 11510 Puerto Real, Cádiz, Spain.

Address correspondence to: Francisco A. Macías, Universidad de Cádiz, c/República Saharaui s/n. Apdo. 40, 11510 Puerto Real, Cádiz, Spain (E-mail: famacias@uca.es).

This research was supported by the Comisión Interministerial de Ciencia y Tecnología (C.I.C.Y.T.), Project No. AGF97-1230-C02-02.

[Haworth co-indexing entry note]: "The Use of Allelopathic Studies in the Search for Natural Herbicides." Macías, Francisco A. et al. Co-published simultaneously in *Journal of Crop Production* (Food Products Press, an imprint of The Haworth Press, Inc.) Vol. 4, No. 2 (#8), 2001, pp. 237-255; and: *Allelopathy in Agroecosystems* (ed: Ravinder K. Kohli, Harminder Pal Singh, and Daizy R. Batish) Food Products Press, an imprint of The Haworth Press, Inc., 2001, pp. 237-255. Single or multiple copies of this article are available for a fee from The Haworth Document Delivery Service [1-800-342-9678, 9:00 a.m. - 5:00 p.m. (EST). E-mail address: getinfo@haworthpressinc.com].

belonging to wild and agricultural ecosystems that have rendered many allelochemicals belonging mainly to several terpene families.

KEYWORDS. Allelopathic studies, natural herbicides, standard bioassay, standard target species, allelochemicals

INTRODUCTION

Projections of World Bank and United Nations indicate that the world population will grow, at least, until 2040 to 10 billion people (Avery, 1995). Increasing demand for food will require an increase in agricultural productivity. One of the keys for this is the control of weeds and pests. Herbicides have helped farmers to increase yields while reducing labor. However, the potential environmental and toxicological costs of herbicides have raised questions about our agricultural dependence on these magic solutions.

Sir Collin Berry (1999) affirms that although the major contaminants in food production are microbiological and occur in food processing, preparation and handling. The chemicals in food including food additives, pesticides and preservatives appear to cause anxiety. The growth in yield must respect human considerations and the environment–in other words, this growth must be sustainable. David A. Evans (1999) has described the criteria that food provision technology must satisfy to be sustainable:

- agronomy — reliance on effective systems
- environment — must conserve the natural balance
- society — must provide net benefit to communities
- economics — must reward all components of production.

Is the development of new herbicides necessary? The answer to this question appears to be a definitive "Yes." In spite of modern weed control, even in developed countries, losses due to weeds in terms of yields and quality are relatively high. Additionally, the use of herbicides has provoked an increasing incidence of resistance in weeds, so new, more efficient and specific herbicides, are needed. Some data concerning this search are very interesting (Petsko, Ringe, and Hogan, 1999):

- Synthesis of 6,200 chemical compounds is required for every drug that makes it to the market.
- The time required to bring a drug to the public is about 13 years.
- The total cost per new commercial compound is well in excess of $350 million.
- The industry has the capacity to test 1,00,000 compounds per year.
- Traditional organic synthesis cannot provide more than 200-300 novel compounds per year.

The industry needs new ideas and methods to obtain new herbicides. For instance, the newest available herbicides have principally four traditional sites of actions, all discovered before 1985 (Hay, 1999):

- acetolactate synthesis inhibitors
- protoporphyrinogen oxidate inhibitors
- *p*-hydroxyphenylpyruvate dioxygenase inhibitors
- phytoene desaturase inhibitors.

A simple deduction of this fact can be made: we need new and effective herbicides affecting new sites of action. But where? and more important, how can we find new herbicides possessing these new sites of action? Duke et al. (1997) gave us one possible answer to this question: compounds isolated from natural sources. They are most likely to provide a multitude of new sites of action on which we can focus research. Up to date, many natural products are commercially used as insecticides, acaricides, fungicides or bactericides, but none of them has been used extensively as herbicides (Copping, 1996).

Another new and important source of compounds for screening is the combinatorial chemistry. It involves the production of every possible molecule from a basis set of modular units. The clearest example is the preparation of all possible peptides from a basic set of amino acids. In spite of new techniques, combinatorial chemistry has several limitations, i.e., the range of chemistries that can be accessed (Petsko, Ringe, and Hogan, 1999). Indeed, not many novel compounds can be expected from combinatorial chemistry. Nevertheless, this technique can provide thousands of similar compounds that lead us to make interesting structure-activity comparisons. Another important limitation is the fact that this technique is, in most cases, non-guided. You can prepare, of course, thousands of compounds. You can test their activity. And you can be lucky, . . . or not. Really, you need more powerful tools that indicate what avenues to pursue.

So, we need new herbicides, with novel structures and new sites of action, which are environmentally friendly and, of course, beneficial. Evans (1999) provides us the profile of the ideal herbicide:

- it works at 500 mg/ha;
- it offers exceptional control (rainfast, broadspectrum, flexible application window, pre- and post-activity) with superior crop safety;
- favorable ecotoxicology (non-oncogenic, reduced risk product);
- cheap to produce on a small dedicated pilot plant (or even in the laboratory); and
- range of formulations to suit all growing conditions.

We know what we want, but do we know how to get it? In the following pages, we will try to indicate how allelopathic studies can help in the search for natural herbicides. First, we note some characteristics of this search:

- These compounds must be safe and the technology sustainable (from different points of view–agronomical, environmental, and economical).
- New sites of action must be explored.
- We need novel compounds in a sufficient number and quantity.
- These compounds must be active at low concentrations and have a wide range of activities.

ALLELOPATHY, LEARNING FROM NATURE

Plants have their own defense mechanisms and allelochemicals are, in fact, natural herbicides. If we knew the ways that plants interacted inter- and intra-specifically within a given ecosystem, we could mimic certain processes and devise applications based on natural herbicides, antibiotics, fungicides, and insecticides. Allelopathic studies develop new compounds which may lead to obtain a new generation of growth inhibitors. These studies have the advantage that natural products may have different target sites than traditional herbicides. Thus, they may be able to control resistant weeds.

One of the main objections to allelopathic studies is the low reproducibility of the results and the difficulty of comparing data from different research groups. We propose that a Standard Bioassay with selected Standard Target Species (belonging to monocotyledonous and dicotyledonous species) be performed. This bioassay should also include a commercial herbicide of known levels of activity. The use of such an internal standard will serve to compare the levels of phytotoxicity shown by the compounds and the levels of a commercial formulation in the same conditions. The second important advantage is that it will also allow comparison of the results of bioassays carried out on different places and situations.

SAFETY AND SUSTAINABLE

Allelopathy includes any process involving secondary metabolites produced in nature that influence the growth and development of plants. This definition does not assure environmental safety. Many potent poisons from natural sources such as hemlock, or more recently, brevetoxins, are known. Nevertheless, we can take some very simple cautions. We can look for those herbicides that are actually present in some extended crops. For instance, crops such as sorghum, alfalfa, wheat, barley, corn, asparagus, coffee, tea, tobacco, and sunflower are allelopathic (Macias et al., 1996c). They have different mechanisms for the incorporation of their allelochemicals into the environment. These allelochemicals are potential herbicides, so we should focus our attention on known systems, as described above. If these compounds could be obtained from parts of the crops actually discarded (mulches, roots, leaves, etc.) they could add economic value to the original crop. They would be cheap and abundant. The development of these kind of herbicides could revalue present byproducts.

The economic implications are obvious. Those farmers cultivating sunflower will be able to collect their leaves for herbicide production. The industries will buy this cheap starting material to produce them. Where is the problem? Why are not there many allelochemicals in the market? Allelopathy is, in many cases, a complex phenomenon and hence, it cannot be assumed that a single compound, or even a small number of them, can be named as being responsible for such a behavior (Macias et al., 1999b). First, we have to understand the particular chemistry of each case, and then aim to optimize the extraction, isolation, and the use of allelochemicals. In our opinion, this is not a limitation, but a challenge. We have to learn from nature, and learn how to make good use of nature.

NEW MODES OF ACTION

Recently, the attention to the mode of action of allelochemicals has increased. Although the primary mode of action has not been conclusively established for almost any allelopathic compound, some physiological actions are known. The quinones, juglone (Hejl, Einhellig, and Rasmussen, 1993) and sorgoleone (Einhellig et al., 1993) inhibit chloroplast oxygen evolution and strongly affect mitochondrial functions (Rasmussen et al., 1992). We could classify these compounds in category C from the HRAC (Herbicide Resistance Action Committee) (Figure 1).

A different case is presented by the monoterpene 1,8-cineole and its commercial derivative cinmethylin. This compound could be considered the first

FIGURE 1. Allelochemicals and related compounds with potential use as herbicide.

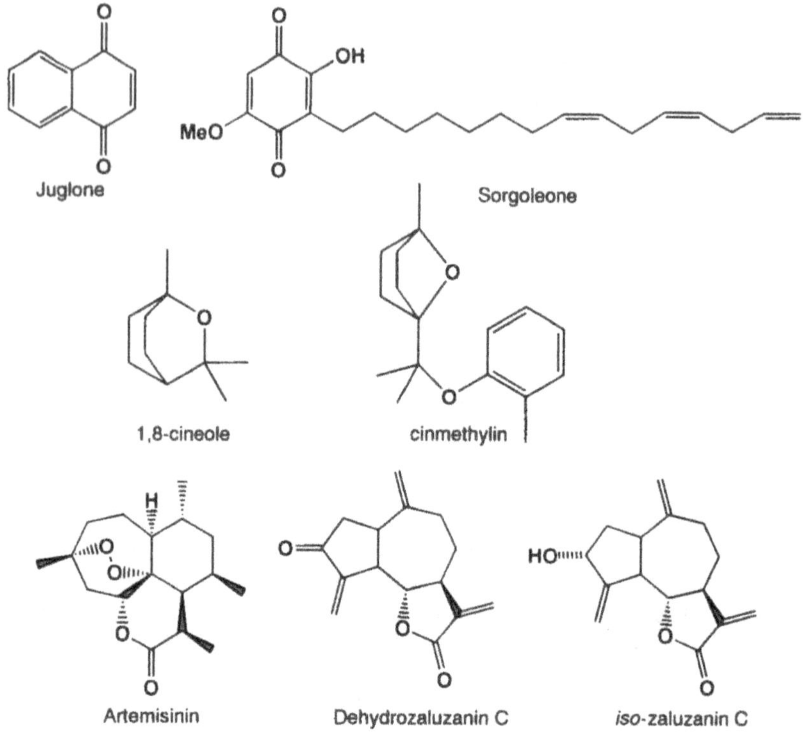

commercialized allelopathic herbicide. It is not a natural product, but its structure is very similar to the natural 1,8-cineole. There are several research groups that are now studying in the interaction of cinmethylin with other traditional herbicides such as chlorsulfuron or metsulfuron-methyl in different crops. Its primary mode of action is still unknown and seems to be new. Some sesquiterpene lactones such as artemisinin, dehydrozaluzanin C, and *iso*zaluzanin C have been studied (Figure 1). Artemisinin is the main allelopathic principle of *Artemisia annua* and it is highly phytotoxic, inhibiting root growth and causing extreme chlorosis. Artemisinin affects mitochondrial oxygen evolution and inhibits mitosis, leading to the appearance of aberrant mitotic phases (Dayan et al., 1999a).

Dehydrozaluzanin C, isolated from the root of many different Compositae species (Bohlman and Le Van, 1977; Bohlman, Brindöpke, and Rastogi, 1978; Bohlman et al., 1980 a,b,c; Bohlman and Zdero, 1982; Bohlman, Singh, and Jakupovic, 1982), causes rapid and pronounced cellular leakage

(Galindo et al., 1999). This compound presents phytotoxic effects on several plant species (lettuce, tomato, carrot, cress, onion, wheat, and barley) in a range between 0.1 and 1 mM (Macias et al., 2000a). Primary effects produced by this compound are still unknown, but some secondary effects such as the disruption of membrane integrity and of amino acid biosynthetic pathways have been reported. This could represent another example of a new mode of action of a natural product.

NOVEL COMPOUNDS AND ALLELOPATHIC STRATEGIES

There exists a great demand of new and novel products. In nature, compounds arise following a mechanism analogous to the combinatorial chemistry that we can call randomized chemistry. Mutations in the DNA lead to synthesis of a multitude of variations of a large number of skeletons. *The Dictionary of Natural Products* edited by Chapman & Hall and CRC Press covers more than 1,40,000 compounds. But, this is like looking for a needle in a haystack. Allelopathy research methodology can help us to find those active principles (Macias, 1995) (Figure 2). The process begins with the selection of the source. We select those species that have shown allelopathic antecedents in the field. Then we have to study the mechanism that the plant uses to excrete the active principles into the environment. The research tries to mimic what happens in nature, so the method of extraction uses water as solvent at room temperature. This extract will be fractionated and bioassayed, and the active fractions will be further studied to isolate and characterize the active principles.

One of the keys of the research process is the correct design of the bioassay and its statistical treatment. Strictly speaking, allelopathy is an ecological process that occurs in a particular ecosystem. If we are making ecological studies we must use those species present in the ecosystem. But, in a broad sense, we can be looking for new active molecules that can be used as herbicides. In such case, we should use weed seedlings. Nevertheless, wild species are genetically more heterogeneous than commercial crops, and their growth under laboratory conditions may be unpredictable. Viability of seeds depends on many conditions, and seeds do not react uniformly to similar treatments at different times. Additionally, weed seeds do not germinate uniformly, probably due to an internal protection mechanism (King, 1966; Cutler and Cole, 1983). Wild seeds present other important problems as the degradation by storage conditions. To solve these problems, a collection of Standard Target Species (STS) of commercial crop seeds have been proposed as models for weeds and crops (Macias, Castellano, and Molinillo, 2000). The dicotyledons *Lactuca sativa* L. (lettuce, Compositae), *Daucus carota* L.

FIGURE 2. Allelopathic strategy for the search for new natural herbicides.

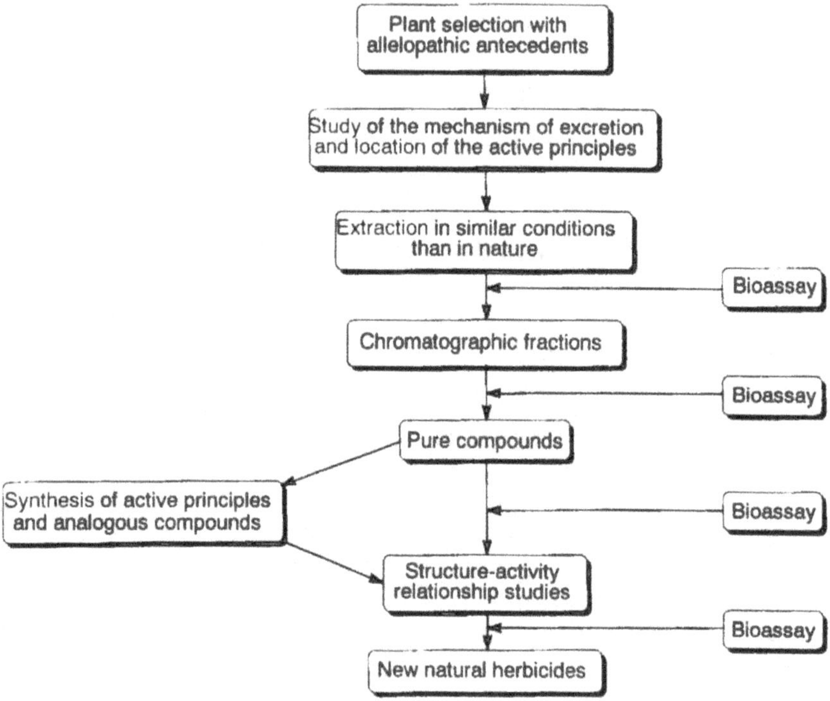

(carrot, Umbelliferae), *Lepidium sativum* L. (cress, Cruciferae), *Lycopersicum esculentum* L. (tomato, Solanaceae); and the monocotyledon *Allium cepa* L. (onion, Liliaceae), *Hordeum vulgare* L. (barley, Gramineae), *Triticum aestivum* L. (wheat, Gramineae), and *Zea mays* L. (corn, Gramineae) have been selected as representative of most common weeds (Holm et al., 1979).

The following steps to standardize these bioassays are proposed as (i) to test the sensitivity of the proposed STS; (ii) to establish the activity profiles of commercial herbicides under standard allelopathic bioassay conditions using nine STS; commercial herbicides have potential to be used as reference data in allelochemical assays to evaluate natural product herbicide templates; and (iii) to propose a particular formulation as an internal standard with reproducible activity to authenticate the phytotoxic responses of experimental allelochemicals.

In order to evaluate the potential of allelopathic agents as new herbicides, a number of bioassays were undertaken with these agents and the results compared with those of commercial herbicides proposed as internal standards provided by NOVARTIS (Macias et al., 1997). Test concentrations ranged between 10^{-2}-10^{-9} M, based on the usual concentration applied on the field (~10^{-2} M). In this standard phytotoxic allelopathic bioassay, herbicides show strong inhibitory activities only at concentrations between 10^{-2}-10^{-3} M and at a lower concentrations this activity disappears or turns stimulatory. Based on the most consistent profile of activity, the mixture terbutryn + triasulfuron (commercialized as Logran Extra; Figure 3) was selected from the nine herbicides tested to be used as an internal standard.

Duke and Abbas (1995) point out some additional problems that arise during the discovery of natural products with potential for use as herbicides: overly complicated chemical structures, extremely small quantity of compound available for bioassays, and rediscovery of known compounds. One of the major problems that we have in this research is the small amounts of compound obtained. Chemistry can help us in this case, because these compounds, belong in many cases, to some of the usual types of natural products (Rice, 1984), as flavonoids, terpenoids, alkaloids and so on. Thus, they can be synthesized using more abundant similar natural products as starting material. This strategy has been used for the preparation of the above mentioned dehydrozaluzanin C.

ACTIVE COMPOUNDS AT LOW CONCENTRATIONS AND WIDE RANGE OF ACTIVITY

Plants usually do not produce big amounts of active natural products, and usually excrete small amounts of them. This disadvantage turns to an advantage from the point of view of the range of activity. Allelochemicals can present really wide ranges of activities. For instance, tricolorin A, isolated

FIGURE 3. Composition of Logran .

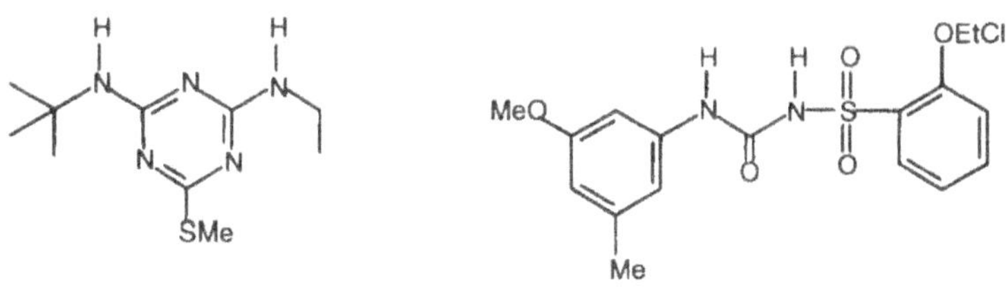

Terbutryn (59.4% w/v)

Triasulfuron (0.6% w/v)

from *Ipomoea tricolor*, presents a value of U_{50} on basal electron flow of 0.33 μM, and therefore, it is a powerful natural uncoupler (Achnine et al., 1999). Brassinosteroids (Figure 4) are a natural product group analogous to brassinolide which was first isolated from rape pollen (Grove et al., 1979). They have been isolated from angiosperms, gymnosperms, an alga and a pteridophyte and may be considered as a new class of plant hormones (Fujioka and Sakurai, 1997). A large number of reports of their bioactivity have been published, but we want to point out the extremely low concentration of brassinolide necessary to inhibit root growth of tomato, 10^{-10} M (Roddick, 1994). Other members of this family of compounds such as 24-epibrassinolide, 22,23,24-triepibrasinolide and 28-homobrassinolide have shown activities within the same range.

Another example of active natural compounds at low concentration is a diterpene group of compounds: quassinoids (Figure 4) have been isolated from several plant species of the simaroubaceae family. Their allelopathic activity (Heisey, 1990; Lawrence, Colwell, and Sexton, 1991) and potential herbicidal activity (Lin et al., 1995) have been reported, as well as some QSAR studies (Dayan et al., 1999b). Quassinoids such as chaparrinune and glaucaxubulone inhibit growth and germination of dicots and monocots at concentration below 10^{-5} M.

Sesquiterpene lactones (SL) are a wide group of compounds with numerous members showing interesting activities at low concentrations. For instance, it has been reported that many SL stimulate witchweed germination at 10^{-9} M (Fischer, Weidenhamer, and Bradow, 1989; Fischer et al., 1990). These compounds were proposed to control witchweed through promoting seed germination in the absence of a suitable host, resulting in death of witchweed seedlings. In the same way, parthenolide and 3,5-dihydroxydehydro-costuslactone (Figure 4) have been recently reported as inductors of the germination of *Orobanche cumana* at 10^{-6} M (Perez et al., 2000). Many other examples of activities at low concentrations of sesquiterpene lactones have been reported (Stevens and Merill, 1985; Fischer, 1986; Francois et al., 1995; Marles et al., 1995) and some SAR studies have been achieved (Macias, Galindo, and Massanet, 1992; Beekman et al., 1997; Macias et al., 1999a).

Chemical studies of *Helianthus* species have shown that this genus is a rich source of terpenoids, and particularly of sesquiterpenoids, with a wide spectrum of biological activities, including potential allelopathic activity (Macias et al., 1999c) (Figure 5). The systematic study of commercial crop varieties of *Helianthus annuus* has afforded a collection of terpenoids (Macias et al., 1999b), including monoterpenes (Macias et al., 1998c), sesquiterpene lactones (Macias et al., 1993a, 1996a), diterpenes (Macias et al., 2000b), triterpenes, and two new families of sesquiterpenoids-heliannanes (Macias et

FIGURE 4. Active natural products at low concentration.

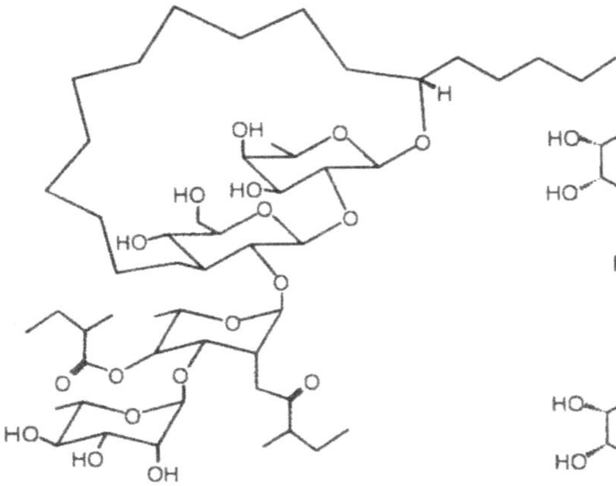

Tricolorin A

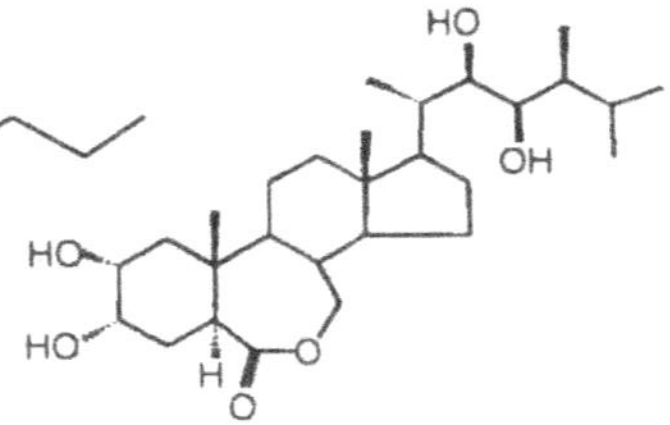

Brassinolide

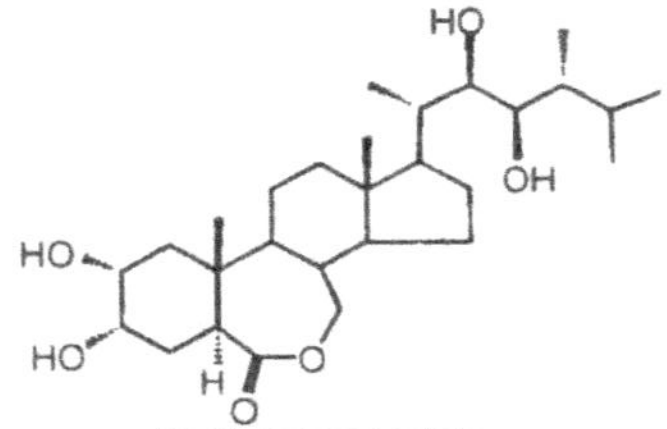

24-Epibrassinolide

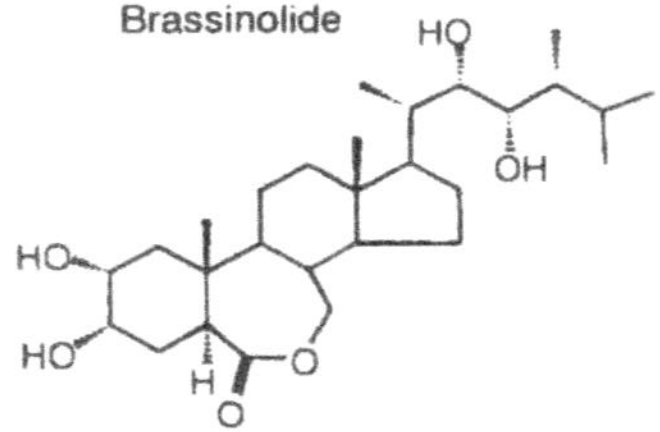

22,23,24-Triepibrassinolide

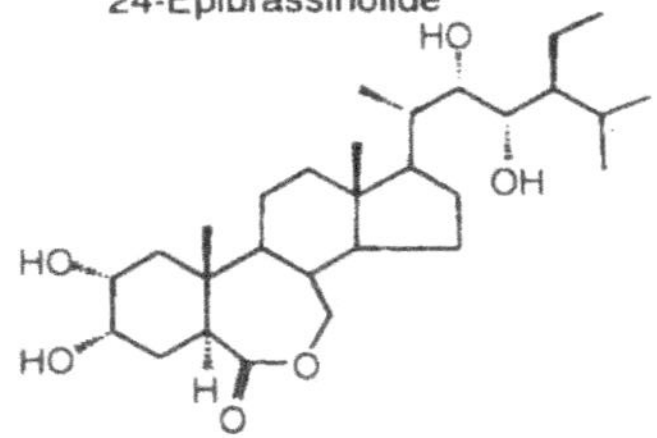

28-Homobrassinolide

Quassinoids

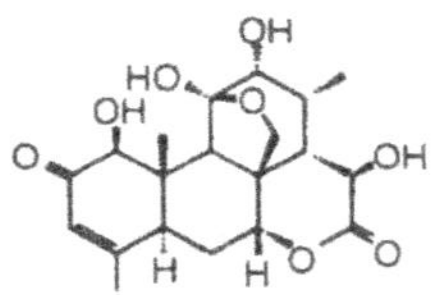

Glaucarubolone

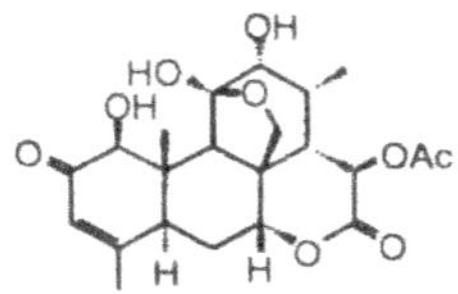

Holacanthone

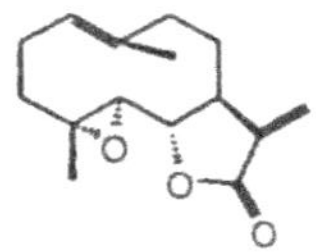

Parthenolide

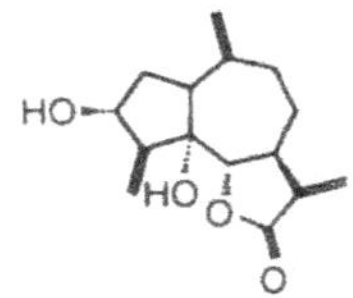

5α-Hydroxy-*iso*zaluzanin C

FIGURE 5. Allelochemicals isolated from *Helianthus annuus*.

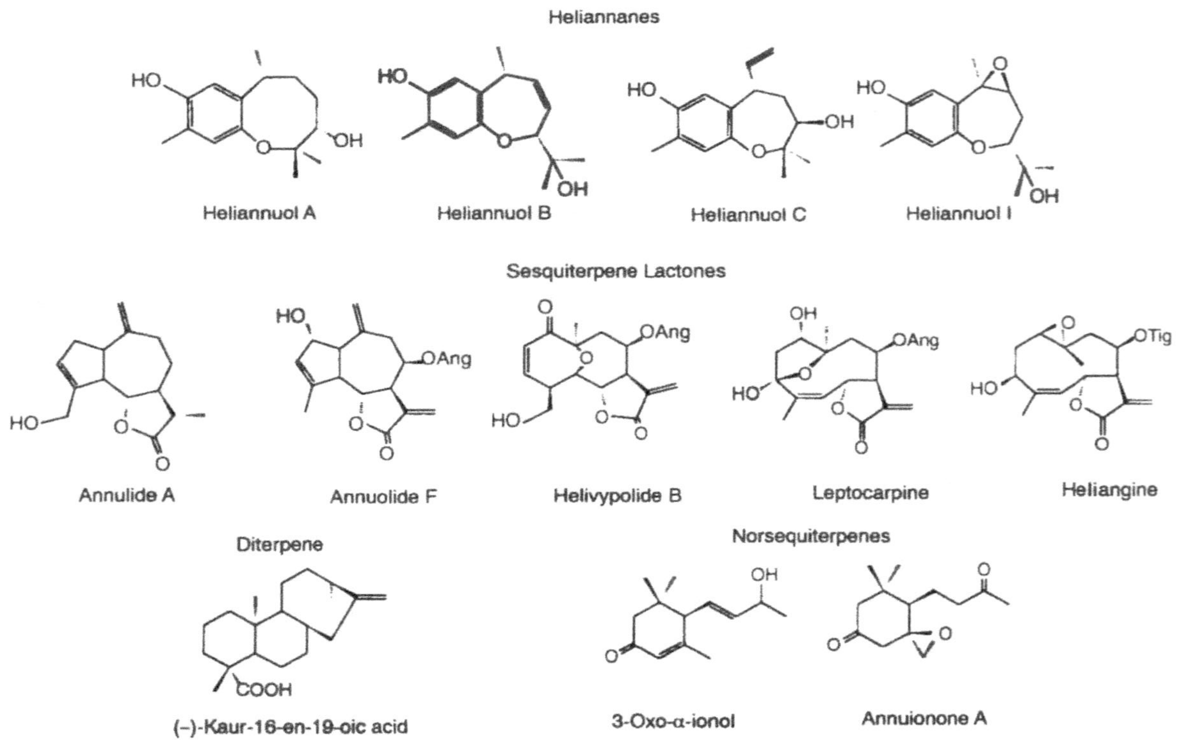

al., 1993b, 1994) and heliespiranes (Macias et al., 1998b) (Figure 5). Some of these compounds have shown excellent levels and profiles of phytotoxicity in comparison with the internal standard used (LOGRAN). As an example we can highlight the inhibition of germination of *Lactuca sativa* shown by the norsesquiterpene helinorbisabone, the heliannanes heliannuol A, C and I, as well as the sesquiterpene lactone annuolide A and helivypolide B at lower concentrations than Logran (down to 10^{-8} M in some cases). Another example could be the inhibition of shoot length caused by annuolide F, heliangine, and leptocarpine on *Lactuca sativa* and by heliannuol B, annuolide A and C, helivypolide B, and (−)-kaur-16-en-19-oic acid on *Lepidium sativum*.

The effects of terpenoids isolated from sunflower species on the growth and germination of monocots is usually the opposite of Logran with stimulation on growth and germination, excepting bisnorsesquiterpenes 3-oxo-α-ionol and annuionone A that inhibit germination of *Allium cepa* and the diterpene (−)-kaur-16-en-19-oic acid that affects the germination and growth of the same species (Macias et al., 1997). We can point out the stimulatory effect presented by annuionone C on root length of both species and the stimulation caused by heliannuol C on root and shoot length of *Hordeum vulgare*.

Finally, the study of *Melilotus messanensis* afforded a number of triterpenes (Figure 6) with a variety of structures (lupane, oleane, norlupane, and gammacerane). Some of these compounds have shown some interesting activities, where we can emphasize the stimulatory effects on the germination of *Allium cepa* by oleananes and the inhibitory activity induced by lupanic acids (Macias, Simonet, and Galindo, 1997; Macias et al., 1996b, 1998a).

CONCLUSIONS

Allelopathic studies are emerging as a challenge for the development of new lead compounds which could provide a new generation of growth inhibitors. In many cases, allelochemicals present excellent activity profiles: wide range of activity and persistence at low doses, high selectivity and sensitivity in terms of species and parameters, and they probably target new sites and present new modes of action. Knowledge of the plant inter- and intra-specific relationships within a specific ecosystem can allow us to mimic certain processes and develop applications of natural herbicides.

FIGURE 6. Allelochemicals isolated from *Melilotus messanensis*.

Lupanic acids

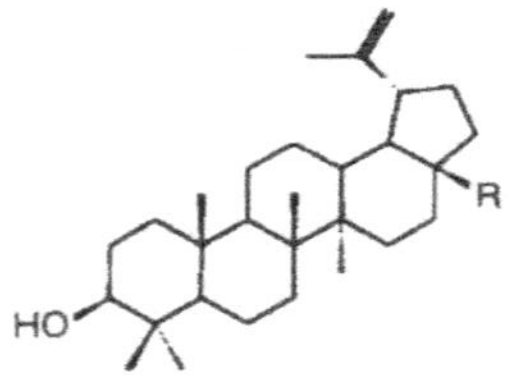

R = COOH, betulinic acid
R = CHO, betulinaldehyde
R = CH_2OH, betuline
R = CH_3, lupeol

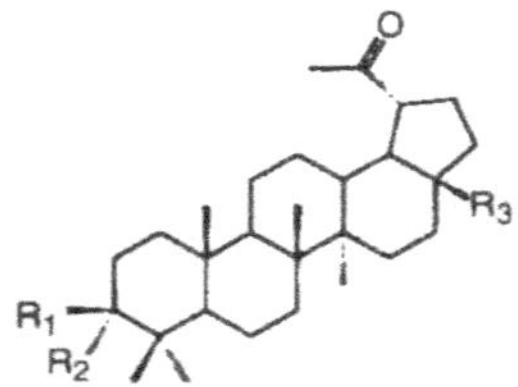

R_1 = OH, R_2 = H, R_3 = CH_2OH; messagenin
R_1 = OH, R_2 = H, R_3 = COOH; platanic acid
R_1 = R_2 = O, R_3 = COOH; 3-oxo-platanic acid

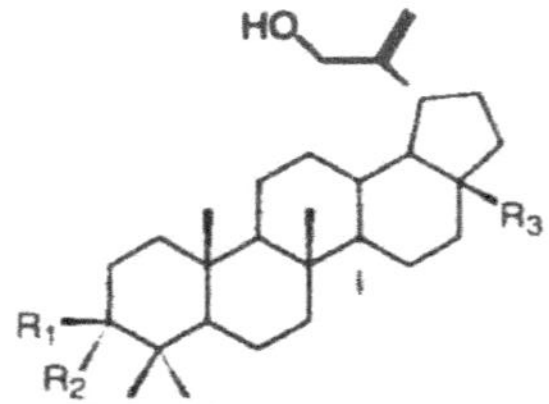

R_1 = OH, R_2 = H, R_3 = COOH; messagenic acid D
R_1 = OH, R_2 = H, R_3 = CH_2OH; lup-20(29)ene-3B,28,30-triol
R_1 = R_2 = 0, R_3 = COOH; messagenic acid G

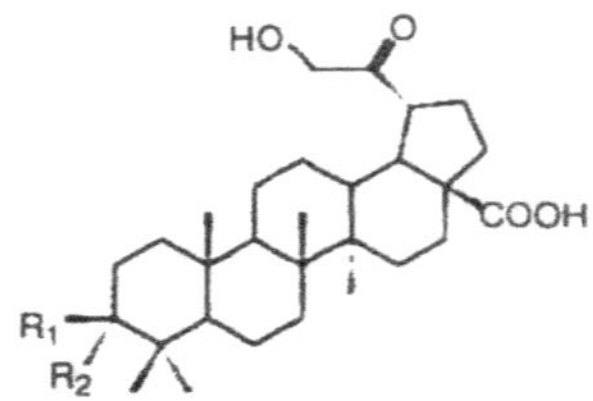

R_1 = R_2 = O; messagenic acid H
R_1 = OH, R_2 = H; messagenic acid

Olenanes

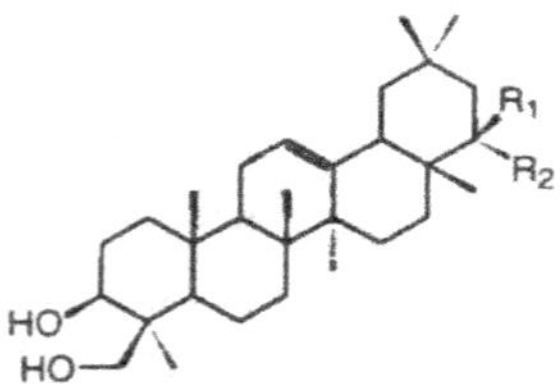

R_1 = OH, R_2 = H; soyasapogenol B
R_1 = R_2 = O; soyasapogenol E

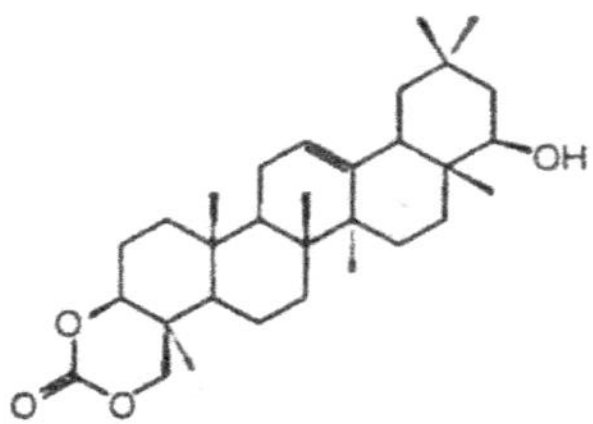

Soyasapogenol G

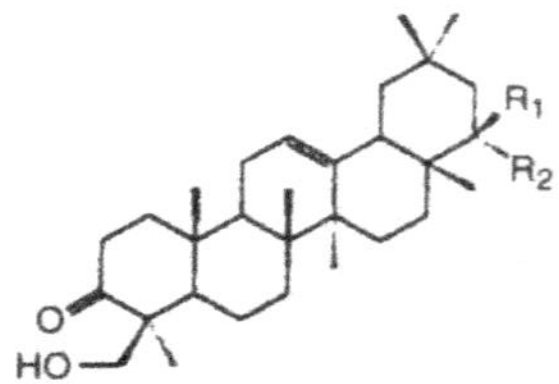

R_1 = OH, R_2 = H; melilotigenin C
R_1 = R_2 = O; melilotigenin B

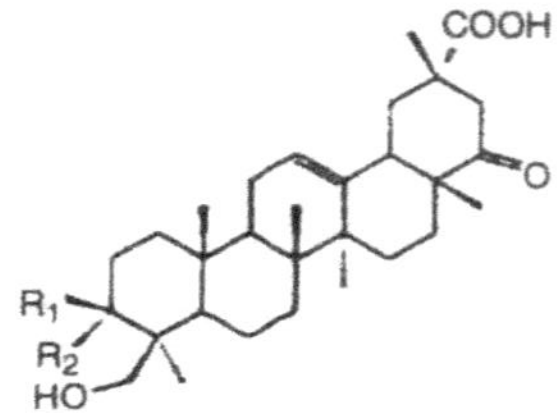

R_1 = OH, R_2 = H; melilotigenin
R_1 = R_2 = O; melilotigenin D

REFERENCES

Achnine, L., M. Bah, R. Pereda-Miranda, R. Mata, and B. Lotina-Hennsen. (1999). Tricolorin A, a potent natural non-protonophoric uncoupler and inhibitor of Q_b redox enzyme on spinach chloroplasts. In *Recent Advances in Allelopathy. Volume I. A Science for the Future*, eds. F.A. Macias, J.C.G. Galindo, J.M.G. Molinillo, and H.G. Cutler. Cádiz, Spain: Servicio de Publicaciones de la Universidad de Cádiz, pp. 205-210.

Avery, D.T. (1995). The environmental triumph of high-yielding farming. In *Saving the Planet with Pesticides and Plastic*. Indianapolis: Hudson Institute, pp. 5.

Beekman, A.C., H.J. Woerdenbag, W.V. Uden, N. Pras, A.W.T. Konings, H.V. Wikstroem, and T.J. Schmidt. (1997). Structure-cytotoxicity relationships of some helenanolide-type sesquiterpene lactones. *Journal of Natural Products* 60: 252-257.

Berry, C. (1999). Caution, precaution or indemnity? The cost of different approaches to regulation. In *Pesticide Chemistry and Bioscience: The Food-Environment Challenge*, eds. G.T. Brooks and T.R. Roberts. Cambridge, UK: The Royal Society of Chemistry, pp. 2530.

Bohlman, F. and C. Zdero. (1982). Naturally occurring terpene derivatives. Part 378. Sesquiterpene lactones from *Brachylaena* species. *Phytochemistry* 21: 647-651.

Bohlman, F., C. Zdero, R.M. King, and H. Robinson. (1980a). Naturally occurring terpene derivatives. Part 287. Sesquiterpene lactones from *Eremanthus* species. *Phytochemistry* 19: 2263-2268.

Bohlman, F., G. Brindöpke, and R.C. Rastogi. (1978). Naturally occurring terpene derivatives. Part 133. A new type of germacranolide from *Vernonia* species. *Phytochemistry* 17: 475-482.

Bohlman, F., J. Jakupovic, R.K. Gupta, R.M. King, and H. Robinson. (1980b). Naturally occurring terpene derivatives. Part 296. Allenic germacranolides, bourbonene-derived lactones, and other constituents from *Vernonia* species. *Phytochemistry* 20: 473-480.

Bohlman, F., L. Müller, R.K. Gupta, R.M. King, and H. Robinson. (1980c). Naturally occurring terpene derivatives. Part 362. Hirsutinolides from *Vernonia* species. *Phytochemistry* 20: 2233-2237.

Bohlman, F. and N. Le Van. (1977). Naturally occurring terpenoid derivatives. Part 96. Sesquiterpene lactones and polyines from the genus *Arctotis*. *Phytochemistry* 16: 487-488.

Bohlman, F., P. Singh, and J. Jakupovic. (1982). Naturally occurring terpene derivatives. Part 439. New germacranolides and other sesquiterpene lactones from *Dicoma* species. *Phytochemistry* 21: 2029-2033.

Copping, L.G. (1996). Preface. In *Crop Protection Agents from Nature: Natural Products and Analogues*. Cambridge, UK: The Royal Society of Chemistry, pp. xvi-xxv.

Cutler, H.G. and R.J. Cole. (1983). Carboxyatractyloside: a compound from *Xanthium strumarium* and *Atractylis gummifera* with plant growth inhibiting properties. The Probable. Inhibitor A. *Journal of Natural Products* 46: 609-613.

Dayan, F.E., A. Hernández, S.N. Allen, R.M. Moraes, J.A. Vroman, M.A. Avery, and

S.O. Duke. (1999a). Comparative phytotoxicity of artemisinin and several sesquiterpene analogs. *Phytochemistry* 50: 607-614.

Dayan, F.E., S.B. Watson, J.C.G. Galindo, A. Hernández, J. Dou, J.D. McChesney, and S.O. Duke. (1999b). Structure-activity relationships of quassinoids in plants. *Pesticide Biochemistry and Physiology* 65: 15-24.

Duke, S.O., F.E. Dayan, A. Hernández, M.V. Duke, and H.K. Abbas. (1997). Natural products as leads for new herbicide modes of action. In *Proceedings of The 1997 Brighton Crop Protection Conference–Weeds. Vol. 2*, eds. K.E. Pallett. Farnham, UK: British Crop Protection Council, pp. 379-386.

Duke, S.O. and H.K. Abbas. (1995). Natural products with potential use as herbicides. In *Allelopathy: Organisms, Processes, and Applications*, eds. Inderjit, K.M.M. Dakshini, and F.A. Einhellig. ACS Symposium Series 582. Washington, DC: American Chemical Society, pp. 348-362.

Einhellig, F.A., J.A. Rasmussen, A.M. Hejl, and I.F. Souza. (1993). Effects of root exudate sorgoleone on photosynthesis. *Journal of Chemical Ecology* 19: 369-375.

Evans, D.A. (1999). How can technology feed the world safely and sustainably? In *Pesticide Chemistry and Bioscience: The Food-Environment Challenge*, eds. G.T. Brooks and T.R. Roberts. Cambridge, UK: The Royal Society of Chemistry, pp. 124.

Fischer, N.H. (1986). The function of mono and sesquiterpenes as plant germination and growth regulators. In *The Science of Allelopathy*, eds. A.R. Putnam and C.S. Tang. New York, USA: Wiley-Interscience, pp. 203-218.

Fischer, N.H., J.D. Weidenhamer, J.L. Riopel, L. Quijano, and M.A. Menelaou. (1990). Stimulation of witchweed germination by sesquiterpene lactones: a structure-activity study. *Phytochemistry* 29: 2479-2483.

Fischer, N.H., J.D. Weidenhamer, and J.M. Bradow. (1989). Dihydroparthenolide and other sesquiterpene lactones stimulate witchweed germination. *Phytochemistry* 28: 2315-2317.

Francois, G., C.M. Passreiter, H.J. Vowedengag, and M. Van Looveren. (1995). Antiplasmodial activities and cytotoxic effects of aqueous extracts and sesquiterpene lactones from *Neurolaena lobata*. *Planta Medica* 62: 126-129.

Fujioka, S. and A. Sakurai. (1997). Brassinosteroids. *Natural Products Report* 14: 1-10.

Galindo, J.C.G., A. Hernández, E. Frank, F.A. Dayan, F.A. Macías, R.N. Paul, and S.O. Duke. (1999). Dehydrozaluzanin C, a natural sesquiterpenolide, causes rapid plasma membrane leakage. *Phytochemistry* 52: 805-813.

Grove, M.D., G.F. Spencer, W.K. Rohwedder, N. Mandava, J.F. Worley, J.D. Warthen, Jr., G.L. Steffens, J.L. Flippen-Anderson, and J.C. Cook, Jr. (1979). Brassinolides, a plant growth-promoting steroids isolated from *Brassica napus* pollen. *Nature* 281: 216-217.

Hay, J.V. (1999). Herbicide discovery in the 21st century–a look into the crystal ball. In *Pesticide Chemistry and Bioscience: The FoodEnvironment Challenge*, eds. G.T. Brooks and T.R. Roberts. Cambridge, UK: The Royal Society of Chemistry, pp. 55-65.

Heisey, R. (1990). Allelopathic and herbicidal effects of extracts from tree of heaven. *American Journal of Botany* 77: 662-670.

Hejl, A.M., F.A. Einhellig, and J.A. Rasmussen. (1993). Effects of juglone on growth, photosynthesis and respiration. *Journal of Chemical Ecology* 19: 559-568.

Holm, L.G., J.V. Pancho, J.P. Herberger, and D.L. Plucknett. (1979). *A Geographical Atlas of World Weeds*. New York: John Wiley & Sons.

King, L.J. (1966). *Weeds of the World, Biology and Control*. London: Leonard Hill Books.

Lawrence, J.G., A. Colwell, and O. Sexton. (1991). The ecological impact of allelopathy in *Ailanthus altissima* (Simaroubaceae). *American Journal of Botany* 78: 948-958.

Lin, L.-J., G. Peiser, B.-P. Ying, F. Mathias, F. Karasina, Z. Wang, J. Itatani, L. Green, and Y.S. Hwang. (1995). Identification of plant growth inhibitory principles in *Ailanthus altissima* and *Castela tortuosa*. *Journal of Agricultural and Food Chemistry* 43: 1708-1711.

Macías, F.A. (1995). Allelopathy in the search for Natural Herbicide Models. In *Allelopathy: Organisms, Processes, and Applications*, eds. Inderjit, K.M.M. Dakshini, and F.A. Einhellig. ACS Symposium Series 582. Washington, DC: American Chemical Society, pp. 310-329.

Macías, F.A., A. Torres, J.M.G. Molinillo, R.M. Varela, and D. Castellano. (1996a). Potential allelopathic sesquiterpene lactones from sunflower leaves. *Phytochemistry* 43: 1205-1215.

Macias, F.A., A.M. Simonet, M.D. Esteban, and J.C.G. Galindo. (1996b). Triterpenoids from *Melilotus messanensis*: Soyasapogenol G, the first natural carbonate derivative. *Phytochemistry* 41: 1573-1577.

Macias, F.A., A.M. Simonet, and J.C.G. Galindo. (1997). Bioactive steroids and triterpenes from *Melilotus messanensis* and their allelopathic potential. *Journal of Chemical Ecology* 23: 1781-1803.

Macias, F.A., A.M. Simonet, J.C.G. Galindo, P.C. Pacheco, and J.A. Sanchez. (1998a). Bioactive polar triterpenoids from *Melilotus messanensis*. *Phytochemistry* 49: 709-717.

Macías, F.A., D. Castellano, and J.M.G. Molinillo. (2000). In the search for a standard phytotoxic bioassays for allelochemicals. selection of standard target species (STS). *Journal of Agricultural and Food Chemistry* (in press).

Macías, F.A., D. Castellano, R.M. Oliva, P. Cross, and A. Torres (1997). Potential use of allelopathic agents as natural agrochemical. In *Proceedings of The 1997 Brighton Crop Protection Conference-Weeds. Vol. 1*, eds. L. Copping. Farnham, UK: British Crop Protection Council, pp. 33-38.

Macias, F.A., J.C.G. Galindo, D. Castellano, and R.F. Velasco. (1999a). Sesquiterpene lactones with potential use as natural herbicide models (I): *trans,trans*-germacranolides. *Journal of Agricultural and Food Chemistry* 47: 4407-4414.

Macias, F.A., J.C.G. Galindo, and G.M. Massanet. (1992). Potential allelopathic activities of several sesquiterpene lactones models. *Phytochemistry* 31: 1969-1977.

Macías, F.A., J.C.G. Galindo, J.M.G. Molinillo, and D. Castellano. (2000a). Dehydrozaluzanin C, a potent plant growth regulator with potential use as a natural herbicide template. *Phytochemistry* 54: 165-171.

Macías, F.A., J.M.G. Molinillo, A. Torres, R.M. Varela, and D. Chinchilla. (2000b). Natural diterpenes from cultivar sunflower leaves with potential use as natural herbicide models. *Journal of Agricultural and Food Chemistry* (Submitted for publication).

Macías, F.A., J.M.G. Molinillo, J.C.G. Galindo, R.M. Varela, A. Torres, and A.M. Simonet, A.M. (1999b). Terpenoids with potential use as natural herbicide templates. In *Biologically Active Natural Products: Agrochemicals*, eds. H.G. Cutler and S.J. Cutler, Boca Raton, USA: CRC Press, pp. 15-31.

Macías, F.A., J.M.G. Molinillo, R.M. Varela, A. Torres, and F.R. Fronczek. (1994). Structural elucidation and chemistry of a novel family of bioactive sesquiterpenes. Heliannuols. *Journal of Organic Chemistry* 59: 8261-8266.

Macías, F.A., J.M.G. Molinillo, R.M. Varela, A. Torres, and J.C.G. Galindo. (1999c). Bioactive compounds from the genus *Helianthus*. In *Recent Advances in Allelopathy. Volume I. A Science for the Future*, eds. F.A. Macias, J.C.G. Galindo, J.M.G. Molinillo, and H.G. Cutler. Cádiz, Spain: Servicio de Publicaciones de la Universidad de Cádiz, pp. 121-148.

Macías, F.A., R.M. Varela, A. Torres, and J.M.G. Molinillo. (1993a). Potential allelopathic guaianolides from cultivar sunflower leaves, var. SH-222. *Phytochemistry* 34: 669-674.

Macías, F.A., R.M. Varela, A. Torres, and J.M.G. Molinillo. (1996c). Field crops as source of natural herbicide models: sunflowers. In *Allelopathy in Pests Management for Sustainable Agriculture. Vol. II*, eds. S.S. Narwal and P. Tauro, Jodhpur, India: Scientific Publishers, pp. 77-88.

Macías, F.A., R.M. Varela, A. Torres, and J.M.G. Molinillo. (1998b). Heliespirona A. The first member of a novel family of bioactive sesquiterpenes. *Tetrahedron Letters* 39: 427-430.

Macías, F.A., R.M. Varela, A. Torres, J.M.G. Molinillo, and F.R. Fronczek, F.R. (1993b). Novel sesquiterpene from bioactive fractions of cultivar sunflower. *Tetrahedron Letters* 34: 1999-2002.

Macías, F.A., R.M. Varela, A. Torres, R.M. Oliva, and J.M.G. Molinillo. (1998c). Bioactive norsesquiterpenes from *Helianthus annuus*. Potential allelopathic activity. *Phytochemistry* 48: 631-636.

Marles, R.J., L. Pazos-Sanou, C.M. Compadre, J.M. Pezzuto, E. Bloszyk, and J.T. Arnanson. (1995). Sesquiterpene lactones revisited: recent developments in the assessment of biological activities and structure relationships. *Recent Advances in Phytochemistry* 29: 333-356.

Perez de Luque, A., J.C.G. Galindo, F.A. Macías, and J. Jorrín. (2000). Sunflower sesquiterpene lactone model induce *Orobanche cumana* seed germination. *Phytochemistry* 53:45-50.

Petsko, G.A., D. Ringe, and J. Hogan. (1999). A new paradigm for the structure-guided pesticide design using combinatorial chemistry. In *Pesticide Chemistry and Bioscience: The FoodEnvironment Challenge*, eds. G.T. Brooks and T.R. Roberts. Cambridge, UK: The Royal Society of Chemistry, pp. 66-70.

Rasmussen, J.A., A.M. Hejl, F.A. Einhellig, and J.A Thomas (1992). Sorgoleone from root exudate inhibits mitochondrial functions. *Journal of Chemical Ecology* 18: 197-207.

Rice, E.L. (1984). *Allelopathy*. 2nd edition. New York: Academic Press.

Roddick, J.G. (1994). Comparative root-growth inhibition activity of four brassinosteroid. *Phytochemistry* 37: 1277-1281.

Stevens, K.L. and G.B. Merrill. (1985). Sesquiterpene lactones and allelochemicals from *Centaurea* species. In *The Chemistry of Allelopathy: Biochemical Interactions Among Plants*, ed. A.C. Thompson. ACS Symposium Series 268. Washington, DC: American Chemical Society, pp. 83-98.

Screening and Future Exploitation of Allelopathic Plants as Alternative Herbicides with Special Reference to Hairy Vetch

Yoshiharu Fujii

SUMMARY. Allelopathic cover crops are one of the most promising application of allelopathy for weed control and reduce the synthetic herbicides input. Fifty three cover crop plant species (26 leguminous, 19 graminaceous and 8 others) were assessed for the allelopathic activity using *Plant Box Method*. It was found that leguminous cover crops such as hairy vetch (*Vicia villosa*) and velvetbean (*Mucuna pruriens*), graminaceous cover crops, such as oat (*Avena sativa*) and rye (*Secale cereale*), certain cultivars of wheat (*Triticum aestivum*) and barley (*Hordeum vulgare*) were promising. By bioassay screening, 23 species were selected for field tests. Fall-sown cover crops such as hairy vetch, rye, wheat, oat, grass pea, and mustard are more effective when compared to spring-sown cover crops. Hairy vetch was most promising for the weed control in abandoned fields because of its ability to die off during summer season to make a thick straw-like mulch. *[Article copies available for a fee from The Haworth Document Delivery Service: 1-800-342-9678. E-mail address: <getinfo@haworthpressinc.com> Website: <http://www.HaworthPress.com> © 2001 by The Haworth Press, Inc. All rights reserved.]*

Yoshiharu Fujii is Head, Allelopathy Laboratory, National Institute of Agro-Environmental Sciences, Tsukuba, Ibaraki, 305-8604, Japan.

Address correspondence to: Yoshiharu Fujii at the above address (E-mail: yfujii@niaes.affrc.go.jp).

[Haworth co-indexing entry note]: "Screening and Future Exploitation of Allelopathic Plants as Alternative Herbicides with Special Reference to Hairy Vetch." Fujii, Yoshiharu. Co-published simultaneously in *Journal of Crop Production* (Food Products Press, an imprint of The Haworth Press, Inc.) Vol. 4, No. 2 (#8), 2001, pp. 257-275; and: *Allelopathy in Agroecosystems* (ed: Ravinder K. Kohli, Harminder Pal Singh, and Daizy R. Batish) Food Products Press, an imprint of The Haworth Press, Inc., 2001, pp. 257-275. Single or multiple copies of this article are available for a fee from The Haworth Document Delivery Service [1-800-342-9678, 9:00 a.m. - 5:00 p.m. (EST). E-mail address: getinfo@haworthpressinc.com].

© 2001 Taylor & Francis.

KEYWORDS. Allelopathy, cover crops, hairy vetch, *Vicia villosa*, weed control, abandoned fields

INTRODUCTION

World over the synthetic herbicides are being used for weed management. Nevertheless, their environmental hazards and economical considerations have stimulated interest in alternative ways of weed management in agroecosystems. In this direction, allelopathic cover crop plants hold a good promise. Cover crops are grown between the periods of regular cropping and reduce soil erosion, reduce nutrient runoff, increase organic matter content besides suppressing weeds (Barnes and Putnam, 1983; Dyck and Liebman, 1994; Dyck, Liebman, and Erich, 1995; Swanton and Murphy, 1996; Teasdale 1996; Weston, 1996). Cover crop plants interfere with weeds by competition and allelopathy (Putnam and DeFrank, 1983; Barnes and Putnam, 1986) and the weed suppressing ability decreases with the increasing biomass (Buhler, 1995) and even their use can replace herbicides use (Yenish, Worsham, and York, 1996).

Our group has been engaged in the search for allelopathic plants in order to determine allelochemicals, its mechanism and the possibility of future exploitation for weed control under field conditions (Fujii, Shibuya, and Yosuda, 1990, 1991; Fujii and Shibuya, 1992b; Fujii, 1994; Fujii, Ono, and Sato, 1994; Fujii, Sato, and Tanahashi, 1995). Under field conditions it is very difficult to exploit allelopathic plants as such for weed suppression because of a number of other competitive factors like light and nutrients. In Japan, some farmers have suggested from their experiences that some ground cover plants suppress weeds drastically. There is a possibility that some cover crops could be used for practical weed suppression under field conditions. In order to explore this, we screened the allelopathic activity of cover plants by plant box method. From these primary selections, and experiments on the fields, we conducted a series of field experiment to select the most promising cover crop for weed control in orchard garden and abandoned fields.

MATERIALS AND METHODS

Procurement of Seeds of Cover Crops and Definition of Hairy Vetch

For the present study, 26 leguminous crops, viz., *Mucuna pruriens* var. *utilis* (Velvet bean) (average of 6 cultivars), *Vicia faba* (Broad Bean) (average of 10 cultivars), *Vicia villosa* (Hairy Vetch) (average of 4 cultivars), *Calopogonium mucunoides* (Calopo), *Melilotus albus* (White Sweet Clover), *Vicia sativa*

(Common Vetch), *Medicago rugosa* (Gama medic), *Canavalia ensiformis* (Jack Bean), *Pueraria phaseoloides* (Tropical Kudzu), *Vigna angularis* (Azuki bean), *Medicago sativa* (Alfalfa), *Trifolium incarnatum* (Crimson Clover), *Tephrosia candida* (White Tephrosia), *Cajanus cajan* (Pigeon Pea), *Lathyrus sativus* (Grass Pea), *Cicer arietinum* (Chickpea), *Vigna radiata* (Mung Bean), *Stylosanthes hamata* (Stylo), *Arachis hypogaea* (Peanut), *Trifolium pratense* (Red Clover), *Crotolaria juncea* (Sunn Hemp), *Astragalus sinicus* (Chinese Milk Vetch), *Lupinus albus* (White Lupine), *Trifolium subterraneum* (Subterranean Clover), *Trifolium repens* (White Clover), *Glycine max* (Soybean), 19 graminaceous plants, viz., *Avena sterilis* (Wild Oat), *Triticum polonicum* (Polish Wheat), *Triticum aestivum* (Wheat, for mulching), *Celosia argentea* (Feather Cockscomb), *Secale cereale* (Rye), *Avena sativa* (Oat), *Setaria italica* (Italian Millet), *Festuca rubra* (Chewing Fescue), *Sorghum bicolor* (Sorghum), *Hordeum vulgare* (Barley), *Panicum maximum* (Guinea Grass), *Festuca arundinacea* (Tall Fescue), *Poa pratense* (Kentucky Bluegrass), *Dactylis glomerata* (Orchard Grass), *Lolium perenne* (Perennial Ryegrass), *Paspalum dilatatum* (Dallis Grass), *Paspalum notatum* (Bahia grass), *Lolium multiflorum* (Italian Ryegrass), *Phalaris arundinacea* (Reed Canary grass), and 8 other plants species, viz., *Amaranthus tricolor* (Ganges Amaranth), *Brassica napus* (Rape), *Brassica alba* (White Mustard), *Helianthus annuus* (Sunflower), *Carthamus tinctorius* (Safflower), *Ricinus communis* (Castor Bean), *Momordica charantia* (Balsam Pear) and *Gossypium barbadense* (Cotton) were selected. Most of the seed used in the present study were obtained from 'Kaneko Seed Co.,' 'The Takii Seed Co.,' 'The Yukijirushi Seed Co.,' and 'Sakta no Tane, Sakata Seed Co.'

The term *'hairy vetch'* is used in Japan both for hairy vetch (*Vicia villosa*) and woolly pod vetch (*Vicia dasycarpa* = *V. villosa* subsp. *dasycarpa*). All seeds of hairy vetch were imported from USA, and some seed companies deny telling the origin, we called them as hairy vetch as a whole. The allelopathic activity and inhibitory activity of both in the field is almost the same.

Screening of Allelopathic Cover Crops by the Plant Box Test

Primary selection of cover crops for allelopathy was done using "Plant Box Method," developed for the assessment of allelopathy (Fujii and Shibuya, 1992a). Young plants were cultivated for one to two months in a sand culture in standing water containing a nutrient solution. The receiver plant used for bioassay was lettuce (var. Great Lakes 366), because of its high sensitivity to bioactive substances.

Inhibitory Activity of Cover Crops

Plant growth inhibitory activities of some cover crops were tested by water and methanol extraction. Dried leaves and shoots (forced air dry at 60°C for over night) were extracted with 150 times of water and 40 times of methanol. For this, 67 g and 250 g of leaves (dry weight) were extracted with 1 liter of water and methanol, respectively. These extracts were mixed with agar (0.5%) to make a stuff bed to support lettuce seeds. After three days in a dark condition at 20°C, germination, radicle and hypocotyl length were measured.

Screening of Cover Crops for Weed Control in the Experimental Fields

In order to explore the weed suppressing ability of cover crops under field conditions field tests were designed. All field trials for screening were done in the experimental field of Shikoku National Agricultural Experimental Station, Japan. For the first screening for spring seeding cover crops, 32 plants were sown on $1 \times 4\ m^2$ quadrats on May 23, 1992. After three months of no weeding, weed density and dry weights were measured on August 25, 1992. For the preliminary screening for fall seeding plants, 23 plant types were sown in $2 \times 2\ m^2$ quadrats on November 5, 1992, and after no weeding for 6 months, dry weight of weeds in each plot was measured on April 20, 1993. On the basis of these preliminary screening tests 11 fall-sown cover crops were selected and seeded in $2 \times 2\ m^2$ quadrats on November 5, 1992 and the weed and crop dry weighs were measured on April 20, 1993. A comparison was made with rice straw mulch (10,000 $kg \cdot ha^{-1}$) and herbicide (Benthicarb plus Prometrin; 40 $kg \cdot ha^{-1}$). Four replications were maintained for each treatment in a completely randomized block design.

Application of Hairy Vetch to Abandoned Paddy Field in the Experimental Station

From the results of screening of cover crops, we focused on hairy vetch, and four trials were conducted on the experimental fields P8 of Shikoku National Agricultural Experiment Station (SNAES), Japan. These were:

1. Large-scale application test of hairy vetch was conducted in comparison with Chinese milk vetch on the uniform paddy field of 1000 m^2 area. This field was divided into 16 blocks and 4 replications (each in a plot of 500 m^2) of 4 different cover crop trials were designed, viz., (i) hairy vetch, (ii) hairy vetch and oat, (iii) Chinese milk vetch, and (iv) no treatment as control. This experiment was started in October 1992 and data from 1992 to 1994 are reported in this paper. The dominant weeds in this field were *Alopecurus aequalis* and *Lamium amplexicaule*.
2. Changing the seeding ratio from standard seeding to four times was examined using a field of 500 m^2, and arranged in a split-split-plot test

with 4 replications. This experiment was started in October 25, 1993 and weed and crop yields were measured on May 20, 1994.

3. Changing the seeding date from October to February was tested using the same field with three replications. This test was started in 1992 and ended in 1994.
4. Effect of soil water content on the growth of hairy vetch was tested in the lysimeter. This lysimeter was designed to know the suitable water ratio for crops, and it is possible to change the water level by stair-step system of over flowing. Each block was about 10 m^2 and there were four stages with two replications.

Soil type in all the above four tests was sandy loam (paddy field converted to upland condition). In addition a fifth trial was conducted on the slope land experimental field of Shikoku National Agricultural Experiment Station on O-asa mountain (slope angle = 9°) using a combination of leguminous cover crops and barley. Seeding was done on November 6, 1993, and crop and weed yield was measured on May 30, 1994. The total area of this field was about 800 m^2 and soil is light volcanic ash soil.

Direct Application Test of Hairy Vetch to Farmer's Field

Direct application and exhibition trial of hairy vetch was done using the fields of cooperative farmers by the courtesy of each district counselors for farmers. (In Japan, there are counselors for farmers in each county. These counselors are public service officers belonging to the Ministry of Agriculture.) Four different farming systems were chosen. These were:

1. Paddy field of Mr. Tsune-kane with an area of 800 m^2 in Man'nou town, Nakatoda county, Kwaga. This paddy field is in a slightly mountainous area, and slightly dry condition. This field has just stopped rice production (referred to as Paddy field I).
2. Paddy field of Mr. Hirata with an area of about 1000 m^2 in Marugame city, Kwaga. This field is in humid condition, close to road, and was abandoned two years ago (referred to as Paddy field II).
3. Grassland with an area of about 0.2 ha owned by Mr. Morichika in Chu-nan town, Nakatoda county, Kwaga. This grassland is in a slope land, and used as an exhibition (referred to as Grassland I).
4. Orchard for Kaki, Japanese persimmon in Kounan town, Oka village and owned by Mr. Oka. The area used for cover crop trial was about 50,000 m^2 in first year, 1992, and extended to 10,000 m^2 in the following year and continued till now (referred to as Orchard I).

RESULTS AND DISCUSSION

Screening of Allelopathic Cover Crops

The results of screening of plants for allelopathic cover crops from leguminous and graminaceous species by Plant Box Method are shown in Tables 1 and 2, respectively. Radicle percentage means the percent radicle length of the young lettuce plants present in the root zone of each donor plant, based on the calculation of radicle length within the root zone controlled by donor plants. Leguminous cover crops such as velvet bean, hairy vetch, yellow sweet clover and white sweet clover have strong allelopathic inhibitory activities (Table 1). Of these legumes, velvetbean, *Crotaralia*, *Canavalia*, *Cajanus*, *Cicer*, *Vigna* and *Glycine* are summer cover crops, and most of others are winter cover crops. *Melilotus* and *Pueraria* are perennial crops. Then it is important to know the characteristic of each cover crop to use for the farmland.

Graminaceous species, such as oat, wheat, millet (*Setaria*), rye and fescue also showed strong inhibitory activity, but most of non-graminaceous plant species such as sunflower, safflower, white mustard, castor bean, etc., showed only medium inhibitory activity by this method (Table 2).

Inhibitory Activity of Cover Crops

To know the existence of allelochemicals in some principal cover crops, we tried their water and methanol extracts (Table 3). Hairy vetch and velvetbean showed the strongest inhibitory activity in both water and methanol extract and the extracts prepared from hairy vetch were more inhibitory compared to that from velvet bean (Table 3). Velvetbean contains an extraordinary high amount of L-3,4-dihydroxyphenylalanine and the inhibitory activity of this compound is rather high in acidic and non-oxidative conditions (Fujii, 1994). However in the case of hairy vetch, there is a possibility of more inhibitory allelochemicals than velvetbean. The results of solvent extraction are consistent with the results from Plant Box test for root exudates in these experiments.

Screening of Cover Crops for Weed Control in the Experimental Fields

To assess the activity of these cover crops, field experiments were performed using the small-scale field in our Experimental Station. Spring sown cover crops were not promising as shown in Table 4. In Japan, we have a rainy season in June, and soon after the beginning of this season, there is a serious infection and vigorous growth of weeds. Some cover crops such as *Helianthus*, *Celosia* and *Panicum* showed relatively strong inhibitory activity

TABLE 1. Assessment of allelopathic activity of 26 leguminous cover crops by the Plant Box method.

Scientific name (English name)	Inhibition (%)[1]	Criteria[2]
Mucuna pruriens var. *utilis* (average of 6 cultivars)	87	***
Vicia faba (Broad Bean) (average of 10 cultivars)	81	**
Vicia villosa (Hairy Vetch) (average of 4 cultivars)	80	**
Calopogonium mucunoides (Calopo)	78	**
Melilotus albus (White Sweet Clover)	77	**
Vicia sativa (Common Vetch)	75	**
Medicago rugosa (Gama medic)	74	**
Canavalia ensiformis (Jack Bean)	72	**
Pueraria phaseoloides (Tropical Kudzu)	71	**
Vigna angularis (Azuki bean)	69	*
Medicago sativa (Alfalfa)	68	*
Trifolium incarnatum (Crimson Clover)	64	*
Tephrosia candida (White Tephrosia)	63	*
Cajanus cajan (Pigeon Pea)	62	*
Lathyrus sativus (Grass Pea)	59	*
Cicer arietinum (Chickpea)	56	*
Vigna radiata (Mung Bean)	53	
Stylosanthes hamata (Stylo)	52	
Arachis hypogaea (Peanut)	51	
Trifolium pratense (Red Clover)	47	
Crotaralia juncea (Sunn Hemp)	43	
Astragalus sinicus (Chinese Milk Vetch)	41	
Lupinus albus (White Lupine)	40	
Trifolium subterraneum (Subterranean Clover)	30	
Trifolium repens (White Clover)	28	
Glycine max (Soybean)	24	

[1] Inhibition (%) means the allelopathic activity measured by the Plant Box Test. All data are with respect to the control and 100 means complete inhibition.
[2] Criteria for the allelopathic activity are shown as *** (means > 85 %), ** (70 to 84%) and * (< 70%).

TABLE 2. Assessment of allelopathic activity of 19 graminaceous and 8 other cover crops by the Plant Box method.

Scientific name (English name)	Inhibition (%)[1]	Criteria[2]
Avena sterilis (Wild Oat)	88	***
Triticum polonicum (Polish Wheat)	87	***
Triticum aestivum (Wheat, for mulching)	82	**
Celosia argentea (Feather Cockscomb)	75	**
Secale cereale (Rye)	71	**
Avena sativa (Oat)	71	**
Setaria italica (Italian Millet)	70	*
Festuca rubra (Chewing Fescue)	70	*
Sorghum bicolor (Sorghum)	64	*
Hordeum vulgare (Barley)	62	*
Amaranthus tricolor (Ganges Amaranth)[3]	61	*
Panicum maximum (Guinea Grass)	60	*
Brassica napus (Rape)[3]	59	*
Festuca arundinacea (Tall Fescue)	55	*
Brassica alba (White Mustard)[3]	53	
Helianthus annuus (Sunflower)[3]	51	
Carthamus tinctorius (Safflower)[3]	45	
Ricinus communis (Castor Bean)[3]	44	
Momordica charantia (Balsam Pear)[3]	42	
Poa pratense (Kentucky Bluegrass)	40	
Dactylis glomerata (Orchard Grass)	38	
Gossypium barbadense (Cotton)[3]	36	
Lolium perenne (Perennial Ryegrass)	36	
Paspalum dilatatum (Dallis Grass)	28	
Paspalum notatum (Bahia grass)	27	
Lolium multiflorum (Italian Ryegrass)	26	
Phalaris arundinacea (Reed Canary grass)	24	

[1] Inhibition (%) means the allelopathic activity measured by the Plant Box Test. All data are with respect to the control and 100 means complete inhibition.
[2] Criteria for the allelopathic activity are shown as *** (means > 85 %), ** (70 to 84%) and * (< 70%).
[3] Means non-graminaceous cover crop.

TABLE 3. Effect of water and methanol extracts of some cover crops.

Cover Crop	Water extract (× 150)			Methanol extract (× 40)		
	G (%)	R (%)	H (%)	G (%)	R (%)	H (%)
<u>*Vicia villosa*</u>	100	<u>12</u>	89	90	<u>18</u>	<u>52</u>
<u>*Mucuna pruriens*</u>	100	<u>17</u>	102	87	<u>26</u>	<u>61</u>
Cajanus cajan	97	41	76	83	<u>13</u>	<u>15</u>
Centrosema pubescens	100	53	118	100	37	80
Crotaralia juncea	100	55	120	97	51	105
Tephrosia candida	100	58	106	100	41	95

G means percent germination; R means percent radicle growth; H means percent hypocotyl growth compared to water treated control.
<u>Underlined</u> means strong inhibition, i.e., > 66 inhibition.

TABLE 4. Effect of spring sown cover crops on the weed growth under field conditions (a primary selection).

Cover Crop	W (%)	Cover Crop	W (%)
Helianthus annuus	85	*Ricinus communis*	14
Celosia argentea	85	*Phaseolus vulgaris*	10
Panicum maximum	83	*Vicia hirsuta*	9
Mucuna pruriens	48	*Medicago sativa*	9
Panicum milliaceum	46	*Luffa cylindrica*	9
Vigna angularis	41	*Glycine ussuriensis*	8
Cassia occidentalis	39	*Memordica charantia*	3
Corchorus olitorius	36	*Canavalia ensiformis*	1
Corchorus capsularis	34	*Vigna angularis*	−2
Gossypium barbadense	33	*Vigna umbellata*	−4
Tephrosia candida	26	*Vigna radiata*	−5
Panicum ramosum	25	*Glycine max*	−12
Amaranthus tricolor	21	*Citrullus lanatus*	−16
Panicum virgotum	19	*Cajanus cajan*	−19
Medicago rugosa	17	*Carthamus tinctorius*	−23
Setaria italica	16	*Crotalaria juncea*	−33

W means percent weed control; 100% means complete inhibition.

to weeds, followed by *Mucuna*, *Vigna* and *Cassia* (Table 4). Competition for light and nutrients must be the most important factors in the fields, and all of these plants have vigorous growth rate and huge leaves. Most of these cover crops were reported as allelopathic, and have relatively strong inhibitory activity in the Plant Box test (Tables 1 and 2). For example, there are many reports on the allelopathy of *Helianthus*, and *Celosia* (Rice, 1984). However, the fields with more than 20% of remaining weed looks like weedy and disagreeable to the farmers and public. In this situation, spring-seeded cover crops in these conditions were not successful under Japanese condition.

The results from the trial for fall seeding are shown in Tables 5 and 6. If cover crops are sown in fall, they tend to grow slow but steadily in winter to make enough biomass in spring and could eliminate the vigorous weeds. It is obvious that dry weight of each cover crop is most important to reduce the growth of weeds. It is true for rye, oat, wheat, woolly pod vetch and Italian ryegrass. Because of the overwhelming canopy of these crops, no weed can survive. But, if we compare the relationship between dry weight and weed suppression activity, *Brassica*, *Vicia*, and *Medicago* do not have enough biomass but show better weed suppression than estimated from their biomass (Tables 5 and 6). Probably, allelopathy plays a role in these cases.

On the basis of this preliminary screening, 11 fall-sown cover crops were selected and tested for their weed-suppressing ability. It was observed that without weeding, hairy vetch, oat, barley, rye, and wheat showed strong inhibitory activity of weeds, but Chinese milk vetch, which is a traditional green manure in Japan and China, showed little weed suppression (Table 6). The weed controlling effect of these cover crops was almost similar to that of rice straw mulch and even herbicide application.

Application of Hairy Vetch to Abandoned Paddy Field in the Experimental Station

The results from large space application of hairy vetch, Chinese milk vetch and mixed planting of vetch and oats are shown in Tables 7 and 8. These data were taken from 1992 to 1994. From these data, it is obvious that hairy vetch almost completely inhibits the growth of weeds in spring. On the other hand, Chinese milk vetch, traditionally used in Japanese paddy field as green manure, could inhibit the weed biomass up to 80% (Tables 7 and 8). Allowing 20% of weeds to grow will make this field weedy next year, and will be abandoned in two or three years without weeding. These results coincide with the observation of farmers that continuous using of Chinese milk vetch causes a serious infection of weeds.

Addition of oat to hairy vetch was aimed to increase the weed suppression ability. In both years, addition of oat decreases the population of weeds, to nearly perfect inhibition of weed until fall (Table 8). Mixed planting of vetch

TABLE 5. Effect of fall-sown cover crops on weed control in the field (a primary selection).

Cover Crop (English Name)	W (%)	Dry Weight ($g \cdot m^{-2}$)
Control (No-weeding)	0	---
Vicia villosa (Hairy Vetch)	100	1171
Secale cereale (Rye)	100	1671
Brassica napus (Rape)	100	998
Lathyrus sativus (Grass Pea)	99	891
Avena sativa (Oat)	99	1426
Hordeum vulgare (Barley)	99	1173
Triticum aestivum (Wheat)	99	1751
Brassica campestris (Field Mustard)	97	834
Pisum sativum (Pea)	95	1359
Brassica alba (White Mustard)	95	416
Vicia dasycarpa (Woolly Pod Vetch)	91	2409
Lolium multiflorum (Italian Ryegrass)	89	1799
Vicia sativa (Common Vetch)	87	1016
Medicago sativa (Alfalfa)	77	384
Phleum pratense (Timothy)	69	445
Lolium perenne (Perennial Ryegrass)	66	516
Lupinus albus (Lupin)	45	341
Vicia angustifolia (Karasuni-endou)	44	473
Festuca arundinacea (Tall Fescue)	44	304
Astragalus sinicus (Chinese Milk Vetch)	31	167
Festuca elatior (Meadow Fescue)	31	372
Vicia hirsuta (Suzumeno-endou)	30	304
Trifolium repens (Ladino Clover)	22	426

W means percent weed control; 100% means complete inhibition.
Dry weight of weeds in the control plot is 381 $g \cdot m^{-2}$ (3810 $kg \cdot ha^{-1}$).
Dominant weeds were *Capsella bursa-pastoris* and *Lamium amplexicaule*.

and oat has companionship and yields of both increased. However, if no care was taken, the outlook of field from spring to summer of mixed cover field was not beautiful because of the remaining stems of oat. On the other hand, hairy vetch kept the stand height of maximum 50 cm, and the outlook of this field was uniform and flat, and free from weeds. Hairy vetch died itself when

TABLE 6. Effect of selected fall-sown cover crops on the weed control.

Cover Crop (English Name)	W (%)	Dry Weight ($g \cdot m^{-2}$)
Control (No-weeding)	0^{a}	---
Secale cereale (Rye)	99^{d}	693
Triticum aestivum (Wheat)	99^{d}	1751
Avena sativa (Oat)	99^{cd}	994
Hordeum vulgare (Barley)	99^{cd}	1173
Vicia villosa (Hairy Vetch)	98^{cd}	816
Brassica campestris (Field Mustard)	97^{cd}	834
Trifolium repens (White Clover)	78^{cd}	356
Medicago sativa (Alfalfa)	77^{cd}	384
Lupinus albus (Lupin)	49^{b}	341
Astragalus sinicus (Chinese Milk Vetch)	36^{b}	167
Melilotus albus (White Sweet Clover)	-16^{a}	30
Rice straw mulch (10000 $kg \cdot ha^{-1}$)	87^{cd}	(1000)
Herbicide (Benthiocarb + Prometrin, 40 $kg \cdot ha^{-1}$)	91^{cd}	---

W means percent of weed control. 100% means complete control.
Similar letters represent insignificant difference at $P > 0.01$ applying Duncan's multiple range test.
Dry weight of weeds in the control plot is 381 $g \cdot m^{-2}$. Dominant weeds in this field are, *Capsella bursa-pastoris* and *Lamium amplexicaule*.

maximum temperature reaches to 30°C. In our experimental field, hairy vetch made a straw like mulch without any work and this mulch protects the field for weed invasion after the death of mother plant. Based on this we concluded that it is better to recommend the farmers to use hairy vetch alone because of the simplicity of sowing and reducing the labor.

Standard seeding rate was enough for the weed control, and increasing the seeding rate beyond 2 times the normal rate made no difference for biomass and weed suppression (Table 9). Not only so, even the application of two seeding rate made no significant difference as regards the weed control and hairy vetch biomass (Table 10). In order to find the optimum seeding date for weed suppression in Japan an experiment was performed where seeding was done on three different dates. It was observed that late seeding allows more weeds to grow and based on this experiment early seeding no later than the 1st week of November is recommended (Table 11).

Some Japanese abandoned paddy fields are suffering from water logging. Resistance of hairy vetch to water logging in paddy field was measured by using water logging lysimeter. It was observed that hairy vetch is resistant to

TABLE 7. Effect of cover crops on weed control in abandoned paddy field. First trial on the experimental station (1992-1993). Seeding date: October 28, 1992, Sampling date: May 7 (First), June 10 (Second), 1993.

Cover Crop (English Name)	W (%)	Crop Yield ($g \cdot m^{-2}$)
First Sampling (May 7, 1993)		
Control (No-weeding)	0 a	---
Astragalus sinicus (Chinese Milk Vetch)	82 b	431 a
Vicia villosa (Hairy Vetch)	99 b	584 ab
A. sativa + *V. villosa* (Oat and Hairy Vetch mix.)	99 b	730 b
Second Sampling (June 10, 1993)		
Control (No-weeding)	0 a	---
Astragalus sinicus (Chinese Milk Vetch)	59 b	135 a
Vicia villosa (Hairy Vetch)	100 c	147 b
A. sativa + *V. villosa* (Oat and Hairy Vetch mix.)	100 c	137 a

W means percent of weed control. 100% means complete control.
Dry weight of weeds in the control plot was 281 (First) and 155 (Second) $g \cdot m^{-2}$.
Similar letters means insignificant difference applying Duncan's multiple range test ($P > 0.05$).

TABLE 8. Effect of cover crops on weed control in abandoned paddy field. Second trial on the experimental station (1993-1994). Seeding date: November 10, 1993, Sampling date: June 2 (First), October 7 (Second), 1994.

Cover Crop (English Name)	W (%)	Crop Yield ($g \cdot m^{-2}$)
First Sampling (June 2, 1994)		
Control (No-weeding)	0 a	----
Astragalus sinicus (Chinese Milk Vetch)	83 b	220 a
Vicia villosa (Hairy Vetch)	100 b	619 ab
A. sativa + *V. villosa* (Oat and Hairy Vetch mix.)	100 b	629 + 367 b
Second Sampling (October 7, 1994)		
Control (No-weeding)	0 a	----
Astragalus sinicus (Chinese Milk Vetch)	77 b	----
Vicia villosa (Hairy Vetch)	90 c	----
A. sativa + *V. villosa* (Oat and Hairy Vetch mix.)	97 c	----

W means percent of weed control. 100% means complete control.
Dry weight of weeds in the control plot was 156 (First) and 247 (Second) $g \cdot m^{-2}$.
Similar letters means insignificant difference applying Duncan's multiple range test ($P > 0.05$).

TABLE 9. Effect of seedling rate of hairy vetch on weed control in SNAES experimental field P19 (Zentsuji, Kagawa) under upland condition. Seeding: October 25, 1993; Survey: May 20, 1994.

Seeding Rate of Hairy Vetch	Dry Weight of Hairy Vetch ($g \cdot m^{-2}$)	Dry Weight of Weed ($g \cdot m^{-2}$)	W (%)
Standard (50 $kg \cdot ha^{-1}$)	563 ± 37 (100)	0	100
× 2 (100 $kg \cdot ha^{-1}$)	620 ± 86 (110)	0	100
× 3 (150 $kg \cdot ha^{-1}$)	633 ± 76 (112)	0	100
× 4 (200 $kg \cdot ha^{-1}$)	665 ± 23 (118)	0	100
Control (No-weeding)	---	98 ± 57	0

W means percent weed inhibition compared to the control (No weeding plot).
± Means the standard deviation. All data were not significant at $P > 0.01$ with Duncan's multiple range test.

TABLE 10. Effect of two seeding rate of Hairy Vetch on the weed control in SNAES experimental field P4 (Zentsuji, Kagawa) under upland condition.

Seeding Rate of Hairy Vetch	Dry Weight of Hairy Vetch ($g \cdot m^{-2}$)	Dry Weight of Weed ($g \cdot m^{-2}$)	W (%)
April 20, 1994			
Standard (50 $kg \cdot ha^{-1}$)	656	0	100
× 3 (150 $kg \cdot ha^{-1}$)	720	0	100
Control (No-weeding)	---	185	0
May 13, 1994			
Standard (50 $kg \cdot ha^{-1}$)	706	0	100
× 3 (150 $kg \cdot ha^{-1}$)	704	0	100
Control (No-weeding)	---	224	0
June 10, 1994			
Standard (50 $kg \cdot ha^{-1}$)	732	0	100
× 3 (150 $kg \cdot ha^{-1}$)	847	0	100
Control (No-weeding)	---	324	0

W means percent weed inhibition compared to the control (No weeding plot).

heavy water logging (Table 12). Water content of around 40% means the heavy water logging, and vetch showed no growth retardation under these conditions. But, in this lysimeter, continuous supply of water is possible, so if there is heavy water logging condition and no movement of water in the field, it might disturb the growth of vetches.

TABLE 11. Effect of seeding date on the weed control in spring in the SNAES experimental field P43 (Zentsuji, Kagawa) under upland condition. Seeding rate; 50 kg·ha^{-1}, standard; Survey: May 23 to June 1, 1994.

Seeding Date and Treatment	Dry Weight of Hairy Vetch (g·m^{-2})	Dry Weight of Weed (g·m^{-2})	W (%)
October 25, 1993			
Hairy Vetch	495 ± 49	1.0	99
Control (No-weeding)		127	0
November 5, 1993			
Hairy Vetch	431 ± 57	1.4	99
Control (No-weeding)		154	0
December 6, 1993			
Hairy Vetch	418 ± 19	11	96
Control (No-weeding)		270	0

W means percent weed inhibition compared to the control (No weeding plot).
± Means the standard deviation.

TABLE 12. Effect of soil moisture content on the growth of hairy vetch in paddy field in SNAES experimental field P41 (Zentsuji, Kagawa). Seeding: November 15, 1993; Survey: May 23, 1994.

Plot Number	1	2	3	4
Water Content Ratio (%)	41.6 ± 3.6	17.8 ± 1.6	15.6 ± 2.0	14.5 ± 3.3
Dry Weight (g.m^{-2})	579	528	470	568

In every plot, there is no weed, almost 100% suppression at the sampling date.

The effect of combination of leguminous cover crops and barley on weed control was examined on the slope land experimental field of Shikoku National Agricultural Experiment Station. The aim of this experiment was to use barley as a cover crop in slope land. Barley, especially known as 'Hadakamugi', a naked barley, is the traditional cultivar and suitable to Shikoku and southeast area of Japan. Hairy vetch slightly reduces the growth of barley, but the weed inhibitory activity of hairy vetch and combination of vetch and barley were the best. This was followed by combination of red clover and barley (Table 13).

TABLE 13. Effect of cover crops on weed growth in the slope of Levee in SNAES slope land experimental field (Slope angle is 9°). Seeding: Novermber 6, 1993; Survey: May 30, 1994.

Combination of Cover Crops	Dry Weight of Legume ($g \cdot m^{-2}$)	Dry Weight of Grass ($g \cdot m^{-2}$)	Dry Weight of Weed ($g \cdot m^{-2}$)	W (%)
Hairy Vetch (HV)	805 ± 148	--	13 ± 18	87
Hairy Vetch + Barley	710 ± 248	186 ± 59	2 ± 2	98
White Clover + Barley	193 ± 118	397 ± 40	110 ± 76	−17
Red Clover + Barley	424 ± 50	363 ± 81	80 ± 34	15
Crimson Clover + Barley	84 ± 41	373 ± 183	151 ± 18	−60
Control (No-weeding)	--	--	94 ± 29	0

W means percent weed inhibition compared to the control (No weeding plot).
± Means the standard deviation.

Direct Application and Exhibition of Hairy Vetch to Farmer's Field

Direct application and exhibition trials of hairy vetch were done using the field of cooperative farmers and these results are summarized in Table 14. In most cases, hairy vetch suppresses weed enough, and the impression of farmers was very good. Before this trial, there was no custom to use hairy vetch as cover crops to control the weeds. Many farmers came to our Experimental Station or exhibition field of cooperative farmers, and most of them started to try hairy vetch. Now, in the middle of Japan, the use of hairy vetch, under Kaki orchard, is spreading to 600 ha, and is still expanding.

Hairy vetch is a well-known green manure and cover crop in the United State and Europe. The origin of hairy vetch was estimated to the area from west Asia to east Mediterranean coast. It was cultivated in England and Germany in the early 19th century then introduced to USA in the middle of 19th century, had a good reputation from USDA, and now widely distributed in the southern part of the USA. Hairy vetch was introduced into Japan in the early 20th century, and had a good result by the Agricultural Experimental Station, but did not distribute until now.

There are some reports on the allelopathy and weed control by hairy vetch. Lazauskas and Balinevichiute (1972) tested the inhibitory activity of extract of seeds to barley, and found that hairy vetch showed the strongest activity. White, Worsham, and Blum (1989) reported that the incorporation of residues of hairy vetch and crimson clover reduced the emergence of solanaceous weeds by 60 to 80%, and water extract of hairy vetch had the stronger inhibitory activity. Johnson, Defelice, and Helsel (1993) reported that the

TABLE 14. Direct application of hairy vetch on farmer's field for weed control.

Treatment	Hairy Vetch Height (cm)	Hairy Vetch Dry wt. ($g \cdot m^{-2}$)	Weed Dry wt. ($g \cdot m^{-2}$)	W (%)
Paddy Field I *(Seeding: December 10, 1993; Survey: June 2, 1994)*				
Seeding after tillage	38.9 ± 7.5	635 ± 141	0	100
Non-till seeding	46.5 ± 7.1	477 ± 62	0	100
Paddy Field II *(Seeding: October 30, 1993; Survey: June 2, 1994)*				
Seeding after tillage	55.0 ± 5.2	724 ± 91	0	100
Control (No weeding)	--	--	152 ± 57	0
Grassland I *(Seeding: November 10, 1993; Survey: June 2, 1994)*				
Tilled after seeding	56.6 ± 6.1	605 ± 112	0	100
Seeding after mowing	60.8 ± 7.8	661 ± 80	6.4	99
Control (No weeding)	--	--	535 ± 243	0
Orchard I *(Seeding rate: 50 kg.ha $^{-1}$; Broadcasting by hand)*				
First year (November 14, 1992 to April 30, 1993)				
No-till seeding	51.2 ± 12.7	228 ± 152	165 ± 146	66
Control (No weeding)	--	--	488 ± 218	0
Second year (October 24, 1993 to May 31, 1994)				
No-till seeding 'A'	50.6 ± 4.5	399 ± 70	29 ± 24	94
No-till seeding 'B'	51.6 ± 4.4	435 ± 77	23 ± 19	95
Control (No weeding)	--	--	514 ± 270	0

W means percent weed inhibition compared to the control (No weeding plot).
A, B means the difference of two seed companies.

mulch made from hairy vetch or rye completely inhibited the weed in non-tillage systems by killing them by herbicide or mechanical cutting. Teasdale and Daughtry (1993) reported that the living mulch of hairy vetch showed better inhibitory activity than the desiccated one. Abdul-Baki and Teasdale (1993) reported a unique system using hairy vetch mulch to compensate for the vinyl plastic film mulch in tomato production. There are so many reports and field observation about the weed suppression of hairy vetch (Fujii and Shibuya, 1992b; Fujii and Ono, 1993), but still now the contribution of allelopathy and its allelochemicals are unknown. We are now going to separate allelochemicals from hairy vetch and will report about the chemical nature.

Hairy vetch has many other merits such as (i) nitrogen fixation to reduce chemical fertilizer, (ii) adds organic materials to reduce chemical fertilizer or soil conditioner, (iii) prevents soil erosion by surface cover, (iv) deep root system promotes the soil porosity, (v) thick plant cover ameliorates the microclimate and reduce maximum temperature and increase minimum temperature, and (vi) induces carnivorous ladybug to reduce the population harmful insects.

CONCLUSIONS

In conclusion, spring-seeded cover crops are not promising in Japanese conditions, but fall-seeded cover crops such as hairy vetch, rye, oat, wheat, and barley have enough inhibitory activity for weed control from spring to early summer. Their inhibitory effects are almost the same as that of traditional way of weeding, such as rice straw mulch and herbicide application. After a series of experiments, it was concluded that hairy vetch is the most promising cover crop for the control of weeds in abandoned fields, grassland, and orchard in the central and southern part of Japan.

REFERENCES

Abdul-Baki, A.A. and J.R. Teasdale. (1993). A no-tillage tomato production system using hairy vetch and subterranean clover mulches. *HortScience* 28: 106-108.

Barnes, J.P. and A.R. Putnam. (1983). Rye residues contribute weed suppression in no-tillage cropping systems. *Journal of Chemical Ecology* 9: 1045-1057.

Barnes, J.P. and A.R. Putnam. (1986). Evidence for allelopathy by residues and aqueous extracts of rye (*Secale cereale* L.). *Weed Science* 34: 384-390.

Buhler, D.D. (1995). Influence of tillage system on weed population dynamics and management in corn and soybean in the central USA. *Crop Science* 35: 1247-1258.

Dyck, E. and M. Liebman. (1994). Soil fertility management as a factor in weed control: the effect of crimson clover residue, synthetic nitrogen fertilizer, and their interaction on emergence and early growth of lambsquarters and sweet corn. *Plant and Soil* 167: 227-237.

Dyck, E., M. Liebman, and M.S. Erich. (1995). Crop-weed interference as influenced by a leguminous or synthetic fertilizer nitrogen source: I. Double-cropping experiments with crimson clover, sweet corn, and lambsquarters. *Agriculture Ecosystem and Environment* 56: 93-108.

Fujii, Y. (1994). Screening of allelopathic candidates by new specific discrimination and assessment methods for allelopathy, and the identification of L-DOPA as the allelopathic substance from the most promising velvetbean. *The Bulletin of the National Institute of Agro-Environmental Sciences* 10: 115-218.

Fujii, Y., K. Sato, and T. Tanahashi. (1995) Weed control in Japanese persimon orchard by hairyvetch. *Weed Research, Japan* 40 (Supplement): 110-111.

Fujii, Y. and S. Ono. (1993). Field trial for the weed control by hairyvetch and assessment of allelopathy. *Weed Research, Japan* 38 (Supplement): 144-145.

Fujii, Y., S. Ono, and K. Sato. (1994). Weed suppression by winter cover crops such as hairy vetch, oats, rye, barley and wheat, and its relation to Plant Box Test. *Weed Research, Japan* 39 (Supplement): 258-259.

Fujii, Y. and T. Shibuya. (1992a). Establishment of a new bioassay for allelopathy; Assessment of allelopathic activity from root exudates by using Plant Box and agar medium. *Weed Research, Japan* 36 (Supplement) 152-153.

Fujii, Y. and T. Shibuya. (1992b). Allelopathy of hairy vetch. Assessment of allelopathic activity and weed control in the field. *Weed Research, Japan* 37 (Supplement): 160-161.

Fujii, Y., T. Shibuya, and T. Yasuda. (1990). L-3,4-dihydroxyphenylalanine as an allelochemical candidate from *Mucuna pruriens* (L.) DC. var. *utilis*. *Agricultural and Biological Chemistry* 55: 617-618.

Fujii, Y., T. Shibuya, and T. Yasuda. (1991). Survey of Japanese medicinal plants for the detection of allelopathic properties. *Weed Research, Japan* 36: 36-42.

Johnson, G.A., M.S. Defelice, and Z.R. Helsel. (1993). Cover crop management and weed control in Corn. *Weed Technology* 7: 425-430.

Lazauskas, P. and Z. Balinevichiute. (1972). Influence of the excretions from *Vicia villosa* seeds on germination and primary growth of some crops and weeds. In *Physiological-Biochemical Basis of Plant Interactions in Phytocenoses*, ed. A.M. Grodzinsky. Vol. 3. Kiev: Naukova Dumka, pp. 76-79 (In Russian, English summary).

Putnam, A.R. and J. DeFrank. (1983). Use of phytotoxic plant residues for selective weed control. *Crop Protection* 2: 173-181.

Rice, E.L. (1984). Manipulated ecosystems: Roles of allelopathy in Agriculture. In *Allelopathy*, 2nd ed. London: Academic Press, pp. 8-12.

Swanton, C.J. and S.D. Murphy. (1996). Weed science beyond the weeds: the role of integrated weed management (IWM) in agroecosystem health. *Weed Science* 44: 437-445.

Teasdale, J.R. (1996). Contribution of cover crops to weed management in sustainable agricultural systems. *Journal of Production Agriculture* 9: 457-479.

Teasdale, J.R. and C.S.T. Daughtry. (1993). Weed suppression by live and desiccated hairy vetch. *Weed Science* 41: 207-212.

Weston, L.A. (1996). Utilization of allelopathy for weed management in agroecosystems. *Agronomy Journal* 88: 860-866.

White, R.H., A.D. Worsham, and U. Blum. 1989. Allelopathic potential of legume debris and aqueous extracts. *Weed Science* 37: 674-679.

Yenish, J.P., A.D. Worsham, and A.C. York. (1996). Cover crops for herbicide replacement in no-tillage corn (*Zea mays*). *Weed Technology* 10: 815-821.

Japanese Research and Developments on Plant Pathogens as Weed Control Agents

Fumio Sugawara

SUMMARY. The Japanese explorers of microbial herbicides with plant pathogens have successfully found that *Exserohilum monoceras, Nimbya scirpicola* and *Xanthomonas campestris* are potential weed control agents against barnyard grass (*Echinochloa crus-galli* and *E. oryzicola*), kuroguwai (*Eleocharis kuroguwai*) and annual bluegrass (*Poa annua*), respectively. This paper reviews the aspects of those potential micoherbicides currently developed in Japan. *[Article copies available for a fee from The Haworth Document Delivery Service: 1-800-342-9678. E-mail address: <getinfo@haworthpressinc.com> Website: <http://www.HaworthPress.com>*

KEYWORDS. Microbial herbicides, *Exserohilum monoceras, Nimbya scirpicola, Xanthomonas campestris,* Camperico®

INTRODUCTION

Plants are often infected with a variety of diseases. These diseases are caused by non-biological (soil and weather conditions, environmental pollu-

---

Fumio Sugawara is Professor of Chemical Biology, Department of Applied Biological Science, Research Institute for Science and Technology, Science University of Tokyo, Noda, Chiba 278-8510, Japan.

Address correspondence to: Fumio Sugawara at the above address (E-mail: sugawara @rs.noda.sut.ac.jp).

The author would like to express his appreciation to Dr. Gohbara of Japan Tobacco Inc. for his helpful information.

[Haworth co-indexing entry note]: "Japanese Research and Developments on Plant Pathogens as Weed Control Agents." Sugawara, Fumio. Co-published simultaneously in *Journal of Crop Production* (Food Products Press, an imprint of The Haworth Press, Inc.) Vol. 4, No. 2 (#8), 2001, pp. 277-285; and: *Allelopathy in Agroecosystems* (ed: Ravinder K. Kohli, Harminder Pal Singh, and Daizy R. Batish) Food Products Press, an imprint of The Haworth Press, Inc., 2001, pp. 277-285. Single or multiple copies of this article are available for a fee from The Haworth Document Delivery Service [1-800-342-9678, 9:00 a.m. - 5:00 p.m. (EST). E-mail address: getinfo@haworthpressinc.com].

tion, damages from agricultural chemicals and biological secondary metabolites) and pathogens (viruses, bacteria and fungi). In the diseases caused by pathogens, various physical and biochemical interactions occur between the host plants and pathogens. In nature, host-parasite interactions are diverse and plants are infected with relatively few microorganisms. Furthermore, a given microorganism generally infects only a few plant species. This host specificity has been major area of research in plant pathology. Currently, pathologists and entomologists are examining the prospects of utilizing insects and pathogens in biocontrol strategies to control weeds in agricultural crops. Plant pathogens, especially fungi and bacteria, induce disease symptoms in their respective host(s) due to the phytotoxins produced by them. Some pathogens also induce symptoms due to lytic enzymes. These compounds may be candidates for novel herbicides or serve as models for herbicide development (Gohbara, 1996).

Weed pathogens have had millennia to evolve with their hosts and devise biochemical mechanisms to kill them or influence their physiology. Now, we should take advantage of this to devise new strategies for weed control. The first step in this process is an examination of the structure and function of phytotoxins from weed pathogens. The second step should be the exploration of plant pathogens, especially weedy pathogens, to use them as microbial herbicides.

The studies on microbial herbicides began in the US in 1960s (Charudattan and Walker, 1982), but Japanese scientists have studied this area for only twelve years (Gohbara, 1996). In 1997, *Xanthomonas campestris* pv. *poae* (Camperico) has been approved as a microbial herbicide against *Poa annua* in Japan, whereas *Colletotrichum gloeosporioides* f. sp. *malvae* (BioMal) against *Malva pussil* in Canada (1992), *Colletotrichum gloeosporioides* (Collego) against *Aeschynomene virginica* in USA (1982), *Phytophthora palmivora* (DeVine) against *Morrenia odorata* in USA (1981) and *Colletotricum gloeosporioides* f. sp. *cuscuta* against *Cuscuta* sp. in China (LuBao 2) have been approved.

JAPANESE RESEARCHES

The Japanese research started with the finding of a novel pathogen *Epicoccosorus nematosorus* (Suzuki, 1987a,b) to biologically control kuroguwai (*Eleocharis kuroguwai*). In the same year, *Alternaria alternata* was used against cocklebur (*Xanthium* spp.). Only six microbial herbicides are approved today yet, however, over 200 microbes in 33 countries are currently investigated (Gohbara, 1998). Current target weeds of microbial herbicides in Japan are *Xanthium strumarium* (pathogen: *Alternaria alternata*), *Xanthium occidentale* (pathogen: *Alternaria alternata*), *Eleocharis kuroguwai* (pathogen: *Dendryphiella* sp., *Epicoccosorus nematosorus*, *Nimbya scirpicola*), *Echinochloa crus-galli* (pathogen: *Bipolaris ravenelli*, *Curvularia lunata*, *Drechslera poae*, *Exserohilum monoceras*, *Fusarium oxysporum*, *Leoptosphaeria salvinii*, *Ustilago trichophora*), *Echinochloa oryzicola* (pathogen: *Bipolaris revenelli*, *Drechslera poae*, *Exserohilum monoceras*), *Echinochloa colona* (pathogen: *Bipolaris revenelli*, *Drechslera poae*, *Exserohilum monoceras*) and *Poa annua* (pathogen: *Xanthomonas campestris*) (Gohbara, 1996). Microbial herbicides will come up in the near future.

Xanthomonas campestris

The phytopathogenic bacterium, *X. campestris* pv. *poae*, isolate JT-P482, has been successfully developed as a bioherbicide targeting annual bluegrass (*Poa annua*) by researches at Japan Tobacco Inc. (JT) of Japan (Imaizumi et al., 1994; Nishino et al., 1995; Nishino and Fujimori, 1996; Imaizumi et al., 1997; Nishino, Morita, and Fujimori, 1997; Imaizumi and Fujimori, 1997a,b; Nishino and Fujimori, 1998a,b; Imaizumi and Fujimori, 1999a,b). In 1997, *X. campestris* has been marketed by researchers at Japan Tobacco Co. of Japan and Mycogen Corp. of US as Camperico® (Nishino, 1995).

During 1991 to 1993, the JT group collected annual bluegrass from the several areas of Japan (Nishino et al., 1995). Leaf and stem sections (about 1cm long) of the weed were harvested with sterile scissors blade. Bacteria were isolated from surface-sterilized samples by maceration in sterile distilled water. Cells from colonies were transferred onto YNA medium (yeast extract, 5 g; nutrient agar, 23 g; distilled water, 1 L). Small and slightly yellow colored colonies developed within 4 days of incubation at 28°C. Cells from colonies were transferred onto modified King's B agar medium (King, Ward, and Raney, 1954) (potato peptone No. 3, 20 g; KH_2PO_4, 15 g; $MgSO_4.7H_2O$, 0.4 g; glycerol, 10 ml; agar, 15 g; distilled water, 1 L). After 4 days of incubation, growth on the medium was suspended in sterile distilled water to prepare inocula. Healthy annual bluegrass seedlings were inoculated by means of leaf clipping inoculation. The inoculated plants were grown for two weeks in a greenhouse at 25°C/20°C (day/night) under 16 h white light (354-670 nm, 6.7W/m^2, FL20L, Toshiba, Japan).

Eighty-nine bacterial isolates that caused wilt symptoms (Roberts, Vargas, and Detweiler, 1985) on annual bluegrass after inoculation were isolated from annual bluegrass found on golf courses, landscapes and roadsides in 20 prefectures in Japan (Nishino et al., 1995). One isolate, JT-P482 demonstrated significant pathogenic effects on inoculated leaves. The leaf blades became constricted and turned dark green within 5 days after inoculation, and then became necrotic and dry. The plants died about 25 days after inoculation under greenhouse condition (Nishino and Fujimori, 1996). Isolate JT-482 exhibited severe effects on annual bluegrass rather than on rough blue grass.

JT-482 is gram negative and its bacteriological properties were the same as those of *X. campestris* pv. *poannua* and *X. campestris* pv. *poae* (Johnson, 1994; Zhou and Niel, 1995). Based on these results, including the bacterial characteristics and host ranges, Nishino et al. (1995) claimed that both JT-482 and *X. campestris* pv. *poannua* should be considered as a single pathovar, *X. campestris* pv. *poae*.

Plant inoculation with the suspension of the strain *X. campestris* pv. *poae* JT-P482 in the range of 1×10^4 to 1×10^8 cfu (colony forming unit)/ml was observed at 30°C/25°C (day/night) (Nishino and Fujimori, 1996). The inoculum concentration at 10^4 cfu/ml was less effective, but severe wilt was observed 7 weeks after with concentration over 10^6 cfu/ml. High (30°C/25°C) and low (10°C/5°C) temperatures did not severely affect the bacterial growth multiplication and disease development more than 10^6 cfu/ml of inocula. Wilt of the inoculated plant at 90% RH (relative humidity) developed more slowly than at 30-40% RH, however, the humidity effect on disease severity seemed to be caused by the predisposition of host plant rather than by the pathogenesis of the isolate JT-P482. A spontaneous mutant (strain Rif-482) of *X. campestris* pv. *poae* JT-P482, tolerant to the antibiotic rifampicin (3-[methylpiperazinyliminomethyl]rifamycin, Sigma-Aldrich, USA), was used. Host ranges, aggressiveness to annual bluegrass and bacteriological properties of both strain JT-P482 and Rif-482 were similar apart from their tolerance limit to rifampicin. When the fate of a strain Rif-482 in the grass turf field soil was examined, the population of strain Rif-482 declined within 7 days after treatment. In contrast to strain Rif-482, the number of soil bacteria and fungi in the turf field soil was not changed by the bacterial application (Nishino, Morita and Fujimori, 1997). From these observations, *X. campestris* pv. *poae* is neither able to persist nor to grow above the detection limit in soil in the natural ecosystem, and *X. campestris* pv. *poae* shows little effect on soil microorganisms. The longevity of this bacterium within the inoculated plants has been studied (Imaizumi and Fujimori, 1999a). Plants approximately 2 months old were inoculated with strain Rif-482 with the range of 10^{10} to 10^4 cfu/g. Seven months later, bacterial numbers in the both green house and field experiments declined to 10^4 cfu/g (Imaizumi and Fujimori, 1999b). The pathogenic mechanism of this strain is not yet known.

Other Possible Microbes in Japan

Nimbya scirpicola, *Dendryphiella* sp. and unidentified fungi have become the focus of comparison studies evaluating the bioherbicide capabilities of each fungus against *Eleocharis kuroguwai*, the most noxious paddy weeds in Japan (Tanaka et al., 1994, 1995). In 1991, field-testing was conducted to determine the capability of the pathogen to cause disease under natural conditions by JT group (Tanaka et al., 1994).

Nimbya scirpicola had been isolated from a diseased paddy field weed, *E. kuroguwai* (Harada et al., 1992). This fungus evoked severe orange spotting on stems of *E. kuroguwai* and showed specific phytotoxicity. A simple leaf-puncture assay on *H. kuroguwai* was used as a guide for isolating potential phytotoxins of *N. scirpicola*, which was grown in a Czapek-Dox medium, 2 weeks at 28°C. The ethyl acetate extract was separated by silica gel chromatography to yield active phytotoxin. From the analytical data of optical rotation, MS (mass spectra), IR (infrared spectra) and NMR (nuclear magnetic resonance), the structure of this phytotoxin was identified as 1 [4S, 5S (8S,9S)-diepoxy-(3R)(10R)-dihydroxyundeca-1,6-diene], which has been isolated from *Alternaria brassicicola* (Matsumoto et al., 1992). A 10 mg dose of this phytotoxin 1 on *H. kuroguwai* stems produced necrotic lesions (5 mm in length) or serious damage.

Dendryphiella sp. was also isolated from a diseased *E. kuroguwai* (Harada et al., 1992). This fungus produced four phytotoxic compounds named dendryol A (2), B (3), C (4) and D (5) (Tanaka et al., 1995) (Figure 1).

Eight fungi, *Exserohilum monoceras*, *E. rostratum*, *Bipolaris sorokiniana*, *Curvularia lunata*, *C. aeria*, *Colletotrichum graminicola*, *Pyricularia grisea* and *Ustilago trichophora*, were isolated from diseased weed, *Echinochloa* spp. (Tsukamoto et al., 1997). After selecting the most virulent isolate of each species, eight isolates were considered for weed control activity against *Echinochloa crus-galli* using spray and drop inoculation. *E. monoceras* and *C. lunata* reduced the dry weight of plants by over 80% using both types of inoculation. Drop inoculation resulted in higher weed control than spray inoculation. *Exserohilum monoceras* exhibited high herbicidal activity against *Echinochloa crus-galli* var. *crus-galli*, *E. crus-galli* var. *formosensis*, *E. oryzicola* and *E. colona*, which grow in Japanese paddy fields. The herbicidal activity of *E. monoceras* correlated with water management practices (Tsukamoto et al., 1998). Increasing the depth and duration of water flooding produced high herbicidal activity of the fungus. When 10^5 conidia/100 cm^2 were introduced into a pot filled with water into a depth of 7 cm or 9 cm, more than 95% plants were killed, and the dry weight of the plants was reduced by over 95% at 21 days after inoculation. A water depth of 7 cm with 7 or more days of flooding gave almost the same results at 21 days after inoculation. When the flooding duration was prolonged to 14 days, the plants were completely destroyed. Shallow depth or short flooding duration was not effective. *E. monoceras* produces a phytotoxin, monocerin 6 (Robeson and Strobel, 1982; Sugawara et al., 1985). However, it is unclear whether monocerin is the active principle of *E. monoceras* in this study.

FIGURE 1

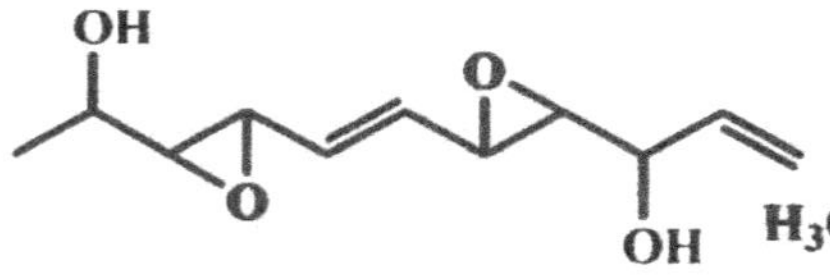

1

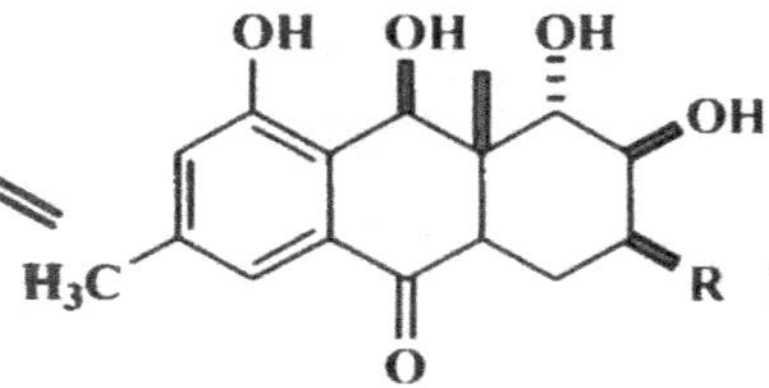

2: R = H

4: R = OH

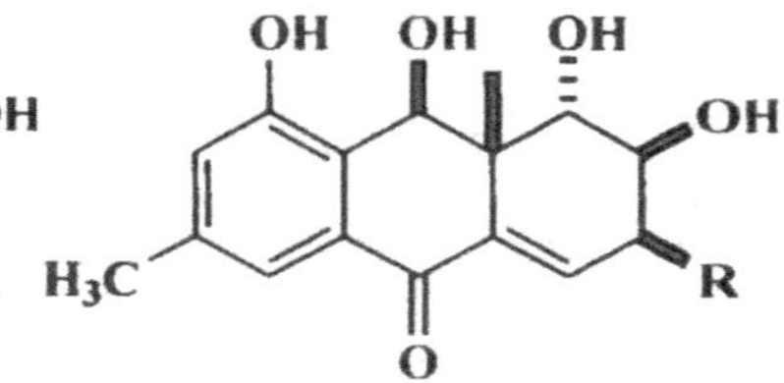

3: R = OH

5: R = H

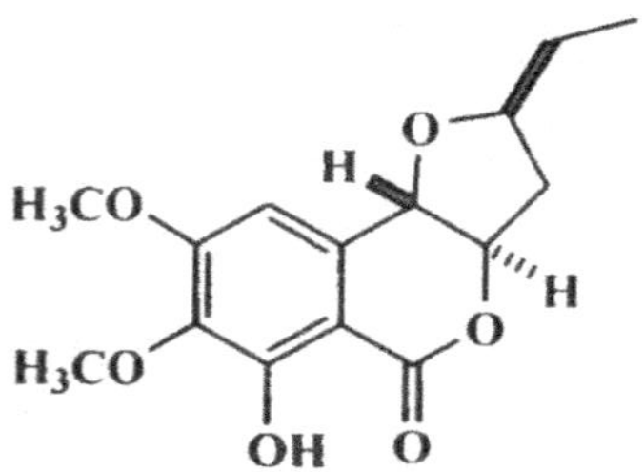

6

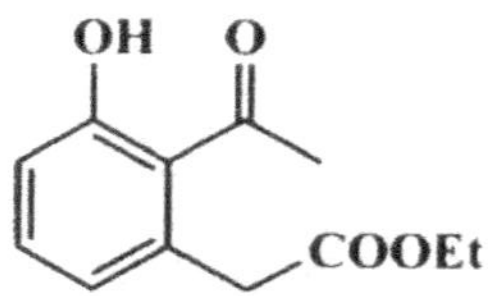

7

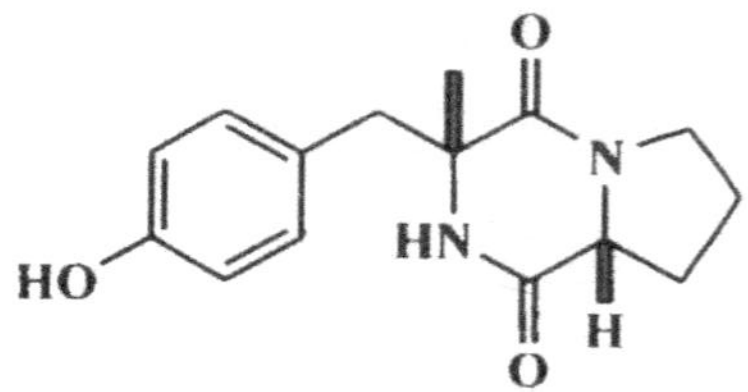

8

FUTURE ASPECTS

The studies on this field are exciting, especially in associations where one microbe produces phytotoxins, which regulate or kill higher plants (Sugawara, 1999). These are examples of pathogenic associations that offer tremendous potential use as selective herbicides. Recent studies have uncovered both novel and previously known phytotoxins that could increase widespread interest as potential herbicides. Curvulin (7) (Kenfield et al., 1988a,b) and maculosin (8) (Stierle, Cardellina, and Strobel, 1988) can be synthesized, and offer possibilities for direct application both as non-specific and as extremely selective herbicides. More complex organic compounds are useful, both as tools for understanding biochemical and physiological processes in plants and as selective, biologically rational agrochemicals. Many other microbes that attack weeds are also candidates for study.

REFERENCES

Charudattan, R. and H.L. Walker. (1982). *Biological Control of Weeds with Plant Pathogens*. New York: John Wiley & Sons.

Gohbara, M. (1996). Use of plant pathogens for the control of weeds in Japan. In *Proceedings of Second International Weed Control Congress* (eds. H. Brown, D.W. Cussans, M. Devine, S.O. Duke, C. Fernandez-Quintanilla, A. Helweg, R.E. Kaurada, M. Landes, P. Kudsk, and J.C. Streibig). Copenhagen, Denmark: Department of weed control and pesticide ecology, pp. 1205-1210.

Gohbara, M. (1998). Current research and development of micoherbicide. *Biocontrol* 2: 20-28 (In Japanese).

Harada, Y., S. Imaizumi, H. Tanaka, H. Negishi, T. Fujimori, M. Yamada, R. Monkura, and Y. Miura. (1992). *Nimbya scirpicola* causing orange stem spot on *Eleocharis kuroguwai*. *Annals of the Phytopathological Society of Japan* 58: 766-768.

Imaizumi, S., R. Houkabe, T. Nishino, Y. Omigawa, and M. Yamada. (1994). Control of annual bluegrass with bioherbicides. (1) Efficacy testing on weed control of annual bluegrass with bacterium, *Xanthomonas campestris*. *Weed Research (Japan)* 39, Supplement I: 42-43.

Imaizumi, S., and T. Fujimori. (1997a). Multiplication and translocation of *Xanthomonas campestris* pv. *poae* (JT-P482) in annual bluegrass (*Poa annua* L.). *Journal of Weed Science Technology* 42: 115-124.

Imaizumi, S., and T. Fujimori. (1997b). Effectiveness of *Xanthomonas campestris* pv. *poae* (JT-P482) in controlling two ecotypes of Japanese bluegrass (*Poa annua* L.). *Journal of Weed Science Technology* 42: 125-134.

Imaizumi, S., and T. Fujimori. (1999a). Persistence of *Xanthomonas campestris* pv. *poae* (JT-P482) in annual bluegrass. *Annals of the Phytopathological Society of Japan* 65: 52-56.

Imaizumi, S. and T. Fujimori. (1999b). Field spread and longevity of *Xanthomonas campestris* pv. *poae* (JT-482) after inoculation to annual bluegrass and mowing. *Annals of the Phytopathological Society of Japan* 65: 57-59.

Imaizumi, S., T. Nishino, K. Miyabe, T. Fujimori, and M. Yamada. (1997). Biological control of annual bluegrass (*Poa annua* L.) with a Japanese isolate of *Xanthomonas campestris* pv. *poae* (JT-P482). *Biological Control* 8: 7-14.

Johnson, B.J. (1994). Biological control of annual bluegrass with *Xanthomonas campestris* pv. *poannua* in bermuda grass. *Horticultural Science* 29: 659-662.

Kenfield, D., G.G. Bunkers, G. Strobel, and F. Sugawara. (1988a). Potential new herbicide-phytotoxins from plant pathogens. *Weed Technology* 2: 519-524.

Kenfield, D.S., Y. F. Hallock, J. Clardy, and G.A. Strobel. (1988b). Curvulin, and *O*-methyl curvulinic acid: phytotoxic metabolites of *Drechslera indica*, which causes necrosis on purslane and spiny amaranthus. *Plant Science* 60: 123-127.

King, E.O., M.K. Ward, and D.E. Raney. (1954). Two simple media for the demonstration of pyocyanin and fluorescein. *Journal of Laboratory of Clinics and Medicine* 44 : 301-307.

Matsumoto, M., S. Matsutani, K. Sugita, H. Yoshida, F. Hayashi, Y. Terui, H. Nakai, N. Uotani, Y. Kawamura, K. Matsumoto, J. Shoji, and T. Yoshida. (1992). Depudecin: a novel compound inducing the flat phenotype of NH3T3 cells doubly transformed by *ras*- and *src*-oncogene, produced by *Alternaria brassiciccola*. *Journal of Antibiotics* 45 : 879-885.

Nishino, J. (1995). *Xanthomonas campestris* (Camperico®). *Nouyaku-Jihou*, 1995: 15-17 (In Japanese).

Nishino, T., K. Morita, and T. Fujimori. (1997). Fate of *Xanthomonas campestris* pv. *poae*, a biological control agent for annual bluegrass, in soil. *Journal of Pesticide Science* 22: 326-330.

Nishino, T., S. Imaizumi, K. Miyabe, M. Yamada, and M. Goto. (1995). *Xanthomonas campestris* pv. *poae* as the causal agent of wilt symptoms on annual bluegrass in Japan. *Annals of the Phytopathological Society of Japan* 61: 555-561.

Nishino, T., and T. Fujimori. (1996). Influence of inoculum concentration, temperature and humidity on the disease of annual bluegrass caused by *Xanthomonas campestris* pv. *poae* JT-P482. *Annals of the Phytopathological Society of Japan* 62: 492-494.

Nishino, T. and T. Fujimori. (1998a). Pathogenic diversity of *Xanthomonas campestris* pv. *poae* within the genus *Poa*. *Annals of the Phytopathological Society of Japan* 64: 1-6.

Nishino, T., and T. Fujimori. (1998b). Population dynamics of *Xanthomonas campestris* pv. *poae*, a biological control agent for annual bluegrass, in turf grass field after application. *Journal of Pesticide Science* 23 : 151-154.

Roberts, D.L., J.M. Vargas, Jr., and R. Detweiler. (1985). Occurrence of bacterial wilt on *Poa annua* and other turf grasses. *Phytopathology* 75: 1289.

Robeson, D.J. and G. Strobel. (1982). Monocerin, a phytotoxin from *Exserohilum turcicum* (=*Drechslera turcicum*). *Agricultural Biological Chemistry* 46: 2681-2683.

Stierle, A.C., J.H. Cardellina II, and G.A. Strobel. (1988). Maculosin, a host specific phytotoxin for spotted knapweed (*Centaurea maculosa*). *Proceedings of National Academy of Science, U.S.A.* 85: 8008-8011.

Sugawara, F. (1999). Phytotoxins as potential herbicides. *Allelopathy Journal* (in press).

Sugawara, K., F. Sugawara, G. Strobel, Y. Fu, H. Cun-Heng, and J. Clardy. (1985).

Exserohilone, a novel phytotoxin produced by *Exserohilum holmii*. *Journal of Organic Chemistry* 50: 5631-5933.

Suzuki, H. (1987a). Biological control of paddy weed, kuroguwai, with a fungal pathogen. In *The Biological Control of Plant Disease. FFTC Book Series No. 42.* Taipei, Taiwan, ROC: Food and Fertilizer Technology Centre for ASPAC Region, pp. 78-87.

Suzuki, H. (1987b). Trials of biological control of kurogwai with a fungal pathogen. *Nougyou oyobi Engei* 63: 741-744.

Tanaka, M., J. Ohra, Y. Tsujino, T. Fujimori, H. Ago, and H. Tsuge. (1995). Dendryol A, B, C, and D, four compounds produced by a weed pathogenic fungus *Dendryphiella* sp. *Zeitschrift fuer Naturforschung* 50c: 751-756.

Tanaka, M., J. Ohra, Y. Tsujino, Y. Sawaji, and T. Fujimori. (1994). Phytotoxin produced by *Nimbya scirpicola*. *Bioscience Biotechnology and Biochemistry* 58: 565-566.

Tsukamoto, H., M. Gohbara, M. Tsuda, and T. Fujimori. (1997). Evaluation of fungal pathogens as biological control agents for the paddy weed, *Echinochloa* species by drop inoculation. *Annals of the Phytopathological Society of Japan* 63: 366-372.

Tsukamoto, M. Tsuda, H., M. Gohbara, and T. Fujimori. (1998). Effect of water management on mycoherbicidal activity of *Exserohilum monoceras* against *Echinochloa oryzae*. *Annals of the Phytopathological Society of Japan* 63: 526-531.

Zhou, T., and J.C. Niel. (1995). Annual bluegrass (*Poa annua*) control with *Xanthomonas campestris* pv. *poannua* in New York State. *Weed Technology* 9: 173-177.

Allelopathic Potential in Rice Germplasm Against Ducksalad, Redstem and Barnyard Grass

Robert H. Dilday
John D. Mattice
Karen A. Moldenhauer
Wengui Yan

SUMMARY. Of 17,279 rice accessions or varieties from 110 countries in the USDA-ARS rice working collection, about 12,000 accessions in the rice collection have been evaluated for allelopathy to ducksalad (*Heteranthera limosa* [Sw.] Willd.) and about 5,000 have been evaluated for allelopathy to redstem (*Ammannia coccinea* Rottb.) and barnyard grass (*Echinochloa crus-galli* [L.] Beauv.). A total of 412 rice accessions were identified in field tests in 1988, 1989 and 1990 that had an area of allelopathic activity > 10 cm to ducksalad and 145 accessions that had the same area of activity to redstem. A total of 94 accessions demonstrated apparent allelopathic activity to barnyard grass. The rice accessions that showed allelopathic activity in the field had 2-3 times more root biomass than cultivars that did not show allelopathic activity.

Robert H. Dilday and Wengui G. Yan are affiliated with USDA-ARS, Dale Bumpers National Rice Research Center, Stuttgart, AR 72160.

John D. Mattice is affiliated with Department of Crop, Soil and Environmental Science, University of Arkansas, Fayetteville, AR 72701.

Karen A. Moldenhauer is affiliated with Rice Research and Extension Center, University of Arkansas, Stuttgart, AR 72160.

Address correspondence to: Robert H. Dilday at the above address (E-mail: rdilday@spa.ars.usda.gov).

[Haworth co-indexing entry note]: "Allelopathic Potential in Rice Germplasm Against Ducksalad, Redstem and Barnyard Grass." Dilday, Robert H., et al. Co-published simultaneously in *Journal of Crop Production* (Food Products Press, an imprint of The Haworth Press, Inc.) Vol. 4, No. 2 (#8), 2001, pp. 287-301; and: *Allelopathy in Agroecosystems* (ed: Ravinder K. Kohli, Harminder Pal Singh, and Daizy R. Batish) Food Products Press, an imprint of The Haworth Press, Inc., 2001, pp. 287-301. Single or multiple copies of this article are available for a fee from The Haworth Document Delivery Service [1-800-342-9678, 9:00 a.m. - 5:00 p.m. (EST). E-mail address: getinfo@haworthpressinc.com].

In barnyard grass infested plots the grain yield reduction of rice accessions that demonstrated allelopathy activity was about 37% as compared to a 60 to 68% reduction in grain yield of non-allelopathic accessions. Chromatograms of allelopathic and non-allelopathic accessions fall into two types, one that is represented by the allelopathic PI 312777 and the other by the non-allelopathic cultivar 'Rexmont'.

KEYWORDS. Allelopathic cultivars, barnyard grass, ducksalad, redstem, reice accessions

INTRODUCTION

Molisch (1937) coined the term allelopathy to define both the detrimental and beneficial biochemical interactions among all classes of plants, including microorganisms. Rice (1984) later modified the definition of allelopathy as "any direct or indirect harmful or beneficial effect by one plant (including microorganisms) on another through the production of chemical compounds that escape into the environment." The chemicals with allelopathic potential are present in virtually all plant tissues, including leaves, flowers, fruits, stems, roots, rhizomes and seeds, and are released by processes such as volatilization, root exudation, leaching, and decomposition of plant residues (Rice, 1984; Putnam and Weston, 1986). Although the term was coined by Molisch in 1937, the concept of allelopathy is a very old component of agricultural science, e.g., Theophrastus (ca. 300 B.C.) referred to 'phytotoxicity' among plants in his book 'Enquiry into Plants'; the Greek philosopher Democritus reported the use of naturally occurring plant products as a practical method of controlling weeds and stated that trees could be killed by treating their roots with a mixture of lupine flowers soaked in hemlock juice (Rizvi and Rizvi, 1992).

Allelopathy is postulated to be one mechanism by which weeds affect crop growth and it occurs widely in natural plant communities (Gressel and Holm, 1964; Whittaker and Feeny, 1971; Bell and Koeppe, 1972). Allelopathic potential of weeds through the release of toxic substances into the environment either through root exudation or from decaying plant material has been demonstrated in about 90 species (Putnam and Weston, 1986). These weeds include quack grass (*Agropyron repens* [L.] Beauv.) (Kommedahl, Kotherimer, and Bernardini, 1959; Gabor and Veatch, 1981), yellow and purple nutsedge (*Cyperus esculentus* L. and *C. rotundus* L.) (Friedman and Horowitz,

1971), Johnson-grass (*Sorghum halepense* [L.] Pers.) (Abdul-Wahab and Rice, 1967), Canada thistle (*Cirsium arvense* [L.] Scop.) (Bendall, 1975), leafy spurge (*Euphorbia esula* L.) (LeTourneau, Failes, and Heggeness, 1956; LeTourneau and Heggeness, 1957), giant foxtail (*Setaria faberii* Herrm.), yellow foxtail (*Setaria glauca* [L.] Beauv.), velvetleaf (*Abutilon theophrasti* Medic.) (Elmore, 1980), and tall fescue (*Festuca arundinacea* Schreb.) (Peters, 1968; Peters and Luu, 1984). In addition to the existence of allelopathy in weeds several workers have reported that crops such as rye (*Secale cereale* L.) and wheat (*Triticum aestivum* L.) (Shilling, Liebl, and Worsham, 1985), sunflower (*Helianthus annuus* L.) (Leather, 1983), oats (*Avena sativa* L.) (Fay and Duke, 1977), barley (*Hordeum vulgare* L.) (McCalla and Haskins, 1964), tobacco (*Nicotiana tabacum* L.) (Patrick, Toussoun, and Snyder, 1963), and rice (*Oryza sativa* L.) (Dilday, Nastasi, and Smith, 1989, 1991; Dilday, Lin, and Yan, 1994) possess allelopathic activity and have weed-suppressing properties. Putnam and Duke (1974) postulated that "wild types" of existing crops may have possessed high allelopathic activity and this character was reduced or lost as they were hybridized and selected for other characteristics.

More than 50 weed species infest direct-seeded rice and cause major losses in U.S. rice production (Smith, Flinchum, and Seaman, 1977). Ducksalad (*Heteranthera limosa* [Sw.] Willd.), an aquatic weed that can reduce rice yield by 27-30% when competing with rice in a water-seeded culture (Smith, Flinchum, and Seaman, 1977; Smith, 1988), is second only to barnyard grass (*Echinochloa crus-galli* [L.] Beauv.) as the most frequently reported weed in rice fields, followed by hemp sesbania (*Sesbania exaltata* [Raf.] Rydb. ex A.W. Hill), bulrushes (*Scirpus* spp.), red rice (*Oryza sativa* L.), broadleaf signalgrass (*Brachiaria platphylla* [Griseb.] Nash) and sprangletop (*Leptochloa* spp.) (Chandler, 1981).

There are 17,279 rice accessions or varieties from 110 countries in the USDA-ARS rice working collection. One objective of our research is to evaluate the USDA-ARS rice collection and identify accessions or varieties with allelopathic properties to weed species, both aquatic and dry seeded weeds, that can be used in variety development programs. About 12,000 accessions in the USDA-ARS rice germplasm collection have been evaluated for allelopathy to ducksalad and about 5,000 accessions have been evaluated for allelopathy to redstem and barnyard grass. Promising germplasm possessing allelopathic properties to ducksalad, redstem and barnyard grass have been identified. The development and utilization of this germplasm can lead to a reduced use of herbicides that may result in improved water quality and less environmental contamination.

The objectives of this study were to evaluate the USDA-ARS rice germplasm collection for apparent allelopathic activity in rice, identify morphological differences and measure the amount of weed suppression and yield potential of rice germplasm that demonstrates the apparent activity.

MATERIALS AND METHODS

History (1985-1999)

Apparent allelopathic activity in rice to ducksalad in field tests was first observed in 1985 and 1986 by R.H. Dilday at the Rice Research and Extension Center, Stuttgart, Arkansas while evaluating accessions in the USDA-ARS world collection for tolerance to alachlor, a herbicide (Dilday, Nastasi, and Smith, 1989, 1991; Dilday, Lin, and Yan, 1994). Field and laboratory experiments were conducted from 1987 to 1999 to identify rice accessions possessing allelopathic properties against ducksalad, redstem, broadleaf signalgrass, rice flatsedge (*Cyperus iria*), sprangletop and barnyard grass as part of the USDA-ARS rice germplasm evaluation project at Stuttgart, AR. Results of experiments involving allelopathy to ducksalad, redstem and barnyard grass are reported.

Ducksalad and Redstem

Approximately 12,000 accessions including checks (approx. 5,000 in 1988 and 1989 and approx. 2,000 in 1990) were seeded in hill plots in April each year in a 0.75 by 0.75 m^2 grid. Each year between 5 and 7 seed were placed in each hill. The field evaluation tests were replicated twice and conducted at the Rice Research and Extension Center (RREC), Stuttgart, AR, on a Crowley silt loam (fine montomorillonitic, thermic Typic Albaqualfs). A natural and uniform infestation of ducksalad occurs at the RREC at Stuttgart and occurred in these tests; therefore, seeding the test with ducksalad was not necessary. Allelopathic activity to ducksalad was recorded each year in July at about the panicle initiation stage for most of the accessions. Two methods were used to record the activity: (i) the radial area (cm) from the base of the rice plant that was affected, and (ii) the percentage of weed control within the affected area based on the number of ducksalad plants in a check plot around a control plant that had no apparent allelopathic activity to ducksalad. In the first method, actual measurements were made from the base of the plant to the outermost edge of the area of activity. The area of activity is defined as the area around the plant where no ducksalad growth appears or a reduced stand of ducksalad is present. If an accession had a radial mean area of activity less than 10 cm, then the accession was rated as having no allelopathic activity. The second method of measuring allelopathic activity is the reduction in the ducksalad population (plants $\cdot$ m^{-2}) compared to a control plant that has slight or no effect on the ducksalad population.

A total of 82 kg $\cdot$ ha^{-1} of N as urea was applied when the seedlings were at the 4th true leaf stage. The remaining 41 kg $\cdot$ ha^{-1} of N was applied in 2 equal increments about 23 days after the first application each year and then 12 days later. The test was irrigated 2 times in May each year to insure uniform seedling emergence. A permanent flood was applied in June. Plant height (cm), days to maturity, plant type, panicle type, hull cover or pubescence, hull color, lemma color, awning, lodging and grain type were recorded for each accession. Plant height was measured in cm from ground level to the tip of the mature panicle. Heading date was determined by calculating the number of days from the date of seedling emergence to the date that 50% of the panicles had emerged. Plant type, hull cover, hull color, lemma color and awning were recorded in the laboratory after threshing. The accessions were characterized as having short (<5.50 mm), medium (5.51-6.60 mm) or long (6.61-7.50 mm) grain. In 1988, the 5,000 accessions including checks were also evaluated for allelopathic activity to redstem because there was a uniform population of volunteer redstem plants.

Barnyard Grass

In 1994, approximately 5,000 rice accessions were evaluated in the field at the RREC for their effect on the growth and development of barnyard grass. Each accession was drill seeded in paired rows about 1.4 m long and 17.0 cm apart in 2 replications. Prior to seeding, the soil was tilled and 9.5 liters of barnyard grass per ha were broadcast with a hand applicator and incorporated with a harrow. Two rice accessions, PI 312777 and PI 338046, that had significantly fewer barnyard grass plants between the paired rows when compared to 2 cultivars, Lemont and Rexmont, were selected for further field evaluations. In 1995, PI 312777, PI 338046, Lemont and Rexmont were drill seeded in 6 row plots 1.4 m long and 17.0 cm apart in 4 replications. A cylinder that was 36 cm in diameter was placed in the center of each plot and the number of barnyard grass plants was recorded approximately 5 weeks after emergence. Each plot was harvested for grain yield at maturity.

Root Development

Sterilized soil was placed in plastic pots that were 20.5 cm in diameter and 20.0 cm in depth. Ten seeds of 20 different rice germplasm accessions were planted in each pot. After the seeds germinated, 10 pots of each accession were thinned to 1 plant/pot and an additional 10 pots were thinned to 5 plants/pot. Therefore, each accession in each stand density was replicated 10 times. When the rice plants were mature they were cut at ground level without damaging the root system. The roots that developed above ground were removed, dried at 50°C for 48 h and weighed. The soil was washed from the roots that developed below the soil level and the roots were dried at 50°C for 48 h and weighed. The data were analyzed by Duncan's Multiple Range Test.

Yield

Seven germplasm accessions that vary in apparent allelopathic activity to barnyard grass based on previous tests and two US semidwarf cultivars, Rexmont and Lemont, plus one blank plot (no rice planted) were evaluated in the field in 1996. Five of the germplasm accessions originated at the International Rice Research Institute (IRRI), Los Banos, Philippines (PI 312777, PI 338046, PI 350468, PI 360150, PI 373026) and two were from China (Gui-Chao and Teqing). The experiment was performed in a randomized block design (four replications) with two treatments, barnyard grass control (clean) and barnyard grass infested. Barnyard grass seeds were broadcast at a rate of 13.7 $lb \cdot ha^{-1}$ to insure a uniform infestation and incorporated with a harrow prior to seeding the rice. After seeding the rice, a second broadcast application of barnyard grass seed (6.8 $lb \cdot ha^{-1}$) was made with a hand applicator. Eighty grams of rice (116 $lb \cdot acre^{-1}$) were seeded in nine-row plots that were 3.8 m long with a 17.0 cm row spacing on 23 May, 1996. A 30 cm diameter cylinder was placed in the center of each plot and the number and dry weight of barnyard grass plants within each cylinder were recorded at maturity. Two 3.2 m rows in each plot were harvested for grain yield.

Chromatograms

Forty accessions were included in the HPLC analysis of rice leaf tissue in 1998. There were three replications per accession and ten rice plants per replication. The leaves were removed from ten day old plants, cut into approximately 1 cm lengths and placed in either 25 or 50 ml erlenmeyer flasks. Methanol was added to make a ratio of 10 mg tissue ml^{-1} methanol, the samples were covered with parafilm and refrigerated overnight. The next morning a volume of the methanol extract was combined with an equal volume of deionized water and the samples were analyzed by HPLC using a C18 column and a 1% acetic acid in water/acetonitrile gradient. Detection was done at 320 nm.

Breeding/Enhancement

PI 338046, an accession that has shown allelopathic activity, was hybridized with Adair, Alan, Katy, Lemont/RA73, M-201/Katy and Newbonnet in

1991. The F_1 hybrids were grown in the greenhouse in the winter of 1991-1992, and the F_2 population was grown at Stuttgart in 1992, the F_3 populations in Puerto Rico in the winter of 1992-1993, the F_4 and F_5 populations at Stuttgart in 1994 and 1995, respectively. Eight F_6 lines were field evaluated for agronomic characteristics such as field and milling yield, days from emergence to heading, plant height, and lodging, etc., in 1996. The central four rows of each plot (1.3 m × 3.8 m) were harvested. The lines were evaluated for allelopathic properties to ducksalad in plastic pots, 37.5 cm in diameter, in the greenhouse in the winter of 1995-1996 and field evaluated for allelopathic properties to barnyard grass in 1998.

Genetics

PI 312777, an accession that has shown allelopathic activity, was hybridized with Lemont in 1994. The F_1 and F_2 populations were grown in the field at Stuttgart, AR in 1995 and 1996, respectively. About 400 F_2 plants were evaluated for allelopathy to ducksalad in 1996. The area of activity around the plant that (as previously described) was the descriptor that was used to determine allelopathic activity.

RESULTS AND DISCUSSION

Ducksalad and Redstem

In 1988, 190 accessions had a radius of activity to ducksalad of greater than 10 cm and these accessions were significantly different from the control, Rexmont. In 1989, another 155 accessions from a different set of 5,000 had a radius of activity to ducksalad of greater than 12 cm. In 1990, 67 accessions from a different set of 2,000 had a radius of activity to ducksalad of greater than 12 cm. A list of these accessions has been reported (Dilday et al., 1998). Accessions that demonstrated allelopathic activity in 1988, 1989 and 1990 originated in 31 countries: Afghanistan, Argentina, Australia, Brazil, Colombia, Dominican Republic, Egypt, France, India, Indonesia, Iran, Iraq, Israel, Italy, Japan, Malaysia, Mali, Mexico, Pakistan, People's Republic of China, Peru, Philippines, Portugal, Republic of Korea, Soviet Union, Spain, Taiwan, Thailand, Turkey, United States and Vietnam.

In a 1988 field experiment, 145 accessions had a radius of activity greater than 10 cm to redstem, and these accessions were significantly different from the control, Rexmont. A list of these accessions has been reported (Dilday et al., 1998). Accessions that demonstrated allelopathic activity to redstem originated in 21 countries: Argentina, Australia, Colombia, Dominican Republic,

Guatemala, Guyana, Hungary, India, Indonesia, Japan, Malaysia, Mexico, Myanmar, Pakistan, People's Republic of China, Philippines, Spain, Sri Lanka, Surinam, Taiwan and U.S. Sixteen of the 145 accessions that showed allelopathy to redstem also demonstrated allelopathy to ducksalad (Table 1). The sixteen accessions originated in four countries: India, Malaysia, Philippines and Taiwan.

Barnyard Grass

In the 1994 field experiment there were 94 accessions that had significantly fewer barnyard grass plants than the controls, Lemont and Rexmont. Two accessions, PI 312777 and PI 338046 and two cultivars were evaluated in the field in 1995. There was no significant difference in the total field yields of PI 312777, PI 338046, Lemont and Rexmont in a weed free environment. However, in the barnyard grass infested plots the cultivars Rexmont and Lemont had a significantly greater number of barnyard grass plants and produced significantly less field yields than PI 312777 and PI 338046 (Table 2). Barnyard grass reduced the grain yield of the two allelopathic accessions, PI 338046 and PI 312777, by 37 and 45%, respectively, whereas the grain yield of the two non-allelopathic cultivars, Lemont and Rexmont was reduced by 60 and 68%, respectively (Table 2).

Root Development

Among the 20 rice germplasm accessions, there were significant differences in both above and below ground root biomass. Furthermore, the pots that contained 5 plants produced significantly more above and below ground root biomass than pots that contained only 1 plant. However, there was no significant difference in the order of root biomass production when the two test (1 and 5 plants pot^{-1}) were compared. PI 312777 and PI 338046, two germplasm accessions that have shown allelopathic activity to barnyard grass and ducksalad in the field, had the greatest root biomass. Two cultivars, Lemont and M-201, which have not shown allelopathic activity in the field had the least root biomass.

Yield

Grain yields of the nine rice accessions were taken in both barnyard grass infested plots and plots free of barnyard grass. The average grain yield reduction in the barnyard grass infested plots was 3822 $kg \cdot ha^{-1}$ or 48.2% lower than the yields from the weed free plots. The average grain yield of the nine rice varieties in the control (no barnyard grass) plots was 7892 $kg \cdot ha^{-1}$ as

TABLE 1. Rice accessions (16) from the USDA-ARS germplasm collection that showed allelopathy against both redstem and ducksalad in a field experiment at Stuttgart, Arkansas, in 1998 and 1989.

Cultivar Name	Weed Control (%) Redstem/Ducksalad	Radius (cm) Redstem/Ducksalad	CI/PI	Origin
B850/Cros 1-7-18-3-2 sein (298965-2)	75/85	13/10	12496	Unknown
Johna 349 (early)	70/85	15/10	247882	India
Mahlar 346	78/85	15/10	247883	India
S 12 DZK-India	70/90	15/13	247887	India
Taichung Native 1 (TN-1)	60/85	15/15	271672	Taiwan
Lui chou 25 108 30	73/85	13/10	303683	Taiwan
IR52 16 7 3	80/60	18/15	312576	Philippines
IR52 30 6 2	80/85	18/13	312580	Philippines
IR238	78/85	15/13	321183	Philippines
CP231/SL017//Nanng mon	80/85	15/15	321184	Philippines
IR781-497-2-3	70/90	13/10	348798	Philippines
IR1317-385	78/90	15/10	348835	Philippines
IR1321-12	78/90	15/10	348842	Philippines
Masrai	68/85	15/10	372764	Malaysia
CR52-3 (India)	70/90	15/10	373020	India
IR788-16-1-1-1	68/90	13/10	373026	Philippines
Rexmont (Control)	0/0	0/0	502968	USA

TABLE 2. Reduction in field yield of PI 3127777, PI 338046, Lemont and Rexmont cultivars due to barnyard grass infestation in a field experiment at Stuttgart, Arkansas, in 1995.

Designation	Weed Free Rice	BYG Infested			Yield Reduction	Plant Height
	(kg·ha^{-1})	Rice (kg·ha^{-1})	BYG Number	BYG Weight (g)	(%)	(cm)
PI 312777	8121	4445	6	15	45	78
PI 338046	7589	4768	10	10	37	71
Lemont	7182	2910	19	30	60	73
Rexmont	7585	2422	30	29	68	68
LSD 0.05	2021	1199	10	8	--	7

compared to 4070 kg · ha^{-1} from the barnyard grass infested plots. Furthermore, two U.S. check cultivars, Rexmont and Lemont, had the most yield reduction (5163 kg · ha^{-1} or 68.1% and 4272 kg · ha^{-1} or 59.5%, respectively). Conversely, two germplasm accessions from the International Rice Research Institute (IRRI), IR 781-92-1-2-1-2-2 (PI 350468) and IR 782-98 (PI 338064), had the least yield reduction of rice (2434 kg ha^{-1} or 36.7% and 2821 kg · ha^{-1} or 37.2%, respectively). Also, the two IRRI accessions, IR 781-92-1-2-1-2-2 and IR 782-98, had the least barnyard grass biomass (2.0 and 10.5 g, respectively) as compared to the two U.S. cultivars Rexmont (29.5 g) and Lemont (30.5 g). The number of barnyard grass plants observed in Gui-Chao (5.0) from China, T65*2/TN1 (Taichung 65*2/Taichung Native 1) (6.5) from IRRI and Teqing (8.5) from China were significantly less when compared to Rexmont (30.0) or the blank plot (41.0).

Chromatograms

There was no difference in chromatograms taken from plants that were 5, 10, or 15 days old; therefore, for this study 10 day old plants were selected. Ten rice plants were used per replication in order to minimize differences between replications in order to make it easier to find differences between accessions. Reproducibility between replications was good. If necessary the analysis can be done on one leaf from a plant so the plant does not have to be sacrificed. The data fall into two types, one which is represented by the allelopathic accession PI 312777 and the other by non-allelopathic Rexmont. In the PI 312777 chromatogram the peaks were substantially higher for the compounds with retention times of 12.2, 12.6 and 13.8 min. The compounds with retention times of 14.1, 14.6 and 15.2 are present in the PI 312777 extract and essentially absent in the extracts of Rexmont (Figure 1). Ongoing work investigating the correlation between bioassay results and chromatographic results indicates that accessions that produce chromatograms similar to those of PI 312777 will inhibit growth of barnyard grass. To date there have been no accessions that produce chromatograms like Rexmont that also show substantial inhibition of barnyard grass.

Breeding/Enhancement

Individual F_2 populations from hybrids involving PI 338046 with Adair, Alan, Katy, Lemont/RA 73, M-201/Katy and Newbonnet were evaluated in the field at Stuttgart in 1992. IR 8 and TN-1 are two parents in the pedigree of PI 338046. Therefore, PI 338046 or IR 8*2//85894 A4-18-1*2/TN-1 has 2 (IR 8 and TN-1) of the 4 common parents identified from the IRRI material which possess allelopathic potential to both ducksalad and redstem. Eight F_6

FIGURE 1. HPLC chromatograms of methanol extracts of rice leaf tissue from PI 31277 and Rexmont from 11.8 min. to 15.8 min.

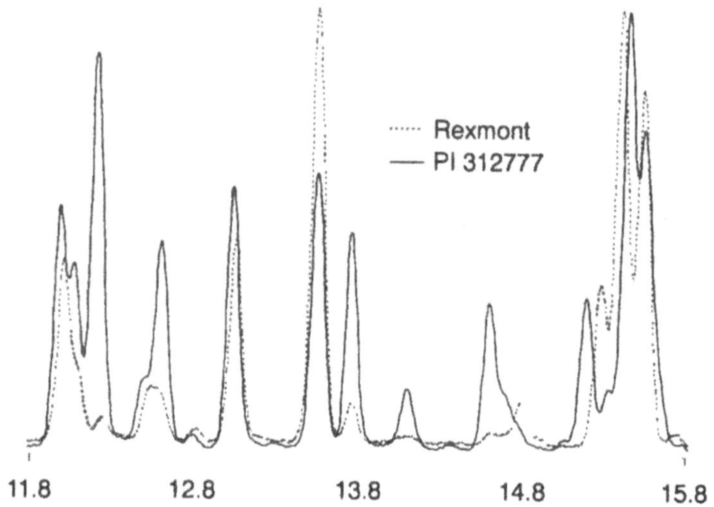

lines from the crosses PI 338046//Lemont/RA 73, PI 338046/Alan, PI 338046//M-201/Katy and PI 338046/Katy were field evaluated in replicated test in 1996. One line from the cross PI 338046/Katy (Stg 94L42-130) was the highest yielding entry (9882 kg/ha) in the Arkansas Preliminary Test in 1995. Stg 94L42-130 exceeded the yield of Katy (7868 kg/ha) by almost 2000 kg/ha. Also, in a greenhouse test in 1995-96 Stg 94L42-130 had significantly fewer ducksalad plants (35) than the non-allelopathic check, Rexmont (97), LSD (0.05) = 31. In a field test in 1998, Stg 95L05-054 had significantly fewer barnyard grass plants than the non-allelopathic check, Rexmont and Katy (Table 3).

Genetics

The F_2 population involving PI 312777/Lemont was distributed normally for allelopathic activity. This indicates that allelopathic activity is quantitatively inherited. Also, our experiences with environmental influences on allelopathic activity support quantitative inheritance. For example, year to year variation, soil type, weed density, and rice plant number can effect the size of the area of allelopathic activity, especially on ducksalad. This appears to agree with Rizvi and Rizvi (1992) comment that "much of the evidence indicates that several chemicals are released together and may exert toxicities

TABLE 3. Number of barnyard grass (BYG) plants in a 36 cm cylinder of 8 F_6 breeding lines compared to an allelopathic accession (PI 338046) and 2 non-allelopathic culitvars (Rexmont and Katy).

Stuttgart Cross No.	Pedigree	BYG Number
Stg 95L05-74	PI 338046//M201/Katy	20
Stg 95L05-032	PI 338046//M201/Alan	18
Stg 95L05-078	PI 338046//M201/Katy	16
Stg95L05-044	PI 338046/Alan	13
Stg94L42-130	PI 338046/Katy	10
Stg95L05-058	PI 338046//LMNT/RA 73	8
Stg95L05-039	PI 338046/Alan	8
Stg95L05-054	PI 338046//LMNT/RA 73	5
Checks		
Rexmont	Newrex/Bellemont	19
Katy	Bonnet 73/CI9722//	18
	Starbonnet/Tetep/3/Lebonnet	
PI 338046	IR 8 *2//B5894A4-18-1 *2/TN 1	5
LSD 0.05		11

in an additive or synergistic manner. To date all cases of alleged allelopathy that have been thoroughly studied appear to involve a complex of chemicals. In no case has one specific phytotoxin been proved to be solely responsible for, or produced as a result of interference by a neighboring plant."

REFERENCES

Abdul-Wahab, A.S. and E.L. Rice. (1967). Plant inhibition by Johnson-grass and its possible significance in old field succession. *Bulletin of the Torrey Botanical Club* 94: 486-497.

Bell, D.T. and D.E. Koeppe. (1972). Noncompetitive effects of giant foxtail on the growth of corn. *Agronomy Journal* 64: 321-325.

Bendall, G.M. (1975). The allelopathic activity of Californian thistle (*Cirsium arvense* L. Scop.) in Tasmania. *Weed Research* 15: 77-81.

Chandler, J.M. (1981). Estimated losses of crops to weeds. In *Handbook of Pest Management in Agriculture*, ed. D. Pimentel. Vol. 1. Boca Raton, FL: CRC Press, Inc., pp. 95-109.

Dilday, R.H., P. Nastasi, and R.J. Smith, Jr. (1989). Allelopathic observation in rice

(*Oryza sativa* L.) to ducksalad (*Heteranthera limosa*). *Proceedings of the Arkansas Academy of Science* 43: 21-22.

Dilday, R.H., P. Nastasi, and R.J. Smith, Jr. (1991). Allelopathic activity in rice (*Oryza sativa* L.) against ducksalad [*Heteranthera limosa* (Sw.) Willd.]. In *Proceedings of Sustainable Agriculture for the Great Plains*, eds. J.N. Hanson, M.J. Shaffer, D.A. Ball, and C. Vern Cole. Beltsville, MD, USA: United States Department of Agriculture, Agricultural Research Service, pp. 193-201.

Dilday, R.H., J. Lin, and W.G. Yan. (1994). Identification of allelopathy in the USDA-ARS rice germplasm collection. *Australian Journal of Experimental Agriculture* 34: 907-910.

Dilday, R.H., W.G. Yan, K.A. Moldenhauer, and K.A. Gravois. (1998). Allelopathic activity in rice for controlling major aquatic weeds. In *Allelopathy in Rice. Proceedings of the Workshop on Allelopathy in Rice*, ed. Maria Olofsdotter. Manila, Philippines: International Rice Research Institute, pp. 7-26.

Elmore, C.D. (1980). Inhibition of turnip (*Brassica rapa*) seed germination by velvetleaf (*Abutilon theophrasti*) seed. *Weed Science* 28: 658-660.

Fay, P.K., and W.B. Duke. (1977). An assessment of allelopathic potential in Avena germplasm. *Weed Science* 25: 224-228.

Friedman, T. and M. Horowitz. (1971). Biologically active substances in subterranean parts of purple nutsedge. *Weed Science* 19: 398-401.

Gabor, W. E., and C. Veatch. (1981). Isolation of a phytotoxin from quackgrass (*Agropyron repens*) rhizomes. *Weed Science* 29: 155-159.

Gressel, J.B., and L.D. Holm. (1964). Chemical inhibition of crop germination by weed seed and the nature of the inhibition by *Abutilon theophrasti*. *Weed Research* 4: 44-53.

Kommedahl, T., J.B. Kotherimer, and J.V. Bernardini. (1959). The effects of quackgrass on germination and seedling development of certain crop plants. *Weeds* 7: 1-12.

Leather, G.R. 1983. Sunflowers (*Helianthus annuus*) are allelopathic to weeds. *Weed Science* 31: 37-42.

LeTourneau, D., G.D. Failes, and M.G. Heggeness. (1956). The effect of aqueous extracts of plant tissue on germination of seeds and growth of seedlings. *Weeds* 4: 363-368.

LeTourneau, D. and M.G. Heggeness. (1957). Germination and growth inhibitors in leafy spurge foliage and quackgrass rhizomes. *Weeds* 5: 12-19.

McCalla, T.M., and F.A. Haskins. (1964). Phytotoxic substances from soil microorganisms and crop residues. *Bacteriological Review* 28: 181-207.

Molisch, H. (1937). *Der Einfluss einer Pflanze auf die andere-Allelopathie*. Fischer, Jena.

Patrick, A.Q., T.A. Toussoun, and A. Snyder. (1963). Phytotoxic substances in arable soils associated with decomposition of plant residues. *Phytopathology* 53: 152-161.

Peters, E.J. (1968). Toxicity of tall fescue to rape and birdsfoot trefoil seed and seedlings. *Crop Science* 8: 650-653.

Peters, E.J. and K.T. Luu. (1984). Allelopathy in tall fescue. In *The Chemistry of Allelopathy*, ed. A.C. Thompson. Washington, DC: American Chemical Society, pp. 273-283.

Putnam, A.R. and L.A. Weston. (1986). Adverse impacts of allelopathy in agricultural system. In *The Science of Allelopathy*, eds. A.R. Putnam and C.S. Tang. New York: John Wiley and Sons.

Putnam, A.R. and W.B. Duke. (1974). Biological suppression of weeds: Evidence for allelopathy in accessions of cucumber. *Science* 185: 370-372.

Rice, E.L. (1984). *Allelopathy*. 2nd edition. Orlando, FL: Academic Press.

Rizvi, S.J.H. and V. Rizvi. (1992). *Allelopathy, Basic and Applied Aspects*. London: Chapman and Hall.

Shilling, D.G., R.A. Liebl, and A.D. Worsham. (1985). Rye (*Secale cereale* L.) and wheat (*Triticum aestivum* L.) mulch: The suppression of certain broadleaf weeds and the isolation and identification of phytotoxins. In *The Chemistry* of *Allelopathy*, eds. A.C. Thompson. Washington, DC: American Chemical Society, pp. 243-271.

Smith, R.J., Jr., W.T. Flinchum, and D.E. Seaman. (1977). *Weed Control in U.S. Rice Production. U.S. Department of Agriculture Handbook 457*. Washington, DC: U.S. Government Printing Office, pp. 78.

Smith, R.J., Jr. (1988). Weed thresholds in southern U.S. rice, *Oryza sativa*. *Weed Technology* 2: 232-241.

Whittaker, R.H. and P.P. Feeny. (1971). Allelochemicals: Chemical interactions between species. *Science* 171: 757-770.

Effects of Application of Alfalfa Pellet on Germination and Growth of Weeds

T. D. Xuan
E. Tsuzuki

SUMMARY. Allelopathy is well known as an interaction among plants (including microbes) through chemicals produced by plants. Recently, many investigations have evaluated the possibility of the use of allelopathic substances as a natural herbicide. Alfalfa (*Medicago sativa* L.) is reported to contain allelopathic chemicals that inhibit the growth of weeds. Experiments were conducted in Laboratory of Crop Science, Experimental Farm of Miyazaki University and Kobayashi city in Miyazaki Prefecture, Japan to determine the allelopathic impact of alfalfa pellet (commercial forage fodder) on germination and growth of lowland weeds, and its potential as a natural herbicide in paddy field. The results indicated that alfalfa pellet significantly inhibited germination and growth of 4 weed species, viz., *Echinochloa orygicola*, *Digitaria ciliaris*, *Cyperus difformis* and *Monocholia vaginalis* in rice paddies. Among the 4 tested weeds, the maximum inhibitory effect of alfalfa pellet was seen against *Cyperus difformis*. The degree of inhibition of weed growth by alfalfa pellet became stronger as the application of concentration increased. The results showed that alfalfa pellet might be used as a natural herbicide in rice paddies and might reduce the amount

T. D. Xuan is Graduate Student, and E. Tsuzuki is Professor, Crop Science and Horticulture Division, Faculty of Agriculture, Miyazaki University, Gakuen Kibanadai Nishi 1-1, Miyazaki City 889-21, Japan.

Address correspondence to: E. Tsuzuki at the above address (E-mail: a01102u@cc.miyazaki-u.ac.jp).

[Haworth co-indexing entry note]: "Effects of Application of Alfalfa Pellet on Germination and Growth of Weeds." Xuan, T. D., and E. Tsuzuki. Co-published simultaneously in *Journal of Crop Production* (Food Products Press, an imprint of The Haworth Press, Inc.) Vol. 4, No. 2 (#8), 2001, pp. 303-312; and: *Allelopathy in Agroecosystems* (ed: Ravinder K. Kohli, Harminder Pal Singh, and Daizy R. Batish) Food Products Press, an imprint of The Haworth Press, Inc., 2001, pp. 303-312. Single or multiple copies of this article are available for a fee from The Haworth Document Delivery Service [1-800-342-9678, 9:00 a.m. - 5:00 p.m. (EST). E-mail address: getinfo@haworthpressinc.com].

of herbicide used in paddy fields.

KEYWORDS. Alfalfa, alfalfa pellet, inhibition, *Medicago sativa*, *Echinochloa orygicola*, *Digitaria ciliaris*, *Cyperus difformis*, *Monocholia vaginalis*, weed control

INTRODUCTION

Alfalfa (*Medicago sativa* L.) has been grown and used in various regions of the world as forage crop. At present, many researchers have verified that alfalfa plants contain water-soluble allelopathic chemicals that are released into the soil environment from fresh leaf, stem, crown tissues, as well as dry hay, old roots and seeds (Nielesen, Cuddy, and Woods, 1960; McElgunn and Heinrichs, 1970; Klein and Miller, 1980; Hall and Henderlong, 1989). Moreover, the growth of alfalfa is also known to be retarded, and markedly damaged by continuous cropping (Webster, Khan, and Moore, 1967; Miller, 1983; Nakahisa, Tsuzuki, and Mitsumizo, 1993). Injury of alfalfa crop by continuous cropping has been studied by a number of investigators and it has been shown to be closely related to allelopathy (Dornbos, Spencer, and Miller, 1990; Ramesh and Miller, 1990; Miller 1983; Nakahisa, Tsuzuki, and Mitsumizo, 1993; Nakahisa et al., 1994; Chung, and Miller, 1995). Nakahisa, Tsuzuki, and Mitsumizo (1993) suggested that allelopathy plays a major role in injury by continuous cropping of alfalfa. Some researchers have evaluated the allelopathic potential of alfalfa as a natural herbicide (Abdul-Rahman and Habib, 1989; Waller, 1989; Dornbos, Spencer, and Miller, 1990; Wyman-Simpson et al., 1991). Decomposed alfalfa roots and their associated soil reduce seed germination and seedling growth of bladygrass (*Imperata cylindrica* [L.] Beauv.) (Abdul-Rahman and Habib, 1989). Saponins in pure alfalfa root inhibited the growth and development of cheatgrass (*Bromus secalinus* L.), barnyard grass (*Echinochloa crus-galli* [L.] Beauv.), pigweed (*Amaranthus* spp.), dandelion (*Taraxacum officinale* L.) and coffeeweed (*Sesbania exaltata* L.) (Waller, 1989). Tsuzuki et al. (1999) suggested that the powder obtained by drying the aerial parts (stem and leaf) of alfalfa significantly inhibited germination of some lowland weeds. The present study was designed to determine the degree of allelopathic effects of alfalfa pellet on the growth and development of weeds and rice plants in paddy fields and to test the possibility of the use of alfalfa pellet as a natural herbicide in rice paddies.

MATERIALS AND METHODS

Alfalfa pellet (commercial forage fodder in Japan) given by Agricultural Co-Operative Association of Japan was used. Experiments were carried out in Faculty of Agriculture, Miyazaki University and Kobayashi city in Miyazaki Prefecture, Japan.

Laboratory Experiments

Seeds of 4 weed species, viz., *Echinochloa orygicola*, *Digitaria ciliaris*, *Cyperus difformis* and *Monocholia vaginalis* were collected from the rice paddies of Miyazaki, Southern Japan, which received no herbicide treatment for the last two years, i.e., 1997 and 1998. Seeds were stored dry at 5°C (water content 12.2% on fresh weight basis). In 1999, before the experiments were carried out, collected seeds were randomly put to germination and it was found to be more than 65%. Powder of 0.14, 0.28, 0.43, and 0.57 g of alfalfa pellet was put in the 9 cm diameter Petri dishes with 15 ml of distilled water and filter paper, respectively. Twenty seeds each of *Echinochloa orygicola*, *Digitaria ciliaris* and *Cyperus difformis* were sown in the dishes, respectively. In addition, thirty seeds of another weed *Monocholia vaginalis* were tested at two concentrations of 0.28 and 0.57 g. The entire set-up was placed in a growth chamber set at 25°C and 4000 lux. After 10 days, number of seeds that germinated were counted, and plant height and root length of all the weeds except *Monocholia vaginalis* were measured. Treatment with distilled water served as the control. All the treatments were arranged in a completely randomized block design with three replications.

Greenhouse Experiment

On the 31st July 1998, paddy soil in the Experiment Farm of Miyazaki University, where early-mature rice (*Oryza sativa* L. var. *Koshihikari*) was cultivated, and previously received no herbicide treatment, was sampled and placed in Wagner pots (1/5000a; 1 a = 100 m^2) and saturated with water for 4 days. Then, the powder of 1.2 and 1.6 g of alfalfa pellet per pot was applied, respectively. After 20 days, number and dry weight of weed plants were observed. The pots without alfalfa pellet treatment served as the control. The experiment was carried out in greenhouse to prevent invading of any weeds into experimental pots. Treatments were arranged in a completely randomized design with two replications.

Field Experiments

Experiment 1

Experiment was conducted in the Experimental Farms of Miyazaki University during May-October in 1999. The experimental rice field was divided into many plots each having an area of 3 × 12 m^2 and the plots received no previous herbicide treatment or any special treatment for the soil. The experiment was conducted in 3 randomly selected plots of rice fields with applications of 100 kg/10 a of alfalfa pellet and 500 ml/10 a (1 a = 100 m^2) of herbicide (*Shizetto Proaburu*, Sankyou Co., Japan), respectively. A non-weed-management was used as the control. *Hinohikari*, a leading rice variety in Japan, was cultivated by conventional methods including water and fertilizer management except weed management. Alfalfa pellet (100 kg/10 ha) was applied in the paddy field at 3 days after the 30 day-old seedlings were transplanted at a density of 20 × 30 cm^2 in the experimental plots. Thirty days after application, number of plants and dry weight of weeds in a randomly sampled block (50 cm^2) in each experimental plot were measured with three replications. Number of tillers and average height of rice plants was investigated with three replications, respectively.

Experiment 2

Experiment was carried out in a paddy field of a farmer in Kobayashi city in Miyazaki Prefecture during May-October in 1999. Alfalfa pellet (160 kg/10 a) was applied at 5 days after seedlings were transplanted. Non-weed treatment was used as the control. Thirty days after alfalfa pellet was applied, measurements were conducted by the same methods as in Experiment 1.

Statistical Analysis

Analysis of variance for the data of the experiments was carried out in the general linear model procedure of the Statistical Analysis System (SAS) program (SAS, 1982). The means were separated on the basis of least significant difference (LSD) at 0.05 and 0.01 probability level.

RESULTS AND DISCUSSION

Laboratory Experiments

It was observed that the germination of all the 4 weed species was reduced in response to the different concentrations of alfalfa pellet (Table 1). The degree of inhibition become stronger as the applied concentration increased. Among the 4 weed species tried, maximum inhibitory effect was observed in case of *Cyperus difformis* where none of the seed could germinate at 0.43 g or higher concentration of alfalfa pellet. On the contrary, very little effect was seen in case of *Echinochloa orygicola*, where about 70% seeds could germinate at the highest concentration (Table 1). Not only the germination but the plant height and root length were also drastically reduced in response to alfalfa pellet. At 0.43 and 0.57 g concentration, there was a significant reduction in the growth of all the three weed species (Table 2). Maximum inhibitory effect of alfalfa pellet was observed in case of *Cyperus difformis* followed by *Digitaria ciliaris* and *Echinochloa orygicola* (Table 2).

Greenhouse Experiments

Application of alfalfa pellet drastically reduced the number and the dry weight of weeds in paddy soil (Table 3). The effect was more pronounced in

TABLE 1. Effects of alfalfa pellet on germination of 4 weed species.

Concentrations (g)	*Echinochloa orygicola*	*Digitaria ciliaris*	*Cyperus difformis*	*Monocholia vaginalis*
Control	100 (100)	75 (100)	65 (100)	81 (100)
0.14	86 (86)	55 (73)	50 (77)	
0.28	86 (86)	20 (27)	50 (77)	37 (46)
0.43	70 (70)	10 (13)	0 (0)	
0.57	70 (70)	5 (7)	0 (0)	14 (17)

Values in parentheses are with respect to control. All values are significant with respect to control.

TABLE 2. Effects of different concentrations of alfalfa pellet on growth of test weeds.

Concentration (g)	*E. orygicola*		*D. ciliaris*		*C. difformis*	
	Plant height	Root length	Plant height	Root length	Plant height	Root length
Control	22.0 (100)	26.0 (100)	6.1 (100)	10.3 (100)	2.2 (100)	2.5 (100)
0.14	24.0 (109)	4.1 (16)**	3.7 (61)*	1.1 (11)**	0 (0)	0.9 (36)**
0.28	16.0 (73)	0.7 (3)**	1.4 (23)**	0.3 (3)**	0 (0)	1.0 (40)**
0.43	11.0 (50)**	0.4 (2)**	0.5 (8)**	0.1 (1)**	0 (0)	0 (0)
0.57	8.2 (37)**	0.3 (1)**	0.2 (3)**	0.02 (0.2)**	0 (0)	0 (0)

Values in parentheses are with respect to control.
*, ** Represent significant difference from the control at $P < 0.05$ and $P < 0.01$, respectively.

response to 1.6 application of alfalfa pellet where the number of weed plants and their dry weight was reduced to nearly 7 and 10% of that in control (Table 3). It showed that alfalfa pellet has the markedly inhibitory effect on the growth of weeds (Figures 1 and 2). In the control pots, *Echinochloa orygicola* was prominent whereas in response to treatment of 1.2 g alfalfa pellet most of the weeds did not appear except a small number of *Monocholia vaginalis*. However, in response to the application of 1.6 g alfalfa pellet none of the weed appeared except a small number of *Echinochloa orygicola*, but the growth of weed plants was significantly reduced (Figures 1 and 2).

TABLE 3. Effect of alfalfa pellet on growth of weeds in paddy soil.

	Amount of alfalfa pellet per pot		
	0 (control)	1.2 g	1.6 g
Number of weed plants (per pot)	224 (100)	35 (16)**	16 (7)**
Dry weight (g) (per pot)	0.62 (100)	0.08 (13)**	0.06 (10)**

Values in parentheses are with respect to control.
** Represent significant difference from the control at $P < 0.01$.

FIGURE 1. Number of weeds of paddy soil in Wagner pots at 20 days after the application of alfalfa pellet with different concentrations. From left to right: Weeds of 1.6 g application of alfalfa pellet, weeds of the control (no application of alfalfa pellet), and weeds of 1.2 g application of alfalfa pellet.

FIGURE 2. Weeds of paddy soil in Wagner pots at 20 days after the application of alfalfa pellet with different concentrations. From left to right: Weeds of 1.6 g application of alfalfa pellet, weeds of the control (no application of alfalfa pellet), the weeds of 1.2 g application of alfalfa pellet.

Field Experiments

Application of alfalfa pellet in the paddy fields of agricultural farms in Miyazaki University reduced the number of weed plants and its dry weight to 39% and 32% as compared with control, respectively (Table 4). The result showed that alfalfa pellet markedly inhibited emergence and growth of weeds. The effect on the dry weight of weeds was almost similar to that of herbicide application (Table 4). On the other hand, the application of alfalfa pellet has no effect on the growth of rice (Table 4). Likewise, the number of weed plants and their dry weight was reduced by application of alfalfa pellet in farmer's field in Kobayashi, Miyazaki, Japan (Table 5). However, there were no differences in plant height and number of tillers of rice plants between alfalfa application and the control (Table 5). The results showed that alfalfa pellet has a markedly inhibitory effects on emergence and growth of weeds in paddy field, agreeing with those in Table 4. It is known that alfalfa contains materials that inhibit the growth of plants in the top and root parts (Lawerence and Kilcher, 1962; Guenzi, Kehr, and McCalla, 1964; Marchaim et al., 1975), and it is heterotoxic as well as autotoxic (Miller, 1983). Oleszek and Jurzysta (1987) isolated saponins from alfalfa roots which were later found to have allelopathic effect against a number of plants in their pure form (Oleszek, 1999). Abdul-Rahman

TABLE 4. Effects of alfalfa pellet application on growth of weeds in paddy field of agricultural farm in Miyazaki University.

Treatment	Weeds		Rice	
	Number of plants	Dry Weight (g)	Plant height (cm)	Number of tillers
Control	24 (100)	1.80 (100)	38.3 (100)	6.1 (100)
Alfalfa	9.3 (39)*	0.58 (32)*	34.5 (90)	6.2 (102)
Herbicide	5.3 (22)**	0.61 (34)*	33.8 (88)	5.5 (90)

Values in parentheses are with respect to control.
** , * Represent significant difference from the control at $P < 0.5$ and $P < 0.1$, respectively.

TABLE 5. Effects of alfalfa pellet application on growth of weeds in paddy field of a farmer in Kobayashi in Miyazaki prefecture.

Treatment	Weeds		Rice	
	Number of plants	Dry Weight (g)	Plant height (cm)	Number of tillers
Control	5.3 (100)	0.07 (100)	42.5 (100)	34.6 (100)
Alfalfa	2.3 (43)*	0.01 (14)**	41.5 (98)	34.6 (100)

Values in parentheses are with respect to control.
**, * Represent significant difference from the control at $P < 0.5$ and $P < 0.1$, respectively.

and Habib (1989) reported the allelopathic effects of decomposed roots and associated soil of alfalfa germination of weed (*Imperata cylindrica*) and identified chemicals such as saponins, caffeic, chlorogenic, isochlorogenic, *p*-coumaric, and ferulic acid from root exudates and residues. Dornbos, Spencer, and Miller (1990) identified medicarpin, 4-methoxymedicarpin, sativan, and 5-methoxysativan, and Nakahisa et al. (1994) identified ferulic and salicylic acids as growth-inhibiting substances in the body of alfalfa. Wyman-Sympson et al. (1991) isolated saponins from the roots of six alfalfa cultivars and tested their biological activity. In the present study also, it is presumed that allelochemicals particularly the saponins are released from the alfalfa pellet and these are responsible for the observed weed-suppressing ability.

CONCLUSIONS

From the present study, it could be concluded that alfalfa pellet has a markedly inhibitory effect on the growth of the weeds in the paddy fields and could be used as a natural herbicide in rice paddies. However, the effect varied with the applied concentration and the type of weed species.

REFERENCES

Abdul-Rahman, A.A and S.A. Habib. (1989). Allelopathic effect of alfalfa (*Medicago sativa*) on bladygrass (*Imperata cylindrica*). *Journal of Chemical Ecology* 15: 2289-2300.

Chung, I.M. and D.A. Miller. (1995). Natural herbicide potential of alfalfa residue on selected weed seed species. *Agronomy Journal* 87: 920-925.

Dornbos, D.L., G.F. Spencer, Jr., and R.W. Miller. (1990). Medicarpin delays alfalfa seed germination and seedling growth. *Crop Science* 30: 162-166.

Guenzi, W.D., W.R. Kehr, and T.M. McCalla. (1964). Water-soluble phytotoxic substances in alfalfa forage; variation, with variety, cutting, year and stage of growth. *Agronomy Journal* 56: 499-500.

Hall, M.H. and P.R. Henderlong. (1989). Alfalfa autotoxic fraction characterization and initial separation. *Crop Science* 29: 425-528.

Klein, R.R. and D.A. Miller. (1980). Allelopathy and its role in agriculture. *Communications in Soil Sciences and Plant Analysis* 11: 43-56.

Lawerence, T. and M.R. Kilcher. (1961). The effect of fourteen root extracts upon germination and seedling length of fifteen plant species. *Canadian Journal of Plant Science* 42: 308-313..

Marchaim, E., E. Birk, A. Dorvat, and T. Berman (1975). Kinetics of the inhibition of cotton seed germination by lucerne saponins. *Plant and Cell Physiology* 16: 857-864.

McElgunn, J.D. and D.H. Heinrichs. (1970). Effects of root temperature and a suspected phytotoxic substance on the growth of alfalfa. *Canadian Journal of Plant Science* 50: 307-311.

Miller, D.A. (1983). Allelopathic effects of alfalfa. *Journal of Chemical Ecology* 9: 1059-1072.

Nakahisa, K., E. Tsuzuki, and T. Mitsumizo. (1993). Study on the allelopathy of alfalfa (*Medicago sativa* L.). 1. Observation of allelopathy and survey for substances inducing growth inhibition. *Japanese Journal of Crop Science* 62: 294-299.

Nakahisa, K., E. Tsuzuki, H. Terao, and S. Kosemura. (1994). Study on the allelopathy of alfalfa (*Medicago sativa* L.). 2. Isolation and identification of allelopathic substance in alfalfa. *Japanese Journal of Crop Science* 63: 278-284.

Nielsen, K.F., T. Cuddy, and W. Woods. (1960). The influence of the extracts of some crops and soil residues on germination and growth. *Canadian Journal of Plant Science* 40: 188-197.

Oleszek, W. (1999). Allelopathic significance of plant saponins. In *Recent Advances in Allelopathy. Volume I. A Science for the Future*, eds. F.A. Macias, J.C.G. Galindo, J.M.G. Molinillo, and H.G. Cutler. Cádiz, Spain: Servicio de Publicaciones de la Universidad de Cádiz, pp. 167-178.

Oleszek, W. and M. Jurzysta. (1987). Allelopathic potential of alfalfa root medicagenic acid glycosides and their fate in soil environments. *Plant and Soil* 98: 67-80.

Ramesh. S.H. and D.A. Miller. (1990). Allelopathy and autotoxicity in alfalfa: characterization and effects of preceding crops and residue incorporation. *Crop Science* 30: 1255-1259.

SAS. (1982). *SAS User's Guide*. Statistical Analysis System Institute.

Tsuzuki. E., M. Miura, N. Sakaki, and T. Yoshino. (1999). Study on the control of weeds by using higher plants. *Report of Kyushu Branch of the Crop Science Society of Japan* 65: 39-40.

Waller, G.R. (1989). Allelochemical action of some natural products. In *Phytochemical Ecology: Allelochemicals, Mycotoxins, Insect Pheromones and Allomones*, eds. C.H. Chou and G.R. Waller. Academia Sinica Monograph 9. Taipei, Taiwan, ROC: Academia Sinica, pp. 129-154.

Webster, G.R., S.U. Khan, and A.W. Moore. (1967). Poor growth of alfalfa (*Medicago sativa* L.) on some Alberta soils. *Agronomy Journal* 59: 37-42.

Wyman-Simpson, C.L., G.R. Waller, M. Jurzysta, J.K. McPherson, and C.C. Young. (1991). Biological activity and chemical isolation of root saponins of six cultivars of alfalfa (*Medicago Sativa* L.). *Plant and Soil* 135: 83-94.

Allelopathic Effects of Some Volatile Substances from the Tomato Plant

Young Sik Kim
Bong-Seop Kil

SUMMARY. The allelopathic effects of chemical substances released from the tomato plant (*Lycopersicon esculentum* Mill.) were investigated on various receptor plants. Bioassays under controlled conditions in a growth chamber showed that germination and seedling growth of *Lactuca sativa* and growth of *Vitis* were reduced when grown near tomato plants. It was also observed that when plants were grown in a U-tube pot with tomato plant, the tomato roots excreted allelochemicals that inhibit growth of tested species. To identify the phytotoxic compounds from tomato plant, a GC/MS method was employed. Forty compounds, such as trans-2-hexenal, α-terpineol, linalool, phenylacetaldehyde, methylsalicylic acid and tetradecanoic acid were identified from the essential oil of tomato plants. The volatile substances from leaves inhibited seed germination and seedling growth of receptor plants in laboratory tests. Inhibitory response varied with the concentration of compounds. These findings suggest that the tomato plant may have an interesting allelopathic potential. *[Article copies available for a fee from The Haworth Document Delivery Service: 1-800-342-9678. E-mail address: <getinfo@haworthpressinc.com> Website: <http://www.HaworthPress.com> © 2001 by The Haworth Press, Inc. All rights reserved.]*

Young Sik Kim is Professor, Wonkwang Health College, Iksan, 570-749, Republic of Korea.

Bong-Seop Kil is Professor, Department of Life Science, Wonkwang University, Iksan, 570-749, Republic of Korea.

Address correspondence to: Bong-Seop Kil at the above address (E-mail: bskil@wonnms.wonkwang.ac.kr).

[Haworth co-indexing entry note]: "Allelopathic Effects of Some Volatile Substances from the Tomato Plant." Kim, Young Sik, and Bong-Seop Kil. Co-published simultaneously in *Journal of Crop Production* (Food Products Press, an imprint of The Haworth Press, Inc.) Vol. 4, No. 2 (#8), 2001, pp. 313-321; and: *Allelopathy in Agroecosystems* (ed: Ravinder K. Kohli, Harminder Pal Singh, and Daizy R. Batish) Food Products Press, an imprint of The Haworth Press, Inc., 2001, pp. 313-321. Single or multiple copies of this article are available for a fee from The Haworth Document Delivery Service [1-800-342-9678, 9:00 a.m. - 5:00 p.m. (EST). E-mail address: getinfo@haworthpressinc.com].

© 2001 Taylor & Francis.

KEYWORDS. *Lycopersicon esculentum*, allelopathic effect, aqueous extract, essential oil, germination, seedling growth

INTRODUCTION

Allelopathy is an important mechanism of plant interference mediated by the addition of plant-produced phytotoxins to the plant environment and competitive strategy of plants (Muller, 1969; Chou and Lin, 1976; Rice, 1984; Fischer et al., 1994; Langenheim, 1994). Allelochemicals are released from plant tissue in a variety of ways including emission of volatile substances from living plant parts, exudation from roots, or leaching from above ground parts by rain, dew, fog, etc. (Rice, 1984). Many researchers have found that inhibitory substances involved in allelopathy are terpenoids, and phenolic substances (Muller, 1965; Kil and Yim, 1983; Weidenhamer et al., 1994; Seigler, 1996).

Tomato plants are known to exhibit phytotoxic effects (Kim and Kil, 1987). Aqueous extracts of leaves, stem and roots of tomato inhibited the germination and seedling growth of test species and the phytotoxicity of the extracts increased with increasing time of extraction, i.e., 24 h $<$ 48 h $<$ 72 h (Kim and Kil, 1987). Bate-Smith (1972) reported the presence of some phenolics in the leaves of tomato plants. Tomato plants are aromatic in nature and spread a strong smell into the surrounding environment. Probably, this is due to the presence of some volatile terpenoids which may be responsible for the observed inhibitory effects of tomato plants. Keeping this in mind, the present study was undertaken to (a) verify the presence of volatile inhibitors in the tomato plants, and (b) identify the chemical substances responsible for the inhibitory effects.

MATERIALS AND METHODS

Plant Material

For the present study tomato plants (*Lycopersicon esculentum* Mill. cv. Seogwang) were chosen as the donor plant. The receptor species selected were: *Lactuca sativa* L. for seed germination and seedling growth test; *Vitis vinifera* L. for growth test near donor species; *Brassica campestris* var. *pekinensis* Makino, *Sorghum bicolor* Moench, *Perilla frutescens* var. *japonica* Hara and *Zea mays* L. for pair test; *Bidens bipinnata* (Lour.) Merr. et Scherff., *Plantago asiatica* L., *Lactuca sativa* L. and *Eragrostis ferruginea* (Thunb.) P. Beauv. for relative seedling elongation test in association with tomato leaves; and *Amaranthus mangostanus* L. for volatile substance bioassay, respectively.

Germination and Growth Studies

Germination and seedling elongation of *L. sativa* were tested near tomato plants. Fifty seeds of *L. sativa* were placed in a 12 cm diameter Petri dish lined with a moistened layer of filter paper. Petridishes were then put into a growth chamber with or without tomato plants and maintained at 20°C for 16 h in light and 15°C for 8 h in dark, respectively. The experiment was extended over a period of two weeks to allow maximum seed germination and growth. Three replicates were maintained for each treatment. After 2 weeks number of germinated seeds, length and dry weight of seedlings were measured.

To find out the effects of volatile substances released by tomato plant, ten tomato plants planted in a pot and ten grapevines in another pot were taken and put apart about 10 cm in a growth chamber. Ten grapevines planted in a pot and placed in a growth chamber without tomato plants served as control. After six months, grapevines were harvested and their growth was measured.

Volatile Substances Test

To study the effect of volatile substances from tomato plants, a 1.8 liter glass chamber (height 10.5 cm and diameter 14 cm) was used (Yun, Kil, and Han, 1993). Fifty seeds each of *Bidens bipinnata*, *Plantago asiatica*, *Lactuca sativa*, and *Eragrostis ferruginea* were placed on a filter paper overlining a thin moistened cotton wad in the glass chamber. Fresh leaves of tomato plants in different quantities (5, 10, 15, 20, 25, and 30 g) were placed in a glass beaker embraced within every glass chamber. Growth studies in chambers without sliced tomato leaves served as control. All the chambers were covered with vinyl wrap and placed at room temperature (~25°C). Four replicates were maintained for each treatment. After 4 days, seedling elongation was measured and presented as with respect to control. The data were analyzed using student's *t*-test.

Cultivation in U-Tube Pot

U-shaped plastic tubes were used for studying the effect of root exudate from paired-plants. For this, *Lycopersicon esculentum-Brassica campestris* var. *pekinensis* pair, *Lycopersicon esculentum-Sorghum bicolor* pair, *Lycopersicon esculentum-Perilla frutescens* var. *japonica* pair, *Lycopersicon esculentum-Zea mays* pair were planted in the ends of U-tube, respectively. These paired plants were cultivated with supplemented 20 ml of Hoagland's solution per week for two months. Four replicates were maintained for each set. A growth interaction index (GII) of seedling growth and dry weight was calculated as follows:

$$GII = \frac{\text{Growth Superior} - \text{Growth Inferior}}{\text{Growth Superior}} \times 100$$

Extraction and Analysis of Essential Oils

Essential oils from the tomato plant were extracted using Karlsruker's apparatus (Stahl, 1973). These were analyzed by a Gas Chromatograph (Hewlett Packard 5890A capillary) as per Kil and Yim (1983). The column temperature was programmed from 100 to 300°C at a rate of 4°C/min. A split ratio of 1:50 was used and 1 µl sample injection was made with a 5 ml SGE syringe employing the solvent flush technique. Identification of chromatographic peaks was made by comparison of their retention times with those of chemical standards (Sigma and Aldrich Co.) and mass spectra analysis.

Bioassay with Volatile Substances

Four volatile chemicals, viz., α-terpineol, linalool, eugenol and tetradecanoic acid were tested on the germination of *Amaranthus mangostanus* under the same set of conditions as for the volatile substance test. The volatile chemicals were applied at concentrations of 10, 20, 30, 40 and 50 ppm. After 72 h the percent germination was measured.

RESULTS

The germination and percent elongation of *L. sativa* grown near tomato plants were almost similar to the control. However, dry weight of *L. sativa* was suppressed by 34.3%. (Table 1). Growth of roots and twigs, and stem after harvest of grapevines planted near the tomato plants was comparatively inhibited (Table 2). The fresh and dry weight of grapevines planted near the tomato plants were 56.5 and 61.2% of the control, respectively.

TABLE 1. Germination and seeding growth of *Lactuca sativa* tested in growth chambers in association with *Lycopersicon esculentum* plants.

	Germination (%)	Elongation S + R (cm)	Fresh Weight (g)	Dry weight (mg)
Control	88.00	20.03	08.97	0.70
Treatment	87.00	18.55	07.64	0.24*
Percent of control	0.988	0.926	0.851	0.343

Control: *L. sativa*; Treatment: *L. sativa* pot near *Lycopersicon esculentum*.
S and R represent shoot and root of seedling.
* Means statistically significant at $p < 0.05$.

In general, relative elongation of receptor plants treated with volatile substances of tomato plant was reduced. Seedling elongation of *Plantago asiatica* and *Bidens bipinnata* were significantly reduced whereas *Eragrostis ferruginea* and *Lactuca sativa* did not respond much, even at higher concentration of tomato leaves (Table 3). Little differences in growth were observed when similar plants were grown in U-tube pots. However, the differences in growth were apparent when one of the plants was tomato, e.g., tomato-corn, tomato-sorghum, etc. (Table 4). This clearly depicts that tomato roots release some chemical substances which were inhibitory to the plants growing at the other side of the U-tube pot (Table 4).

Gas chromatography of the essential oils extracted from tomato plants exhibits the presence of 40 components of diverse chemical nature such as alkane, alcohol, aldehyde, ester, carboxylic acid, phenol, terpenoid, ketone, epoxide and tricosine. The major constituents of leaf essential oil were *trans*-2-hexenal, *cis*-hexen-1-ol, eugenol and hexadecanoic acid (Figure 1). Based on this authentic samples of α-terpineol, linalool, eugenol and tetrade-

TABLE 2. Growth of *Vitis vinifera* planted near the *Lycopersicon esculentum* plants in growth chamber.

	Root (cm)	Stem (cm)	Twig (cm)	Fresh weight of stem before planted (g)	Stem after harvest (g) Fresh weight	Dry weight
Control	39.90	21.04	103.44	16.75	118.46	30.02
Treatment	24.30	18.70	78.17	20.84	66.95	18.37
Percent of control	0.609*	0.888	0.756	1.244	0.565*	0.612*

Control: *V. vinifera*; Treatment: *V. vinifera* pot near *Lycopersicon esculentum*.
* Means statistically significant at $p < 0.05$.

TABLE 3. Relative seedling elongation of receptor plants grown in growth chamber with different quantities of tomato leaves.

Species	Amount of tomato leaves (g)					
	5	10	15	20	25	30
Bidens bipinnata	85	69	58	59	48*	46*
Plantago asiatica	96	87	73	50*	49*	49*
Lactuca sativa	77	78	80	74	73	73
Eragrostis ferruginea	84	82	83	75	77	75

* Represents significantly different from control at $p < 0.05$.

TABLE 4. Comparison of the growth of paired-plants grown in U-tube pots.

Pair	Length (m)	GII of length	Dry weight (g)	GII of weight
Tomato-*Brassica* pair				
+–*Lycopersicon esculentum*	0.52	0.6	10	0.9
+–*Brassica campestris* var. *pekinensis*	0.18		1	
+–*Brassica campestris* var. *pekinensis*	0.21	0.1	0.7	0.2
+–*Brassica campestris* var. *pekinensis*	0.25		0.5	
Tomato-*Sorghum* pair				
+–*Lycopersicon esculentum*	0.45	0.2	10	0.9
+–*Sorghum bicolor*	0.36		1	
+–*Sorghum bicolor*	0.30	0.1	0.8	0.2
+–*Sorghum bicolor*	0.35		1	
Tomato-*Perilla* pair				
+–*Lycopersicon esculentum*	0.39	0.4	7	0.9
+–*Perilla frutescens* var. *japonica*	0.22		0.7	
+–*Perilla frutescens* var. *japonica*	0.22	0.1	0.4	0
+–*Perilla frutescens* var. *japonica*	0.19		0.4	
Tomato-*Zea* pair				
+–*Lycopersicon esculentum*	0.41	0.2	7	0.8
+–*Zea mays*	0.30		1	
+–*Zea mays*	0.40	0.1	1	0.1
+–*Zea mays*	0.45		0.9	

canoic acid were tested using *Amaranthus mangostanus* as a bioassay plant. Seed germination of *A. mangostanus* was reduced at all the tested concentration of pure chemicals and the inhibitory effect increased with increasing concentration of oils (Table 5).

DISCUSSION

The results clearly show that tomato plants release volatile substances that act as growth inhibitors. The germination, growth, and dry weight accumulation of the plants grown in vicinity of tomato was reduced. However, the degree of toxicity was dependent on the receptor plant and concentration.

FIGURE 1. Gas chromatograms of essential oil of tomato leaf.

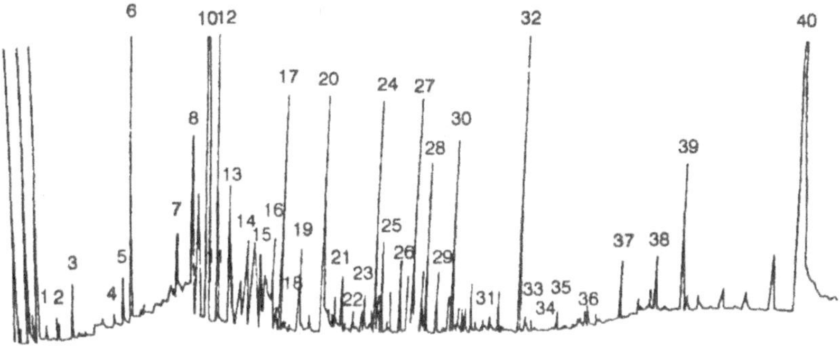

Keys to numbers: 1. Methylbenzene; 2. Decane; 3. Hexanal; 4. Dodecane; 5. *iso*-amyl alcohol; 6. *trans*-2-Hexenal; 7. 1-hexen-3-ol; 8. *n*-Hexanol; 9. *cis*-3-Hexenyl acetate; 10. *cis*-Hexen-1-ol; 11. Tetradecane; 12. *trans*-2-Hexen-1-ol; 13. Linalooloxide; 14. Acetic acid; 15. Pentadecane; 16. Benzaldehyde; 17. Linalool; 18. Hexadecane; 19. β-Caryophyllene; 20. Phenylacetaldehyde; 21. α-Terpineol; 22. Methylbenzoate; 23. Ethylphenylacetate; 24. Methylsalicylate; 25. Geraniol; 26. Guaiacol; 27. Benzylalcohol; 28. 2-Phenylethylalcohol; 29. β-Caryophyllene epoxide; 31. Heneicosane; 32. Eugenol 33. Thymol; 34. Docosane; 35. Tricosine; 36. Tetracosane; 37. Pentacosane; 38. Hexacosane; 39. Tetradecanoic acid; 40. Hexadecanoic acid. Components are listed only if their presence was confirmed by mass spectral analysis.

TABLE 5. Effect of α-terpineol, linalool, eugenol and tetradecanoic acid on germination of *Amaranthus mongostanus*.

Compound	Concentration (ppm)					
	Control	10	20	30	40	50
α-Terpineol	59ab	44ab	43ab	42b	41b	40b
Linalool	60a	40b	42b	30c	28c	2c
Eugenol	80a	51ab	47bc	40bc	40bc	23c
Tetradecanoic acid	83a	80ab	60bc	60bc	60bc	50c

Values followed by the different letters in a row are significantly different at $p < 0.05$.

These findings are in agreement with those of Halligan (1976), Heisey and Delwiche (1983), Oleszek (1987) and Yun (1990). Kim and Kil (1987) reported that aqueous extracts from tomato leaves, roots, and plants are highly toxic and later on they identified five phenolic acids, viz., tannic acid, hydroquinone, *p*-hydroxybenzoic acid, vanillic acid and ferulic acid from tomato leaves by HPLC (Kim and Kil, 1989).

Phenolic compounds are a main inhibitor of germination and seedling growth (Olmsted and Rice, 1970; Whittaker and Feeny, 1971; Tinnin and Muller, 1972; Horsley, 1977; Carballeira, 1980). In the present study essential oils from the tomato leaves were found to contain forty components. Among them *trans*-2-hexenal, *cis*-hexen-1-ol, *trans*-2-hexen-1-ol, eugenol and hexadecanoic acid were the major constituents. These were found to be very toxic against *A. mangostanus*. Such an observation is not new, as a number of reports are available in literature which point to the presence of active volatile components in the plants responsible for their phytotoxic/inhibitory effects (Muller, 1965; Weaver and Klarich, 1977; Kil, 1992; Weidenhamer et al., 1994). As regards the mode of action of these volatile allelochemicals very few studies are available. Muller (1965) observed that terpenes reduce cell division and elongation leading to reduced growth of cucumber radicle and hypocotyl. Kohli, Batish, and Singh (1998) reported that volatile oils from *Eucalyptus* spp. reduced the chlorophyll content and respiratory activity in *Parthenium hysterophorus*.

It is, therefore, concluded that tomato plants exhibit allelopathy due to the presence of volatile chemical substances.

REFERENCES

Bate-Smith, E.C. (1972). Attractants and repellents in higher animals. In *Phytochemical Ecology*, ed. J.B. Harborne. New York: Academic Press, pp. 45-56.

Carballeira, A. (1980). Phenolic inhibitors in *Erica australis* L. and the associated soil. *Journal of Chemical Ecology* 6: 593-596.

Chou, C.H. and H.J. Lin. (1976). Autointoxication mechanism of *Oryza sativa*. I. Phytotoxic effects of rice residues in soil. *Journal of Chemical Ecology* 2: 353-367.

Fischer, N.H., G.B. Williamson, J.D. Weidenhamer, and D.R. Richardson. (1994). In search of allelopathy in the Florida scrub: the role of terpenoids. *Journal of Chemical Ecology* 20: 1355-1380.

Halligan, J.P. (1976). Toxicity of *Artemisia californica* to four associated herb species. *American Midland Naturalist* 95: 406-421.

Heisey, R.M. and C.C. Delwiche. (1983). A survey of California plants for water-extractable and volatile inhibitors. *Botanical Gazette* 144: 382-390.

Horsley, S.B. (1977). Allelopathic interference among plants. II. Physiological modes of action. In *Proceedings of the Fourth North American Forest Biology Workshop*, eds. H.E. Wilcox and A.F. Hamer. Syracuse, NY: State University of New York, pp. 93-136.

Kil, B.S. and Y.J. Yim. (1983). Allelopathic effects of *Pinus densiflora* on undergrowth of red pine forest. *Journal of Chemical Ecology* 9: 1135-1151.

Kil, B.S. (1992). Effect of pine allelochemicals on selected species in Korea. In *Allelopathy: Basic and Applied Aspects*, eds. S.J.H. Rizvi and V. Rizvi. London: Chapman and Hall, pp. 205-242.

Kim, Y.S. and B.S. Kil. (1987). A bioassay on susceptivity of selected species to phytotoxic substances from tomato plants. *Korean Journal of Botany* 30: 59-67.

Kim, Y.S. and B.S. Kil. (1989). Identification and growth inhibition of phytotoxic substances from tomato plants. *Korean Journal of Botany* 32: 41-49.

Kohli, R.K., D.R. Batish, and H.P. Singh. (1998). Eucalypt oils for the control of parthenium (*Parthenium hysterophorus* L.). *Crop Protection* 17: 119-122.

Langenheim, J.H. (1994). Higher plant terpenoids: a phytocentric overview of their ecological roles. *Journal of Chemical Ecology* 20: 1223-1280.

Muller, W.H. (1965). Volatile materials produced by *Salvia leucophylla*: effect on seedling growth and soil bacteria. *Botanical Gazette* 126: 195-200.

Muller, C.H. (1969). Allelopathy as a factor in ecological process. *Vegetatio* 18: 348-357.

Olmsted, C.E. and E.L. Rice. (1970). Relative effects of known plant inhibitors on species from first two stages of old field succession. *Southwestern Naturalist* 15: 165-173.

Oleszek, W. (1987). Allelopathic effects of volatiles from some *Cruciferae* species on lettuce, barnyard grass and wheat growth. *Plant and Soil* 102: 271-274.

Rice, E.L. (1984). *Allelopathy*. 2nd edition. NY: Academic Press.

Seigler, D.S. (1996). Chemistry and mechanism of allelopathic interactions. *Agronomy Journal* 88: 876-885.

Stahl, E. (1973). *Thin-Layer Chromatography*. 2nd ed. London: George Allen and Unwin Ltd.

Tinnin, R.O. and C.H. Muller. (1972). The allelopathic influence of *Avena fatua*: The allelopathic mechanism. *Bulletin of the Torrey Botanical Club* 99: 287-292.

Weaver, R.J. and D. Klarich. (1977). Allelopathic effects of volatile substances from *Artemisia tridentata* Nutt. *The American Midland Naturalist* 97: 508-512.

Weidenhamer, J.D., M. Menelaou, F.A. Macias, N.H. Fischer, D.R. Richardson, and G.B. Williamson. (1994). Allelopathic potential of menthofuran monoterpens from *Calamintha ashei*. *Journal of Chemical Ecology* 20: 3345-3359.

Whittaker, R.H. and P.P. Feeny. (1971). Allelochemics: chemical interaction between species. *Science* 171: 757-770.

Yun, K.W. (1990). Allelopathic effects of chemical substances in *Artemisia princeps* var. *orientalis* on selected species. *Doctoral dissertation. Department of Biology, Wonkwang University, Iri, Korea.*

Yun, K.W., B.S. Kil, and D.M. Han. (1993). Phytotoxic and antimicrobial activity of volatile constituents of *Artemisia princeps* var. *orientalis*. *Journal of Chemical Ecology* 19: 2757-2766.

Growth Stage of *Cyperus esculentus* Influences Its Allelopathic Effect on Ectomycorrhizal and Higher Plant Species

C. F. Reinhardt
S. R. Bezuidenhout

SUMMARY. Live plants and plant material in various forms have important roles in ecological processes. Weeds may be involved in interactions with desirable species at different trophic levels through the production and release of allelochemicals. The allelopathic potential of the weed *Cyperus esculentus* (yellow nutsedge) was assessed by preparing aqueous extracts from tubers and foliage from immature or mature plants, and testing their effects on an ectomycorrhiza and two higher plant species. The extracts from tubers and both foliage types significantly reduced the colony diameter of the ectomycorrhiza *Boletus maxaria*, which was isolated from roots of the forestry species *Pinus patula*, on agar medium. Extracts (2% m/v) from *C. esculentus* tubers and the foliage of immature plants inhibited the germination of lettuce (*Lactuca sativa*) seeds significantly. At 5% concentration, the extracts from foliage of immature weed plants completely inhibited the germination of lettuce seeds at the end of the 7-day trial period. The

C. F. Reinhardt is Professor of Weed Science, and S. R. Bezuidenhout is Postgraduate student (Weed Science), Department of Plant Production and Soil Science, University of Pretoria, Pretoria 0002, Republic of South Africa.

Address correspondence to: C. F. Reinhardt at the above address (E-mail: creinhard@postino.up.ac.za).

[Haworth co-indexing entry note]: "Growth Stage of *Cyperus esculentus* Influences Its Allelopathic Effect on Ectomycorrhizal and Higher Plant Species." Reinhardt, C. F., and S. R. Bezuidenhout. Co-published simultaneously in *Journal of Crop Production* (Food Products Press, an imprint of The Haworth Press, Inc.) Vol. 4, No. 2 (#8), 2001, pp. 323-333; and: *Allelopathy in Agroecosystems* (ed: Ravinder K. Kohli, Harminder Pal Singh, and Daizy R. Batish) Food Products Press, an imprint of The Haworth Press, Inc., 2001, pp. 323-333. Single or multiple copies of this article are available for a fee from The Haworth Document Delivery Service [1-800-342-9678, 9:00 a.m. - 5:00 p.m. (EST). E-mail address: getinfo@haworthpressinc.com].

emergence of corn (*Zea mays*) was retarded in soil where *C. esculentus* tubers were planted 28 days before planting the crop, irrespective of whether the weeds continued to grow or were physically removed at the time the corn was sown. In contrast, where weed tubers and corn seeds were planted at the same time, emergence of corn was not affected. It is suggested that the growth stage of weeds be considered in the assessment of their direct and indirect allelopathic effects, and in predictions for their impacts on crop development and yield.

KEYWORDS. Allelopathy, crops, *Cyperus esculentus*, *Pinus patula*, weeds

INTRODUCTION

Cyperus esculentus (yellow nutsedge) is a herbaceous, perennial species that is considered to be one of the world's worst weeds (Holm et al., 1977). It is a serious problem of cropping systems in tropical and temperate climates where large losses in crop yields are ascribed to its pronounced competitiveness (Holm et al., 1977). This weed is characterized by prolific vegetative growth producing a complex underground system of rhizomes that forms basal bulbs and tubers through which it spreads asexually (Wills, Hoagland, and Paul, 1980; Stoller and Sweet, 1987; Gifford and Bayer, 1995). Tubers are recognized as the primary dispersal unit, because flowering is variable and many populations do not flower during a particular growing season. According to Stoller and Sweet (1987) mature viable seeds do develop occasionally, but seedlings from seed lack the vigor required for survival.

Several researchers have reported the allelopathic and growth-inhibiting effect of *Cyperus esculentus* on various crop species (Tames, Getso, and Vieites, 1973; Stoller, Wax, and Slife, 1979; Drost and Doll, 1984). Meissner, Nel, and Smit (1979) reported similar plant responses to the serious weed *C. rotundus* (purple nutsedge). In general, allelochemicals are reportedly present in nearly all plant tissues, i.e., leaves, fruit, stems, rhizomes and roots, and are released by such processes as volatilization, root exudation, leaching and decomposition of plant residues (Rice, 1984; Putnam, 1985). Leaves appear to be the most consistent source, and most studies indicate that fewer and less potent toxins occur in roots, or are released from them (Rice, 1995). Methodologies for isolation of allelochemicals often involve extraction in an appropriate solvent or collection on an adsorbent (Putnam, 1985; Rice, 1995). A commonly used solvent is water or an aqueous methanol solution in which dried or living plant material is soaked. After extracting the material for varying lengths of time, the solution or suspension is usually filtered or centrifuged prior to bioassay. More drastic procedures involve maceration or pulverization of plant material prior to solvent extraction (Putnam, 1985).

Mycorrhizal fungi (literally fungus-root) form symbiotic relationships with most terrestrial plants. Ectomycorrhizae predominate on the roots of forest species, mostly in the form of mushroom- and puffball-producing basidiomycetes and ascomycetes. It has been shown that mycorrhizae help plants to acquire mineral nutrients from the soil (Marx, 1969; Marais 1974; Schenck, 1981; Graham, 1988) and even protect trees from root pathogen infection (Marx, 1973). Parke, Linderman, and Black (1983) found that the fungal symbiont increased the drought resistance of plants by increasing water uptake and/or altering the physiological status.

The aim of this study was to investigate the role of plant growth stage in the allelopathic effects that *C. esculentus* have on ectomycorrhizal growth, as well as on seed germination and early growth of crop seedlings.

MATERIALS AND METHODS

Effect of Aqueous Extracts of C. esculentus on Mycorrhizal Growth

Foliage and tubers were collected from *C. esculentus* plants occurring at the Giant's Castle estate (29°59′S; 29°12′E; altitude 1400 m) of Mondi Forests (Pty.) Ltd. in the KwaZulu-Natal province of South Africa. The ectomycorrhiza *Boletus maxaria* that served as indicator species in the study was isolated from an established *Pinus patula* plantation at the same field site. Foliage from immature and mature *C. esculentus* plants were collected and kept separately. The material was kept fresh by cooling it, and cut into 20-mm segments a day after it was collected; then mixed with distilled water in the ratio 50 g:1000 ml, and immediately macerated in a Waring Commercial blender for 30 seconds. The suspension was filtered through Whatmann no. 1 filter paper and the supernatant was used for testing allelopathic potential.

A modified Melin-Norkrans (MMN) agar medium (Marx, 1969) was prepared for mycorrhizal growth. After autoclaving the mixture for 20 minutes at 121°C, the pH of the medium was adjusted to between 5.5 and 5.7. Petri dishes (90 mm in diameter) were each filled with 30 ml of MMN medium. Two methods were used for determining possible inhibition of mycorrhizal growth. One technique involved application of the ectomycorrhiza inoculum in the center of the Petri dish, and 1ml extract placed around it in four holes made in the agar medium. With the second method, 1 ml extract was spread evenly over the agar surface after the inoculum was placed in the center. For the control in each case, 1 ml distilled water was applied instead of the inoculum. The Petri dishes were incubated at 25°C in the dark. After three weeks, the colony diameter was measured, diagonally and across. Each treatment was replicated five times in a completely randomized design.

Effect of C. esculentus Extracts on the Germination of Lettuce

Two extract concentrations of 2 and 5% (mass/volume) were prepared from tubers, mature or immature plants as described above. Thirty lettuce seeds were placed on Whatmann no. 1 filter paper in a covered Petri dish, 90 mm in diameter, and 5 ml of each extract solution applied. The control treatment was distilled water. The trial consisted of four treatments: (1) control, (2) tuber extract, (3) immature plant extract, and (4) mature plant extract. Each treatment was replicated five times in a completely randomized design. Petri dishes were incubated at a day/night temperature regime of 30/20°C, with 12/12 h day/night intervals. On each day of a 7-day period the number of seeds that had germinated in each treatment was recorded and expressed as a percentage of the total number of seeds per Petri dish. Germination was considered to have occurred when the radicle length was at least 2 mm.

Effect of C. esculentus on the Rate of Emergence of Corn Seedlings

Pots (195 mm diameter × 200 mm deep) with holes in the base for drainage, were filled with 2.5 kg of a sandy loam soil collected from the University of Pretoria's Hatfield experimental farm (25°45′S; 28°15′E; altitude 1320 m). One treatment involved the sowing of corn (*Zea mays* L. cv. PNR 6346) and the planting of *C. esculentus* tubers at the same time; for the second treatment, tubers were planted 28 days before (= day minus 28) the day (= day 0) plants were physically removed and the corn seed sown immediately thereafter. In the third treatment, weed plants were not removed prior to the seeding of the crop. No weeds or weed material were present in soil of the control treatment at any stage. All pots received 250 ml of a complete nutrient solution (Nitsch, 1972) on alternate days for the duration of the trial. Crop emergence in each treatment was measured by counting the number of coleoptiles that had emerged from the soil. Measurement commenced eight days after seeding of the crop, i.e., on the day the first coleoptiles emerged from the soil, and were made on each of the next three days. Data were expressed as the percentage seedlings that emerged out of the total number of seeds sown in each pot. Each treatment was replicated 10 times in a completely randomized design.

Data for all three experiments were subjected to analysis of variance. Mean separation was done with the Least Significant Difference (LSD) test at the 5% level of significance.

RESULTS AND DISCUSSION

Effect of C. esculentus Extracts on Ectomycorrhizal Growth

The first method used involved application of extract solutions in holes made in the agar medium. This procedure was considered unreliable because the ectomycorrhizae grew over the holes. Data obtained with the second method described under Materials and Methods is presented in Table 1. Compared to control treatment, the growth of ectomycorrhiza species was significantly inhibited by all the weed extracts. The inhibition caused by the extract from mature *C. esculentus* plants was significantly greater than that by the tuber extract (Table 1).

The growth inhibition reported here for aqueous extracts of *C. esculentus* on a mycorrhiza species associated with *Pinus patula* could at least partly explain the problems which the forestry industry have with the establishment of new pine plantations on former crop fields that are heavily infested with the weed. The phenomenon of failure of forestry species on former crop fields is referred to as the "oldland syndrome" (Noble and Schumann, 1993). Symptoms on seedlings thus affected are: poorly developed root systems, stunting, lack of apical dominance, chlorosis and necrosis of fascicles, and dying-off of the growth tips. The inhibition of the growth of *Boletus maxaria* in the present study was less pronounced than that reported by Reinhardt, Khalil, and Bezuidenhout (1996) for unidentified mycorrhizal types on the roots of *P. patula* seedlings, which were exposed to *C. esculentus* and two other agronomic weeds. They contended that the inhibition of mycorrhizae, which are considered essential for the successful establishment of pine species (Marais, 1974), represents an indirect allelopathic effect on the forestry species that, although not crippling on its own, may be crucial in determining their resistance to other stress factors.

TABLE 1. Effect of different *Cyperus esculentus* extracts on ectomycorrhiza (*Boletus maxaria*) growth on agar medium.

Treatment (Material extracted)	Diameter (mm) of Ectomycorrhiza Colony
Control	7.20 a
Tubers	6.33 b
Foliage of immature plants	6.00 bc
Foliage of mature plants	5.83 c
Standard error	0.14
CV (%)	9.92

Means followed by the same letter do not differ significantly at $P = 0.05$.

Mycorrhizae are reported to promote the uptake by plants of immobile elements such as phosphorus (P), zinc (Zn) and copper (Cu) (Schenck, 1981; Tinker, 1984; Graham, 1988); increase the efficiency of mineral uptake, thus enhancing plant growth (Marx, 1969). Parke, Linderman, and Black (1983) concluded that mycorrhizae can also increase water uptake and/or alter the plant's physiology to reduce stress responses to drought. Other roles for the symbiont are to reduce transplant injury and help plants withstand high temperatures (Marx and Bryan, 1971), as well as to provide protection against pathogens (Marx, 1973).

Bioassays using microorganisms as indicator species have been employed by others to evaluate allelochemical effects. Chu-Chou (1978) found that water extracts of roots of old *Pinus radiata* completely inhibited the growth of a mycorrhizal fungus associated with seedlings of the species, and that it also caused root necrosis and wilting of the seedlings. Pellissier (1993) concluded that phenolic acids produced by plant species in under-storey vegetation of a spruce forest inhibited the respiration of two spruce mycorrhizal species, *Laccaria laccata* and *Cenococcum graniforme*.

Effect of C. esculentus Extracts on the Germination of Lettuce

Germination rates for lettuce seed exposed to different aqueous extracts are shown in Figures 1 and 2. After seven days of exposure to the extracts, seed germination in response to 5% concentration of all the extracts was significantly less than in the control, while at the 2% level only the mature plant extract did not reduce seed germination significantly (Table 2). At the 5% level, the extracts from immature *C. esculentus* plants completely inhibited germination (Figure 2). This suggests that immature *C. esculentus* plants contain more allelochemicals, or higher concentrations of particular compounds than mature plants or the tubers.

Effect of C. esculentus Extracts on the Rate of Emergence of Corn

Emergence of corn seedlings was significantly delayed at the two treatments where the weed was grown for 28 days and then either totally removed or left prior to sowing the crop (Table 3). This suggests that allelochemicals, which were released into the soil by actively growing *C. esculentus* plants, reduced the rate of emergence and probably the rate of germination as well. The finding demonstrates the need to recognize the risk of poor crop establishment on fields with current or recent infestations of the weed. Data in Table 3 show that the inhibitory effects were transient, and that emergence at all treatments was similar four days after the first seedlings emerged and the first measurement was made. It is impossible to extrapolate the results to field

FIGURE 1. Germination of lettuce seeds exposed to different *Cyperus esculentus* extracts at the 2% concentration level.

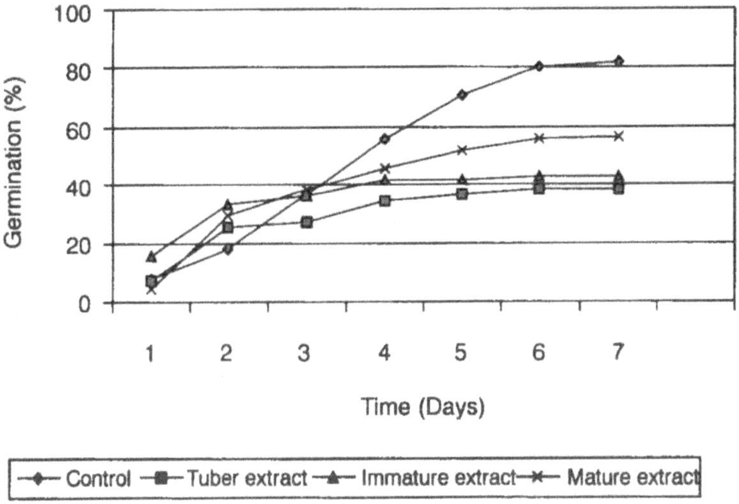

FIGURE 2. Germination of lettuce seeds exposed to different *Cyperus esculentus* extracts at the 5% concentration level.

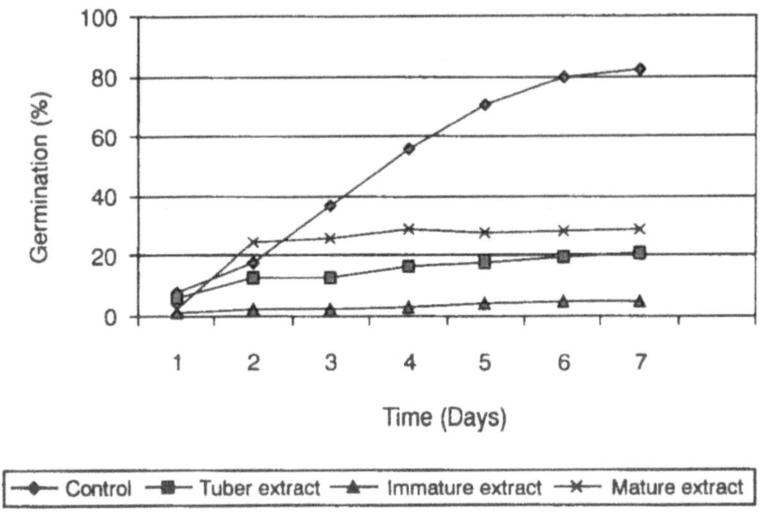

TABLE 2. Germination percentage of lettuce seed after a 7-day exposure to 2 and 5% (m/v) concentrations of different aqueous extracts from *Cyperus esculentus*.

Treatment (Material extracted)	Germination % 2% Extract conc.	5% Extract conc.
Control	82 a	82 a
Tuber	38 b	21 bc
Foliage of immature plants	43 b	5 c
Foliage of mature plants	61 ab	29 b
Standard error	0.23	5.5
CV (%)	13	36

Means followed by the same letter do not differ significantly at $P = 0.05$.

TABLE 3. Emergence (%) of corn seedlings on soil that was either infested with *C. esculentus* plants prior to the sowing of the crop, or where tubers of the weed and crop seeds were planted at the same time.

Planting of weed tubers relative to sowing crop seed	Weed presence	Days after emergence started			
		1	2	3	4
Control	Absent	90 a	95 a	95 a	95 a
Day minus 28*	Undisturbed growth	13 b	38 b	70 b	93 a
Day minus 28	Removed on day 0	18 b	35 b	75 b	93 a
Day 0**	Undisturbed growth	95 a	98 a	98 a	98 a
Day 0	Removed on day 0	83 a	98 a	98 a	98 a
Standard error		7	8	7	-
CV (%)		36	35	26	-

* Day minus 28: Planting of weed tubers, 28 days prior to crop seeding.
** Day 0: Sowing of crop seed, and planting of weed tubers.

conditions, but it is conceivable that under conditions favoring the production and release of allelochemicals in high concentrations by *C. esculentus*, corn seedlings would be placed under chemical stress that might weaken their resistance to other, more well-known environmental stress factors.

Stoller, Wax, and Slife (1979) identified a strong inverse relationship between *C. esculentus* density (shoots/m^2) and percentage reduction in corn yields. An 8% yield reduction was reported for weed increments in the order of 100 shoots/m^2, with 41% loss occurring when no weed control was done on a field initially infested with 1200 shoots/m^2. Drost and Doll (1984) concluded that extracts and residues of *C. esculentus* affect the growth of *Glycine max* (soybean) and corn. Tames, Getso, and Vicitez (1973) found compounds in *C. esculentus* tubers that were inhibitory to oat (*Avena sativa*) coleoptiles and seed germination of other crops. Horowitz and Friedman (1971) reported that dried *C. esculentus* tubers mixed in soil significantly reduced the root and top growth of barley (*Hordeum vulgare*). Meissner, Nel, and Smit (1979) grew *C. rotundus* on sterilized, well-fertilized soil and showed that the growth of barley, cucumber (*Cucumis sativus*) and tomato (*Lycopersicon esculentum*) were considerably reduced even though the weed was removed prior to the crops being seeded.

CONCLUSIONS

Results suggest that the allelopathic potential of the weed *C. esculentus* varies with growth stage, and that both primitive and higher plant species may be sensitive to the allelochemicals produced by this weed. Although allelopathic effects on desirable species may be transient, even minor impediments of growth and development at critical stages, especially during early development of a species, could weaken its resistance to the more well-known environmental stresses. The growth-inhibiting effect of *C. esculentus* on the mycorrhizal species *Boletus maxaria* could explain, in part at least, why seedlings of *Pinus patula* often fail to establish where new plantations are attempted on former crop fields. Allelopathy researchers should consider that plants probably differ in their allelopathic potential from one growth stage to another. This information is also relevant for crop producers, since knowledge of the inherent and transient allelopathic potentials of weeds would benefit practices aimed at avoiding or limiting weed interference with crops.

REFERENCES

Chu-Chou, M. (1978). Mycorrhizal fungi of *Pinus radiata* in New Zealand. *Soil Biology and Biochemistry* 11: 557-562.

Drost, D.C. and J.D. Doll. (1980). The allelopathic effect of yellow nutsedge (*Cyperus esculentus*) on corn (*Zea mays*) and soybeans (*Glycine max*). *Weed Science* 28: 229-233.

Gifford, E.M. and D.E. Bayer. (1995). Developmental anatomy of *Cyperus esculentus* (yellow nutsedge). *International Journal of Plant Sciences* 156: 622-629.

Graham, J.H. (1988). Interactions of mycorrhizal fungi with soil-borne plant pathogens and other organisms: an introduction. *Phytopathology* 78: 365-366.

Holm, L.G., D.L. Plucknett, J.V. Pancho, and J.P. Herberger. (1977). *The World's Worst Weeds: Distribution and Biology.* Hawaii, Honolulu: University Press, pp. 125-133.

Horowitz, M. and J. Friedman. (1971). Biological activity of subterranean residues of *Cynodon dactylon* L., *Sorghum halepense* L. and *Cyperus rotundus* L. *Weed Research* 11: 88-93.

Marais, L.J. (1974). Mycorrhizal effects on *Pinus patula* Schlecht. and Cam. in South Africa. *M.Sc. Thesis. University of Pretoria, Pretoria.*

Marx, D.H. (1969). The influence of ectotrophic mycorrhizal fungi on the resistance of pine roots to pathogen infections. I. Antagonism of mycorrhizal fungi to root pathogenic fungi and soil bacteria. *Phytopathology* 59: 153-163.

Marx, D.H. (1973). Mycorrhizae and feeder root disease. In *Ectomycorrhizae*, eds. G.S. Marks and T.T. Kozlowski. London: Academic Press, pp. 351-382.

Marx, D.H. and W.C. Bryan. (1971). Influence of ectomycorrhizae on survival and growth of aseptic seedlings of loblolly pine at high temperatures. *Forest Science* 17: 37-41.

Meissner, R., P.C. Nel, and N.S.H. Smit. (1979). Influence of red nutgrass (*Cyperus rotundus*) on growth and development of some crop plants. In *Proceedings 3rd National Weed Conference*, South Africa, pp. 39-52.

Nitsch, J.P. (1972). Phytotrons: Past achievements and future needs. In *Crop Processes in Controlled Environments*, eds. A.R. Rees, K.E. Cockshull, D.W. Hand, and R.G. Hurd. London: Academic Press, pp. 33-55.

Noble, A.D. and A.W. Schumann. (1993). The amelioration of *Pinus patula* mortality on former agricultural sites through fertilization: A bioassay and greenhouse study. *South African Forestry Journal* 164: 35-41.

Parke, J.L., R.G. Linderman, and C.H. Black. (1983). The role of ectomycorrhizas in drought tolerance of Douglas-fir seedlings. *New Phytologist* 95: 83-95.

Pellissier, F. (1993). Allelopathic effect of phenolic acids from humic solutions of two spruce mycorrhizal fungi: *Cenococcum graniforme* and *Laccaria laccata*. *Journal of Chemical Ecology* 19: 2105-2114.

Putnam, A.R. (1985). Weed allelopathy. In *Weed Physiology. Vol. 1. Reproduction and Ecophysiology*, ed. S.O. Duke. Boca Raton, FL: CRC Press, pp. 131-155.

Reinhardt, C.F., S. Khalil, and S.R. Bezuidenhout. (1996). Bioassay techniques in assessing the allelopathic effect of weeds in crops and plantation species. In *Recent Advances in Allelopathy. Vol. 1*, eds. F.A. Macias, J.C.G. Galindo, J.M.G. Molinillo, and H.G. Cutler. University of Cadiz: Servicio de Publicaciones, pp. 29-46.

Rice, E.L. (1984). *Allelopathy*. 2nd edition. New York: Academic Press.

Rice, E.L. (1995). Allelopathy in forestry. In *Biological Control of Weeds and Plant Diseases: Advances in Applied Allelopathy*, ed. E.L. Rice. Norman: University of Oklahoma Press, pp. 317-378.

Schenck, N.C. (1981). Can mycorrhizae control root disease? *Plant Disease* 65: 230-234.

Stoller, E.W. and R.D. Sweet. (1987). Biology and life cycle of purple and yellow nutsedge (*Cyperus rotundus* and *Cyperus esculentus*). *Weed Technology* 1: 66-73.

Stoller, E.W., L.M. Wax, and F.W. Slife. (1979). Yellow nutsedge (*Cyperus esculentus*) competition and control in corn (*Zea mays*). *Weed Science* 27: 32-37.

Tames, R.S., M.D.V. Getso, and E. Vieitez. (1973). Growth substances isolated from tubers of *Cyperus esculentus* var. *aureus*. *Physiologia Plantarum* 28: 195-200.

Tinker, P.B. (1984). The role of micro-organisms in mediating and facilitating the uptake of plant nutrients from soil. *Plant and Soil* 76: 77-91.

Wills, G.D., R.N. Hoagland, and R.N. Paul. (1980). Anatomy of yellow nutsedge (*Cyperus esculentus*). *Weed Science* 28: 432-437.

Autotoxic Potential of Cucurbit Crops: Phenomenon, Chemicals, Mechanisms and Means to Overcome

Jing Quan Yu

SUMMARY. The present review deals with the phenomemon of autotoxicity in cucurbit crops which are known for their serious problem of soil sickness in commercial production. There are great differences in the autotoxic potential of cucurbit crops. Some species such as watermelon, melon and cucumber showed autotoxic potential, but others did not. Some phenolic acids have been isolated and identified from their root tissue and root exudates as autotoxins. These autotoxins affect ion uptake, membrane permeability, photosynthesis and phytohormone balance. Plants are more susceptible to the effects of autotoxins under stress. Autotoxicity in cucurbits could be overcome or alleviated by selecting resistant genotype, proper rotation, proper soil and plant residues management, grafting, adsorbent as well as microbial degrading. *[Article copies available for a fee from The Haworth Document Delivery Service: 1-800-342-9678. E-mail address: <getinfo@haworthpressinc.com> Website: <http://www.HaworthPress.com> © 2001 by The Haworth Press, Inc. All rights reserved.]*

KEYWORDS. Autotoxicity, cucurbits, autotoxins, soil sickness, mechanism, means to overcome

---

Jing Quan Yu is Professor, Horticultural Department, Huajiachi Campus, Kaixuan Road 268, Zhejiang University, Hangzhou 310029 P.R. China.

Address correspondence to: Jing Quan Yu at the above address (E-mail: yu@public.hz.zj.cn).

The author is thankful to National Natural Science Foundation of China for financial support for this study.

[Haworth co-indexing entry note]: "Autotoxic Potential of Cucurbit Crops: Phenomenon, Chemicals, Mechanisms and Means to Overcome." Yu, Jing Quan. Co-published simultaneously in *Journal of Crop Production* (Food Products Press, an imprint of The Haworth Press, Inc.) Vol. 4, No. 2 (#8), 2001, pp. 335-348; and: *Allelopathy in Agroecosystems* (ed: Ravinder K. Kohli, Harminder Pal Singh, and Daizy R. Batish) Food Products Press, an imprint of The Haworth Press, Inc., 2001, pp. 335-348. Single or multiple copies of this article are available for a fee from The Haworth Document Delivery Service [1-800-342-9678, 9:00 a.m. - 5:00 p.m. (EST). E-mail address: getinfo@haworthpressinc.com].

INTRODUCTION

A reduction in both crop yield and quality often occurs when same crop or its relative species are cultivated on same soil successively. This phenomenon is usually called soil sickness. It has been recognized with a long history and was even cited in a famous Chinese book *Jiminyaoshu* which was written more than 1500 years ago. However, scientific work was initiated at the beginning of this century (Schreiner and Reed, 1907). In recent years, the extension of protected cultivation has made the problem more serious and it is one of the factors limiting the sustainable development of agriculture. It is a complicated natural phenomenon whose causes are not yet fully understood. Various factors such as the buildup of pests in soil, disorder in physico-chemical properties of soil, autotoxicity and other unknown factors are believed to be involved (Börner, 1960; Tsuchiya, 1990; Yu, 1999). Autotoxicity is a special kind of allelopathy, which occurs within same species. Autotoxic plant releases autotoxins to affect the growth of adjacent plants or subsequent plants. Plant residues, root exudates, leachates of plant and volatile oil are main sources of autotoxins (Börner, 1960). Autotoxicity has been reported in both the natural and managed ecosystems and is observed in a number of crops including rice, asparagus, wheat, corn, sugarcane, apple, peach, citrus, grape, tea, pea, tomato, eggplant, cucumber, etc. (Singh, Batish, and Kohli, 1999). Cucurbit crops are one of the vulnerable species (Tsuchiya, 1990).

Cucurbit crops are widely cultivated around the world. In their commercial production, poor growth is frequently observed in the monoculture of some cucurbit crops. This leads farmer to change cropping soil year by year. The problem is especially serious in watermelon production and a 6-7 year rotation is recommended. In most cases, soil-borne pest is believed to be the main cause. Soil fumigation, however, did not eliminate this phenomenon in some cases, suggesting that factors other than soil-borne pest are partly responsible for soil sickness (Yu, 1999). It is widely recognized that plant growth is poor in the subsequent cropping than that of first cropping under protected cultivation. In the hydroponic culture of cucumber and tomato, poor plant growth together with root death was frequently observed when the plants were carrying a lot of fruits (Rahman and Newton, 1984). This phenomenon was not so apparent when the nutrient solution was renewed periodically thereby suggesting the possible involvement of autotoxicity in their production (Yu, Lee, and Matsui, 1993; Yu et al., 2000).

AUTOTOXIC POTENTIAL IN CUCURBIT CROPS

Cucurbit crops are believed to have autotoxic potential. However, there is a great degree of variation in autotoxic potential of cucurbit crops. Yu et al.

(2000) investigated the autotoxic potential in cucurbit crops and found that root extracts of watermelon, melon and cucumber showed great phytotoxicity to the radicle elongation and seed germination of respective species even when they were extracted at 1:100 (w/v). The phytotoxicity increased with the concentration of root extracts. However, root extracts of gourd, snake gourd and bitter gourd did not show autotoxicity until the root extracts concentrations were higher than 1:20. The species with autotoxic potential in germination test were used to explore autotoxic potential of their root exudates and root residues. The root exudates collection system developed by Tang (1986) was employed to investigate the autotoxic potential of watermelon, melon and cucumber. Seedling growth was apparently inhibited by root exudates from adult plants of respective species. The inhibition rate of dry matter production were 16% for watermelon, 15% for melon, and 21% for cucumber, respectively (Yu et al., 2000). In addition, root exudates greatly inhibited seed germination of respective plants. The root exudates of cucumber, however, did not inhibit seed germination and radicle elongation of figleaf gourd, which is commonly used to solve soil sickness problem and to increase their tolerance to low temperature in the commercial production of some cucurbit crops. They further investigated the autotoxic potential of root extracts in rhizosphere by using an improved root exudates trapping system in which Amberlite XAD-4 was replaced by a root-tissue-filled column at a rate of one gram per seedling. It was found that seedling growth of watermelon, melon and cucumber were apparently poor in culture system with the root column than that without the column, suggesting that root residues produced autotoxins during the decomposition.

There are several studies about the autotoxic potential of cucurbit crops in hydroponic culture. In the hydroponic production of cucumber, root death is often observed at the reproductive stage especially in NFT-grown plants. In fact, this phenomenon was also observed in other crops such as tomato (Rahman and Newton, 1984). In addition to nutritional disorder and accumulation of pathogens, autotoxins are also believed to be involved in this phenomenon. Accordingly, farmers usually renew the nutrient solution periodically. Yu and Matsui (1994) found that addition of activated charcoal had no effect on the inorganic composition, but it greatly improved the growth and the yield of cucumber plants. The root exudates collected by a continuous root exudate collection system showed greatest phytotoxicity when the plants carried a lot of fruits. Recently, Asao et al. (1998) confirmed the results of Yu and Matsui (1994) and revealed the genotype varieties in autotoxic potential. Some genotypes, however, did not show autotoxic potential. Putnam and Duke (1974), Lockerman and Putnam (1979, 1981a,b) also emphasized differences in allelopathic potential of cucumber accessions. Moreover, the autotoxic potential is greatly influenced by environmental condition. Usually, plant exerts greater autotoxicity under stress condition such as low pH and low oxygen supply (Einhellig, 1987; Yu and Matsui, 1997; Asao et al., 1999c), which will be discussed later.

AUTOTOXINS

Since soil sickness is a serious problem in the commercial production of some cucurbit crops such as watermelon, the identification of the autotoxins has been the research target for many scientists. Hatsuda et al. (1961) isolated salicylic acid from the water extracts of watermelon root and this compound greatly inhibited the radicle elongation of watermelon and lettuce (Tsuchiya, 1990). When watermelon seedling was incubated in a salicylic acid solution, seedling showed similar symptoms as caused by *Fusarium* (Niuchi and Koutsu, 1959). Gaidmak (1971) detected the presence of phenolic compounds and amino acids in the nutrient solution after hydroponic culture of cucumber and tomato, but they did not identify the phytotoxic compounds. Yu and Matsui (1994) collected the root exudates of cucumber by a continuously root exudates collection system (Tang and Young, 1982). They divided the root exudates into basic and neutral fraction, acidic fraction and water soluble substances fraction. Both the basic and neutral fraction and acidic fraction were phytotoxic with the acidic fraction showing greater phytotoxicity. In the basic and neutral fraction, they identified *p*-thiocyanatephenol as the autotoxin and it showed a much greater phytotoxicity than *p*-hydroxybenzoic acid. It induced a net ion leakage at a concentration less than 0.1 mM (Yu and Matsui, 1993b). In the acidic fraction, nine compounds, viz., benzoic, 3-phenylpropionic, cinnamic, *p*-hydroxybenzoic, 2,5-dihydroxybenzoic, myristic, *p*-hydroxycinnamic, palmitic and stearic acids were identified (Yu and Matsui, 1994). Recently, Asao et al. (1999a) identified 2,4-dichlorobenzoic acid from the activated charcoal, which was added to the nutrient solution for the hydroponic culture of cucumber. This compound showed much greater phytotoxicity than other compounds they identified. It is worth noting that chloro-containing compounds are rarely present in plant.

MECHANISMS

Recently, the action mode of autotoxicity has received wide attention. Blum and his colleagues carried out a series of works on the effects of phenolic acid on cucumber seedling (Blum, Dalton, and Rawlings, 1984; Blum, Dalton, and Shann, 1985a,b; Blum, Weed, and Dalton, 1987; Klein, and Blum, 1990a; Hollapa and Blum, 1991; Blum, 1998). Their work greatly deepened our understanding on autotoxicity of cucurbit crops. It is, however, difficult to summarize the mechanisms of autotoxicity at present stage. Among the cited effects, we can underline the following.

Phytohormonal Level

It has been suggested that allelochemicals affect plant growth by interacting with phytohormone (Einhellig, 1986). There are evidences that phytohormone level can be strongly affected by autotoxins. For example, abscisic acid level in cucumber seedling increased greatly in several hours after ferulic acid treatment (Hollapa and Blum, 1991), which was identified from root exudates of cucumber and tomato (Yu and Matsui, 1993a, 1994).

Membrane Permeability

Increased ion leakage by autotoxins or allelochemicals has been documented in several crops including cucumber (Politycka, 1996; Yu and Matsui, 1997), asparagus (Hartung, Putnam, and Stephens, 1989) and soybean (Baziramakenga, Leroux, and Simard, 1995). In cucumber, ion leakage was greatly increased by root exudates, root extracts and autotoxins. There was a close relationship between the increased ion leakage and the Log P (octanol-water partition coefficient) (Yu and Matsui, 1997). It has also been found that these autotoxic agents increased MDA (malondiadehyde) content in cucumber root, which is an index of membrane peroxidation (Yu, unpublished). Effects of autotoxins on the active oxygen species will be discussed in following section.

Stomatal Aperture

Treatment of seedling with root exudates, extracts and autotoxins decreased leaf conductance in cucumber in our study. Ferulic acid decreased leaf water potential and turgor pressure in cucumber seedlings. However, decreased water potential and turgor pressure are not thought to be caused by changes in osmotic properties of symplast or stomatal conductance (Booker, Blum, and Fiscus, 1992).

Nutrient Uptake

The effects of phenolic acids on ion uptake has been intensively studied (Lyu and Blum, 1990; Booker, Blum, and Fiscus, 1992; Yu and Matsui, 1993b, 1994). These studies showed that root exudates and phenolic acid decreased ion uptake. The inhibition magnitude depended on their concentration, pH, etc. Inhibition of ion uptake was related linearly to the proportion of root in contact with phenolic acid (Lyu and Blum, 1990; Klein and Blum, 1990a; Lehman, Blum, and Gerig, 1994) and root contact appears to be more important than root uptake (Blum, Shafer, and Lehman, 1999).

Active Oxygen Species (AOS)

Recently, the role of AOS in stressed plants has received wide attention in ecophysiology studies. Plants generate more AOS under stress condition. AOS may damage lipids, proteins, chlorophyll and nucleic acids, thereby disrupting the cell. Meanwhile, plants have evolved several mechanisms to prevent or alleviate injuries from AOS (Rice-Evans, Diplock, and Symons, 1991). These include scavenging by natural antioxidants and enzymatic antioxidant systems such as superoxide dismulase (SOD), catalase, and peroxidases (POD). Some workers have pointed that allelochemicals may exert more significant effects under stress condition (Tang et al., 1995, Reigosa, Sanchez-Moreiras, and Gonzalez, 1999). Recent works by Politycka (1996) and our laboratory (Yu, unpublished data) showed that there were apparent increases in POD and SOD activities after autotoxin treatment, indicating the increased generation of AOS. However, increased POD and SOD activities did not match the generation of AOS. At last, an increase in MDA was also observed in autotoxin-treated plants (Yu et al., unpublished). Moreover, phenolic acids also influence the activities of phenylalanine ammonia-lyase, phenol-beta-glucosyltransferase and beta-glucosidase in cucumber roots (Politycka, 1998).

Photosynthesis and Transpiration

Photosynthesis can be greatly affected by autotoxic agents. In our study, net assimilate rate decreased 4 h from the onset of treatment. Similarly, transpiration was also greatly inhibited (Blum, Dalton, and Shann, 1985a; Yu, unpublished data).

In conclusion, mechanism of autotoxicity is complicated. The primary measurable effect of phenolic acids, the main autotoxins identified from cucurbit crops until now, is supposed to be a reduction in hydraulic conductivity (i.e., water uptake) and net nutrient uptake. Increased ABA level, and reduced rates of photosynthetic rate, carbon translocation to roots, transpiration and leaf expansion all appear to be secondary effects attributable to reduced uptake of water and nutrients (Blum, Shafer, and Lehman, 1999).

FACTORS AFFECTING AUTOTOXICITY

Biotic and abiotic factors not only influence the production of autotoxins, but also modify the effects of autotoxins. Although there are increasing works on environmental effects, most of them were done with phenolics.

Light

Despite some works on the effects of light on phenolics production (Zucker, 1969), the role of light on the production of autotoxins is still unclear. Research is being carried out to investigate autotoxicity of cucumber plants grown under different photoperiod and temperature (Pramanik, in press).

Nutrients

Nutrient deficits tend to increase the production of phenolic compounds. In addition, autotoxins or allelochemicals exert greater detrimental effects on plants under nutrient stress (Klein and Blum, 1990b; Vaughan and Ord, 1990). The presence of carbohydrates, amino acids, other organic molecules can increase the inhibitory activities of phenolic acid (Blum, Shafer, and Lehman, 1999).

Water Stress

There are some reports about the improved production of allelochemicals under water deficit condition, little, however, is known about the effects of water deficits on the phytotoxicity of allelochemicals or autotoxins (Reigosa, Sanchez-Moreiras, and Gonzalez, 1999). As stated earlier, one of the effects of autotoxin is achieved by the inhibition of water uptake, hence, water deficits are supposed to enhance phytotoxicity.

Temperature Stress

Temperature stress can increase the synthesis of autotoxins. Tomato plants, for example, released more phytotoxins in rhizosphere at 40°C than at 25°C (Yu, unpublished data).

Plant Age

Exudation of phytotoxic substances greatly depended on plant growth stage. In cucumber plant, root exudates at reproductive stage were more phytotoxic than at vegetative stage (Yu and Matsui, 1994).

Genotypes

Significant difference in autotoxic potential was found among genotypes of cucumber. Some accessions or cultivars showed significant autotoxic or allelopathic potential, but others did not (Lockerman, and Putnam, 1981a; Asao et al., 1998).

Soil Type

Autotoxicity is greatly influenced by soil type (Blum, Weed, and Dalton, 1987). Soil physico-chemical properties such as organic matter content, pH, microbial activity, adsorption ability all affect the toxicity and the availability of autotoxins (Dalton, Blum, and Weed, 1983; Blum, Dalton and Shann, 1985a; Blum, Weed and Dalton, 1987; Blum, Shafer and Lehman, 1999).

Soil Microbes

In some cases, autotoxins such as phenolic acids in rhizosphere are a carbon source for many microorganisms and can cause the selective induction and/or reproduction of microbes capable of utilizing them (Blum, Dalton, and Rawlings, 1984; Blum, Dalton, and Shann, 1985; Shafer and Blum, 1991; Blum, 1998; Blum, Shafer, and Lehman, 1999). In fact, a higher proportion of phenolic acids is expected to be utilized by soil microbes than fixation by soil and uptake by root (Shann and Blum, 1987). This provides plants a strong barrier between autotoxins in rhizosphere and the root surface. The resulting microbial products will likely differ in their phytotoxicity (Blum, Shafer, and Lehman, 1999).

pH and Aeration

It has been well established that phenolic acids exert more significant effects on ion uptake and leaf expansion at low pH than at relatively high pH (Blum, Dalton, and Shann, 1985b; Yu and Matsui, 1997). Decreased rate of uptake with increased alkalinity is also expected, suggesting that uptake of phenolic acids occurs primarily in the protonated form (Blum, Shafer, and Lehman, 1999). Oxygen supply also seems important in phytotoxicity. Plant under low oxygen supply usually produced more phytotoxic substances than under normal condition (Yu, unpublished data). In addition, cucumber plants were more sensitive to autotoxins at low oxygen supply than at higher oxygen supply (Asao et al., 1999c).

MEANS TO OVERCOME AUTOTOXICITY

Recently, there is increasing awareness of the importance of overcoming autotoxicity. Until now, however, there are no ideal and practical ways which could readily overcome autotoxicity. Some possible ways to overcome autotoxicity are listed below.

Selecting Resistant Genotype

As discussed above, there are great differences in the autotoxic potential among genotypes. In the breeding program, we should pay more attention to select cultivars with poor autotoxic potential.

Proper Rotation

Proper crop rotation is preferred so that there is enough time for the degradation of autotoxins.

Proper Soil and Plant Residues Management

Since phenolic acid showed phytotoxicity only at acidity (Yu and Matsui, 1997), a proper soil neutralization would eliminate and alleviate autotoxicity. Moreover, removal of residues should be recommended since they are main source of autotoxins present in soil.

Grafting

Grafting is widely used for the successful establishment of fruit trees. Asao et al. (1999b) reported that grafting on some rootstocks of cucumber species extended the harvest period of cucumber. Our recent studies have showed that rootstocks like figleaf gourd also showed tolerance to autotoxins of watermelon in additional to its tolerance to low temperature (Yu et al., 2000). However, further experiments are needed.

Using Adsorbent

Activated charcoal is very effective to eliminate autotoxicity of plants grown by hydroponics. There was a 10-30% increase in fruit yield in cucumber and tomato plants (Yu, Lee, and Matsui, 1993; Yu and Matsui, 1994). Unfortunately, this method could not be extended to field.

Microbial Degradation

As described above, some microbes could use phenolic acid as carbon source (Blum, 1998). This provides us a useful way to overcome autotoxicity if their product are not phytotoxic to plants (Vaughan, Sparling, and Ord, 1983; Vaughan, Jones, and Ord, 1993; Zucconi, 1993). Recently, the usefulness of mycorrhizae is also suggested since additional effects such as improving ion and water uptake could also be expected (Tobar, 1994; Mallik and Zhu, 1995; Trimble, 1995; Blum, Shafer, and Lehman, 1999). Further studies should be encouraged in this direction.

CONCLUSION AND FUTURE PROSPECTS

From the foregoing discussion, it is clear that autotoxicity plays a significant role in agricultural field of cucurbit crops. Although a large body of literature has accumulated on the autotoxic or allelopathic potential of cucurbit crops, there have been few studies that followed a special criterion similar to Koch's postulates. In fact, it is very difficult to fully implement these criteria (Blum, Shafer, and Lehman, 1999). There is a little information about the behavior of identified autotoxins in rhizosphere. Only if we have adequate knowledge about autotoxins in rhizosphere, we can understand or explain autotoxicity reasonably. It is also worth studying the effects of autotoxins on incidence of some soil-borne diseases since autotoxins show significant synergism with soil-borne pathogens (Hartung and Stephen, 1983; Peirce and Miller, 1993). This secondary effect might be greater than its primary effects on plants in some cases. More attention should be paid to the use of beneficial microbes which utilize autotoxins as carbon source or alleviate the inhibition of water and nutrients by autotoxins.

REFERENCES

Asao, T., M. Umeyama, K. Ohta, T. Hosoki, N. Ito, and H. Ueda. (1998). Decrease of yield of cucumber by non-renewal of the nutrient solution and its reversal by supplementation of activated charcoal. *Journal of Japanese Society for Horticultural Science* 67: 99-105.

Asao, T., M.H.R. Pramanik, K. Tomita, Y. Ohba, K. Ohta, T. Hosoki, and Y. Matsui. (1999a). Influences of phenolics isolated from the nutrient solution nourishing growing cucumber plants on fruit yield. *Journal of Japanese Society for Horticultural Science* 68: 847-853.

Asao, T., N. Shimidu, K. Ohta, and T. Hosoki. (1999b). Effects of rootstocks on the extension of harvest period of cucumber (*Cucumis sativus* L) grown in non-renewal hydroponics. *Journal of Japanese Society for Horticultural Science* 68: 598-602.

Asao, T., Y. Ohba, K. Tomita, K. Ohta, and T. Hosoki. (1999c). Effects of activated charcoal and dissolved-oxygen levels in the hydroponic solution on the growth and influences of phenolics isolated from the nutrient solution. *Journal of Japanese Society for Horticultural Science* 68: 1194-1196.

Baziramakenga, R., G.D. Leroux, and R.R. Simard. (1995). Effects of benzoic and cinnamic acids on growth, mineral composition and chlorophyll content of soybean. *Journal of Chemical Ecology* 20: 2821-2833.

Blum, U. (1998). Effects of microbial utilization of phenolic acid and their phenolic acid breakdown products on allelopathic interactions. *Journal of Chemical Ecology* 24: 685-708.

Blum, U., B.R. Dalton, and J.O. Rawlings. (1984). Effects of ferulic acid and some of its microbial metabolic products on radicle growth of cucumber. *Journal of Chemical Ecology* 10: 1169-1191.

Blum, U., Dalton, B.R., and J.R. Shann. (1985a). Effects of various mixtures of ferulic acid and some of its microbial metabolic products on cucumber leaf expansion and dry matter in nutrient culture. *Journal of Chemical Ecology* 11: 619-641.

Blum, U., B.R. Dalton, and J.R. Shann. (1985b). Effects of ferulic and *p*-coumaric acids in nutrient culture on cucumber leaf expansion as influenced by pH. *Journal of Chemical Ecology* 11: 1567-1582.

Blum, U., R. Shafer, and M.E. Lehman. (1999). Evidence for inhibitory allelopathic interactions involving phenolic acids in field soils: Concepts vs. an experimental model. *Critical Reviews in Plant Sciences* 18: 673-693.

Blum, U., S.B. Weed, and B.R. Dalton. (1987). Influence of various soil factors on the effects of ferulic acid on leaf expansion of cucumber seedlings. *Plant and Soil* 98: 111-130.

Booker, F.L., U. Blum, and E. Fiscus. (1992). Short-term effects of ferulic acid on ion uptake and water relations in cucumber seedlings. *Journal of Experimental Botany* 43: 649-655.

Börner, H. (1960). Liberation of organic substances from higher plants and their role in the soil sickness problem. *Botanical Review* 26: 393-424.

Dalton, B.R., U. Blum, and S.B. Weed. (1983). Allelopathic substances in ecosystems: effectiveness of sterile soil components in altering recovery of ferulic acid. *Journal of Chemical Ecology* 9: 1185-1201.

Einhellig, F.A. (1986). Mechanisms and modes of action of allelochemicals. In *The Science of Allelopathy*, eds. A.R. Putnam and C.S. Tang. NY: John Wiley and Sons, pp.171-188.

Einhellig, F.A. (1987). Interaction of allelochemicals and other stress factors of plant environment In *Allelochemicals: Role in Agriculture and Forestry*, eds. G.R. Waller. ACS Symposium series No. 330. Washington, DC: American Chemical Society, pp. 343-351.

Gaidmak, V.M. (1971). Biologically active substances in nutrient solution after cucumbers and tomatoes were grown on pure and multiple broken brick. In *Physiological and Biological Basis of Plant Interactions in Phytocenoses*, ed. A.M. Grodzinsky. Vol. 2. Kiev: Naukova-Dumka, pp. 55-60.

Hartung, A.C. and C.T. Stephens. (1983). Effects of allelopathic substances produced by asparagus on the incidence and severity of Fusarium crown rot. *Journal of Chemical Ecology* 9: 1163-1174.

Hartung, A.C., A.R. Putnam, and C.T. Stephens. (1989). Inhibitory activity of asparagus root tissue and extracts on asparagus seedlings. *Journal of American Society for Horticultural Science* 114: 144-148.

Hatsuda, Y., S. Murao, S. Nishimura, and T. Manasaki. (1961). Biochemical studies on soil sickness. 3. On the toxic substances in watermelon roots. *Nippon Nogeikagaku Kaishi* 35: 1107-1108.

Hollapa, L.D. and U. Blum. (1991). Effects of exogenously applied ferulic acid, a potential allelopathic compound, on leaf growth, water utilization, and endogenous abscisic acid levels of tomato, cucumber and bean. *Journal of Chemical Ecology* 17: 865-886.

Klein, K. and U. Blum. (1990a). Inhibition of cucumber leaf expansion by ferulic acid in split-root experiments. *Journal of Chemical Ecology* 16: 455-463.

Klein, K. and U. Blum. (1990b). Effects of soil nitrogen level on ferulic acid inhibition of cucumber leaf expansion. *Journal of Chemical Ecology* 16: 1371-1383.

Lehman, M., U. Blum, and T.M. Gerig. (1994). Simultaneous effects of ferulic and *p*-coumaric acids on cucumber leaf expansion in split-root experiments. *Journal of Chemical Ecology* 20: 1773-1782

Lockerman, R.H. and A.R. Putnam. (1979). Evaluation of allelopathic cucumber as an aid to weed control. *Weed Science* 27: 54-57.

Lockerman, R.H. and A.R. Putnam. (1981a). Mechanisms for differential interference among cucumber (*Cucumis sativus* L.) accessions. *Botanical Gazette* 142:427-430.

Lockerman, R.H. and A.R. Putnam. (1981b). Growth inhibitors in cucumber plants and seeds. *Journal of American Society for Horticultural Science* 106: 418-422.

Lyu, S.W. and U. Blum. (1990). Effects of ferulic acid, an allelopathic compound, on net P, K, and water uptake by cucumber seedlings in a split-root system. *Journal of Chemical Ecology* 16: 2429-2439.

Mallik, A.U. and H. Zhu. (1995). Overcoming allelopathic growth inhibition by mycorrhizal inoculation. In *Allelopathy: Organisms, Processes and* Applications, eds. Inderjit, K.M.M. Dakshini, and F.A. Einhellig. ACS Symposium Series 582. Washington, DC: American Chemical Society, pp. 39-57.

Niuchi, S. and N. Koutsu (1959). Studies on pathological chemistry of Fusarium wilt in watermelon. 12. Screening phenolic compounds causing vascular bundle browning. *Annals of Phytopathological Society of Japan* 24: 139-144.

Peirce, L.C. and H.G. Miller. (1993). Asparagus emergence in Fusarium-treated and sterile media following exposure of seeds or radicles to one or more cinnamic acids. *Journal of American Society for Horticultural Science* 118: 23-28.

Politycka, B. (1996). Peroxidase activity and lipid peroxidation in roots of cucumber seedlings influenced by derivatives of cinnamic and benzoic acids. *Acta Physiologia Plantarum* 18: 365-370.

Politycka, B. (1998). Phenolics and the activities of phenylalanine ammonia-lyase, phenol-beta-glucosyltransferase and beta-glucosidase in cucumber roots as affected by phenolic allelochemicals. *Acta Physiologia Plantarum* 24: 405-410.

Putnam, A.R. and W.B. Duke. (1974). Biological suppression of weeds: Evidence foe allelopathy in accessions of cucumber. *Science* 185: 370-372.

Rahman, A.R.M. and P. Newton. (1984). Death of roots of tomato plants grown with nutrient film technique. *Proceedings of the International Society of Soilless Culture* 6: 59-70.

Reigosa, M.J., A. Sanchez-Moreiras, and L. Gonzalez. (1999). Ecophysiological approach in allelopathy. *Critical Reviews in Plant Sciences* 18: 577-608.

Rice-Evans, C.A., A.T. Diplock, and M.C.R. Symons. (1991). *Techniques in Free Radical Research*. Amsterdam: Elsevier Science.

Schreiner, O. and H.S. Reed. (1907). The production of deleterious excreations by roots. *Bulletin of the Torrey Botany Club* 34: 279-303.

Shafer, S.R. and U. Blum. (1991). Influence of phenolic acids on microbial populations in the rhizosphere of cucumber. *Journal of Chemical Ecology* 17: 369-389.

Shann, J.R. and U. Blum. (1987). The uptake of ferulic acid and *p*-hydroxybenzoic acids by *Cucumis sativus*. *Phytochemistry* 26: 2959-2964.

Singh, H.P., D.R. Batish, and R.K. Kohli. (1999). Autotoxicity: concept, organisms and ecological significance. *Critical Reviews in Plant Sciences* 18: 757-772.

Tang, C.S. (1986). Continuous trapping techniques for the study of allelochemicals from higher plants. In *The Science of Allelopathy*, eds. A.R. Putnam and C.S. Tang. New York: John Wiley and Sons, pp. 113-131.

Tang, C.S., W.F. Cai, K. Kohl, and R.K. Nishimoto. (1995). Plant stress and allelopathy. In *Allelopathy: Organisms, Processes and* Applications, eds. Inderjit, K.M.M. Dakshini, and F.A. Einhellig. ACS Symposium Series 582. Washington, DC: American Chemical Society, pp. 142-157.

Tang, C.S. and C.C. Young. (1982). Collection and identification of alllelopathic compounds from disturbed root system of Bigalta Limpograss (*Hermarthria altissima). Plant Physiology* 69: 155-160.

Tobar, R. (1994). Improved N uptake and transport from ^{15}N-labelled nitrate by external hyphae of AM under water stressed condition. *New Phytologist* 126: 119-122.

Trimble, M.R. (1995). Influence of P nutrition and vericular-AM fungi on growth and yield of greenhouse cucumber. *Canadian Journal of Plant Science* 75: 251-259.

Tsuchiya, K. (1990). Problem on allelopathy in vegetable cropping. *Agriculture and Horticulture* 65: 9-16.

Vaughan, D., D. Jones, and B.G. Ord. (1993). Amelioration by *Voltuella ciliata* of the phytotoxicity of vanallic acid towards to the growth of *Pisum sativum* L. *Soil Biology and Biochemistry* 25: 11-17.

Vaughan, D. and B.G. Ord. (1991). Extraction of potential allelochemicals and their effects on root morphology and nutrient contents. In *Plant Root Growth: An Ecological Perspective*. British Ecological Society Special Publication No. 10. Oxford: Blackwell Scientific Publications, pp. 399-421.

Vaughan, D., G.P. Sparling, and B.G. Ord. (1983). Amelioration of the phytotoxicity of phenolic acids by some soil microbes. *Soil Biology and Biochemistry* 15: 613-614.

Yu, J.Q. (1999). Autotoxic potential of vegetable crops. In *Allelopathy Update: Basic and Applied Aspect*, ed. S.S. Narwal. NY: Science Publishers, Inc., pp. 149-162.

Yu, J.Q., K.S. Lee, and Y. Matsui. (1993). Effects of the addition of activated charcoal on the growth of tomato grown in the hydroponic culture. *Soil Science and Plant Nutrition* 39: 13-22.

Yu, J.Q. and Y. Matsui. (1993a). Extraction and identification of phytotoxic substances accumulated in the nutrient solution for the hydroponic culture of tomato. *Soil Science and Plant Nutrition* 39: 691-700.

Yu, J.Q. and Y. Matsui. (1993b). *p*-Thiocyanateophenol as a novel allelopathic compound in the exudation from root of cucumber. *Chem. Express* 8: 577-580.

Yu, J.Q. and Y. Matsui. (1994). Phytotoxic substances in root exudates of cucumber (*Cucumis sativus* L.). *Journal of Chemical Ecology* 20: 21-31

Yu, J.Q. and Y. Matsui. (1997). Effects of root exudates of cucumber (*Cucumis sativus* L.) and allelochemicals on ion uptake by cucumber seedling. *Journal of Chemical Ecology* 23: 817-827.

Yu, J.Q., S.Y. Shou, Y.R. Qian, and Z.J. Zhu. (2000). Autotoxic potential of some cucurbit crops. *Plant and Soil* 223: 149-153.

Zucconi, F. (1993). Allelopathies and biological degradation in agricultural soils: an introduction to the problem of soil sickness and other soil-born diseases. *Acta Horticulturae* 324: 11-21.

Zucker, M. (1969). Induction of phenylalanine ammonialyase in *Xanthium* leaf discs. *Plant Physiology* 44: 912-922.

Effects of Alfalfa Saponins on *In Vitro* Physiological Activity of Soil and Rhizosphere Bacteria

Robert E. Hoagland
Robert M. Zablotowicz
Wieslaw A. Oleszek

SUMMARY. *In vitro* effects of purified alfalfa (*Medicago sativa* L.) saponins (soyasaponin I, medicagenic acid [MA-3,28Na], MA-3glu, 28Na, and MA-3,28glu) on rhizosphere bacteria cell suspensions were investigated. Triphenyltetrazolium chloride (TTC)-dehydrogenase and fluorescein diacetate (FDA) hydrolytic activities were strongly inhibited (90-95%) by MA-3,28Na and MA-3glu,28Na in *Bacillus thuringiensis* strains (HD-2 and UZ404). In *Curtobacterium flacumafaciens* (JM1011), TTC and FDA activities were reduced only by MA-3,28Na, while little or no effects were observed on the gram-negative strains *Pseudomonas fluorescens* (RA-2) and *Agrobacterium tumefaciens* (A-136). Soyasaponin I decreased FDA hydrolysis in HD-2 and JM1011, but increased FDA activity by 2-fold in RA-2. MA-3glu,28Na increased protein exudation or leakage by 5 to 10-fold in all strains, and MA increased leakage in all strains except RA-2. Soyasaponin in-

Robert E. Hoagland is Research Chemist and Robert M. Zablotowicz is Research Microbiologist, USDA-Agricultural Research Service, Southern Weed Science Research Unit, P.O. Box 350, Stoneville, MS 38776 USA.

Wieslaw A. Oleszek is Professor, Department of Biochemistry, Institute of Soil Science and Plant Cultivation, Osada Palacowa, 24-100 Pulaway, Poland.

Address correspondence to: Robert E. Hoagland at the above address (E-mail: rhoaglan@ag.gov).

[Haworth co-indexing entry note]: "Effects of Alfalfa Saponins on *In Vitro* Physiological Activity of Soil and Rhizosphere Bacteria." Hoagland, Robert E., Robert M. Zablotowicz, and Wieslaw A. Oleszek. Co-published simultaneously in *Journal of Crop Production* (Food Products Press, an imprint of The Haworth Press, Inc.) Vol. 4, No. 2 (#8), 2001, pp. 349-361; and: *Allelopathy in Agroecosystems* (ed: Ravinder K. Kohli, Harminder Pal Singh, and Daizy R. Batish) Food Products Press, an imprint of The Haworth Press, Inc., 2001, pp. 349-361. Single or multiple copies of this article are available for a fee from The Haworth Document Delivery Service [1-800-342-9678, 9:00 a.m. - 5:00 p.m. (EST). E-mail address: getinfo@haworthpressinc.com].

creased protein leakage about 2-fold in JM1011, RA-2 and A-136, but reduced extracellular protein in HD-2 and UZ404. Results suggest that MA-3glu,28Na can deleteriously affect rhizobacteria, and in most cases, the aglycone (MA-3,28Na) is as biologically active as the saponin MA-3glu,28Na.

KEYWORDS. Alfalfa, saponin, antibacterial, allelopathy, rhizobacteria, dehydrogenase, esterase, protein leakage, sodium salts of the alfalfa root saponins

INTRODUCTION

Saponins are comprised of plant glycosides which contain a polar moiety; i.e., water-soluble sugar(s) attached to a lipophilic moiety of steroid or triterpenoid nature. These compounds are found in roots, shoots, seeds, and flowers of many plant species. They are of agronomic interest because of allelopathic interference with plant growth (Oleszek and Jurzysta, 1987; Waller, Jurzysta, and Thorne, 1995). The ecological significance of plant saponins is of interest and has recently been reviewed (Oleszek, Hoagland, and Zablotowicz, 1999). Various saponins exhibit antifungal activity, including toxicity to plant pathogenic fungi. Saponins can influence the growth of soil microflora, and the fungus *Trichoderma* spp. was found to be extremely sensitive to low saponin concentrations (Mishustin and Naumova, 1955; Zimmer, Pedersen, and McGuire, 1967). The effects of β-escin and several other saponins on the growth and activity of several genera of rhizosphere bacteria have also been reported (Zablotowicz, Hoagland, and Wagner, 1996). A wide range of sensitivity to various saponins was observed among bacterial species. Certain bacteria, especially those of the Rhizobiaceae (*Bradyrhizobium*, *Agrobacterium* and *Rhizobium* species) were most susceptible to growth inhibition, while *Bacillus* sp. were most prone to a loss of metabolic activity.

Alfalfa (*Medicago sativa* L.) produces allelopathic saponins which have been implicated as a cause of yield reductions in subsequent crops. Chemical structures of many of these compounds have been determined, and differential biological activity has been associated with differing structures (Waller, Jurzysta, and Thorne, 1995). Alfalfa plant parts differ greatly in the type and quantity of aglycones and saponins present in these tissues (Nowacka and Oleszek, 1994). Among these, medicagenic acid glycosides have been shown to be the most biologically active saponins. They are dominant in alfalfa roots

(Oleszek, Hoagland, and Zablotowicz, 1999). Other alfalfa saponins such as soyasaponin I occur in several genera and species of legumes. The effects of saponins from alfalfa and other plants have been studied in both phytopathogenic and nonpathogenic fungi and found to be inhibitory to a variety of fungal species (Zimmer, Pedersen, and McGuire, 1967; Levy et al., 1989, Oleszek et al., 1990, Oleszek, 1993). Certain alfalfa saponins have also been shown to be inhibitory to some phytopathogenic bacteria (Timbekova, Isaev, and Abubakirov, 1996). Although, alfalfa saponins are allelopathic to plants and inhibitory to certain fungi, their effects on bacteria have not been as widely studied and are poorly understood. Thus, our objectives of these experiments were to critically assess the effects of highly purified alfalfa saponins on a variety of rhizosphere bacteria using *in vitro* parameters that have been shown to be useful indicators of saponin toxicity.

MATERIALS AND METHODS

Chemicals

Saponins were isolated from alfalfa roots using methanol extraction and purified using high-resolution preparative chromatography as described elsewhere (Oleszek, 1993; Nowacka and Oleszek, 1994). Certain compounds, MA-3,28Na and MA-3glu,28Na were used as di- and mono-sodium salts, respectively, to enhance solubility for these bioassays. The chemical structures of alfalfa saponin used in the present study are summarized in Figure 1. Sodium salts were prepared by dissolving the purified saponin or sapogenin (100 mg) in methanol (5 mL). Then, sodium hydroxide (10 mL, 5%) was added with continuous stirring. After equilibration, excess water (100 mL) was added to the solution and it was loaded on C-18 column (2 cm × 3 cm), conditioned with distilled water. The column was then washed with water until the eluate pH was neutral. The sapogenin/saponin sodium salt was eluted from the column with methanol, and evaporation of the methanol yielded the sodium salt powder. All other chemicals were obtained from Sigma Chemicals (St. Louis, MO).

Bacterial Strains and Treatment

Bacterial strains used in these studies (Table 1) have been described elsewhere (Zablotowicz, Hoagland, and Wagner, 1996). Cultures were maintained as frozen glycerol stocks. For experimental protocols, all strains were grown on tryptic soy broth on a rotary-shaking incubator (30°C at 100 rpm). After 24 h growth, cells were harvested by centrifugation (8,000 × g, 10

FIGURE 1

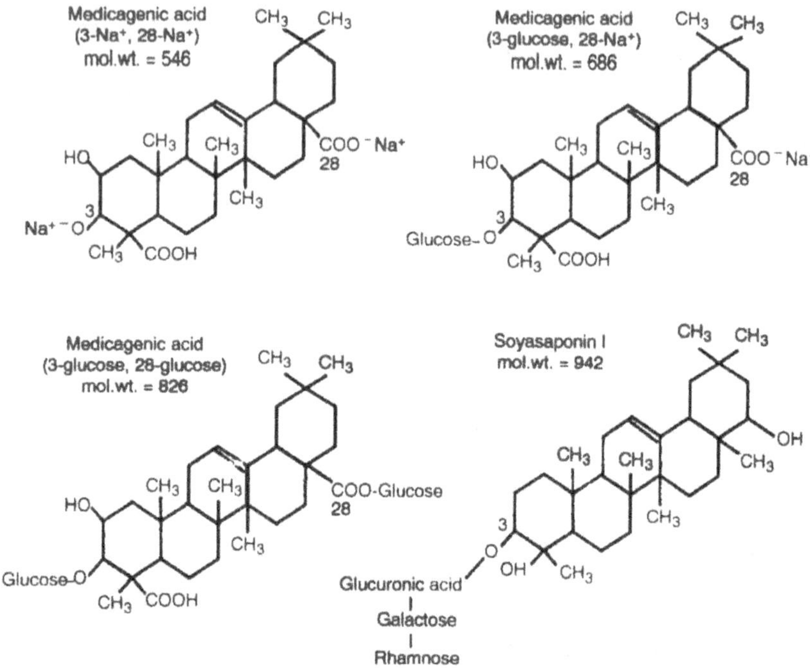

min), washed three times in potassium phosphate buffer (KPi, 50 mM, pH 6.8), and resuspended to yield a concentrated suspension of optical density = 8.0 (660 nm). Cell suspensions (3.8 mL) were placed in sterile 25 mL Corex centrifuge tubes (four replicate tubes per treatment) and treated with ethanolic solutions (80% ethanol, 10 mM) of the saponins to attain a concentration of 500 mM in 2% ethanol. Controls received only ethanol. Following treatment, cell suspensions were incubated (30°C at 100 rpm) for 24 h and then assayed for enzyme activity and protein release.

Enzymological and Physiological Assays

Based upon previous research (Zablotowicz, Hoagland, and Wagner, 1996) we investigated three parameters: fluorescein diacetate (FDA) hydrolysis; triphenyltetrazolium chloride (TTC) dehydrogenase activity, and cellular protein leakage after 24 h exposure to the saponins. Cellular leakage or exudation of protein was determined in cell supernatants following centrifugation (14,000 × g, 10 min). Each supernatant was carefully removed and assayed

TABLE 1. Species of rhizosphere organisms tested.

Genus species	Strain	Source
Gram-Negative Bacteria		
Agrobacterium tumefaciens	A-136	Rhizosphere bacterium; Walter Ream, Oregon State University.
Pseudomonas fluorescens	RA-2	Rice rhizosphere (Hoagland and Zablotowicz, 1995)
Gram-Positive Bacteria		
Bacillus thuringiensis	HD-2	Soil bacterium; USDA-ARS NRRL, Peoria, IL
Bacillus thuringiensis	UZ404	Soybean rhizosphere; Zablotowicz et al., 1991
Curtobacterium flacumafaciens	JM1011	Cotton Rhizosphere; J. McInroy, Auburn University

for FDA activity and protein content. Protein in supernatants was determined using the Bradford reagent (Bradford, 1976).

FDA is a general substrate for several hydrolytic enzymes and serves as substrate for esterases, lipases and certain proteases (Guilbault and Kramer, 1964; Medzon and Brady, 1969). Its activity was determined spectrophotometrically measuring the product of hydrolysis (fluorescein). The assay consisted of 750 μL of cells (or extracellular supernatant), 750 μL KPi buffer (0.1 M, pH 7.6) and 40 μL FDA (2 $mg \cdot mL^{-1}$ in acetone). The assay mixture was vortexed and incubated at 30°C for 5 to 60 min depending on activity. The assay was terminated by extraction with acetone (1.5 mL), followed by centrifugation (14,000 × g, 5 min). The optical density of each supernatant was measured at 490 nm and the total amount of product formed was calculated, based upon an extinction coefficient of 80.3 $mM^{-1} \cdot cm^{-1}$.

TTC-dehydrogenase activity can be used as an estimation of respiratory activity or electron transport system potential. Cells (1.0 ml) plus 50 μl TTC (10%) were incubated for 1 h at 30°C. The reactions were terminated/extracted with 9.0 ml MeOH, and the optical density at 485 nm was determined (Casida, Klein, and Santoro, 1964). The effects of various concentrations (50 to 500 μM) of medicagenic acid (3, 28 Na salt) on TTC-dehydrogenase and FDA-hydrolytic activity of *Bacillus thuringiensis* strain HD-2 was assessed using methodology described above.

RESULTS AND DISCUSSION

Protein Leakage

All saponin treatments increased extracellular protein content, except Soyasaponin I (Table 2). The greatest effects were caused by MA-3,28Na and by two of its glucosylated derivatives, MA-3glu,28Na and MA-3,28glu. In general the aglycone (MA-3,28Na) was as active as the mono-glucosylated derivative (MA-3glu,28Na). However, the di-glucosylated derivative (MA-3,28glu), generally had less activity than sodium salt of the aglycone (MA-3,28Na). Aglycones are formed when the sugar has been removed and Na added. Extreme leakage of cellular protein as evidenced in both *Bacillus* strains may be indicative of cell death, although sporulation may be a possible explanation for protein leakage in this bacterial genera. Although statistically significant effects on extracellular protein were observed in response to Soyasaponin I, these effects were minimal in comparison to MA and its derivatives (Table 2). In other studies glycyrrhetic acid, and to a lesser extent β-escin elicited cellular leakage from several rhizobacterial strains described in this study, while betulin and hecogenin had no effect (Zablotowicz, Hoagland, and Wagner, 1996). The magnitude of leakage elicited by MA-3,28Na and MA-3glu,28Na was of a similar magnitude to that elicited by glycyrrhetic acid.

TABLE 2. Effect of alfalfa saponins (500 μM) on extracellular protein production by rhizosphere bacteria after 24 h exposure.

	Untreated	Soyasaponin I	MA 3,28glu	MA 3glu,28Na	MA 3,28Na
	Protein (ng · mL^{-1})				
P. fluorescens RA-2	76 d[1]	62 e	102 c	141 a	111 b
A. tumefaciens A-136	5 e	11 d	17 c	53 a	35 b
B. thuringiensis HD-2	94 d	66 e	166 c	489 b	558 a
B. thuringiensis UZ-404	67 d	28 e	112 c	458 b	507 a
C. flacumafaciens JM-1011	12 d	24 c	27 c	92 a	59 b

[1] Means of 4 replicates, means followed by the same letter do not differ at the 95% confidence level.

TCC-Dehydrogenase Activity

Saponins had little effect on TCC-dehydrogenase activity in the two gram-negative strains (RA-2 and A-136), and in the gram-positive strain *C. flacumafaciens* JM1011 (Table 3). MA-3,28Na and its mono-glucosylated derivative (MA-3glu,28Na) elicited drastic reductions of dehydrogenase activity in both *B. thuringiensis* strains. This reduction represents an overall loss of respiratory activity, and the substantial reduction in the two *Bacillus* strains is further evidence for cell death.

FDA Hydrolytic Activity (Esterase, Lipase, Protease)

Whole-cell FDA activities in *B. thuringiensis* strains were the most affected by saponins; i.e., MA-3glu,28Na and MA-3,28Na reduced activity in whole cells by 80-95% (Table 4). In *C. flacumafaciens*, significant reductions of whole-cell FDA activity was observed in cells treated with soyasaponin I, and both glycosylated MA derivatives. Treatment of RA-2 with soyasaponin I doubled cellular FDA activity compared to untreated cells. In other studies, β-escin increased FDA activity in strain RA-2 by 3.5 to 5-fold compared to untreated cell activity (Zablotowicz, Hoagland, and Wagner, 1996).

In both *B. thuringiensis* strains, extracellular FDA activity was most affected by MA-3,28Na and its glucosylated derivatives (Table 4). In these

TABLE 3. Effect of alfalfa saponins (500 μM) on triphenyltetrazolium (TTC) activity in rhizosphere bacteria after 24 h exposure.

	Untreated	Soyasaponin I	MA 3,28glu	MA 3glu,28Na	MA 3,28Na
	Whole Cells Activity ($nmol \cdot mL^{-1} \cdot h^{-1}$)				
P. fluorescens RA-2	56 a[1]	55 a	58 a	58 a	60 a
A. tumefaciens A-136	38 b	39 b	41 b	41 b	50 a
B. thuringiensis HD-2	1050 a	1028 a	882 b	191 c	9 d
B. thuringiensis UZ-404	465 b	705 a	557 b	13 c	2 c
C. flacumafaciens JM-1011	92 b	105 ab	117 a	89 c	44 d

[1] Means of 4 replicates, means followed by the same letter do not differ at the 95% confidence level.

TABLE 4. Effect of alfalfa saponins (500 μM) on cellular and extracellular fluorescein diacetate (FDA) hydrolysis by rhizosphere bacteria after 24 h exposure.

	Untreated	Soyasaponin I	MA 3,28glu	MA 3glu,28Na	MA 3,28Na
	Cellular Activity (nmol·mL^{-1}·h^{-1})				
P. fluorescens RA-2	552 b[1]	1034 a	560 b	581 b	586 b
A. tumefaciens A-136	786 a	750 a	817 a	815 a	819 a
B. thuringiensis HD-2	982 a	704 c	1044 a	169 d	48 f
B. thuringiensis UZ-404	4536 a	4269 ab	4070 b	212 c	187
C. flacumafaciens JM-1011	405 a	303 c	354 b	364 b	390 a
	Extracellular Activity (nmol·mL^{-1}·h^{-1})				
P. fluorescens RA-2	36 b	8 c	39 b	49 a	53 a
A. tumefaciens A-136	2 c	3 b	4 a	4 a	3 b
B. thuringiensis HD-2	24 d	14 d	44 c	203 b	437 a
B. thuringiensis UZ-404	349 d	183 e	545 c	1724 b	1884 a
C. flacumafaciens JM-1011	59 b	29 c	35 c	35 c	167 a

[1] Means of 4 replicates, means followed by the same letter do not differ at the 95% confidence level.

strains the highest levels of extracellular FDA hydrolysis were caused by MA-3,28Na, moderate levels by the mono-glucosylated, and the lowest levels by the di-glucosylated MA. Extracellular FDA activity was also about 2.5-fold greater than in untreated *C. flacumafaciens* cells incubated with MA-3,28Na, while levels lower than in untreated cells occurred by treatments with di- and mono-glucosylated MA, or soyasaponin I. FDA activity in either whole-cell or extracellular assays of *Agrobacterium tumefaciens* strain A136 was not affected by any of the four compounds tested (Table 4). In our studies, only a minimal effect of alfalfa saponins was observed on enzymatic activity of resting cells of *A. tumefaciens*. In other studies, the growth of *A. tumefaciens* was severely inhibited by β-escin, but enzymatic activity was not altered (Zablotowicz et al., 1996).

Effect of MA-3,28Na Concentration on Bacillus thuringiensis strain HD-2

The effects of a 24 h exposure of various concentrations of disodium MA extracellular protein and enzyme activities of *B. thuringiensis* strain HD are presented in Table 5. Extracellular protein increased about 2-fold at 50 μM MA-3,28Na, and was about 4-fold greater at concentrations from 125 to 500 μM. TTC-dehydrogenase activity was more sensitive to this compound than cellular FDA activity, i.e., a 59% and 25% loss in activity, respectively, occurred at 50 μM. At 500 μM, virtually no dehydrogenase activity was present, while cellular FDA hydrolytic activity was about 18% of that in untreated cells of this strain. A linear increase in extracellular FDA hydrolytic activity was exhibited at concentrations of 50 to 500 μM MA-3,28Na (Table 5).

CONCLUSIONS

These studies demonstrate differential physiological responses of various bacterial genera to structurally different alfalfa saponins. Leakage/exudation

TABLE 5. Extracellular protein, TTC-dehydrogenase activity, and cellular and extracellular fluorescein diacetate (FDA) hydrolytic activity of *Bacillus thuringiensis* HD-2 suspensions as affected by various concentrations of medicagenic acid (3,28 sodium salt), after 24 h exposure.

Concentration (μM)	Protein ($ng \cdot mL^{-1}$)	TTC-dehydrogenase $nmol \cdot mL^{-1}h^{-1}$	FDA-hydrolysis $nmol \cdot mL^{-1}h^{-1}$	
	extracellular	cellular	cellular	extracellular
0	127 a[1]	532 a	2226 a	298 a
50	277 b	223 b	1668 b	296 a
125	458 c	35 c	1227 c	377 b
250	498 c	14 c	653 d	469 c
500	547 c	9 c	390 e	613 d

[1] Mean four replicates, means followed by the same letter do not differ at the 95% confidence level.

of protein was observed in all five strains in response to medicagenic acid (MA) and its derivatives. The extraordinarily high cellular leakage of protein in strains UZ404 and HD-2 (both gram-positive, *B. thuringiensis*), correlated directly with loss of their whole-cell enzyme activities. Increased protein excretion may be associated with either autolysis or exudation. However, considering the magnitude of the protein release from these *Bacillus* strains, death, or perhaps sporulation (a distinctive characteristic of *Bacillus*) may have occurred. During the sporulation process, several extracellular enzymes are known to be excreted, e.g., proteases, amylases, RNA- and DNA-degrading phosphodiesterases, and cell wall-lytic enzymes (Schaeffer, 1969). This may explain the increased extracellular FDA-hydrolytic activity found in *B. thuringiensis* following exposure to MA-3,28Na and MA-3glu, 28Na.

A major site of action of saponins is their interference with membrane integrity, especially interactions with sterols (Bangham and Horne, 1962). Thus, bacteria and certain fungi (Phycomycetes, e.g., *Phytophthora*) should be less sensitive to saponins since their membranes are low in sterols. Recent studies comparing medicagenic acid-sensitive and acid-resistant *Trichoderma viride* strains indicated that fatty acid composition is as important as sterol content with regard to saponin sensitivity (Gruiz, 1996). Although the mechanism is unknown, based upon data presented here it is clear that certain saponins-glycosides and their aglycones (such as MA and its derivatives) can affect membrane leakage of gram-negative and gram-positive bacteria.

Various *in vitro* studies have demonstrated effects of saponins from alfalfa on other crops and microorgansims. Some saponins exhibit fungal inhibition at very low concentrations. For example, certain MA glycosides inhibited *Trichodemra viride* growth by 100% at a saponin concentration of 1.6 ppm (Oleszek et al., 1990), and were also highly inhibitory to several species of medically important yeasts at 3 to 15 ppm (Polacheck et al., 1986). Although the *B. thuringiensis* strains were the most sensitive of the bacteria we evaluated, only moderate inhibition of vital activity occurred at 50 μM MA-3,28Na. Other alfalfa saponins (13 glycosides from roots; 10 from leaves) were examined for growth inhibition of several phytopathogenic bacteria (Timbekova, Isaev, and Abubakirov, 1996). Almost all alfalfa saponins, except a hederagenin analog, suppressed growth of *Corynebacterium michiganense* and *A. tumefaciens*. Growth of the alfalfa pathogen *C. insidiosum*, was suppressed by saponin mixtures from alfalfa roots, purified MA-3glu, and a hederagenin analog. None of the saponins affected the soft rot bacterium *Erwinia carotovora*, and only leaf saponins were inhibitory to *Xanthomonas campestris*.

Allelopathic interactions of saponins in nature are comprised of a multitude of factors. Thus, the effects of saponins on both plants and soil/rhizosphere microorganisms must be considered. For example, treatment of soil with the saponin β-escin, had no significant effect on the total bacterial population, but populations of gram-negative bacteria and fluorescent pseudomonads increased (Zablotowicz, Hoagland, and Wagner, 1996). This indicates that gram-positive bacteria are more sensitive to β-escin, compared to gram-negative bacteria. In these same studies, fungal propagules decreased after incubation with β-escin. In similar studies, soil applied β-escin dramatically inhibited emergence and growth of several species, including wheat and the weed barnyard grass (*Echinochloa crus-galli* L. Beauv.) (Hoagland, Zablotowicz, and Reddy, 1996). Nodulation of hemp sesbania (*Sesbania exaltata* [Raf.] Rybd.) was unaffected by β-escin, but soybean (*Glycine max* [L.] Merr.) nodulation was reduced in treated soil. Somewhat contrary to this, the growth of *Bradyrhizobium japonicum* and *Rhizobium meliloti* was drastically inhibited by β-escin relative to other bacterial species tested (Zablotowicz, Hoagland, and Wagner, 1996). Incorporation of alfalfa roots or shoots into soil inhibited wheat (*Triticum aestivum* L.) growth, and soil extracts from this alfalfa-amended soil also inhibited *T. viride* growth (Oleszek and Jurzysta, 1987). This inhibition of both species diminished over time, and was associated with microbial detoxification/metabolism of the saponins.

The overall effects of saponins on various species are complex due to the wide variation of their chemical structures, and the broad range of sensitivity to these compounds exhibited by plants and microorganisms. These data indicate that soil and rhizosphere bacteria can be affected by saponins. The ecological impact of saponins produced by crop plants and weeds on microbial populations and on certain microbial processes in soil warrants further study.

REFERENCES

Bangham, A.D. and R.W. Horne. (1962). Action of saponin on biological cell membranes. *Nature* 196: 952-954.

Bradford, M.M. (1976). A rapid and sensitive method for the quantitation of microgram quantities of protein utilizing the principle of protein-dye binding. *Analytical Biochemistry* 72: 248-254.

Casida, L.E., D.A. Klein, and T. Santoro. (1964). Soil dehydrogenase activity. *Soil Science* 98: 371-376.

Guilbault, G.G. and D.N. Kramer. (1964). Fluorometric determination of lipase, acylase, alpha- and gamma-chymotrypsin and inhibitors of these enzymes. *Analytical Chemistry* 36: 409-412.

Gruiz, K. (1996). Fungitoxic activity of saponins: practical use and fundamental principles. In *Saponins Used in Traditional and Modern Medicine*, eds. G.R. Waller and K. Yamasaki. New York: Plenum Press, pp. 527-537.

Hoagland, R.E. and R.M. Zablotowicz. (1995). Rhizobacteria with exceptionally high aryl acylamidase activity. *Pesticide Biochemistry and Physiology* 52: 190-200.

Hoagland, R.E., R.M. Zablotowicz, and K.N. Reddy. (1996). Studies of the phytotoxicity of saponins on weed and crop plants. In *Saponins Used in Food and Agriculture*, eds. G.R. Waller and K. Yamasaki. New York: Plenum Press, pp. 57-73.

Levy, M., N. Zehavi, N. Naim, and I. Polachek. (1989). Isolation, structure determination, synthesis, and antifungal activity of a new alfalfa root saponin. *Carbohydrate Research* 193: 115-123.

Medzon, E.L. and M.L. Brady. (1969). Direct measurement of acetylesterase in living protist cell. *Journal of Bacteriology* 97: 402-415.

Mishustin, E.N. and A.M. Naumova. (1955). Secretion of toxic substances by alfalfa and their effect on cotton and soil microflora. *Akademiya Nauk SSSR Izvestiya Seriya Biologicheskaya* 6: 3-9 (In Russian with English summary).

Nowacka, J., and W. Oleszek. (1994). Determination of alfalfa (*Medicago sativa*) saponins by high performance liquid chromatography. *Journal of Agricultural and Food Chemistry* 42: 727-730.

Oleszek, W. (1993). Allelopathic potentials of alfalfa (*Medicago sativa*) saponins: their relation to antifungal and hemolytic activities. *Journal of Chemical Ecology* 19: 1063-1074.

Oleszek, W., R.E. Hoagland, and R.M. Zablotowicz. (1999). Ecological significance of plant saponins. In *Principles and Practices in Plant Ecology: Allelochemical Interactions*, eds. Inderjit, K.M.M. Dakshini, and C.L. Foy. Boca Raton, FL: CRC Press, pp. 451-465.

Oleszek, W. and M. Jurzysta. (1987). The allelopathic potential of alfalfa root medicagenic acid glycosides and their fate in soil environment. *Plant and Soil* 98: 67-80.

Oleszek, W., K.R. Price, I.J. Colquhoun, M. Jurzysta, M. Ploszynski, and G.R. Fenwick. (1990). Isolation and identification of alfalfa (*Medicago sativa* L.) root saponins: Their activity in relation to a fungal bioassay. *Journal of Agricultural and Food Chemistry* 38: 1810-1817.

Polacheck, U., M. Zehavi, M. Naim, M. Levy, and R. Evron. (1986). Activity of compound G2 isolated from alfalfa roots against medically important yeasts. *Antimicrobial Agents and Chemotherapy* 30: 290-294.

Schaeffer, P. (1969). Sporulation and products of antibiotics, exoenzymes and exotoxins. *Bacteriological Reviews* 33: 48-71.

Timbekova, A.E., M.I. Isaev, and N.K. Abubakirov. (1996). Chemistry and biological activity of triterpenoid glycosides from *Medicago sativa*. In *Saponins Used in Food and Agriculture*, eds. G.R. Waller and K. Yamasaki. New York: Plenum Press, pp. 171-182.

Waller, G.R., M. Jurzysta, and R.L. Thorne. (1995). Root saponins from alfalfa (*Medicago sativa* L.) and their allelopathic activity on weeds and wheat. *Allelopathy Journal* 2: 21-30.

Zablotowicz, R.M., C.D. Bierle, A.H. Bosworth, and J.C. Gates. (1991). Soybean rhizosphere colonization by bacterial inoculants in midwestern soils. *Phytopathology* 81: 1215.

Zablotowicz, R.M., R.E. Hoagland, and S.C. Wagner. (1996). Effects of saponins on the growth and activity of rhizosphere bacteria. In *Saponins Used in Food and Agriculture*, eds. G.R. Waller and K. Yamasaki. New York: Plenum Press, pp. 83-95.

Zimmer, D.E., M.W. Pedersen, and C.F. McGuire. (1967). A bioassay for alfalfa saponins using the fungus *Trichoderma viride* Pers. ex Fr. *Crop Science* 7: 223-224.

Activity and Persistence of Sorgoleone, a Long-Chain Hydroquinone Produced by *Sorghum bicolor*

L. A. Weston
M. A. Czarnota

SUMMARY. *Sorghum bicolor* (L.) Moensch and other related Sorghum species produce sorgoleone–a long chain hydroquinone compound (MW = 358), which is exuded by living root systems. This compound exhibits potent activity as a photosynthetic inhibitor. A diverse group of sorghum germplasm was evaluated for sorgoleone production. Production was quite variable, with certain accessions producing up to 15 mg sorgoleone/g fresh root weight. The root exudate composition among accessions was less variable, with sorgoleone being the major constituent (76-99%) within the extract. The potential binding of sorgoleone to the D1 protein of the PSII complex was evaluated in triazine resistant and susceptible redroot pigweed (*Amaranthus retroflexus* L.) thylakoids. Sorgoleone, metribuzin and diuron exhibited competitive binding with atrazine in susceptible thylakoids, while no competition was evident in resistant thylakoids. Sorgoleone has an intermediate affinity between that of diuron and metribuzin from estimated binding constants. Computer-aided design programs have prov-

L. A. Weston is Professor, and M. A. Czarnota is Graduate Student, Department of Horticulture, Cornell University, Ithaca, NY 14853.

Address correspondence to: L. A. Weston at the above address (E-mail: law20@cornell.edu).

The authors would like to thank S. C. Weller and C. N. Yerkes of Purdue University, West Lafayette, IN, for providing R and S pigweed used in this study and R. Harmon for assistance in data collection and analysis.

[Haworth co-indexing entry note]: "Activity and Persistence of Sorgoleone, a Long-Chain Hydroquinone Produced by *Sorghum bicolor*." Weston, L. A., and M. A. Czarnota. Co-published simultaneously in *Journal of Crop Production* (Food Products Press, an imprint of The Haworth Press, Inc.) Vol. 4, No. 2 (#8), 2001, pp. 363-377; and: *Allelopathy in Agroecosystems* (ed: Ravinder K. Kohli, Harminder Pal Singh, and Daizy R. Batish) Food Products Press, an imprint of The Haworth Press, Inc., 2001, pp. 363-377. Single or multiple copies of this article are available for a fee from The Haworth Document Delivery Service [1-800-342-9678, 9:00 a.m. - 5:00 p.m. (EST). E-mail address: getinfo@haworthpressinc.com].

en useful to further evaluate structural activity relationships for sorgoleone, and PSII inhibitors. Sorgoleone at concentrations up to 40 ppmw significantly reduced shoot fresh weight in several broadleaf weed species when incorporated in soil, but had limited effect on root development. Sorgoleone at the same concentrations applied postemergence also caused significant growth reduction to a number of weed species. Recovery of soil impregnated with sorgoleone was most effective (up to 85%) when acetonitrile:water (80:20 v/v) was used as the extractant, in comparison to methanol:water or water alone. Recovery declined over a 42 day period after incorporation, but detectable sorgoleone levels were observed after 7 weeks.

KEYWORDS. Allelopathy, *Sorghum bicolor*, root exudates, sorgoleone, photosynthesis, photosystem II, soil activity

INTRODUCTION

The use of allelopathic traits of crop species for weed suppression in agroecosystems is an idea first suggested by A. R. Putnam in the 1980's (Putnam and DeFrank, 1983; Putnam, DeFrank, and Barnes, 1983). More recently, the weed suppressive nature of cover crop species or green manures has been utilized to aid in weed management in both horticultural and agronomic cropping systems (Weston, 1996). Weed suppression provided by these covers can be variable, and is dependant upon factors such as inherent weed pressure encountered, climate, cover crop, amount and longevity of residue, soil microbial populations and water availability. The interactions of higher plants within an agroecosystem are often poorly understood and strong evidence for allelopathic interference is typically unavailable or poorly documented in the literature (Inderjit and Weston, 2000). However, the interference and weed suppressive potential exhibited by *Sorghum bicolor* (L.) Moensch in the field and by the natural product sorgoleone in the laboratory present a unique opportunity to assess the physiological basis of plant growth interference through allelopathy.

In the U.S., sorghum is cultivated upon more than 600,000 ha for grain, with greater than 350,000 ha used for forage, green manures and in the production of sorghum syrup (Heath, Barnes, and Metcalfe, 1985). Sorghum is often chosen as a summer annual cover crop because of its rapid growth and ability to suppress weeds (Forney, Foy, and Wolf, 1985). In Michigan sorghum covers are killed and used as weed suppressive residues in orchards.

The allelopathic effect may last for several weeks after sorghum is killed (Putnam and DeFrank, 1983; Putnam, DeFrank, and Barnes, 1983). In Mississippi, Smeda has shown that spring-planted sorghum residues provide up to 90% reductions in weed biomass for 6 to 8 weeks in no-till summer-planted soybeans (personal communication). When sorghum was incorporated as a green manure, it strongly suppressed annual weeds in succeeding alfalfa crops (Forney, Foy, and Wolf, 1985). Einhellig and Rasmussen (1989) have shown that the inhibitory effects of grain sorghum on surrounding weed growth occurred through the following growth season.

The ability of sorghum residues or green manures to suppress weeds has been discussed in the literature for years, but until recently, the chemistry of the potential inhibitor(s) remained unknown. Several sorghum species including *S. halepense*, *S. vulgare*, *S. sudanese*, and *S. bicolor* have shown allelopathic interference with weed and crop growth (Einhellig and Souza, 1992). Sorghums produce and release cyanogenic glycosides, and a number of phenolic breakdown products of these glycosides contribute to short-term plant growth suppression (generally less than 8 weeks) in field and greenhouse experiments (Einhellig and Rasmussen, 1989; Guenzi and McCalla, 1966; Nicollier, Pope, and Thompson, 1983; Weston, Harmon, and Mueller, 1989). The toxicity of sorghum to livestock has long been associated with the release of HCN from dhurrin (*p*-hydroxy-(S)-mandelonitrile β-D-glucopyranoside) (Figure 1), which is present in sorghum herbage. Although dhurrin was not phytotoxic in selected assays, two compounds identified from sudex (*S. bicolor* × *S. sudanese*) herbage by Weston, Harmon, and Mueller (1989), *p*-hydroxybenzoic acid and aldehyde (Figure 1), are breakdown products of dhurrin. These simple phenolics possessed significant activity in bioassays of germinating seedlings and hydroponic systems (Weston, unpublished data). Using a modified Parker bioassay that simulated field conditions, Weston, Harmon, and Mueller (1989) determined that herbage of young seedlings (2 to 4 weeks of age) was most phytotoxic and produced greater levels of these products on a per g tissue basis than older herbage (6 to 8 weeks of age).

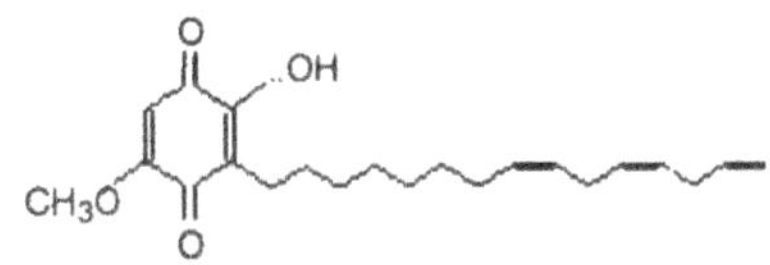

Sorgoleone MW = 358

Sorghum roots, herbage and germinating seeds all release phytoinhibitors. Panasuik, Bills, and Leather (1986) established that germinating sorghum seeds inhibited germination and growth of grass and broadleaf species. In our own work, under closely controlled conditions for light and moisture avail-

FIGURE 1. Double reciprocal plots for binding of ^{14}C atrazine in competition with sorgoleone, diuron, or metribuzin to thylakoids isolated from triazine-resistant (R) and triazine-susceptible (S) redroot pigweed (Nimbal et al., 1996b, reprinted with permission).

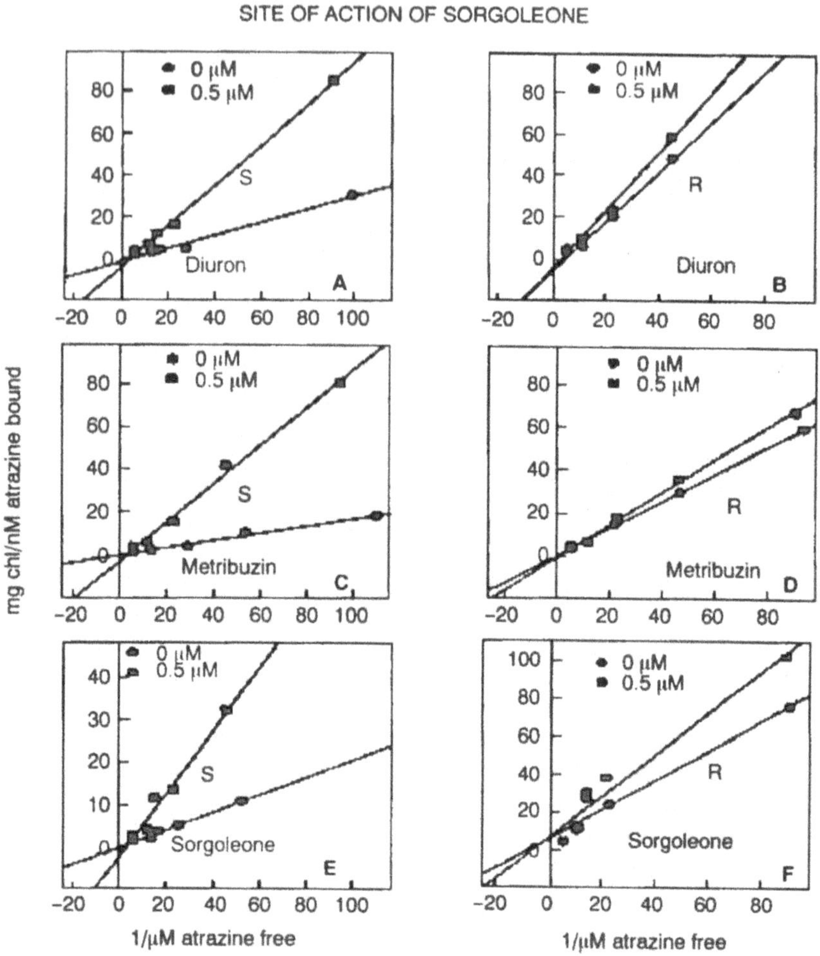

ability, germinating sorghum was also inhibitory to growth of several weeds (Hoffman et al., 1996). Forney and Foy (1985) found that rhizosphere products of hydroponically grown sudex were more phytotoxic than compounds from other plant parts and toxicity increased with increasing plant age up to 6 weeks. They also noted a yellow pigmentation associated with the rhizosphere product. Our own work with ornamentals planted into living sudex root systems showed that seedlings developed chlorosis rapidly in the presence of sudex roots, and exhibited minimal growth over a 7 week period. This inhibition could not be overcome by the addition of water-soluble fertilizer. The living root system of sudex was most likely associated with the greatest phytotoxicity to ornamental seedlings (Geneve and Weston, 1988).

Netzley and Butler (1986) first reported the presence of sorgoleone in the root exudates of *S. bicolor*. They isolated sorgoleone, the oxidized quinone form of a hydrophobic *p*-benzoquinone from sorghum root exudates. Sorgoleone is the major constituent of sorghum root exudates (Table 1) and is easily extracted from roots in methylene chloride plus a small amount of acetic acid (Netzley and Butler, 1986; Nimbal et al., 1996a). Sorgoleone possesses remarkable phytotoxicity in numerous plant growth assays (Einhellig and Souza, 1992; Nimbal et al., 1996a) and is primarily an inhibitor of plant growth, apparently through inhibition of photosynthesis (Einhellig and Souza, 1992; Einhellig et al., 1993; Nimbal et al., 1996b) and respiration (Rasmussen et al., 1992).

The mechanism of action of many herbicides inhibiting photosynthesis involves the blockage of electron transport in photosystem II (PS II) by binding to the Q_B electron acceptor at the D1 protein. This characteristic type of inhibition has been reported in triazines, phenylureas, ureas, and uracils among others (Mets and Thiel, 1989). Binding studies have shown the existence of a common binding niche at Q_B to which these herbicides bind and a close correlation between electron transport inhibition and herbicide binding (Nimbal et al., 1996b). Plants resistant to triazines serve as a useful tool to examine the concept of different but overlapping binding sites at the 32 kDa protein of the PSII complex in diuron-type herbicide inhibitors. Studies in our laboratory with sorgoleone have shown that sorgoleone is a potent inhibitor of photosynthesis and its site of inhibition is within the PSII complex. The activity of sorgoleone compared well with that of diuron (Nimbal et al., 1996a; Gonzalez et al., 1997). Further studies were conducted to evaluate the ability of diverse sorghum germplasm to produce sorgoleone, the specific mode of action and binding characteristics of sorgoleone within PSII, and soil persistence of sorgoleone over time, key questions related to the importance of sorgoleone as an allelochemic and its physiological role in the interference mechanism of *Sorghum bicolor*.

MATERIALS AND METHODS

Screening of Sorghum Germplasm for Sorgoleone Production

A diverse collection of grain sorghum germplasm (Table 1) was screened for sorgoleone production with 5-day-old seedling roots. Germination per-

TABLE 1. Sorgoleone production by various sorghum genotypes[a]

Sorghum genotype	Root fresh wt (g)	Sorgoleone (mg)	Sorgoleone mg/g root wt	% Purity
Rtx433	0.15 ± 0.03	0.10 ± 0.00	0.67	91.8
R NB9040	0.13 ± 0.01	1.00 ± 0.25	7.70	89.5
IS 3723C	0.29 ± 0.01	1.23 ± 0.18	4.24	85.7
IS 8266C	0.22 ± 0.01	1.10 ± 0.10	5.00	87.2
B N122	0.35 ± 0.01	2.00 ± 0.10	5.71	75.9
Rtx7078	0.15 ± 0.01	1.80 ± 0.21	12.00	98.7
B Martin	0.32 ± 0.03	1.73 ± 0.13	5.40	78.2
IS 8160C	0.22 ± 0.03	2.00 ± 0.15	9.10	83.8
Rtx415	0.20 ± 0.02	1.87 ± 0.03	9.35	92.4
Rtx430	0.08 ± 0.04	0.30 ± 0.10	3.75	94.9
IS 1318C	0.17 ± 0.03	2.43 ± 0.47	14.20	87.0
Rtx7000	0.25 ± 0.02	2.30 ± 0.32	9.20	78.9
Greenleaf	0.06 ± 0.01	0.73 ± 0.47	11.40	99.1
IS 7333C	0.27 ± 0.01	1.63 ± 0.27	6.00	83.4
EH-Sart	0.30 ± 0.04	1.60 ± 0.21	5.33	87.1
IS 5893C	0.19 ± 0.02	0.33 ± 0.07	1.74	82.4
Btx3042	0.19 ± 0.02	0.30 ± 0.06	1.58	77.7
B Redlan	0.15 ± 0.03	2.67 ± 1.31	17.80	80.2
R N97	0.25 ± 0.00	1.70 ± 1.15	6.80	81.6
Piper	0.17 ± 0.01	0.17 ± 0.06	1.00	98.6
IS 1269C	0.20 ± 0.03	0.77 ± 0.15	1.10	83.2
IS 7041C	0.37 ± 0.02	0.60 ± 0.31	1.62	87.9
IS 1098C	0.08 ± 0.01	0.20 ± 0.10	2.50	88.2
IS 12611C	0.27 ± 0.02	0.67 ± 0.03	2.48	84.1
B Wheatland	0.33 ± 0.06	0.83 ± 0.23	2.50	78.3

[a]The data are means (and SD) of three replicates of 25 seedlings each.
Nimbal et al., 1996a, reprinted with permission, ©1996 American Chemical Society.

centage and root fresh weight was recorded for each accession, and total quantity of sorgoleone produced was determined. The extraction procedure for sorgoleone was performed as described in Nimbal et al. (1996a), where sorghum seeds were germinated in a seed germination chamber at 29°C for 5 days in the dark. Seedling roots were excised and dipped in methylene chloride plus 1% glacial acetic acid. The crude extracts were filtered and then subjected to HPLC (reverse phase Nova-Pak C_{18} column, 3.9 mm × 150 mm). The mobile phase was 75% acetonitrile/25% acidified water. Water was acidified with glacial acetic acid at 2.5%. Sorgoleone was detected as reported in Nimbal et al. (1996a) using a Waters absorbance detector at 280 nm after 20 µl of the acetonitrile solubilized crude extract sample was injected. Flow rate was 2 ml/min with a 10 minutes total run time. For further experimentation with sorgoleone, Pioneer 8333 seeds (Pioneer Seed Co.) were utilized and extracted. Pioneer 8333 was utilized because of its wide spread use and ready availability at the time. Sorgoleone (MW = 358) constituted more than 90% of the extract.

Binding of Sorgoleone and other PSII Inhibitor Herbicides

Experiments were conducted using labeled atrazine and unlabeled herbicide inhibitors including sorgoleone (obtained from extraction as described above), metribuzin, and diuron (Nimbal et al., 1996b). The ability of sorgoleone to compete for a common binding site in PSII was evaluated in comparison to other inhibitors by varying the ^{14}C atrazine concentration from 0.01 to 0.4 µM in the presence or absence of 0.5 or 1.0 µM concentration of inhibitors. All binding experiments were conducted as reported in Nimbal et al. (1996b), at room temperature and reduced light. Thylakoid membranes were isolated from resistant and susceptible redroot pigweed seedlings (*Amaranthus retroflexus*) obtained from the laboratory of S. Weller with permission of R. Smeda, following the protocol of Smeda et al. (1993). Chloroplasts equivalent to 50 µg Chl/ml were suspended in 1 ml of binding medium containing ^{14}C atrazine. Inhibitors were then added to each tube and samples were incubated for 3 min after vortexing. Finally, samples were centrifuged at 15,000 g for 5 min and an aliquot of 0.5 ml from the supernatant was used for radioactivity measurements to determine the amount of unbound atrazine. The amount of bound ^{14}C atrazine was calculated by subtracting unbound atrazine in the supernatant from total added to the chloroplast suspension. Competitive binding was evaluated by plotting the total concentration of unlabeled inhibitor versus the concentration bound ^{14}C atrazine per mg chlorophyll. Besides sorgoleone, other inhibitors evaluated for binding ability as measured by atrazine displacement included bentazon, metribuzin and diuron. The mechanism of competition for each inhibitor was evaluated using double reciprocal plots. Competitive binding is indicated by a common intercept on the ordinate. The binding constant (K_b) for unlabeled inhibitors was computed by replotting the slopes of the double reciprocal plot versus the total concentration of unlabeled inhibitor. The abscissa intersection of the straight line fitted for this new data plot indicates the binding constant for that particular inhibitor (Nimbal et al., 1996b).

The binding of sorgoleone to the Q_B binding site of the D1 protein was further analyzed using 3 dimensional computer aided graphics programs such as MOPAC 93 (Quantum Chemistry, Bloomington, IN) and Chem-X (Chemical Design Limited, Oxfordshire, UK). Plastoquinone, sorgoleone, PSII inhibiting herbicides and 12 benzoquinones were evaluated using various software packages for bulk, electronic and energy properties. Electrostatic and lysophilic field potentials were evaluated, among other properties and correlated with biological activity of these compounds and their binding potential within the Q_B binding site of the D1 protein.

Soil Activity and Persistence of Sorgoleone

Since sorgoleone is a major component of root exudates produced by sorghum which are released into soil by living sorghum root systems, we wanted to evaluate the soil activity of sorgoleone against weed species. Assays were performed by impregnating a sand soil mixture (50% sand/50% Maury sterilized Maury silt loam) with sorgoleone at concentrations ranging from 10 to 80 ppmw (parts per million by weight). Sorgoleone was applied by dissolving in a trace amount of acetone first, and final suspension in milli-Q water. Selected weed species were then grown for a 3 week period in plastic cones filled with treated soil. Root and shoot fresh weights were evaluated at experimental termination. Weeds evaluated included common purslane (*Portulaca oleracea*), velvetleaf (*Abutilon theophrasti*), sicklepod (*Cassia obtusifolia*), redroot pigweed (*Amaranthus retroflexus*), large crabgrass (*Digitaria sanguinalis*) and green foxtail (*Setaria viridis*). Seedlings were grown in a growth chamber at 28°C under a 14 h light 10 h dark regime with 80% relative humidity. Cones were seeded with 3 weed seeds per cone and treatments were replicated 5 times and arranged as a completely randomized design within the growth chamber. Weed seed was purchased from Azlin Seed Company. Data on seedling size including shoot fresh weight was collected 21 days after treatment. Percent inhibition for each treatment was calculated by comparison with an untreated soil control. Additionally, other weed seedlings were raised in a growth chamber as described above for a 2 week period. At 14 days of age, seedlings were treated with 0.6 kg/ha sorgoleone, by postemergence application of sorgoleone to the foliage of weed seedlings, applied with a small microsprayer containing sorgoleone dissolved in water with 5% acetone. Spray volume utilized for this application was 80 L/ha. Controls were treated with the same water solution containing 5% acetone as well. Ten days after treatment, fresh weight of seedlings was collected. Five replicates were performed per treatment and the experiment was repeated in time. Percent inhibition as compared to the control was calculated for all data collected.

A stock solution of 10 mM sorgoleone was prepared in acetonitrile and added to triplicate 100 g soil samples in 500 ml glass bottles and mixed by shaking to achieve 25 ppmw concentration. Sorgoleone fortified soil was then incubated at room temperature. Sorgoleone persistence over time was evaluated by collecting samples of 10 g soil from each triplicate set at 0, 7, 14, 28 and 42 days after incubation. Extraction was performed using 20 ml of acetonitrile/water (80/20 v/v) and shaking on a rotary shaker at 180 rpm for 6 hours. Soil extracts were filtered using a .2 μm syringe and evaporated to dryness. Extracts were redissolved in acetonitrile (1 ml) and analyzed by HPLC using gradient analysis. HPLC was Waters system with a C_{18} Nova-Pak column as described previously and mobile phase was acetonitrile/acidified water (2.5% glacial acetic acid), 80:20 v/v for 1 minute, 75:25 v/v 1-3 min and 60:40 v/v 3-6 min. The flow rate was 2 ml/min. Sorgoleone standard was obtained by extracting sorghum seedling roots and purification by HPLC. Purified extract contained 98% sorgoleone or greater. Preliminary experiments to evaluate extractability of sorgoleone from soil with acetonitrile/water and methanol/water were performed, as well as extraction time for efficient recovery.

RESULTS AND DISCUSSION

Screening of sorghum germplasm for sorgoleone production indicated that considerable variability exists among genotypes with regard to the amount of sorgoleone produced (Table 1) (see also Nimbal et al., 1996a). Rtx433 produced only 0.67 mg/g sorgoleone/g fresh root weight, while B Redlan and IS 1318C produced 17.8 and 14.2 mg/g, respectively. Most other genotypes ranged between 1.5 and 10 mg/g among cultivars. In another collection, Hess, Ejeta, and Butler (1992) showed little difference among production of mg sorgoleone per g of root fresh weight, but indicated production was sensitive to environmental conditions. Our studies, which were performed under identical controlled environmental conditions with all sorghum seed harvested at the same time and produced under identical field conditions (direct communication, J. Pedersen, USDA sorghum breeder) showed that production and secretion of sorgoleone may also be dependent on inherent genetic differences among *Sorghum* genotypes. On average, sorgoleone constituted 85-90% of the root exudate composition of the germplasm evaluated, but the range varied from 76 to 99%. This range in purity from sorgoleone analyzed by HPLC under identical conditions also suggests that differential allelopathic interference among genotypes may exist. More recent studies with different species and accessions of sorghum raised on capillary mats to extract larger quantities of sorgoleone from greater quantities of sorghum roots also show differences in level of sorgoleone produced, again suggesting that inherent genetic differences among accessions and species of sorghum account for sorgoleone variation, since all environmental conditions were kept identical throughout the experiment (unpublished data, Czarnota).

To date, all sorghum species tested which included *Sorghum bicolor*, *Sorghum sudanese*, and *Sorghum halepense*, produce significant levels of sorgoleone, although minor related constituents in the extract may vary.

Since direct binding studies for sorgoleone onto thylakoid membranes are not possible in the absence of ^{14}C sorgoleone, we determined the potential binding of sorgoleone to the D1 protein of the PS II complex by evaluation of competitive binding of sorgoleone versus atrazine. The binding behavior was then evaluated in isolated triazine susceptible (S) and triazine resistant (R) redroot pigweed thylakoids. Sorgoleone, diuron, and metribuzin did not show competitive binding to thylakoids of R-pigweed, even at higher concentrations (Figure 1 and also Nimbal et al., 1996b). A quantitative analysis of competitive binding data in which data are transformed to linear relationships is possible in double reciprocal plots (Nimbal et al., 1996b). When atrazine concentration was varied in the presence or absence of 0.5 μM concentrations of the 3 inhibitors (Figures 1A, 1C, and 1E), regression lines generated for plots with and without inhibitor had similar ordinate but different abscissa intercepts. Identical results were obtained with concentrations of 1.0 μM inhibitors and are not presented. These findings confirm that all 3 inhibitors competed effectively with atrazine for binding to S pigweed thylakoids. No competition between atrazine and these inhibitors was evident in R thylakoids. Binding constants were calculated and were similar to values reported in past literature with diuron and metribuzin (Buman, Gealy, and Fuerst, 1992). The affinity for binding, from the estimated binding constants, indicates that sorgoleone has an intermediate affinity between that of diuron and metribuzin (data not presented, see Nimbal et al., 1996b).

The use of 3D computer aided design to evaluate structural activity relationships among herbicides and potential herbicide binding sites is widely utilized to predict potential biological activity. An evaluation of sorgoleone in the 3D binding pocket of the Q_B site at the D1 protein has shown that the electrostatic charge distribution of sorgoleone is similar to that of metribuzin and other PSII inhibitors and placement of the molecule within the binding pocket is similar to that of other quinone inhibitors. The actual alignment of sorgoleone within the D1 binding pocket is almost identical to that of plastoquinone, allowing sorgoleone to competitively displace plastoquinone and leading to the subsequent disruption in electron transport (Czarnota et al., 1999). The similarity in charge distribution of sorgoleone to that of plastoquinone may account for the strong binding of sorgoleone at the PS II reaction center but it is more likely that close similarities in chemical structure and alignment are responsible for competitive displacement. Further structural activity work is currently in progress to evaluate the binding potential of sorgoleone in comparison to other related quinones.

To determine if sorgoleone had soil activity against a number of weed species, bioassays were conducted by impregnating soil with sorgoleone at concentrations ranging from 10 to 80 ppmw. After a 3 week growth period, sorgoleone presence resulted in inhibition of shoot growth with little or no effect on root development. A concentration dependent inhibition of growth was observed in selected species, including lettuce, common purslane and redroot pigweed (Table 2). Shoot fresh weight as well as shoot length of common purslane, velvetleaf, sicklepod and pigweed were reduced at concentrations 40 ppmw or less while crabgrass and green foxtail appeared to be less susceptible to inhibition at these concentrations (data not presented for insensitive species). Einhellig and Rasmussen (1989) also noted in field experiments in which grain sorghum residues were located the previous year that the inhibitory effects of grain sorghum were primarily on broadleaf weeds, with little activity observed on grass weeds. They observed these effects the following year, indicating that the allelopathic potential of the crop and or chemical(s) may persist. Our experimentation with soil extraction of sorgoleone-inoculated soil over time attempted to evaluate soil persistence (Nimbal and Weston, 2000). The hydrophobic nature of sorgoleone makes it difficult to extract from soil using aqueous extraction. However, extraction with acetonitrile and water (80:20 v/v) improved recovery and extraction for short periods of time (1 h) gave recovery rates of up to 85% when compared to longer extraction periods (24 h) where recoveries averaged 45%. Methanol:water extraction (80:20 v/v) gave low rates of recovery at all extraction times. Recovery of parent sorgoleone was highest initially, and declined substantially over a 6 week period (data not presented). Thin layer chromatography of soil extracts indicated the presence of sorgoleone-derived metabolites, but structure of these metabolites is currently unknown.

The results of application of 0.6 kg ai/ha sorgoleone to 14 day-old weed seedlings are shown in Table 3. In this case, sorgoleone application was evaluated 10 days after treatment when seedlings were 24 days of age and resulted in significant growth reduction as measured by shoot growth in several broadleaf and grass species, with greatest injury and growth reduction

TABLE 2. Response of selected weeds to soil applied sorgoleone 21 days after treatment (DAT). I means inhibition with respect to no sorgoleone control.

Shoot weight (g)											
Lettuce				Common Purslane				Pigweed			
0 ppm	10 ppm	40 ppm	80 ppm	0 ppm	10 ppm	40 ppm	80 ppm	0 ppm	10 ppm	40 ppm	80 ppm
0.29	0.32	0.21	0.12	0.20	0.19	0.17	0.14	0.13	0.12	0.09	0.07
%I	0	29	58	%I	5	14	32	%I	9	25	44

TABLE 3. Response of selected 2-week-old weeds to foliar application of sorgoleone at 0.6 kg ai/ha. Data collected 10 DAT.

Shoot weight (g) at harvest																	
Nightshade		Pigweed		Lambsquarter		Ragweed		Giant Foxtail		Sicklepod		Common Purslane		Large Crabgrass		Velvetleaf	
check	trt	check	trt	check	trt	check	trt	check	trt	check	trt	check	trt	check	trt	check	trt
2.2	0.2	5.8	1.1	11.9	8.8	5.5	0.6	5.7	5.0	1.4	0.5	2.3	1.1	2.1	1.2	0.5	0.3
%I = 90		%I = 82		%I = 26		%I = 88		%I = 12		%I = 60		%I = 53		%I = 43		%I = 40	

I represents percent inhibition; check means control; trt means treatment.

noted for broadleaf species. In particular, black nightshade, redroot pigweed and giant ragweed were most inhibited 10 days after treatment (inhibition greater than 80%) while giant foxtail and large crabgrass exhibited less inhibition (less than 45%) at this rate. Affected species exhibited severe chlorosis followed by necrosis, similar to an application of a photosynthetic inhibitor herbicide such as atrazine or diuron. Bleaching and growth inhibition was visually observed days after treatment in sensitive species, and significant stunting of growth measured by 10 days after treatment. In most sensitive species, necrosis and death eventually resulted while others exhibited regrowth. Concentration of sorgoleone applied was similar to that of atrazine or other synthetic herbicides which are also photosynthetic inhibitors, indicating this natural product, if properly formulated is highly active both pre- and post-emergence, when applied to the soil or to seedling foliage.

Allelopathic crops offer potential for development of model herbicides as well as providing a source of germplasm that could be manipulated to enhance weed suppression in an environmentally compatible manner (Weston, 1996). As we learn more about the mechanisms of allelochemical selectivity, mode of action and regulation of biosynthesis, we should be able to successfully manipulate our germplasm resources to select for enhanced weed suppression. The strong weed suppressive ability of *Sorghum* cover crops and phytotoxicity of sorgoleone offer interesting possibilities for effective biorational approaches to weed management.

REFERENCES

Buman, R.A., D.R. Gealy, and E.P. Fuerst. (1992). Relationship between temperature and triazinone on herbicide activity. I. Herbicide binding to thylakoid membranes. *Pesticide Biochemistry and Physiology* 43: 22-28.

Czarnota, M.A., F. Dayan, A. Rimando, and L.A.Weston. (1999). Sorgoleone: Production, localization and mode of action of a potent inhibitor produced by *Sorghum* spp. In *Second World Congress on Allelopathy*. Lakehead University, Thunderbay, Canada. Abstract pp. 77.

Einhellig, F.A. and J.A. Rasmussen. (1989). Prior cropping with grain sorghum inhibits weeds. *Journal of Chemical Ecology* 15: 951-960.

Einhellig, F.A., J.A. Rasmussen, J.A. Hejl, and I.F. Souza. (1993). Effects of root exudate sorgoleone on photosynthesis. *Journal of Chemical Ecology* 19: 369-375.

Einhellig, F.A. and I.F. Souza. (1992). Allelopathic activity of sorgoleone. *Journal of Chemical Ecol*ogy 18: 1-11.

Forney, D.R. and C.L. Foy. (1985). Phytotoxicity of products from rhizospheres of a sorghum-sudangrass hybrid (*S. bicolor* × *S. sudanese*). *Weed Science* 33: 597-604.

Forney, D.R., C.L. Foy, and D.D. Wolf. (1985). Weed suppression in no-till alfalfa (*Medicago sativa*) by prior cropping of summer annual forage grasses. *Weed Science* 33: 490-497.

Geneve, R.L. and L.A. Weston. (1988). Growth reduction of eastern redbud (*Cercis canadensis* L.) seedlings caused by interaction with a sorghum-sudangrass hybrid (sudex). *Journal of Environmental Horticulture* 6: 24-26.

Gonzalez, V., C.I. Nimbal, L.A. Weston, and G.M. Cheniae. (1997). Inhibition of a photosystem II electron transfer reaction by sorgoleone, a natural product. *Journal of Agriculture and Food Chemistry* 45: 1415-1421.

Guenzi, W.D. and T.M. McCalla. (1966). Phenolic acids in oats, wheat, sorghum and corn residues and their phytotoxicity. *Agronomy Journal* 58: 303-304.

Heath, M.E., R.F. Barnes, and D.S. Metcalfe. (1985). Summer Annual Grasses. In *Forages: The Science of Grassland Agriculture*, eds. M.E. Heath, R.F. Barnes, and D.S. Metcalfe. Ames, Iowa: Iowa State University Press, pp. 278-290.

Hess, D.E., G. Ejeta, and L.G. Butler. (1992). Selection of sorghum genotypes expressing a quantitative biosynthetic trait that confers resistance to *Striga*. *Phytochemistry* 31: 493-497.

Hoffman, M.L., L.A. Weston, J.C. Snyder, and E.E. Regnier. (1996). Interference mechanisms between germinating seeds and between seedlings: Bioassays using cover crop and weed species. *Weed Science* 44: 402-407.

Inderjit and L.A. Weston. (2000). Are laboratory bioassays for allelopathy suitable for prediction of field responses? *Journal of Chemical Ecology*. 26: 2111-2118.

Mets, L. and A. Thiel. (1989). Biochemistry and genetic control of the photosystem II herbicide target site. In *Target Sites of Herbicide Action*, eds. P. Boger and G. Sandmann. Boca Raton, FL: CRC Press, pp. 1-24.

Nicollier, J.F., D.F. Pope, and A.C. Thompson. (1983). Biological activity of dhurrin and other compounds from johnsongrass (*Sorghum halepense*). *Journal of Agriculture and Food Chemistry* 31: 744-748.

Netzley, D.H and L.G. Butler. (1986). Roots of sorghum exude hydrophobic droplets containing biologically active components. *Crop Science* 26: 776-778.

Nimbal, C.I., J. Pedersen, C.N. Yerkes, L.A. Weston, and S.C. Weller. (1996a). Activity and distribution of sorgoleone in grain sorghum germplasm. *Journal of Agriculture and Food Chemistry* 44: 1343-1347.

Nimbal, C.I., C.N. Yerkes, L.A. Weston, and S.C. Weller. (1996b). Herbicidal activity and site of action of the natural product sorgoleone. *Pesticide Biochemistry and Physiology* 54: 73-83.

Nimbal, C.I. and L.A. Weston. (2000). Selectivity of sorgoleone applied pre- and postemergence and soil extraction. *Journal of Chemical Ecology* (submitted for publication).

Panasuik, O., D.D. Bills, and G.R. Leather. (1986). Allelopathic influence of *Sorghum bicolor* on weeds during germination and early development of seedlings. *Journal of Chemical Ecology* 12: 1533-1543.

Putnam, A.R. and J. DeFrank. (1983). Use of phytotoxic plant residues for selective weed control. *Crop Protection* 2: 173-181.

Putnam, A.R., J. DeFrank, and J.P. Barnes. (1983). Exploration of allelopathy for weed control in annual and perennial cropping systems. *Journal of Chemical Ecology* 9: 1001-1010.

Rasmussen, J.A., A.M. Hejl, F.A. Einhellig, and T.A. Thomas. (1992). Sorgoleone from root exudate inhibits mitochondrial functions. *Journal of Chemical Ecology* 18: 197-207.

Smeda, R., P.M. Hasegawa, P.B. Goldbrough, N.K. Singh, and S.C. Weller. (1993). A serine-to-threonine substitution in the triazine herbicide binding protein in potato cells results in atrazine resistance without impairing productivity. *Plant Physiology* 103: 911-917.

Weston, L.A. (1996). Utilization of allelopathy for weed management in agroecosystems. *Agronomy Journal* 88: 860-866.

Weston, L.A., R. Harmon, and S. Mueller. (1989). Allelopathic potential of sorghum-sudangrass hybrid (sudex). *Journal of Chemical Ecology* 15: 1855-1865.

Role of Benzoxazinones in Cereals

Annette Friebe

SUMMARY. Heterocyclic 1,4-benzoxazin-3-ones and related benzoxazolinones are important natural products of the cereal crops maize, wheat and rye. Much research has focused on the role of these compounds as defensive agents against plant diseases and pests. Studies of a wide variety of biological effects on herbivores and nematodes, plant pathogenic fungi and bacteria are reviewed in this article. Phytotoxic activity of the compounds is also considered with respect to allelopathic interaction with higher plants. Recent investigations of molecular aspects of interactions of cyclic hydroxamic acids and related benzoxazolinones with acceptor organisms have demonstrated different effective detoxification strategies and tolerance mechanisms. This research has greatly advanced our knowledge about the multiple roles that these allelochemicals play in ecological systems. *[Article copies available for a fee from The Haworth Document Delivery Service: 1-800-342-9678. E-mail address: <getinfo@haworthpressinc.com> Website: <http://www.HaworthPress.com> © 2001 by The Haworth Press, Inc. All rights reserved.]*

KEYWORDS. Benzoxazinone, benzoxazolinone, hydroxamic acid, allelochemical, allelopathy, detoxification, DIBOA, DIMBOA, BOA, MBOA

Annette Friebe is Professor, Institute of Agricultural Botany, University of Bonn, Meckenheimer Allee 176, 53115 Bonn, Germany (E-mail: friebe@uni-bonn.de).

The author gratefully thanks Kerstin Gross for critical reading and Isabel Wieland for help with the manuscript. The author was financially supported by Lise-Meitner grant of the Ministerium für Wissenschaft und Forschung NRW.

[Haworth co-indexing entry note]: "Role of Benzoxazinones in Cereals." Friebe, Annette. Co-published simultaneously in *Journal of Crop Production* (Food Products Press, an imprint of The Haworth Press, Inc.) Vol. 4, No. 2 (#8), 2001, pp. 379-400; and: *Allelopathy in Agroecosystems* (ed: Ravinder K. Kohli, Harminder Pal Singh, and Daizy R. Batish) Food Products Press, an imprint of The Haworth Press, Inc., 2001, pp. 379-400. Single or multiple copies of this article are available for a fee from The Haworth Document Delivery Service [1-800-342-9678, 9:00 a.m. - 5:00 p.m. (EST). E-mail address: getinfo@haworthpressinc.com].

© 2001 Taylor & Francis.

INTRODUCTION

In the late 1950s and the 1960s a series of benzoxazinoide compounds were identified as natural products of important cereals. Since their discovery these substances have attracted considerable attention because of their diverse range of biological activities. Intense effort has gone into investigating the role of cyclic hydroxamic acids as chemical resistance factors against herbivore attack, plant diseases and pathogen infections. Numerous studies have demonstrated a wide range of biological activities including effects against aphids, pathogenic fungi, and bacteria, as well as allelopathic influences on higher plants. Nevertheless, attempts to correlate the level of hydroxamic acids in plants with plant field resistance to different diseases and pests often gave controversial results. In recent years, studies were carried out to unveil the molecular aspects of the interactions of benzoxazinoide allelochemicals with herbivores, fungi and higher plants. The increasing insight into effective detoxification strategies and tolerance mechanisms has greatly advanced our understanding of the multiple role that cyclic hydroxamic acids and related derivatives play in ecological systems.

NATURAL OCCURRENCE AND STRUCTURAL DIVERSITY

Heterocyclic 1,4-benzoxazin-3-ones were found to be secondary natural products of the crops maize, wheat and rye but they were not detected in cultivated barley, oat and rice (Niemeyer, 1988). In addition, cyclic hydroxamic acids occur constitutively in a wide variety of wild Poaceae belonging to the Triticeae (for instance genera *Agropyron*, *Hordeum* and *Elymus*) and Avenae (genera *Deschampsia* and *Phalaris*) (Copaja, Barria, and Niemeyer, 1991; Friebe et al., 1995; Gianoli and Niemeyer, 1998), and *Echinochloa crus-galli* (Pheto, 1993). These substances have also been found in three dicot families–Acanthaceae, Ranunculaceae and Scrophulariaceae (Wolf, Spencer, and Plattner, 1985; Özden et al., 1992; Pratt, Kumar, and Chilton, 1995).

The biosynthesis of benzoxazinoide compounds in plants depends on the tryptophan pathway. Recently, the Bx1 gene encoding a tryptophan synthase α and four cytochrome P-450-dependent monooxygenase genes, which are required for the biosynthesis of cyclic hydroxamic acids, have been cloned from maize (Frey et al., 1997). During biosynthesis, 2,4-dihydroxy-2*H*-1,4-benzoxazin-3(4*H*)-ones are derived from their lactam precursors *Ia-c* (Figure 1). These 4-deoxy derivatives of cyclic hydroxamic acids normally only show a weak biological activity (Hashimoto and Shudo, 1996). In living plants, 2,4-dihydroxy-2*H*-1,4-benzoxazin-3(4*H*)-ones are sequestered in the forms of their biologically inactive (2R)-2-β-D-glucosides *IIa-c* (Figure 2). The glu-

FIGURE 1. Chemical structure of 2-hydroxy-2*H*-1,4-benzoxazin-3(4*H*)-ones (**Ia-c**), lactam precursors of cyclic hydroxamic acids.

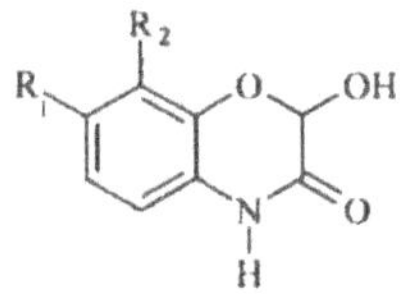

R_1 = H, R_2 = H	**Ia** HBOA
R_1 = OCH_3, R_2 = H	**Ib** HMBOA
R_1 = OCH_3, R_2 = OCH_3	**Ic** HM_2BOA

FIGURE 2. Hydrolysis of (2R)-β-D-glucopyranosyloxy-4-hydroxy-2*H*-1,4-benzoxazin-3(4*H*)-ones (**IIa-c**) to 2,4-dihydroxy-2*H*-1,4-benzoxazin-3(4*H*)-ones (**IIIa-c**) and further decomposition to benzoxazolinones (**IVa-c**).

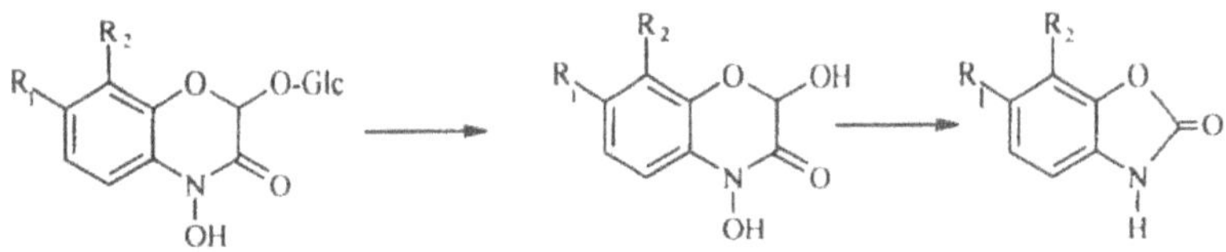

R_1 = H, R_2 = H	**IIa** GDIBOA	**IIIa** DIBOA	**IVa** BOA
R_1 = OCH_3, R_2 = H	**IIb** GDIMBOA	**IIIb** DIMBOA	**IVb** MBOA
R_1 = OCH_3, R_2 = OCH_3	**IIc** $GDIM_2BOA$	**IIIc** DIM_2BOA	**IVc** M_2BOA

cosides are enzymatically hydrolyzed in response to tissue damage or pathogen infection and the more toxic aglycons are released. In soil or aqueous solution the cyclic hydroxamic acids *IIIa-c* subsequently decompose to give the respective ring contracted benzoxazolinones *IVa-c* (Figure 2) (Woodward et al., 1978; Yenish, Worsham, and Chilton, 1995).

DIBOA (*IIIa*) was found to be the predominant benzoxazinoid in rye (Barnes and Putnam, 1987), whereas DIMBOA (*IIIb*) is the most abundant derivative in maize and wheat. Cyclic hydroxamic acids have been discovered in both aerial parts and roots of investigated cereals (Niemeyer, 1988). DIBOA was also isolated from pollen extracts of rye (Zhang et al., 1995). The level of these compounds increase after germination, reaching a maximum after few days and then decrease with increasing age of the plant. Intensive studies on the benzoxazinoid content in cereals demonstrated that the level is strongly species and cultivar dependent. In addition, the concentration is highly influenced by environmental growth conditions. In this context, investigations have shown increasing levels of hydroxamic acids caused by UV-light (Bergvinson, Larsen, and Arnason, 1995), water deficiency (Richardson and Bacon, 1993), higher temperature (Gianoli and Niemeyer, 1997a), and by high fertility regimes (Mwaja, Masiunas, and Weston, 1995). Relatively high concentrations of cyclic hydroxamic acids have been detected in wheat and maize. Maximum concentrations up to 8.07 mmol/ kg fresh weight were measured in shoots of wheat seedlings of different cultivars (Nicol et al., 1992). The maximum recorded levels of hydroxamic acids in roots of various lines and cultivars of maize ranged from 4.4 (Xie et al., 1992) to 9.98 mmol/kg fresh weight (Friebe et al., 1998).

Studies on the structural diversity of 1,4-benzoxazinones and related benzoxazolinones lead to the identification of a wide variety of biologically active derivatives in cereals. In this context glucosides of 7-hydroxylated cyclic hydroxamic acids and lactams have been described (Hofman, Hofmanova, and Hanus, 1969; Woodward et al., 1979; Nagao et al., 1985). The highly active and labile 4,7-dimethoxy-1,4-(2*H*)-benzoxazin-3-one and the respective glucoside were detected in maize (Hofman and Hofmanova, 1970; Hedin, Davis, and Williams, 1993), wheat (Kluge, Grambow, and Sicker, 1997; Grambow et al., 1986), and *Coix lachryma* (Nagao et al., 1985). Structural diversity has been shown not only for benzoxazinoids, but also for related benzoxazolinones. The antifungal 4-acetyl-benzoxazolin-2-one was isolated from kernels of a *Fusarium*-resistant hybrid of *Zea mays* (Fielder et al., 1994). Interestingly, the chlorinated derivatives, 4-chloro-6,7-dimethoxy-2-benzoxazolinone (Anai et al., 1996) and 5-chloro-6-methoxy-2-benzoxazolinone (Kato-Noguchi, Kosemura, and Yamamura, 1998), have also been described for maize.

EFFECTS ON HERBIVORES AND NEMATODES

Cyclic hydroxamic acids have been implicated in the resistance of cereals to herbivore attacks for a long time. Intensive efforts have been made to unveil the role of these allelochemicals in plant interactions with cereal aphids (Niemeyer, 1990). Numerous studies have demonstrated a positive correlation between hydroxamate levels of plants and resistance to aphid infestation (Long et al., 1977; Beck et al., 1983; Thackray et al., 1990; Barria, Copaja, and Niemeyer, 1992; Nicol et al., 1992). Although, cyclic hydroxamates have been considered as preformed defense chemicals of cereals, an accumulation of these compounds in primary leafs of wheat after an infestation with *Rhopalosiphum padi* has also been shown (Gianoli and Niemeyer, 1997b). Nevertheless, results of broader arrays of different genotypes with respect to their field resistance were controversial. For instance, a comprehensive study of the relationship between the concentration of DIMBOA and infestation of 21 maize lines by *Rhopalosiphum padi* showed no positive correlation (Rustamani et al., 1996). The addition of benzoxazinoids to artificial diets provoked two effects on aphids: antibiosis at lower concentrations and feeding deterrence at higher concentrations (Argandona et al., 1983; Niemeyer, Givovich, and Copaja, 1993). Detailed investigations of deterrent effects were carried out by monitoring the feeding behavior of aphids by electrical penetration graphs. These studies with the Russian wheat aphid (*Diuraphis noxia*) showed that higher levels of benzoxazinoids cause a delay in attaining a sustained phloem ingestion and produced lower mean relative growth rates (Givovich and Niemeyer, 1996; Mayoral, Tjallingii, and Castanera, 1996). However, no significant correlation was found between the DIMBOA content and the total time of phloem feeding. Using this technique different species of cereal aphids could be subdivided into sensitive (*Rhopalosiphum padi, Schizaphis graminearum, Sitobion avenae, Metopolophium dirhodum*) and insensitive (*Rhopalosiphum maidis*) species to the feeding deterrent effect of hydroxamic acids (Givovich and Niemeyer, 1995). The insensitivity of *Rhopalosiphum maidis* to the allelochemical seemed to be due to a feeding strategy of avoiding contact with the respective compounds, by decreasing the number of cellular punctures in live tissue.

Insect digestive enzymes have been suggested as possible targets for the action of plant defense chemicals. Inactivation of thiol and serin proteinases, for example, was shown for α-chymotrypsin and papain (Perez and Niemeyer, 1989; Cuevas, Niemeyer, and Perez, 1990). The reaction of hydroxamic acids with nucleophilic residues of enzymes such as cysteine residues at the active site has been directly related to the mechanism of enzyme inhibition. Advances in understanding the role of hydroxamates in the interaction of cereals with aphids also came from experimental studies of different aphid detoxification enzymes (Leszczynski and Dixon, 1992; Leszczynski, Matok, and Dixon, 1992; Leszczynski et al., 1993). There was a negative correlation between the activity of UDP-glucose transferases in *Sitobion avenae* and the concentration of DIMBOA *IIIb* in wheat fed on by this aphid. In contrast, a highly positive correlation was found between the activity of glutathione-S-transferases in aphids and the total DIMBOA levels of the host plants. Thus, the induction of this important detoxification enzyme might enable the insects to overcome toxic effects of hydroxamic acids.

Much research has also focused on the role of cyclic hydroxamic acids as a resistance factor in maize against different corn borer species. The European corn borer (*Ostrinia nubilalis*), the Asian corn borer (*Ostrinia furnacalis*), and the Southwestern corn borer (*Diatraea grandiosella*) are major insect pests of maize, affecting production through physiological yield reduction and harvest losses. DIMBOA *IIIb* and its breakdown product MBOA *IVb* are recognized as important leaf-feeding resistance factors in maize. In addition, the 2-hydroxy-4,7-dimethoxy-1,4-benzoxazin-3-one, a DIMBOA derivative which was found in maize whorl surface waxes, was thought to be involved in the resistance of maize to the Southwestern corn borer (Hedin, Davis, and Williams, 1993). DIMBOA and MBOA were demonstrated to deter feeding in both field and laboratory studies (Robinson, Klun, and Brindley, 1978; Guthrie et al., 1986). Inhibitory effects on growth and digestive processes of European corn borer have been shown by Campos et al. (1989) and Houseman et al. (1992). An investigation of the effects of DIMBOA on several important enzymatic systems of the Asian corn borer revealed an induction of higher detoxification enzyme activities of cytochrome P-450 monooxygenase and glutathione S-transferase, and it is hypothesized that these enzymes may reduce the toxicity of DIMBOA (Fengming et al., 1995). It was also confirmed by Campos et al. (1989) that the European corn borer excretes DIMBOA very quickly. Thus, detoxification and excretion have to be recognized as possible defense mechanisms of insects. It was demonstrated by a detailed investigation of the feeding preference of corn borer larvae in relation to the leaf profile of DIMBOA that the larvae consumed immature tissue at higher rates than more mature tissues, despite significantly higher levels of DIMBOA in immature tissue (Bergvinson, Hamilton, and Arnason, 1995). Therefore, DIMBOA content has to be considered only as a part of a more comprehensive plant defense mechanism. For instance, fiber content and phenolic fortification of cell walls are proposed as additional resistance components in corn against corn borer infection.

Development parameters, including survival, weight and head capsule width of Western corn rootworm (Coleoptera: Chrysomelidae) larvae were negatively correlated with hydroxamic acid levels of maize roots (Xie et al., 1992). Assabgui, Hamilton, and Arnason (1995) compared variations in root hydroxamic acid content and field resistance of 4- and 16-week old maize plants. The time of maximal root damage, as well as the level of recovery, could be correlated with the changes in total content of hydroxamic acids over time. Therefore, they concluded that these allelochemicals play a role in field resistance of the investigated maize inbred lines to larvae of Western corn rootworms. Interestingly, 6-methoxy-2-benzoxazolinone has been identified as semiochemical for host location by Western corn rootworm larvae (Bjostard and Hibbard, 1992; Abou Fakhr, Hibbard, and Bjostad, 1994).

The release of biologically active allelochemicals into soil is of special importance for interaction of plants with the soil ecosystem. Root exudates have long been known to induce hatching of certain plant parasitic nematodes and are also of significance to the host location. The influence of root exudates from *Zea mays* on the behavior of the root lesion nematode *Pratylenchus zeae* was investigated in detail (Friebe et al. 1998). Attractive and repellent influences of the host plant exudate, as well as isolated constituents, on the larvae were determined by a modified bioassay according to Castro et al. (1989). For concentrated aqueous exudates a repulsion of nematodes was found. In contrast, at lower concentrations maize root exudate showed a high attraction. Different benzoxazinoid allelochemicals as well as respective benzoxazolinones have been identified as main components in root exudates and have been tested in the bioassay. The investigation of selected derivatives of cyclic hydroxamic acids on their effects on the behavior of *Pratylenchus zeae* clearly demonstrated attractant effects of DIMBOA (*IIIb*) at low concentrations, similar to concentrations found in root exudates under natural conditions. A screening of different maize cultivars showed no correlation between DIMBOA level in root tissue and nematode infection. Nevertheless, cultivars with high concentrations of hydroxamic acids up to 9.98 μmol/g fresh weight exhibited high levels of root lesion nematode infection. Therefore, an important role for these compounds as defense chemicals against *Pratylenchus zeae* seems unlikely. Rather, it was demonstrated by this study that cyclic hydroxamic acids can act as signals for host recognition by nematodes.

BIOLOGICAL ACTIVITY AGAINST PLANT PATHOGENIC FUNGI AND BACTERIA

Since cyclic hydroxamates were first described in plants, antibacterial and antifungal effects of cyclic hydroxamic acids and related benzoxazolinones have been demonstrated (Virtanen and Hietala, 1955; Wahlroos and Virtanen, 1958; Whitney and Mortimore, 1959, 1961; Elnaghy and Shaw, 1966; Dabler, Pappelis, and Bemiller, 1969). Since these early studies, an abundance of research has focused on the role of these allelochemicals as defensive agents against a wide variety of plant diseases.

The bacteriostatic, but not bacteriocidal activity of DIMBOA to soft rooting *Erwinia* species was demonstrated by Corcuera et al. (1978) and Woodward et al. (1978). Because glucosidic precursor molecules are not biologically active, the 2-hydroxy group of cyclic hydroxamic acids has been considered to be important for biological activity. In addition, the activity of the 4-deoxy derivatives, lactam precursors of the cyclic hydroxamic acids, is very weak. Structure-activity relationships indicated that the reactivity of benzoxazinoids with nucleophiles could be the chemical basis for their biological activity, at least in part (Hashimoto and Shudo, 1996). In this context, the presence of the strong electron-donating methoxy group greatly enhances the reactivity and therefore the cyclic hydroxamic acid DIMBOA (*IIIb*) often shows a higher biological activity in comparison to the non-methoxylated derivative DIBOA (*IIIa*). For natural benzoxazolinones (*IVa-c*) both the electrophilic character of the nitrogen atom of the heterocyclic ring and the lipophilic character of substituents of the aromatic ring are responsible for bioactivity (Bravo, Copaja, and Lazo, 1997). In soil, the cyclic hydroxamic acids decompose rapidly to give the respective ring-contracted benzoxazinones. The decomposition of allelochemicals from rye cover crop residue under field simulated conditions was investigated by Yenish, Worsham, and Chilton (1995). These experiments demonstrated that the total concentrations of DIBOA-glucoside, DIBOA and BOA declined to 50% of their initial level approximately 10 to 12 days after clipping. The disappearance of benzoxazolinones in soil has been directly related to the microbial metabolization of these compounds. For instance, *Acinetobacter calcoaceticus*, a gram-negative bacterium isolated from field soil, was found to be able to metabolize BOA (Chase et al., 1991).

The transformation products of BOA, MBOA and M_2BOA in soil have been identified as 2-amino-3*H*-phenoxazin-3-one *Va* and methoxylated aminophenoxazinones *Vb-c*, respectively (Figure 3) (Gagliardo and Chilton, 1992; Kumar, Gagliardo, and Chilton, 1993). In addition, the metabolization of BOA to 2-amino-3*H*-phenoxazin-3-one has been observed for root-colonizing bacteria of *Avena sativa* and other cereal crops. In this case, *o*-aminophenol has been detected as an intermediate compound and a further acetylation of 2-amino-3*H*-phenoxazin-3-one to 2-acetylamino-3*H*-phenoxazin-3-one *VI* occurs (Figure 4) (Friebe, Wieland, and Schulz, 1996). Although, phytotoxic

FIGURE 3. Chemical structure of 2-amino-3*H*-phenoxazin-3-ones (**Va-c**), soil microbial metabolites of benzoxazolinones (**IVa-c**).

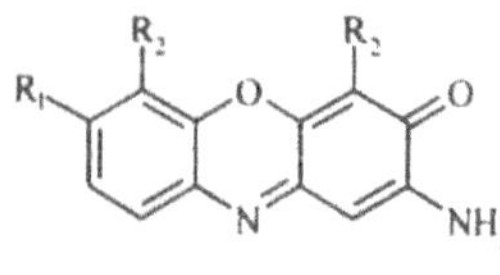

R_1 = H, R_2 = H	**Va** 2-amino-3*H*-phenoxazin-3-one
R_1 = OCH_3, R_2 = H	**Vb** 2-amino-7-methoxy-3*H*-phenoxazin-3-one
R_1 = OCH_3, R_2 = OCH_3	**Vc** 2-amino-4,6,7-trimethoxy-3*H*-phenoxazin-3-one

FIGURE 4. Microbial metabolization of BOA (**IVa**) to 2-amino-3*H*-phenoxazin-3-one (**Va**) and 2-*N*-acetylamino-3*H*-phenoxazin-3-one (**VI**) by root-colonizing bacteria of oat (Adapted from Friebe, Wieland, and Schulz, 1996).

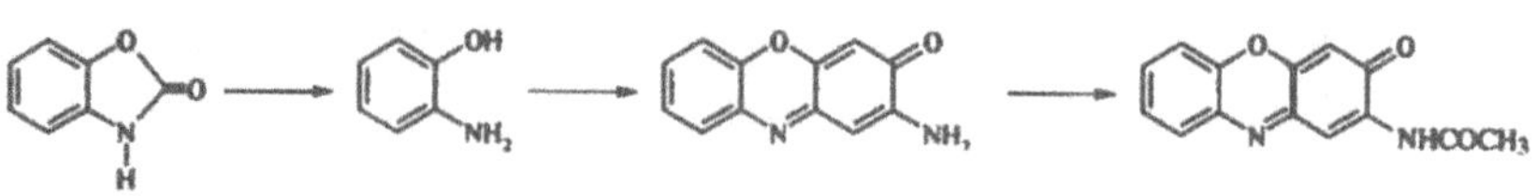

IVa *o*-aminophenol **Va** **VI**

influences of aminophenoxazinones have been described (Gagliardo and Chilton, 1992; Kumar, Gagliardo, and Chilton, 1993; Friebe, Wieland, and Schulz, 1996) radicle elongation tests with *Avena sativa* showed that microbial transformation contributes to a higher tolerance of oat to BOA.

2-Aminophenoxazin-3-one and its acetyl derivative are natural occurring antibiotics, known as questiomycin and N-acetylquestiomycin. Gerber and Lechevalier (1964) isolated and identified both compounds as microbial products from the actinomycete *Waksmania aerata* sp. *nov.* and *Pseudomonas iodina*. Questiomycin and its acetyl derivative belong to the group of actinomycin antibiotics. Their antimicrobial effects may be of special interest with respect to the development of species dependent bacteria populations in the rhizosphere and their biological significance to the respective plants. The ability of plant-associated microorganisms to produce substances with inhibitory effects against pathogens may be of outstanding interest for biological control of plant diseases. For instance, phenazine antibiotic producing *Pseudomonas fluorescens* have received attention as biocontrol agents for the suppression of take-all of wheat (Thomashow and Weller, 1988; Weller, Howie, and Cook, 1988; Fujimoto, Weller, and Thomashow, 1995). Interestingly, it has been demonstrated that the phenazine-producing bacterium *Brevibacterium iodinum* shows a close biosynthetic relationship between phenazine and aminophenoxazinone biosynthesis, possibly including a common intermediate in the biosynthesis of both metabolites (Herbert, Mann, and Römer, 1982).

Take-all, caused by the pathogenic fungus *Gaeumannomyces graminis* var. *tritici*, is one of the most critical root diseases of the hydroxamate-containing cereal wheat. Take-all decline presents a natural form of biological control, whereby the severity of take-all declines following continuous monoculture of wheat. Soil from fields, on which take-all decline occurred is highly suppressive to *Gaeumannomyces graminis* var. *tritici* and fluorescent pseudomonads are thought to be involved in this inhibition. Root exudates contribute to the ability of bacteria to colonize roots (Weller, 1988), and therefore it seems possible that exuded cyclic hydroxamic acids play a role in natural take-all decline in wheat monocultures. Because of their antifungal effects benzoxazinones have been considered as constitutive defense chemicals of cereal plants (Gross, 1989). Additionally, benzoxazinoid compounds have also been described as phytoalexins. After the infection of wheat with *Puccinia graminis* and of resistant maize with *Exserohicum turcicum*, an accumulation of 2-hydroxy-4,7-dimethoxy-2*H*-1,4-benzoxazin-3(4*H*)-one (Bücker and Grambow, 1990) and a significant increase in DIMBOA levels (Zhu, He, and Liu, 1994) have been reported, respectively.

Inhibition of spore germination of the Northern corn leaf blight-producing fungus *Helminthosporium turcicum* was demonstrated using DIMBOA at concentrations of 1 to 10 ppm (Couture, Routley, and Dunn, 1971). In maize a significant correlation was found between hydroxamate concentrations in seedling stem tissue and leaf tissue, and field resistance to this fungus (Long, Dunn, and Routley, 1975, 1978). Additionally, benzoxazinoid compounds have been implicated in the resistance of maize to stalk rot, *Diplodia zeae* (BeMiller and Pappelis, 1965), and the resistance of wheat to stem rust, *Puccinia graminis* (Elnaghy and Linko, 1962). Inhibitory effects of benzoxazinones and related benzoxazolinones on mycelial growth of *Microdochium nivale* (Wahlroos and Virtanen, 1959), *Fusarium moniliforme* (Richardson and Bacon, 1995), *Fusarium culmorum*, *Gauemannomyces graminis* (Friebe et al., 1998), and *Fusarium subglutinans* (Vilich et al., 1999) have been observed. Generally, preformed chemical defense compounds are not considered as major determinants in pathogenesis. They rather contribute to a general and fundamental resistance of plants to a great number of potential pathogens. During the natural evolutionary process only few fungi have developed strategies to overcome these plant specific chemical defense barriers. In these cases the ability of phytopathogenic fungi to detoxify the antifungal compounds can determine their host range.

Different *Gaeumannomyces* isolates have been investigated with respect to their ability to degrade the wheat allelochemicals BOA and MBOA (Friebe et al., 1998). Strong species-dependent differences in the biotransformation of the benzoxazolinones have been observed, following incubation of *G. graminis* var. *tritici*, *G. graminis* var. *graminis*, and *G. graminis* var. *avenae* in media containing 1mM of the respective compounds (Figure 5). After seven days of incubation only slight reductions of BOA (*IVa*) and MBOA (*IVb*) concentrations were found in the presence of *G. graminis* var. *avenae* and *G. graminis* var. *tritici*. In contrast, *G. graminis* var. *graminis* showed a considerably higher degradation potential with nearly complete metabolization after an incubation time of 7 days. During this period an accumulation of metabolites in the media was detected. The main degradation products were subsequently identified as N-(2-hydroxyphenyl)-malonamic acid *VIIa* and *N*-(2-hydroxy-4-methoxyphenyl)-malonamic acid *VIIb* (Figure 6). These are also produced by metabolization of cyclic hydroxamic acids DIBOA and DIMBOA. 2-Amino-3*H*-phenoxazinone *Va* was detected as a by-product of BOA biotransformation.

In contrast to BOA and MBOA, no significant inhibition of mycelial growth could be found when exposing them to the fungal metabolites *N*-(2-hydroxyphenyl)-malonamic acid and *N*-(2-hydroxy-4-methoxyphenyl)-malonamic acid. Therefore, this degradation leads to an effective detoxification of the plant defense compounds. The tolerance of the investigated *Gaeumannomyces* isolates to BOA and MBOA could be correlated with their detoxification ability. Further, the ability to cause root rot symptoms in wheat

FIGURE 5. Degradation of 1 mM BOA (above, solid symbols) and 1 mM MBOA (below, open symbols) by *Gaeumannomyces* spp. (Ggt ◆◇, Ggg ●○, Gga ▲△). Values are means of duplicate experiments with 3 replicates. Different letters indicate significant differences at 168 h (modified Figure from Friebe et al., 1998).

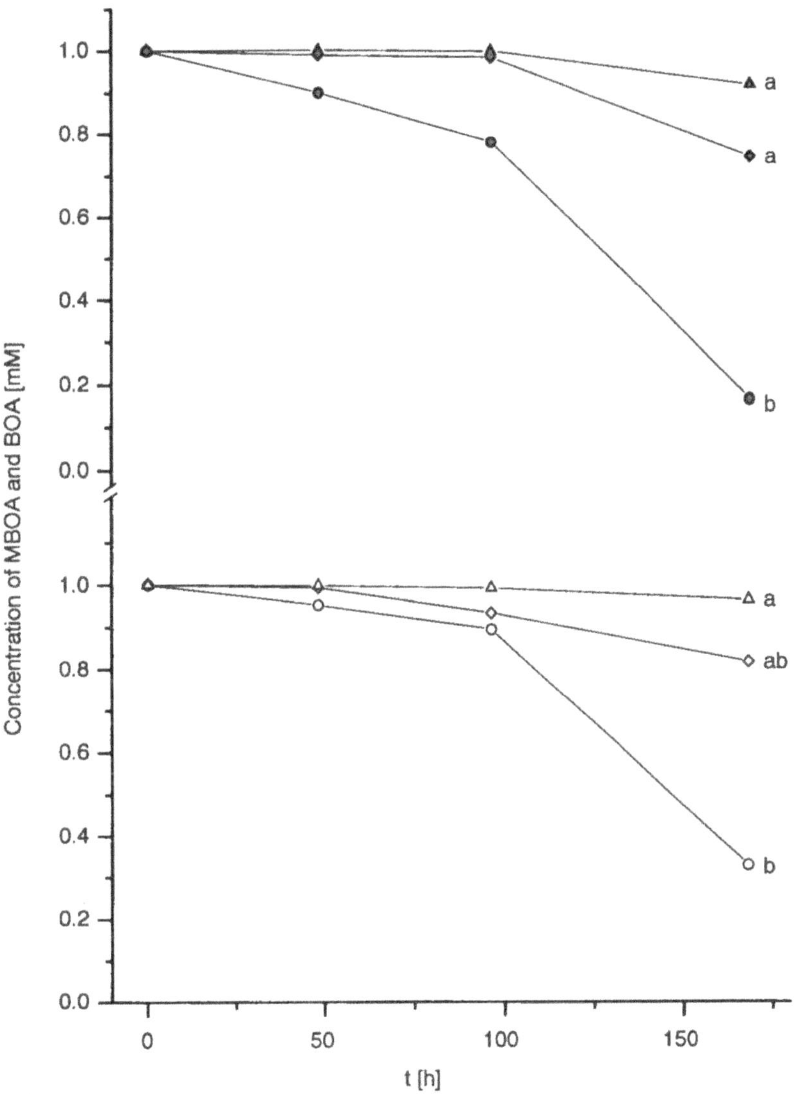

paralleled their potential to degrade the respective wheat allelochemicals (Friebe et al., 1998; data not shown). *N*-(2-hydroxyphenyl)-malonamic acids have also been identified as products of the fungal metabolization of BOA and MBOA by maize isolates of *Fusarium moniliforme* (Yue, Bacon, and Richardson, 1998) and *Fusarium subglutinans* (Vilich et al., 1999). A detailed investigation of the biotransformation of BOA by this *F. subglutinans* isolate indicated that the first metabolization step involves a cleavage of the heterocyclic ring, leading to an intermediate formation of *o*-aminophenol. This compound was rapidly converted into *N*-(2-hydroxyphenyl)-malonamic acid (Figure 7) catalyzed by a *N*-malonyltransferase (Friebe, unpublished results). After further incubation of *F. subglutinans* the malonamic acid was also completely degraded. Miller et al. (1996) isolated 4-acetyl-benzoxazolin-2-one from a Giberrella ear rot-resistant corn hybrid. At a concentration of 4 μM this compound causes a 50% inhibition of the mycotoxin 3-acetyl deoxynivalenol production by *Fusarium culmorum* (Miller et al., 1996). Interestingly, our studies showed that *F. subglutinans* is not able to detoxify this acetylated benzoxazolinone, in contrast to BOA and MBOA (Friebe et al.; unpublished results). A comprehensive study on the detoxification capacity of different *Fusarium* spp. indicated that the metabolization of benzoxazolinones seems to be a common fungal detoxification mechanism against these plant allelochemicals. BOA and MBOA were degraded to malonamic acids by *Fusarium avenaceum*, *F. graminearum*, and *F. oxysporum*-all isolated from maize. Nevertheless, the investigated fungi differed greatly in their

FIGURE 6. Chemical structure of hydroxyphenylmalonamic acids (**VIIa** and **VIIb**), fungal metabolites of benzoxazolinones (**IVa-c**).

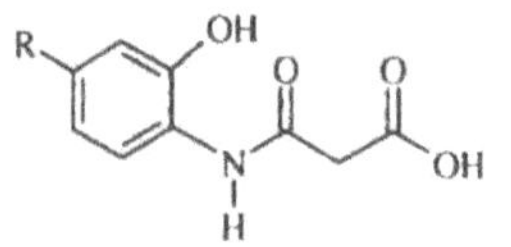

R = H — **VIIa** *N*-(2-hydroxyphenyl)-malonamic acid
R = OCH_3 — **VIIb** *N*-(2 hydroxy-4-methoxyphenyl)-malonamic acid

FIGURE 7. Molecular mechanism of the BOA metabolization by *Fusarium subglutinans*.

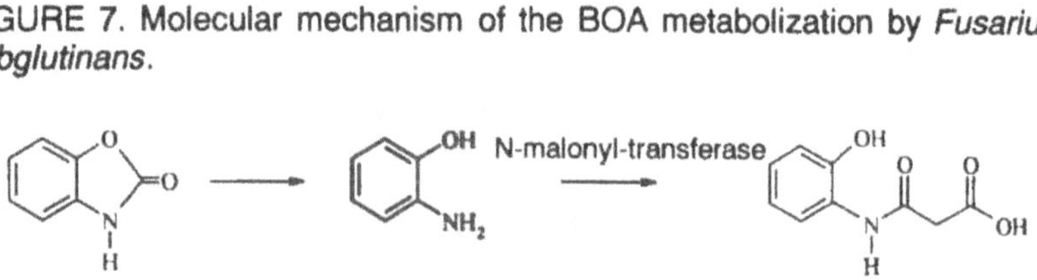

metabolic change activity. For *Fusarium poae* no metabolic change was found during incubation periods up to 20 days (Friebe, unpublished results).

ALLELOPATHIC INTERACTION WITH HIGHER PLANTS

Rye and maize are cereals of high allelopathic potential, and inhibitory effects on higher plants have been observed for a long time. For instance, there are many early descriptions of phytotoxic effects of rye and its residues (Cubbon, 1925; Hill, 1926; Osvald, 1953). Nevertheless, molecular mechanisms of these allelopathic interactions remained unclear for many years. Increasing insight in the nature of active compounds came from the isolation of hydroxamic acids and the studies on their influences on germination and growth of other plants (Barnes and Putnam, 1987). Barnes and Putnam (1987) demonstrated a significant reduction of the seedling root and shoot elongation of large crabgrass, barnyard grass, proso millet, tomato, and lettuce caused by DIBOA isolated from rye. The decomposition product BOA also inhibited seedling development of the test plants. Additionally, both compounds caused a reduction of the chlorophyll production in *Chlamydomonas rheinhardtii.* At low concentrations of the allelochemicals stimulatory effects are often observed. Surprisingly, 5-chloro-6-methoxy-2-benzoxazolinone isolated from maize seedlings also showed a high allelopathic potential (Kato-Naguchi, Kosemura, and Yamamura, 1998).

The phytotoxic benzoxazinoide allelochemicals are released into soil by the decomposition of plant residues (Yenish, Worsham, and Chilton, 1995). On the basis of these effects, different strategies of weed control have been developed, using cereals as mulches and cover crops (Putnam and DeFrank, 1983; Shilling, Liebl, and Worsham, 1985). Additionally, cyclic hydroxamic acids are exuded by intact roots and therefore, they are thought to be responsible for the allelopathic potential of living plants. Pérez and Ormeño-Nuñez (1991) demonstrated the presence of DIBOA in root exudates of rye cv. Forrajero Baer. In contrast, respective allelochemicals have not been detected in exudates of the hydroxamic acid containing wheat cv. Andifén. Therefore, the authors concluded that exudation of hydroxamic acids is an active process. Later, a higher weed suppressing activity of this rye cultivar, in comparison to wheat, was shown in field experiments. At similar crop density, spacing, and total crop biomass production, dry weight of wild oats and broad-leaved weeds in rye were 81% and 94%, respectively, less than in wheat (Pérez and Ormeño-Nuñez, 1993).

Cyclic hydroxamic acids and related compounds have been identified as allelochemicals not only of crops, but also of monocotyledonous weeds. For instance, phytotoxic benzoxazinoid compounds were detected in tissues of barnyard grass (Pheto, 1993). Quackgrass is one of the most aggressive perennial weeds in northern temperate parts of the world. Inhibitory effects of living quackgrass on growth of cress were demonstrated using a root exudate recirculating system (Schulz et al., 1994). DIBOA was detected in both root exudates and tissues and therefore, it was assumed to be, at least partially, responsible for the high allelopathic potential of this weed (Friebe et al., 1995; Schulz et al., 1994).

To understand the molecular effects that can account for the inhibitory activities, the influence of benzoxazinones and related benzoxazolinones on the activities of different enzymatic systems was investigated. Inhibitory effects on the metabolism of chloroplasts were described (Queirolo et al., 1981, 1983; Massardo et al., 1994). In this context, a strong inhibition of the chloroplast ATPase coupling factor by DIMBOA was found, whereas no effects were observed for the respective glucoside (Massardo et al., 1994). In addition, inhibitory influences of cyclic hydroxamic acids on the activity of the plasma membrane H^+-ATPase were demonstrated. At low concentration the allelochemicals caused a stimulation of this enzyme which paralleled the stimulation of root elongation of *Avena sativa* (Friebe et al., 1997). MBOA was identified as an auxin inhibiting substance of maize (Hasegawa et al., 1992). Modification of auxin-receptor interactions could be a reason for this biological activity of benzoxazolinones (Venis and Watson, 1978; Hoshi-Sakoda et al., 1994). Inhibitory effects of cyclic hydroxamic acids and related compounds were shown to be strongly species dependent (Barnes and Putnam, 1987; Pérez, 1990; Schulz et al., 1994). In this context, a higher sensitivity of dicots to DIBOA, in comparison to monocots, was demonstrated (Barnes and Putnam, 1987; Schulz et al., 1994). Allelopathic tolerance of plants is mainly due to their ability to detoxify phytotoxic natural compounds (Schulz and Friebe, 1999). A relatively high tolerance of *Avena sativa* to cyclic hydroxamic acids and benzoxazoliones was shown (Pérez, 1990). A detailed investigation of the molecular reasons for this tolerance revealed a pronounced detoxification capacity of oat (Wieland et al., 1998; Schulz and Friebe, 1999). In oat roots, BOA absorbed from aqueous media is enzymatically hydroxylated (Figure 8). Subsequently this intermediate compound is glucosylated to form BOA-6-O-β-D-glucoside. The BOA-N-glucosylation was shown to be an additional detoxification pathway in oat (Wieland et al., 1998).

CONCLUDING REMARKS

An abundance of research in the last 50 years clearly demonstrates, that cyclic hydroxamic acids and related compounds play an important role in interactions of cereals with the environment. The variety of biological activities including antiherbivore, antifungal and phytotoxic effects is immense

FIGURE 8. Different pathways of BOA (**IVa**) detoxification by *Avena sativa* (Adapted from Wieland et al., 1998).

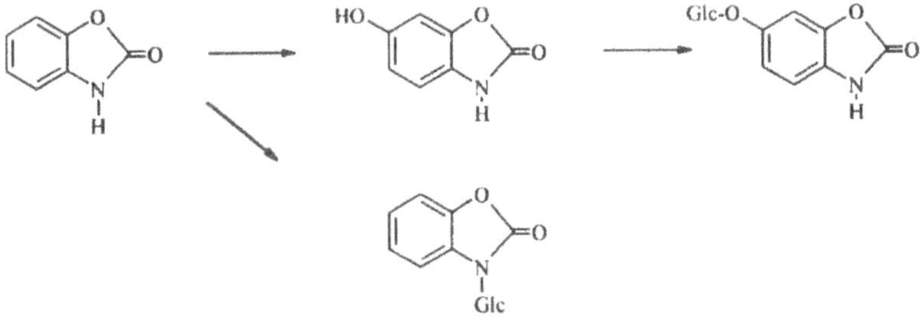

and, therefore, the significance of these allelochemicals as defense chemicals of cereal crops must not be underestimated. Nevertheless, as wide is the spectrum of effects, so diverse are the strategies of detoxification and avoidance of inhibitory effects by other organisms. The influence of benzoxazinones is an example of the perfectly organized ecological equilibrium that has been developed during a long evolutionary process.

REFERENCES

Abou Fakhr, E.M., B.E. Hibbard, and L.B. Bjostad. (1994). Tolerance of western corn rootworm larvae (Coleoptera: Chrysomelidae) to 6-methoxy-2-benzoxazolinone, a corn semiochemical for larval host location. *Journal of Economic Entomology* 87: 647-652.

Anai, T., H. Aizawa, N. Ohtake, S. Kosemura, S. Yamamura, and K. Hasegawa. (1996). A new auxin-inhibiting substance, 4-Cl-6,7-dimethoxy-2-benzoxazolinone, from light-grown maize shoots. *Phytochemistry* 42: 273-275.

Argandona, V.H., L.J. Corcuera, H.M. Niemeyer, and B.C. Campbell. (1983). Toxicity and feeding deterrency of cyclic hydroxamic acids from Gramineae in synthetic diets against the greenbug, *Schizaphis graminum*. *Entomologia Experimentalis et Applicata* 34: 134-138.

Assabgui, R.A., R.I. Hamilton, and J.T. Arnason. (1995). Hydroxamic acid content and plant development of maize (*Zea mays* L.) in relation to damage by the Western corn rootworm, *Diabrotica virgifera virgifera* LeConte. *Canadian Journal of Plant Science* 75: 851-856.

Barnes, J.P. and A.R. Putnam. (1987). Role of benzoxazinones in allelopathy by rye (*Secale cereale* L.). *Journal of Chemical Ecology* 13: 889-906.

Barria, B.N., S.V. Copaja, and H.M. Niemeyer. (1992). Occurrence of DIBOA in wild *Hordeum* species and its relation to aphid resistance. *Phytochemistry* 31: 89-91.

Beck, D.L., G.M. Dunn, D.G. Routley, and J.S. Bowman. (1983). Biochemical basis of resistance in corn to the corn leaf aphid. *Crop Science* 23: 995-998.

BeMiller, J.N. and A.J. Pappelis. (1965). 2,4-Dihydroxy-7-methoxy-1,4-benzoxazin-3-one glucoside in corn. I. Relation of water soluble, 1-butanol soluble glycoside fraction content of pith cores and stalk rot resistance. II. Isolation of 2-methoxy-2(3)-benzoxazolinone fraction as a measure of glucoside content and tissue differences of glucoside content. *Phytopathology* 55: 1237-1243.

Bergvinson, D.J., J.S. Larsen, and J.T. Arnason. (1995). Effect of light on changes in maize resistance against the European corn borer, *Ostrinia nubilalis* (Hübner). *Canadian Entomologist* 127: 111-122.

Bergvinson, D.J., R.I. Hamilton, and J.T. Arnason. (1995). Leaf profile of maize resistance factors to European corn borer, *Ostrinia nubilalis*. *Journal of Chemical Ecology* 21: 343-354.

Bjostard, L.B. and B.E. Hibbard. (1992). 6-Methoxy-2-benzoxazolinone: A semiochemical for host location by Western corn rootworm larvae. *Journal of Chemical Ecology* 18: 931-944.

Bravo, H. R., S.V. Copaja, and W. Lazo. (1997). Antimicrobial activity of natural 2-benzoxazolinones and related derivatives. *Journal of Agricultural Food Chemistry* 45: 3255-3257.

Bücker, C. and H.J. Grambow. (1990). Alterations in 1,4-benzoxazinone levels following inoculation with stem rust in wheat leaves carrying various alleles for resistance and their possible role as phytoalexins in moderately resistant leaves. *Zeitschrift fuer Naturforschung* 45c: 1151-1155.

Campos, F., J. Atkinson, J.T. Arnason, B.J.R. Philogene, P. Morand, N.H. Werstiuk, and G. Timmins. (1989). Toxicokinetics of 2,4-dihydroxy-7-methoxy-1,4-benzoxazin-3-one (DIMBOA) in the European corn borer, *Ostrinia nubilalis* (Hübner). *Journal of Chemical Ecology* 15: 1989-2001.

Castro, C.E., N.O. Belser, H.E. McKinney, and I.J. Thomason. (1989). Quantitative bioassay for chemotaxis with plant parasitic nematodes. Attractant and repellent fractions for *Meloidogyne incognita* from cucumber roots. *Journal of Chemical Ecology* 15: 1297-1309.

Chase, W.R., M.G. Nair, A.R. Putnam, and S.K. Mishra. (1991). 2,2′-oxo-1,1ázobenzene: microbial transformation of rye (L.) allelochemical in field soils by *Acinetobacter calcoaceticus*: III. *Journal of Chemical Ecology* 17: 1575-1584.

Copaja, S.V., B.N. Barria, and H.M. Niemeyer. (1991). Hydroxamic acid content of perennial Triticeae. *Phytochemistry* 30: 1531-1534.

Corcuera, L.J., M.D. Woodward, J.P. Helgeson, A. Kelman, and C.D. Upper. (1978). 2,4-Dihydroxy-7-methoxy-2H-1,4 benzoxazino-3*(4H)*-one, an inhibitor from *Zea mays* with differential activity against soft rotting *Erwinia* species. *Plant Physiology* 61: 791-795.

Couture, R.M., D. Routley, and G.M. Dunn. (1971). Role of cyclic hydroxamic acids in monogenic resistance of maize to *Helminthosporium turcicum*. *Physiological Plant Pathology* 1: 515-521.

Cubbon, M.H. (1925). Effect of a rye crop on the growth of grapes. *Journal of American Society of Agronomy* 17: 568-577.

Cuevas, L., H.M. Niemeyer, and F.J. Perez. (1990). Reaction of DIMBOA, a resistance factor from cereals, with α-chymotrypsin. *Phytochemistry* 29: 1429-1432.

Dabler, J.M., A.J. Pappelis, and J.N. Bemiller. (1969). Effects of phenolic acids and corn extracts upon spore germination of *Diplodia zeae*. *Phytopathology* 69: 1098-1101.

Elnaghy, M.A. and M. Shaw. (1966). Correlation between resistance to stem rust and the concentration of a glucoside in wheat. *Nature* 23: 17-18.

Elnaghy, M.A. and P. Linko. (1962). The role of 4-O-glucosyl-2,4-dihydroxy-7-methoxy-1,4-benzoxazin-3-one in resistance of wheat to stem rust. *Physiologia Plantarum* 15: 764-771.

FengMing, Y., X. ChongRen, L. SongGang, L. ChangShan, and L. JuHuai. (1995). Effects of DIMBOA on several enzymatic systems in Asian corn borer, *Ostrinia furnacalis* (Guenee). *Journal of Chemical Ecology* 21: 2047-2056.

Fielder, D.A., F.W. Collins, B.A. Blackwell, C. Bensimin, and J.W. Apsimon. (1994). Isolation and characterization of 4-acetyl-benzoxazolin-2-one (4-ABOA), a new benzoxazolinone from *Zea mays*. *Tetrahedron Letters* 35: 521-524.

Frey, M., P. Chomet, E. Glawischnig, C. Stettner, S. Grün, A. Winklmair, W. Eisenreich, A. Bacher, R.B. Meeley, S. Briggs, K. Simcox, and A. Gierl. (1997). Analysis of a chemical plant defense mechanism in grasses. *Science* 277: 696-699.

Friebe, A., I. Wieland, and M. Schulz. (1996). Tolerance of *Avena sativa* to the allelochemical benzoxazolinone. Degradation of BOA by root-colonizing bacteria. *Angewandte Botanik* 70: 150-154

Friebe, A., M. Schulz, P. Kück, and H. Schnabl. (1995). Phytotoxins from shoot extracts and root exudates of *Agropyron repens* seedlings. *Phytochemistry* 38: 1157-1159.

Friebe, A., U. Roth, P. Kück, H. Schnabl, and M. Schulz. (1997). Effects of 2,4-dihydroxy-1,4-benzoxazin-3-ones on the activity of plasma membrane H^+-ATPase. *Phytochemistry* 44: 979-983.

Friebe, A., V. Vilich, L. Hennig, M. Kluge, and D. Sicker. (1998). Detoxification of benzoxazolinone allelochemicals from wheat by *Gaeumannomyces graminis* var. *tritici*, *G. graminis* var. *graminis*, *G. graminis* var. *avenae*, and *Fusarium culmorum*. *Applied Environmental Microbiology* 64: 2386-2391.

Friebe, A., W. Klever, R. Sikora, and H. Schnabl. (1998). Allelochemicals in root exudates of maize: Effects on root lesion nematode *Pratylenchus zeae*. In *Phytochemical Signals and Plant-Microbe Interactions. Recent Advances in Phytochemistry Vol. 32*, eds. J.T. Romeo, K.R. Downum, and R. Verpoorte. New York: Plenum Press, pp. 71-93.

Fujimoto, D.K., D.M. Weller, and L.S. Thomashow. (1995). Role of secondary metabolites in root disease suppression. *Allelopathy-Organisms, Processes and Applications*, eds. Inderjit, K.M.M. Dakshini, and F.A. Einhellig, ACS Symposium Series 582. Washington, DC: American Chemical Society, pp. 330-347.

Gagliardo, R.W. and W.S. Chilton. (1992). Soil transformation of 2-(3*H*)-benzoxazolone of rye into phytotoxic 2-amino-3H-phenoxazin-3-one. *Journal of Chemical Ecology* 18: 1683-1691.

Gerber, N.N. and M.P. Lechevalier. (1964). Phenazines and phenoxazinones from *Waksmania aerata* sp. *nov.* and *Pseudomonas iodina*. *Biochemistry* 3: 598-602.

Gianoli, E. and H.M. Niemeyer. (1997a). Environmental effects on the accumulation of hydroxamic acids in wheat seedlings: The importance of plant growth rate. *Journal of Chemical Ecology* 23: 543-551.

Gianoli, E. and H.M. Niemeyer. (1997b). Characteristics of hydroxamic acid induction in wheat triggered by aphid infestation. *Journal of Chemical Ecology* 23: 2695-2705.

Gianoli, E. and H.M. Niemeyer. (1998). DIBOA in wild Poaceae: sources of resistance to the Russian wheat aphid (*Diuraphis noxia*) and the greenbug (*Schizaphis graminum*). *Euphytica* 102: 317-321.

Givovich, A. and H.M. Niemeyer. (1995). Comparison of the effect of hydroxamic acids from wheat on five species of cereal aphids. *Entomologia Experimentalis et Applicata* 74: 115-119.

Givovich, A. and H.M. Niemeyer. (1996). Role of hydroxamic acids in the resistance of wheat to the Russian wheat aphid, *Diuraphis noxia* (Mordvilko) Hom. (Aphidideae). *Journal of Applied Entomology* 120: 537-539.

Grambow, H.J., J. Lückge, A. Klausner, and E. Müller. (1986). Occurrence of 2-(2-hydroxy-4,7-dimethoxy-2*H*-1,4-benzoxazin-3-one)-β-D-glucopyranoside in *Triticum aestivum* leaves and its conversion into 6-methoxy-benzoxazolinone. *Zeitschrift fuer Naturforschung* 41c: 684-690.

Gross, D. (1989). Antimicrobial defense compounds in the Gramineae. *Journal of Plant Diseases and Protection* 96: 535-552.

Guthrie, W.D., R.L. Wilson, J.R. Coats, J.C. Robbins, C.T. Tseng, J.L. Jarvis, and W.A. Russell. (1986). European corn borer (Lepidoptera: Pyralidae) leaf-feeding resistance and DIMBOA content in maize in inbred lines of dent maize grown under field versus greenhouse conditions. *Journal of Economic Entomology* 79: 1492-1496.

Hasegawa, K., S. Togo, M. Urashima, J. Mizutani, S. Kosemura, and S. Yamamura. (1992). An auxin-inhibiting substance from light-grown maize shoots. *Phytochemistry* 31: 3673-3676

Hashimoto, Y. and K. Shudo. (1996). Chemistry of biologically active benzoxazinoids. *Phytochemistry* 43: 551-559.

Hedin, P.A., F.M. Davis, and W.P. Williams. (1993). 2-Hydroxy-4,7-dimethoxy-1,4-benzoxazin-3-one (*N-O*-Me-DIMBOA), a possible toxic factor in corn, to the southwestern corn borer. *Journal of Chemical Ecology* 19: 531-542.

Herbert, R.B., J. Mann, and A. Römer. (1982). Phenazine and phenoxazinone biosynthesis in *Brevibacterium iodinum*. *Zeitschrift fuer Naturforschung* 37c: 159-164.

Hill, H.H. (1926). Decomposition of organic matter in soil. *Journal of Agricultural Research* 33: 77-99.

Hofman, J. and O. Hofmanova. (1970). 1,4-Benzoxazinone derivatives in plants. A new type of glucoside from *Zea mays*. *Tetrahedron Letters* 3213-3214.

Hofman, J., O. Hofmanova, and V. Hanus. (1969). 1,4-Benzoxazinone derivatives in plants. A new glucoside derivative from *Zea mays*. *Tetrahedron Letters* 5001-5002.

Hoshi-Sakoda, M., K. Usui, K. Ishizuka, S. Kosemura, S. Yamamura, and K. Hasegawa. (1994). Structure-activity relationships of benzoxazolinones with respect to auxin-induced growth and auxin-binding protein. *Phytochemistry* 37: 297-300.

Houseman, J.G., F. Campos, N.M.R. Thie, B.J.R. Philogene, J. Atkinson, P. Norand, and J.T. Arnason. (1992). Effect of maize-derived compounds DIMBOA and

MBOA on growth and digestive processes of European corn borer (Lepidoptera: Pyralidae). *Journal of Economic Entomology* 85: 669-674.

Kato-Noguchi, H., S. Kosemura, and S. Yamamura. (1998). Allelopathic potential of 5-chloro-6-methoxy-2-benzoxazolinone. *Phytochemistry* 48: 433-435.

Kluge, M., H.J. Grambow, and D. Sicker. (1997). (2R)-2-β-D-glucopyranosyloxy-4, 7-dimethoxy-2*H*-1,4-benzoxazin-3(4H)-one from *Triticum aestivum*. *Phytochemistry* 44: 639-641.

Kumar, P., R.W. Gagliardo, and W.S. Chilton. (1993). Soil transformation of wheat and corn metabolites MBOA and DIM_2BOA into aminophenoxazinones. *Journal of Chemical Ecology* 19: 2453-2461.

Leszczynski, B., A. Urbanska, H. Matok, and A.F.G. Dixon. (1993). Detoxifying enzymes of the grain aphid. *Bulletin of the International Organization for Biological Control of Noxious Animals and Plants, West Palaearctic Regional Station (or Organisation Interionationale de Lutte Biologique Contre les Animaux et les Plantes Nuisibles, Section R'egionale Quest Pal'earctique)* 16: 165-172.

Leszczynski, B. and A.F.G. Dixon. (1992). Resistance of cereals to aphids: The interaction between hydroxamic acids and glutathione S-transferases in the grain aphid *Sitobion avenae* (F.) (Homoptera: Aphididae). *Journal of Applied Entomology* 113: 61-67.

Leszczynski, B., H. Matok, and A.F.G. Dixon. (1992). Resistance of cereals to aphids: the interaction between hydroxamic acids and UDP-glucose transferases in the aphid Sitobion avenae (Homoptera: Aphididae). *Journal of Chemical Ecology* 18: 1189-1200.

Long, B.J., G.M. Dunn, and D.G. Routley. (1975). Relationship of hydroxamic acid content in maize and resistance to northern corn leaf blight. *Crop Science* 15: 333-335.

Long, B.J., G.M. Dunn, and D.G. Routley. (1978). Relationship of hydroxamate concentration in maize and field reaction to *Helminthosporium turcicum*. *Crop Science* 18: 573-574.

Long, B.J., G.M. Dunn, J.S. Bowman, and D.G. Routley. (1977). Relationship of hydroxamic acid content in corn and resistance to the corn leaf aphid. *Crop Science* 17: 55-58.

Massardo, F., G.E. Zuniga, L.M. Perez, and L.J. Corcuera. (1994). Effects of hydroxamic acids on electron transport and their cellular location in corn. *Phytochemistry* 35: 873-876.

Mayoral, A.M., W.F. Tjallingii, and P. Castanera. (1996). Probing behavior of *Diuraphis noxia* on five cereal species with different hydroxamic acid levels. *Entomologia Experimentalis et Applicata* 78: 341-348.

Miller, J.D., D.A. Fielder, P.F. Dowd, R.A. Norton, and F.W. Collins. (1996). Isolation of 4-ABOA and diferoylputrescine from an extract of *Gibberella* ear rot-resistant corn that blocks mycotoxin biosynthesis, and the insect toxicity of 4-ABOA and related compounds. *Biochemical Systematics and Ecology* 24: 647-658.

Mwaja, V.N., J.B. Masiunas, and L.A. Weston. (1995). Effects of fertility on biomass, phytotoxicity, and allelochemical content of cereal rye. *Journal of Chemical Ecology* 21: 81-96.

Nagao, T., H. Otsuka, H. Kohda, T. Sato, and K. Yamasaki. (1985). Benzoxazinones from *Coix Lachryma-Jobi* var. *Ma-Yuen*. *Phytochemistry* 24: 2959-2962.

Nicol, D., S.V. Copaja, S.D. Wratten, and H.M. Niemeyer. (1992). A screen of worldwide wheat cultivars for hydroxamic acid levels and aphid antixenosis. *Annals of Applied Biology* 121: 11-18.

Niemeyer, H.M. (1988). Hydroxamic acids (4-hydroxy-1,4-benzoxazin-3-ones), defence chemicals in the gramineae. *Phytochemistry* 27: 3349-3358.

Niemeyer, H.M. (1990). The role of secondary plant compounds in aphid-host interactions. In *Aphid-Plant Genotype Interactions*, eds. R.K. Campbell and R.D. Eikenbary. Amsterdam: Elsevier, pp. 187-205.

Niemeyer, H.M., A. Givovich, and V. Copaja. (1993). Hydroxamic acids: chemical defences in wheat against aphids. In *Pest control and Sustainable Agriculture* ed. S.A. Corey, D.J. Dall, and W.M. Milne, Australia: CSIRO, pp. 39-43.

Özden, S., T. Özden, J. Attila, M. Kücükislamoglu, and A. Okatan. (1992). Isolation and identification via high performance liquid chromatography and thin layer chromatography of benzoxazolinone precursors from *Consolida orientalis* flowers. *Journal of Chromatography* 609:402-406.

Osvald, H. (1953). On antagonism between plants. In *Proceedings of the 7th International Botanical Congress, Stockholm, 1950*, eds. Alquist and Wiksell. Stockholm, pp. 167-171.

Perez, F.J. (1990). Allelopathic effect of hydroxamic acids from cereals on *Avena sativa* and *Avena fatua*. *Phytochemistry* 29: 773-776.

Pérez, F.J. and H.M. Niemeyer. (1989). Reaction of DIMBOA, a resistance factor from cereals, with papain. *Phytochemistry* 28: 1597-1600.

Pérez, F.J. and J. Ormeño-Nuñez. (1991). Differences in hydroxamic acids content in roots and root exudates of wheat (*Triticum aestivum* L.) and rye (*Secale cereale* L.): possible role in allelopathy. *Journal of Chemical Ecology* 17: 1037-1043.

Pérez, F.J. and J. Ormeño-Nuñez. (1993). Weed growth interference from temperate cereals: The effect of a hydroxamic-acids exuding rye (*Secale cereale* L.) cultivar. *Weed Research* 33: 115-119.

Pheto, M. (1993). Occurrence of cyclic hydroxamic acids in the tissues of barnyard grass (*Echinochloa cruz-galli*/L./P. B.), and their role in allelopathy. *Acta Agronomica Hungarica* 42: 197-202.

Pratt, K., P. Kumar, and W.S. Chilton. (1995). Cyclic hydroxamic acids in dicotyledonous plants. *Biochemical Systematics and Ecology* 23: 781-785.

Putnam, A.R. and J. De Frank. (1983). Use of phytotoxic plant residues for selective weed control. *Crop Protection* 2: 173-181.

Queirolo, C.B., C.S. Andreo, R.H. Vallejos, H.M. Niemeyer, and L.J. Corcuera. (1981). Effects of hydroxamic acids isolated from Gramineae on adenosine 5′-triphosphate synthesis in chloroplasts. *Plant Physiology* 68: 941-943.

Queirolo, C.B., C.S. Andreo, H.M. Niemeyer, and L.J. Corcuera. (1983). Inhibition of ATPase from chloroplasts by a hydroxamic acid from the Gramineae. *Phytochemistry* 22: 2455-2458.

Richardson, M.D. and C.W. Bacon. (1993). Cyclic hydroxamic acid accumulation in corn seedlings exposed to reduced water potentials before, during, and after germination. *Journal of Chemical Ecology* 19: 1613-1624.

Richardson, M.D. and W.C. Bacon. (1995). Catabolism of 6-methoxy-benzoxazolinone and benzoxazolinone by *Fusarium moniliforme*. *Mycologia* 87: 510-517.

Robinson, J.F., J.A. Klun, and T.A. Brindley. (1978). European corn borer: A nonpreference mechanism of leaf feeding resistance and its relationship to 1,4-benzoxazin-3-one concentration in dent corn tissue. *Journal of Economic Entomology* 71: 461-465.

Rustamani, M.A., K. Kanehisa, H. Tsumuki, and T. Shiraga. (1996). The relationship between DIMBOA concentration in corn lines and resistance to aphids. *Bulletin of the Research Institute for Bioresources, Okayama University* 4: 33-42.

Schulz, M., A. Friebe, P. Kück, M. Seipel, and H. Schnabl. (1994). Allelopathic effects of living quackgrass (*Agropyron repens* L.). Identification of inhibitory allelochemicals exuded from rhizome borne roots. *Applied Botany* 68: 195-200.

Schulz, M. and A. Friebe. (1999). Detoxification of allelochemicals and enzymes involved. In *Principles and Practices in Plant Ecology: Allelochemical Interactions*, eds. Inderjit, K.M.M. Dakshini, and C.L. Foy. Boca Raton, FL: CRC-Press LLC, pp. 383-400.

Shilling, D.G., R.A. Liebl, and A.D. Worsham. (1985). Rye (*Secale cereale* L.) and wheat (*Triticum aestivum* L.) mulch: The suppression of certain broad-leaved weeds and the isolation and identification of phytotoxins. In *The Chemistry of Allelopathy: Biochemical Interactions Among Plants*, ed. A.C. Thompson. ACS Symposium Series 268. Washington, DC: American Chemical Society, pp. 243-271.

Thomashow, L.S. and D.M. Weller. (1988). Role of phenazine antibiotic from *Pseudomonas fluorescens* in biological control of *Gaeumannomyces graminis* var. *tritici*. *Journal of Bacteriology* 170: 3499-3508.

Thackray, D.J., S.D. Wratten, P.J. Edwards, and H.M. Niemeyer. (1990). Resistance to the aphids *Sitobion avenae* and *Rhopalosiphum padi* in Gramineae in relation to hydroxamic levels. *Annals of Applied Biology* 116: 573-582.

Venis, M.A. and P.J. Watson. (1978). Naturally occurring modifiers of auxin-receptor interaction in corn: Identification as benzoxazolinones. *Planta* 142: 103-107.

Vilich, V., B. Löhndorf, B., R.A. Sikora, and A. Friebe. (1999). Metabolization of benzoxazolinone allelochemicals of *Zea mays* by *Fusarium subglutinans*. *Mycological Research* 103: 1529-1532.

Virtanen A.I. and P.K. Hietala. (1955). 2-(*3H*)-Benzoxazolinone, an anti-Fusarium-factor in rye seedlings. *Acta Chemica Scandinavica* 9: 1543-1544.

Wahlroos, Ö. and A.I. Virtanen. (1958). On the antifungal effect of benzoxazolinone and 6-methoxybenzoxazolinone, respectively on *Fusarium nivale*. *Acta Chemica Scandinavica* 12: 124-128.

Wahlroos, Ö. and A.I. Virtanen. (1959). The precursors of 6-methoxybenzoxazolinone in maize and wheat plants, their isolation and some of their properties. *Acta Chemica Scandinavica* 13: 1906-1908.

Weller, D.M. (1988). Biological control of soilborne plant pathogens in the rhizosphere with bacteria. *Annual Review of Phytopathology* 26: 379-407.

Weller, D.M., W.J. Howie, and R.J. Cook. (1988). Relationship between in vitro inhibition of *Gaeumannomyces graminis* var. *tritici* and suppression of take-all of wheat by fluorescent pseudomonads. *Phytopathology* 78: 1094-1100.

Whitney, N.J. and C.G. Mortimore. (1959). Isolation of the antifungal substance

6-methoxybenzoxazolinone from field corn (*Zea mays* L.) in Canada. *Nature* 184: 1320.

Whitney, N.J. and C.G. Mortimore. (1961). Effect of 6-methoxybenzoxazolinone on the growth of *Xanthomonas stewartii* (Erw. Smith) Dowson and its presence in sweet corn (*Zea mays* var. *Saccharata bailey*). *Nature* 189: 596-597.

Wieland, I., M. Kluge, B. Schneider, J. Schmidt, D. Sicker, and M. Schulz. (1998). 3-β-D-Glucopyranosyl-benzoxazolin-2*(3H)*-one–a detoxification product of benzoxazolin-2*(3H)*-one in oat roots. *Phytochemistry* 49: 719-722.

Wolf, R.B., G.F. Spencer, and R.D. Plattner. (1985). Benzoxazolinone, 2,4-dihydroxy1,4-benzoxazin-3-one, and its glucoside from *Acanthuus mollis* seeds inhibit velvetleaf germination and growth. *Journal of Natural Products* 48: 59-63.

Woodward, M.D., L.J. Corcuera, J.P. Helgeson, A. Kelman, and C.D. Upper. (1978). Decomposition of 2,4-dihydroxy-7-methoxy-2H-1,4-benzoxazin-3(4H)-one in aqueous solution. *Plant Physiology* 61: 796-802.

Woodward, M.D., L.J. Corcuera, H.K. Schnoes, J.P. Helgeson, and C.D. Upper. (1979). Identification of 1,4-benzoxazin-3-ones in maize extracts by gas-liquid chromatography and mass spectrometry. *Plant Physiology* 63: 9-13.

Xie, Y., J.T. Arnason, B.J.R. Philogene, H.T. Olechowski, and R.I. Hamilton. (1992). Variation of hydroxamic acid content in maize roots in relation to geographic origin of maize germplasm and resistance to western corn rootworm (Coleoptera: Chrysomelidae). *Journal of Economic Entomology* 85: 2478-2485.

Yenish, J.P., A.D. Worsham, and W.S. Chilton. (1995). Disappearance of DIBOA-glucoside, DIBOA, and BOA from rye (*Secale cereale* L.) cover crop residue. *Weed Science* 43: 18-20.

Yue, Q., C.W. Bacon, and M.D. Richardson. (1998). Biotransformation of 2-benzoxazolinone and 6-methoxy.benzoxazolinone by *Fusarium moniliforme*. *Phytochemistry* 48: 451-454.

Zhang, X., F.K. Habib, M. Ross, U. Burger, A. Lewenstein, K. Rose, and J.-C. Jaton. (1995). Isolation and characterization of a cyclic hydroxamic acids from a pollen extract, which inhibits cancerous cell growth in vitro. *Journal of Medicinal Chemistry* 38: 735-738.

Zhu, Y.L., R. He, and J.L. Liu. (1994). Changes of DIMBOA contents in resistant and susceptible near-isolines of maize during infection with *Exserohicum turcicum*. *Acta Agronomica Sinica* 20: 653-657.

Allelopathic Research in Permanent Improved Pastures of Inland Temperate Australia

J. E. Pratley

SUMMARY. Agricultural productivity is very dependent upon the successful establishment of crops and pastures. High populations of vigorous seedlings enable the crop or pasture to withstand competition from weeds and infection from diseases as well as providing good utilisation of nutrient inputs and water. Residues of weed and pasture species can interfere with crop establishment and pasture regeneration. Research at Charles Sturt University (CSU) and elsewhere in Australia has focused on a better understanding of these relationships and the management of each situation to advantage the pasture and to minimise the impact of the allelopathic effects. Such effects include poor germination and early seedling growth and are influenced by the variety and species of agricultural plants being grown. *[Article copies available for a fee from The Haworth Document Delivery Service: 1-800-342-9678. E-mail address: <getinfo@haworthpressinc.com> Website: <http://www.HaworthPress.com> © 2001 by The Haworth Press, Inc. All rights reserved.]*

KEYWORDS. Allelopathy, pasture regeneration, temperate Australia, crop establishment

---

J. E. Pratley is affiliated with Farrer Centre, Charles Sturt University, Wagga Wagga, NSW 2678, Australia (E-mail: jpratley@csu.edu.au).

[Haworth co-indexing entry note]: "Allelopathic Research in Permanent Improved Pastures of Inland Temperate Australia." Pratley, J. E. Co-published simultaneously in *Journal of Crop Production* (Food Products Press, an imprint of The Haworth Press, Inc.) Vol. 4, No. 2 (#8), 2001, pp. 401-414; and: *Allelopathy in Agroecosystems* (ed: Ravinder K. Kohli, Harminder Pal Singh, and Daizy R. Batish) Food Products Press, an imprint of The Haworth Press, Inc., 2001, pp. 401-414. Single or multiple copies of this article are available for a fee from The Haworth Document Delivery Service [1-800-342-9678, 9:00 a.m. - 5:00 p.m. (EST). E-mail address: getinfo@haworthpressinc.com].

INTRODUCTION

Pasture production in inland temperate Australia can be summarised as:

i. extensive grazing by sheep, cattle, native and feral animals on the rangelands. These areas are categorised by low and unreliable rainfall and native perennial species augmented by ephemerals when conditions allow. No understanding of the role of allelopathy exists in these areas and therefore is not considered further;
ii. short-term pastures which are part of a pasture/crop rotation. These pastures are usually legume-based and play an important role in soil structure rehabilitation and soil fertility improvement through nitrogen fixation by the pasture legume. The species involved are commonly annual legumes such as subterranean clover (*Trifolium subterraneum*), annual medics (*Medicago* spp.) and lucerne (*Medicago sativa*). Allelopathy is considered to be playing an important role in these pastures particularly at establishment and with weed interactions. However, these issues are not discussed further in this chapter;
iii permanent improved pastures, which usually have a perennial grass growing in combination with an annual or perennial legume. It is only in recent times that the role allelopathy might be playing in such systems has been considered. This chapter outlines some of the research that has been undertaken to increase the understanding of the allelopathic processes involved in such permanent improved pastures.

THE PERMANENT IMPROVED PASTURE SYSTEM

Background

In southern temperate Australia–the so-called higher rainfall areas (>550 mm average annual rainfall), a sown perennial grass such as phalaris (*Phalaris aquatica*), tall fescue (*Festuca arundinacea*), cocksfoot (*Dactylis glomerata*) or perennial ryegrass (*Lolium perenne*) is commonly combined with an annual legume such as subterranean clover (*Trifolium subterraneum*) or a perennial legume such as white clover (*Trifolium repens*) or lucerne (*Medicago sativa*). Often there is also the presence of other weedy cool-season annual grasses such as *Bromus*, *Hordeum*, *Vulpia* and *Holcus* spp.

It is important to note in this context that the summer period from December to March is relatively dry in southern Australia and so the pastures respond by providing carryover dry residues. The perennial grasses, particularly phalaris, exhibit a summer dormancy. The annual legume survives this period as seed and germinates with the autumn rains, as do the annual grasses. The sustainability of the grass/legume combination, therefore, depends very importantly on the initial establishment of the legume and the autumn regeneration of the annual legume. Subterranean clover ideally contributes 30 to 60% of the feed offered in the cooler months, especially spring, but in more established pastures this contribution commonly declines to less than 10% regardless of grazing regime (Morley, Bennett, and Mckinney, 1969; Leigh, Halsall, and Holgate, 1995).

A decline in legume content is reflected in poorer animal performance because of the higher protein and digestibility of those species. Several researchers have demonstrated the clear relationship between legume proportion and livestock performance (Ulyatt, Lancashire, and Jones, 1977; Freer and Jones, 1984; Kenry and Reed 1984; Askin, Powan, and Lucas, 1987). Several reasons advanced for legume decline have included low seed reserves carried over from the previous season (Dear and Loveland, 1985), effects of grazing such as treading and ingestion of seed pods (Carter, 1987), soil structure decline reducing seed burial of subterranean clover (Taylor, 1985) and the presence of plant residues, weeds and diseases.

Direct Allelopathic Effects Between Pasture Species

Research conducted at CSIRO Canberra demonstrated that one major factor in legume decline was the dry carryover litter of the phalaris. Leigh and Halsall (1994) and Leigh et al. (1995) describe that in longstanding pastures comprising phalaris (cv. Australian) and subterranean clover (cv. Mt. Barker), the presence of phalaris litter impeded the regeneration of the subterranean clover. Where the litter was reduced or removed (by raking between the phalaris tussocks, application of paraquat or by burning) the numbers of seedlings that established and survived increased (Figure 1a) with dry matter production of the clover almost doubling (Figure 1b). The addition of phalaris litter also had a negative impact on the subterranean clover establishment that was replaced effectively by increased production of phalaris or weeds.

Further investigation under controlled laboratory conditions (Halsall et al., 1995) showed that under sterile conditions to separate allelopathy from the effects of decomposition or infection by plant pathogens in the residue environment, cold water extracts from a range of species produced inhibition effects in the test species. In particular, phalaris extracts reduced germination of all 11 lines of legumes. Medics were particularly sensitive. Radicle lengths were reduced by up to 30% of controls and root tips were swollen and thickened. Nodulation of the legumes was also affected in all cases except for vetch (*Vicia dasycarpa* cv. Namoi).

Halsall et al. (1995) extended the study to evaluate the effects of cold water extracts from a range of species on a number of commonly grown

FIGURE 1. Effects of the removal or addition of phalaris litter on the number of subterranean clover seedlings (a) and clover dry matter production in spring (b) (Leigh et al., 1995, reprinted with permission).

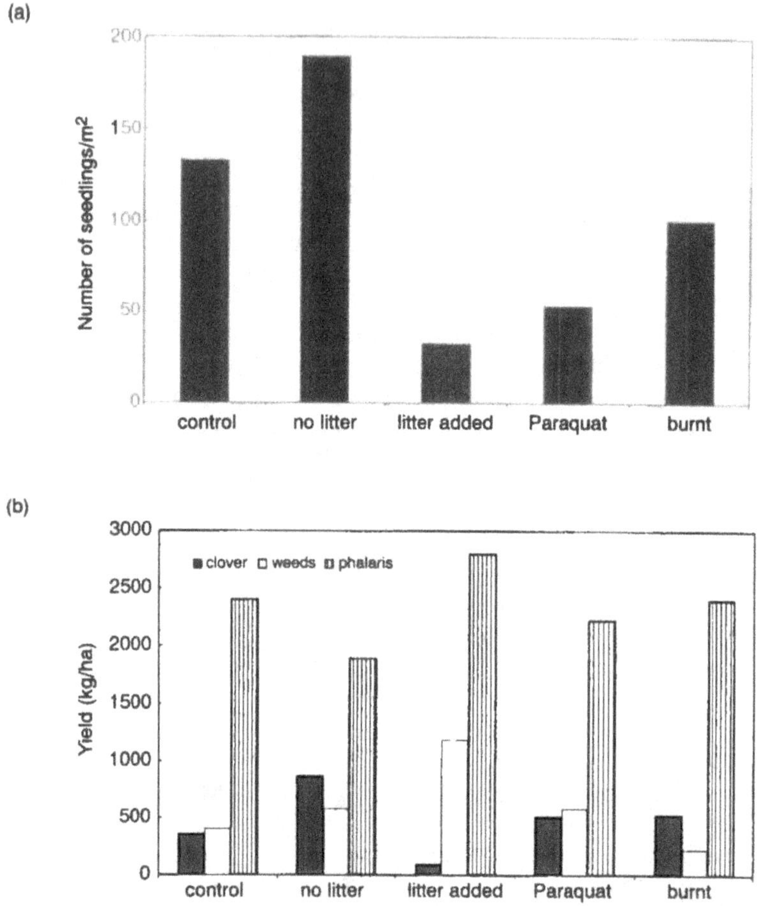

legumes. These trials revealed that white clover (cv. Grasslands Hua), barrel medic (*Medicago truncatula* cv. Jemalong) and balansa clover (*Trifolium balansae*) were also sensitive to three cultivars of phalaris extracts, as well as to *Eragrostis curvula* and perennial ryegrass (*Lolium perenne*), though not to cocksfoot (*Dactylis glomerata*). Of particular interest from all this research was that considerable variation in potency existed between donor cultivars as well as in sensitivity between receiver cultivars (Figure 2). Leigh and Halsall (1994) established a breeding program to produce subterranean clover plants

FIGURE 2. Rankings for tolerance to phalaris residues for subclover lines (Leigh and Halsall, 1994, reprinted with permission).

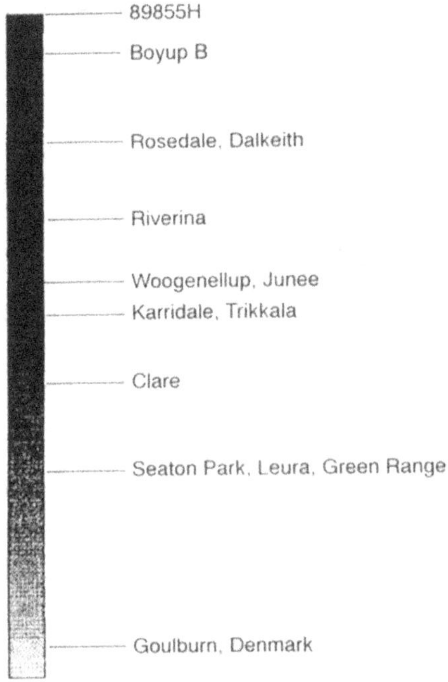

resistant to the allelochemicals present in the phalaris residues. Following use of irradiation techniques to induce mutation, a comprehensive screening process to select tolerant individuals was undertaken and a tolerant line obtained.

Indirect Allelopathic Effects

The allelopathic effect can sometimes be induced by another organism. Two examples are discussed here, namely the presence of an endophyte and crown rust infestation.

Endophyte Presence in Perennial Ryegrass

Foot et al. (1987) reported that the presence of an endophyte (*Acremonium lolii*) in perennial ryegrass substantially reduced the density of subterranean clover in a field experiment at Hamilton, Victoria. Similar effects had been reported in New Zealand (Sutherland and Hoglund, 1989). Further work by Snell and Quigley (1993) in pots compared the effects on establishment of subterranean clover (cv. Trikkala) of a chopped mulch of senescent perennial ryegrass obtained either from a high endophyte site or a low endophyte site. A third treatment comprised a 1:4 mixture while the control treatment had a mulch of inert polypropylene particles. Although no effects on subterranean clover germination were obtained, there was a progressive reduction in nodule score with increased endophyte. Root length and root weight was also reduced.

Quigley and McFarlane (1993) report two field experiments in which the sowing rate of perennial ryegrass was varied within a uniform nutrition regime or variable phosphate nutrition with constant sowing rate. Both experiments were split for plus and minus endophyte presence in the sown ryegrass seed. The high endophyte seed (E+) resulted in higher ryegrass plant densities and lower subterranean clover dry matter production (Table 1). These outcomes could predispose the subterranean clover to progressive decline in persistence in the sward through increased competition and reduced seed set. Quigley et al. (1990) demonstrated in a bioassay study that extracts from milled perennial ryegrass culms reduced the root lengths of young subterranean clover seedlings where endophyte was present. Whether such chemicals would be leached under field conditions, however, was not established although the effects produced in the field suggest allelochemical involvement.

Crown Rust Infestation Effects

Mattner (1998) investigated the impact of crown rust (*Puccinia coronata*) on perennial ryegrass (*Lolium perenne*) on white clover (*Trifolium repens*)

TABLE 1. Effects of *Acremonium* endophyte (E+) on perennial ryegrass establishment and subterranean clover production (adapted from Quigley and McFarlane, 1993).

	Experiment 1			Experiment 2		
Measurement	E+	E−	S.E. of diff.	E+	E−	S.E. of diff.
Proportion of seed containing endophyte	84	5		84	5	
Proportion of plants containing endophyte	75	0		70	0	
Established ryegrass plants (no./m^2)	249	188	15.6	311	266	17.1
Clover density (number/m^2)	92	86	4.0	56	71	11.0
Clover dry matter production (mg/plant)	60	68	3.5	51	55	5.6

pastures, a combination which predominates in improved pastures in south-eastern Australia. Crown rust is a serious fungal pathogen of ryegrass and can reduce its yield by more than 50%. In mixture with white clover, however, the loss of ryegrass production does not result in an increase in clover biomass. Mattner (1998) used bioassays involving soil from pots growing either rusted ryegrass or non-rusted ryegrass. Plus and minus nutrient and steam-sterilisation treatments were incorporated into the design to overcome differences in nutrient content and microflora. The data (Table 2) clearly showed that the biomass of clover grown in soil from rusted ryegrass was significantly less than that of the control or in soil from non-rusted ryegrass. Where leachate was collected from pots of each treatment and applied to clover, growth of the clover receiving soil leachate from rusted ryegrass was less than that of plants watered from non-rusted ryegrass or control. This investigation eliminated the impacts due to soil microorganisms and to nutrient

TABLE 2a. The production of white clover in response to leachate derived from rusted or non-rusted perennial ryegrass. Bioassays were performed prior to and after the inoculation of ryegrass with rust (Adapted from Mattner, 1998).

White clover growth parameter (units)	Leachate source			LSD	
	Control	Non-rusted ryegrass	Rusted ryegrass	$p < 0.05$	$p < 0.01$
Pre-inoculation bioassay					
Total biomass (g)	1.10	0.76	0.78	0.16	0.22
Post-inoculation bioassay					
Total biomass (g)	2.86	3.11	2.28	0.53	0.75
Leaf area (cm^2)	203.88	196.2	145.4	49.9	ns
Stolon number	4.71	5.00	4.00	0.76	ns
Leaf number	36.5	34.12	25.5	8.78	ns
Nodulation index (#/g root dry weight)	1222.81	1121.11	1143.78	ns	ns

ns means not significant.

TABLE 2b. Biomass production of perennial ryegrass donors

Ryegrass Biomass (g)	
Non-rusted	Rusted
43.8	14.8
$p < 0.001$	

effects. It provides strong evidence for ryegrass, under stress through disease, of enhanced allelopathy from the disease thereby reducing the production of its competitors and improving prospects for survival of the ryegrass.

DIRECT IMPACT OF WEEDS

The principle of carryover of dry material applies also to weeds. The impact has been demonstrated clearly by work at Charles Sturt University, Wagga Wagga, Australia. Pratley (1989) initially collected dry residues of silvergrass (*Vulpia* spp.) at the end of the growing season (December-early summer), at crop and pasture establishment time (May-late autumn) and mid-season (August-late winter) from the same site. These residues were applied at different rates to wheat sown in pots and the effects on establishment evaluated. Whilst emergence by the December and August collections had little impact, the residues collected in May resulted in decimation of the establishing seedlings (Figure 3). Further work (Pratley and Ingrey, 1990) showed that the December-collected material could be made potent by exposure to ultraviolet light. These results demonstrated that there was a process of degradation of the material either physically or chemically from the end of one season to the beginning of the next through the dry, hot summer period. Providing this period remained dry prior to the establishment period, allelochemicals were present in sufficient quantity to leach out of the material to interfere with plant establishment.

FIGURE 3. Effect of different residue burdens of *Vulpia* at different degradation states on wheat growth (Adapted from Pratley, 1989).

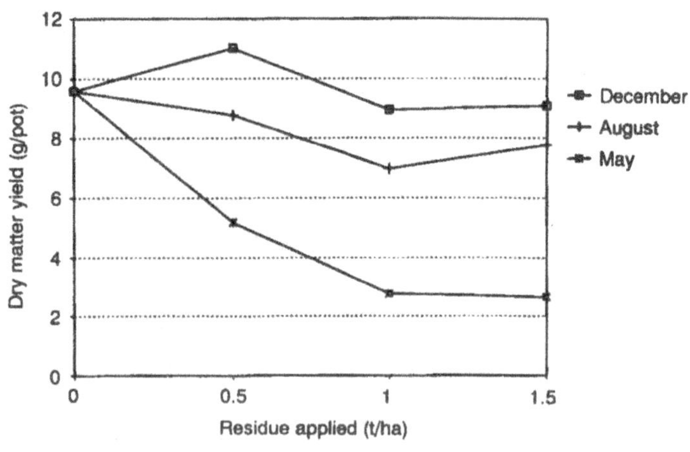

An, Pratley, and Haig (1993) then proceeded to mix the aqueous extracts from the *Vulpia* in a flask with increasing quantities of soil. All flasks were incubated in the dark at 20°C for 6 days after which the filtered and centrifuged supernatants were used for germination bioassays in petri dishes. Whereas the leachate by itself remained potent for up to 60 days (Figures 4 and 5) (An, Pratley, and Haig, 1997), the presence of soil reduced the potency period. Such effects occurred up to a ratio of 1:5 residue:soil (Figures 6 and 7).

FIGURE 4. Phytotoxicity time course of *Vulpia* residue on seed germination as affected by germinating time (Adapted from An, 1996).

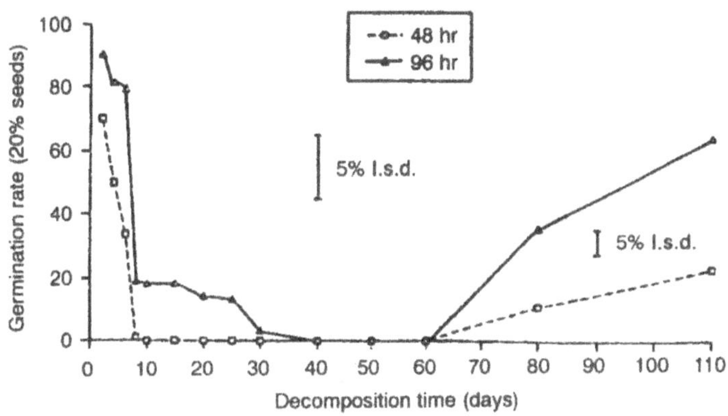

FIGURE 5. Phytotoxicity dynamics of *Vulpia* residue extracts as measured by seedling lengths of test species (wheat cv. Vulcan). Only the first extract results are shown (Adapted from An, 1996).

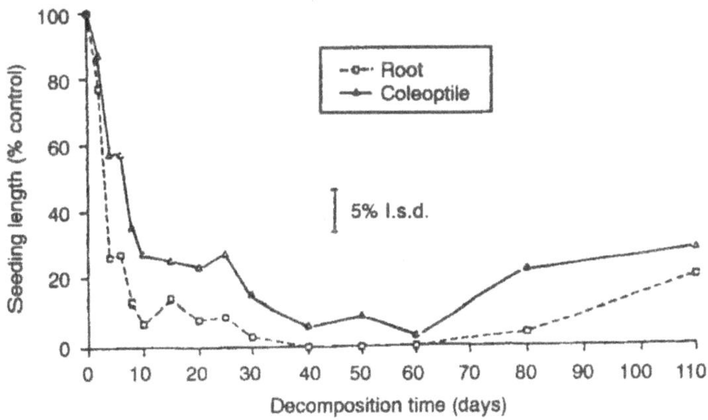

FIGURE 6. Phytotoxicity of *Vulpia* residues on germination of wheat as affected by the addition of soil (An, Pratley, and Haig, 1993).

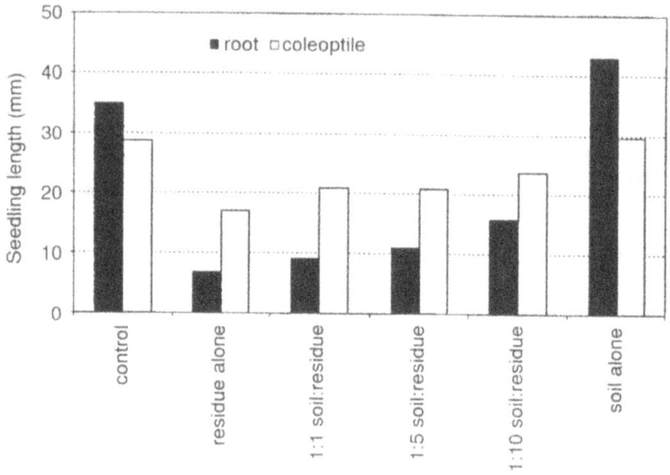

Figure 7. Phytotoxicity of *Vulpia* residues on seedling length of wheat as affected by addition of soil (An, Pratley, and Haig, 1993).

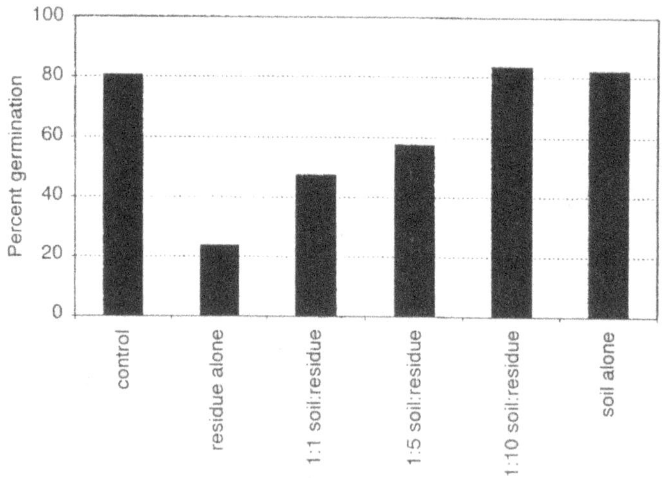

The work proceeded to demonstrate differences in potency on wheat seedlings between *Vulpia* species (An, Pratley, and Haig, 1996b) as shown in Figure 8, *V. bromoides* being stronger in effect than *V. myuros*. Also reported were the differences between plant parts in terms of their impact, data showing that shoots were marginally more toxic than roots but seeds *per se* were not particularly toxic (Figure 9). Greatest effect was observed on radicle growth with a lesser effect obtained on coleoptiles. Responses to the vulpia leachates were shown to vary with species and between varieties within species (An et al., 1997) confirming the work of Leigh and Halsall (1994). Generally, cocksfoot, phalaris, canola and medics were relatively tolerant whilst lupins, barley and wheat were most susceptible. Within subterranean clovers, there was a range of tolerance similar to that demonstrated by Leigh and Halsall (1994).

Application of these findings to the field (An, Pratley, and Haig, 1996a) was undertaken where a farmer wished to establish a permanent pasture into a dense vulpia stand. His initial attempts involved the application of a non-selective herbicide at the start of the growing season in autumn, and then direct drilling pasture comprising cocksfoot, subterranean clover and lucerne. This produced an outcome of vigorous silvergrass. The alternate approach was to move away from traditional sowing time to a midwinter period, involving a non-selective herbicide application, a cultivation to mix any plant debris with the soil to enhance degradation and a delay of 3 weeks to move past the most toxic period. A successful pasture establishment outcome was achieved.

FIGURE 8. Relative phytotoxicity of decomposing residues of different *Vulpia* species on seedling elongation in wheat. Means identified by the same letter are not significantly different at the 5% level (Duncan's new multiple-range test) (An, Pratley, and Haig, 1996b, reprinted with permission).

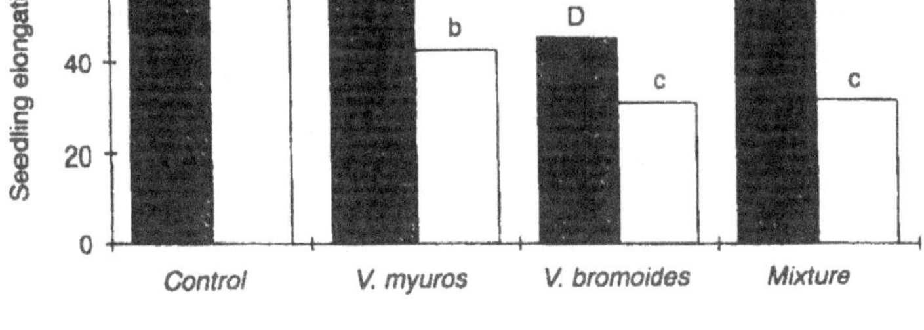

FIGURE 9. Relative phytotoxicity of decomposing residues from plant parts of *Vulpia* on seedling elongation in wheat. Means identified by the same letter are not significantly different at the 5% level (Duncan's new multiple-range test) (An, Pratley, and Haig, 1996b, reprinted with permission).

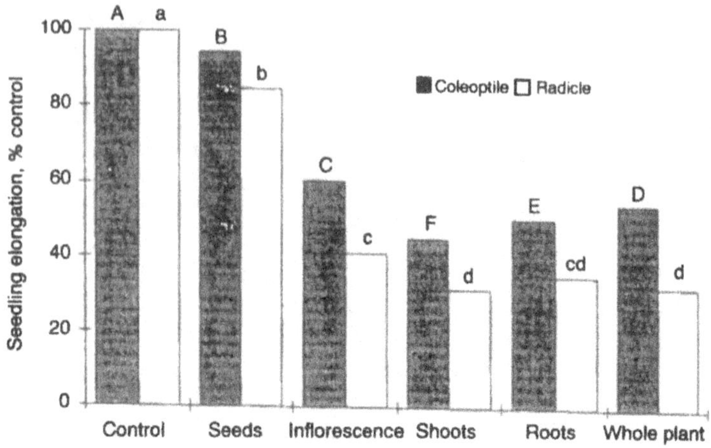

CONCLUSIONS

This chapter has demonstrated that allelopathy is playing a significant role in the establishment and productivity of pasture communities. Understanding of the processes will increasingly allow management of the pastures to minimise the effects. Several principles have emerged from the research:

- that allelopathy is genetically controlled both in terms of allelochemical production and in susceptibility. There is significant variation both between species and between varieties within species for both potency and tolerance;
- that dry residues carried over from one growing season to another are an impediment to the regeneration of species at the break of the new season in autumn. Management, therefore, needs to ensure that such carryover is minimised for sustainability of pasture composition;
- that stresses such as disease can induce an allelopathic relationship. Again, pasture management can influence the extent to which the impact is realised.

This account of current research indicates that sufficient attention needs to be given to allelopathy as a contributing factor to the productivity and botanical composition of improved pastures. Ignoring allelopathy is likely to result in lower production.

REFERENCES

An, M. (1996). Allelopathy in *Vulpia* residues. *Ph.D. thesis, Charles Sturt University, Wagga Wagga, Australia.*

An, M., J.E. Pratley, and T. Haig. (1993). The effect of soil on the phytotoxicity of residues of *Vulpia myuros*. In *Proceedings 7th Australian Agronomy Conference, Adelaide, South Australia*. The Australian Society of Agronomy, pp. 162-164.

An, M., J.E. Pratley, and T. Haig. (1996a). Field application of *Vulpia* phytotoxicity management: a case study. In *Proceedings 8th Australian Agronomy Conference, Toowoomba*. The Australian Society of Agronomy, pp. 616.

An, M., J.E. Pratley, and T. Haig. (1996b). Differential phytotoxicity of *Vulpia* species and their plant parts. *Allelopathy Journal* 3: 185-194.

An, M., J.E. Pratley, T. Haig, and P. Jellett. (1997). Genotypic variation of plant species to the allelopathic effects of vulpia residues. *Australian Journal of Experimental Agriculture* 37: 647-660.

Askin, D., D.B. Powan, and R.J. Lucas. (1987). The effects of clover content and herbage mass on lamb growth rate in the autumn. In *Herbivores Nutrition Research*, eds. J.B. Hacker and J.M. Ternmouth. Sydney: Australian Society of Animal Production, pp. 233-234.

Carter, E.D. (1987). Establishment and natural regeneration of annual pastures. In *Temperate Pastures, Their Production, Use and Management*, eds. J.L. Wheeler, C.J. Pearson, and G.E. Robards. Australia: CSIRO, pp. 35-57.

Dear, B.S. and B. Loveland. (1985). A survey of seed reserves of subterranean clover pastures on the Southern Tablelands of New South Wales. In *Proceedings 3rd Australian Agronomy Conference, Hobart*. The Australian Society of Agronomy, pp. 214.

Foot, J., C. Lenghaus, K.F.M. Reed, and W.K. Yong. (1987). Perennial ryegrass staggers–current research at Hamilton. *Australian Advances in Veterinary Science* 9: 77-80.

Freer, M. and D.B. Jones. (1984). Feeding value of subterranean clover, lucerne, phalaris and Wimmera ryegrass for lambs. *Australian Journal of Experimental Agriculture and Animal Husbandry* 24: 156-164.

Halsall, D.M., J.H. Leigh, S.E. Gollasch, and M. Holgate. (1995). The role of allelopathy in legume decline in pastures. II. Comparative effects of pasture, crop and weed residues on germination, nodulation and root growth. *Australian Journal of Agricultural Research* 46: 189-207.

Kenry, P.T. and K.F.M. Reed. (1984). Effect of pasture type on the growth and wool production of weaner sheep during summer and autumn. *Australian Journal of Experimental Agriculture and Animal Husbandry* 24: 322-332.

Leigh, J. and D.M. Halsall. (1994). Ecological implications of grazing systems: litter as a factor affecting the balance between grass, clover and weeds. In *Proceedings 9th Annual Conference of Grassland Society of New South Wales*, pp. 16-24.

Leigh, J.H., D.M. Halsall, and M.D. Holgate. (1995). The role of allelopathy in legume decline in pastures. I. Effects of pasture and crop residues on germination and survival of subterranean clover in the field and nursery. *Australian Journal of Agricultural Research* 46: 179-188.

Leigh, J.H., D.M. Halsall, W.J. Muller, R.N. Oram, J.P. Edlington, and A.E. Gerssler.

(1995). The role of allelopathy in legume decline in pastures. III. Tolerance of subterranean clover lines to the allelopathic compounds released by phalaris residues. *Australian Journal of Agricultural Research* 46: 1039-1049.

Mattner, S.W. (1998). *The impact of crown rust on interference between perennial ryegrass and white clover.* Ph.D. Thesis, University of Melbourne, Australia.

Morley, F.H.W., D. Bennett, and G.T. McKinney. (1969). The effect of rotational grazing with breeding ewes on phalaris-subterranean clover pastures. *Australian Journal of Experimental Agriculture and Animal Husbandry* 9: 74-84.

Pratley, J.E. (1989). Silvergrass residue effects on wheat. In *Proceedings 5th Australian Agronomy Conference, Perth.* The Australian Society of Agronomy, p. 472.

Pratley, J.E. and J.D. Ingrey. (1990). Silvergrass allelopathy on crops and pastures. In *Proceedings 9th Australian Weeds Conference, Adelaide*, pp. 436-439.

Quigley, P.E., F.J. Snell, P.J. Cunningham, and W. Frost. (1990). Effects of endophyte infected ryegrass on the establishment, persistence and production of mixed pastures in Australia. Proceedings *1st International Symposium on Acremonium/Grass Interactions Louisiana, USA*, pp. 49-50.

Quigley, P.E. and N.M. McFarlane. (1993). Effects of endophyte in perennial ryegrass on pasture establishment and botanical composition in Victoria. In *Proceedings XVII International Grassland Congress*, pp. 329-331.

Snell, F.J. and P.E. Quigley. (1993). Allelopathic effects of endophyte in perennial ryegrass residues on young subterranean clover plants. In *Proceedings XVII International Grassland Congress*, pp. 343-345.

Sutherland, B.L. and J.H. Hoglund. (1989). Effect of ryegrass containing the endophyte *Acremonium lolii* on the performance of associated white clover and subsequent crops. *Proceedings New Zealand Grassland Association* 50: 265-269.

Taylor, G.B. (1985). Effect of tillage practices on the fate of hard seeds of subterranean clover in a ley farming system. *Australian Journal of Experimental Agriculture* 25: 568-573.

Ulyatt, M.J., J.A. Lancashire, and W.T. Jones. (1977). The nutritive value of legumes. *Proceedings New Zealand Grassland Association* 38: 107-118.

Allelopathy in Agroecosystems in Spain

Luis González
Manuel J. Reigosa

SUMMARY. This paper reviews the effects of allelopathic interactions in agroecosystems in Spain on plant physiological activity and their ecological advantages. The phenological stage of growth of donor plants and the effective allelochemicals in the soil solution while studying the role of phenolic compounds were highlighting. Finally possible future prospects and conclusions regarding weed control by allelochemicals under integrated crop management strategies are discussed. *[Article copies available for a fee from The Haworth Document Delivery Service: 1-800-342-9678. E-mail address: <getinfo@haworthpressinc.com> Website: <http://www.HaworthPress.com> © 2001 by The Haworth Press, Inc. All rights reserved.]*

KEYWORDS. Allelopathy, microorganisms, allelochemicals, natural herbicides, ecological significance, pepper, sunflower

INTRODUCTION

Nobody will argue about the enormous importance of agriculture in Spain. To protect our crops from competition by weeds in the 1950s traditional

---

Luis González is Assistant Professor, and Manuel J. Reigosa is Professor, Laboratory of Plant Ecophysiology, Department of Plant Biology and Soil Science, Faculty of Sciences of Vigo, University of Vigo, Aptdo. 874, 36200 Vigo, Spain.

Address correspondence to: Manuel J. Reigosa at the above address (E-mail: mreigosa@uvigo.es).

The authors wish to thank Catherine Clarke for her kind correction of the manuscript.

[Haworth co-indexing entry note]: "Allelopathy in Agroecosystems in Spain." González, Luis, and Manuel J. Reigosa. Co-published simultaneously in *Journal of Crop Production* (Food Products Press, an imprint of The Haworth Press, Inc.) Vol. 4, No. 2 (#8), 2001, pp. 415-432; and: *Allelopathy in Agroecosystems* (ed: Ravinder K. Kohli, Harminder Pal Singh, and Daizy R. Batish) Food Products Press, an imprint of The Haworth Press, Inc., 2001, pp. 415-432. Single or multiple copies of this article are available for a fee from The Haworth Document Delivery Service [1-800-342-9678, 9:00 a.m. - 5:00 p.m. (EST). E-mail address: getinfo@haworthpressinc.com].

methods of farming were integrated with chemical methods. Nowadays many herbicides are used mainly due to the current intensive cultivation because of the build-up of resistance in some weeds. It is becoming necessary to use more herbicides during each application, which increase the problem of pollution and toxicity.

The increase in competition within the European market as new countries join has forced farmers to introduce changes in their traditional agriculture systems that most of the times led them to use chemicals as crop protectant. This fact sometime causes unwanted side effects such as lack of selectivity, development of weed resistance, toxicity to man and accumulation in the environment. It is, therefore, necessary to reduce the weed population in crops, but in line with the new European perspectives and objectives, by means that could not be considered hazardous. In this sense biological control agents can best provide an integrated strategy. The phytochemical relations between plants will play an important role in this integrated strategy.

There are still some discussions about the relative importance of the allelopathic phenomena under field conditions. Carral, Reigosa, and Carballiera (1988) have proposed a method to measure the effect of allelopathy in herbaceous plants when they are growing in a slope. It is suggested that when a plant releases allelochemicals in a slope, some asymmetry will arise due to this fact in the distribution pattern of the target species. This could lead to several combinations of alterations in the species distribution as shown in Figure 1. Sampling and calculating association indexes or measuring specific biomass uphill and downhill of the supposed allelopathic producer plant can assess the distribution pattern. Carballeira, Carral, and Reigosa (1988) have demonstrated such kind of heterogeneity due to allelopathy in the case of the weed *Rumex* sp. Some of these results are presented in Figure 2. This methodology has been used to assess the effects of two weeds, ***Polygonum persicaria*** and *Conyza canadensis* (Reigosa, Sánchez, and González, 1999) (Figures 3 and 4).

ECOLOGICAL ADVANTAGES

Chemical interaction between plants in a crop system is mediated by a lot of factors; nutrients, light, temperature and even humidity are known to affect the toxicity of the phytochemicals. The level of interaction is also dependent on the species and stage of growth of both the crop and the weed. It is known that many agroecological interactions are much more complex than was imagined a few years ago. The ecological aspects of allelopathic compounds should be kept in mind when cultivating or breeding crops for various purposes mainly in a sustainable environment. Our results with pepper plants show a clearly different phenolic profile, which relates to the variety studied

FIGURE 1. Expected situations when a plant releases allelochemicals to the environment when it is situated in a slope. Uphill and downhill response of the target species can vary.

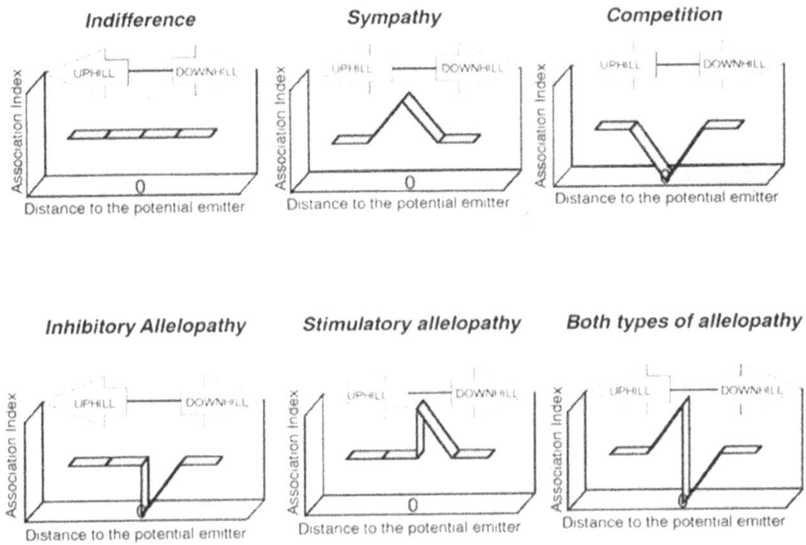

or the phenological stage of growth (González, Souto, and Reigosa, 1993). These findings are of high agroecological importance, mainly when considered with our studies which show the potential of decomposing pepper plant residues to release allelochemicals for up to ninety days (González, Souto, and Reigosa, 1997). For the allelopathic process to be agroecologically real, it needs an effective concentration of allelochemicals in the soil solution that injures weeds over a long period of time (we must not forget that one of the strategies of weed success is their ability to stagger germination).

Allelochemicals released from the pepper plant into the soil, undergo complex and unknown processes in which soil microflora plays an important ecological role. Hence, relationships between phenols released by pepper plants and soil microbial activity were studied. Our group has studied the effect of those allelochemicals on a wide range of microorganisms (proteolitic, *Nitrosomonas*, nitrogen fixers, ammonifiers, *Nitrobacter*, algae, amilolitic, bacteria, fungi and cellulolitic). We can highlight the results obtained from the microorganisms involved in the nitrogen cycle in the soil because of their importance in maintaining soil nutrition (Table 1). Phenolic compounds are among the most frequently utilized chemicals of plants to communicate with microorganisms. They are involved in many chemically mediated interac-

FIGURE 2. Results of *Rumex* species when it is growing in a slope. Biomass of the different target species is shown.

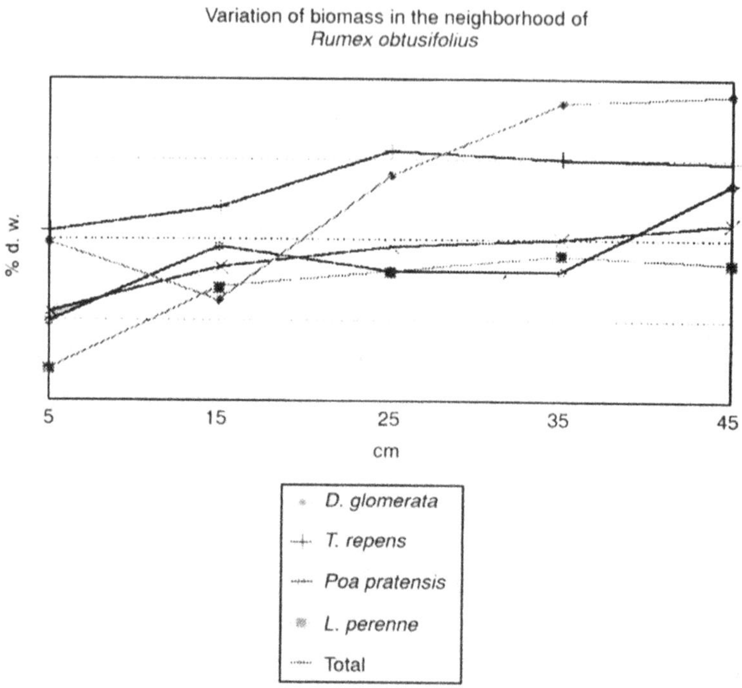

tions between plants and root-infecting microorganisms including bacteria and fungi (Siqueira et al., 1991). Many of the pepper diseases in the Northwest of Spain are due to fungal infection. Therefore, increases in fungal populations, as shown in our experiments, must be taken into account considering best crop management.

In a sustainable agroecological context, it is not necessary to look for a phytochemical that either prevents the germination of the invasive flora or kills the individuals growth. Sometimes, it is necessary to prevent seed formation, good root development, the correct shoot development or any physiological activity that avoids the loss in productivity yield. Thus, most of the compounds reported in this review could be successfully used as new herbicides if we found the appropriate crop-weed system where its use would be profitable. We must not forget the importance of the phenomenon of synergism that was demonstrated by others in relation to allelopathy. This work was checked by our group on a large number of the most important

FIGURE 3. Presence in the field of *Dactylis glomerata*, *Trifolium repens*, *Medicago sativa* and bare soil against *Polygonum persicaria*. The discontinuous line represents the inhibition of the presence of the test species with respect to the expected values ($p \leq 0.05$) and the continuous line indicates the stimulation of the presence of the species ($p \leq 0.05$). In the graphic for the presence of bare points the marks are as above. The horizontal line represents the number of test plants expected in each case and calculated with respect to the control.

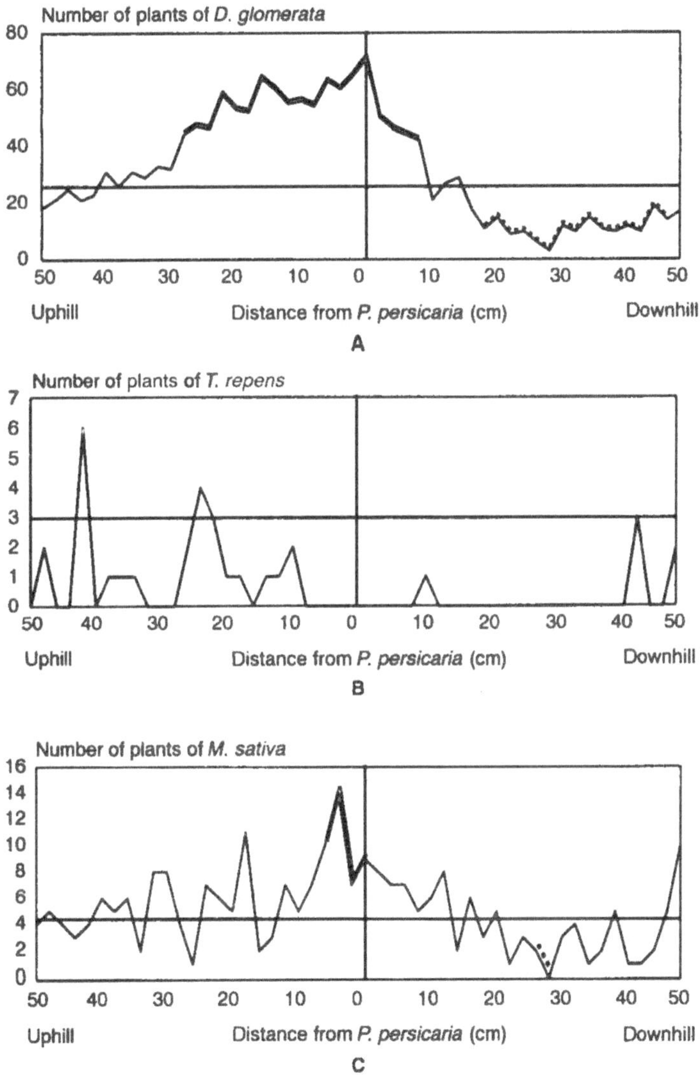

FIGURE 3 (continued)

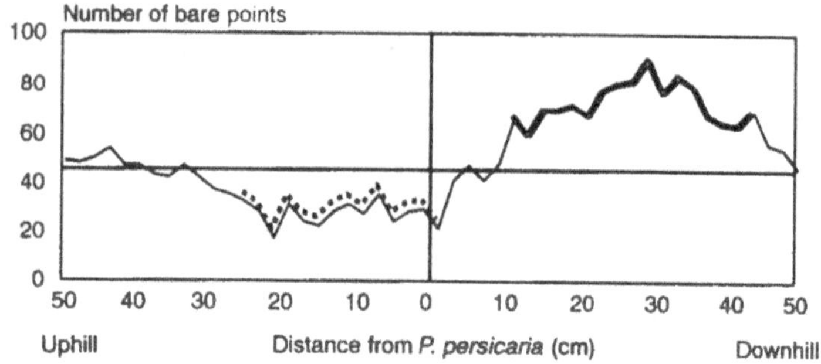

D

weeds in the Northern Spain using the standards from the allelochemicals released by the pepper plant during decomposition (Reigosa, González, and Souto, 1999). These results showed that in different bioassays, carried out in the laboratory, the effect on both growth and germination was usually different than that expected from individual compounds. Thus, we found synergistic and antagonistic phenomena (González, Souto, and Reigosa, 1993; Reigosa, González, and Souto, 1999). In nature, plant chemical interactions almost never depend on a single compound, but instead depends on a mixture of compounds. Often these mixtures are made up of one or more series of structurally closely related compounds, and crude plant extracts are frequently more bioactive than their most active constituent, once isolated and purified (Isman, Matsuura, and Mackinnon, 1996).

Low molecular weight phenolic compounds and sesquiterpene lactones were found in pepper and sunflower and studied as related compounds. The compounds within these families usually only differ in the patterns of hydroxylation, methylation or other ester group. Isman, Matsuura, and Mackinnon (1996) pointed out the significance of these mixtures of related compounds. They either equip the plant with a diverse arsenal capable of defense against a wide array of potential enemies or the mixtures act in a synergistic fashion (a multicomponent mixture is more efficacious than a single compound). This later hypothesis has been tested several times (Berenbaum, 1985) and it is supported from our results (Reigosa, González, and Souto, 1999).

In an agroecological perspective the use of related mixtures might have important consequences for the practical use of crop extracts for weed control. It has been suggested that the development of resistance to crop extracts in weeds may be less probable than that to commercial herbicides. The use of non-toxic phenolic compounds to enhance the toxic effect of others could be

FIGURE 4. Presence in the field of *Dactylis glomerata*, *Trifolium repens*, *Medicago sativa* and bare soil against *Conyza canadensis*. The discontinuous line represents the inhibition of the presence of the test species with respect to the expected values ($p \leq 0.05$) and the continuous line indicates the stimulation of the presence of the species ($p \leq 0.05$). In the graphic for the presence of bare points the marks are as above. The horizontal line represents the number of test plants expected in each case and calculated with respect to the control.

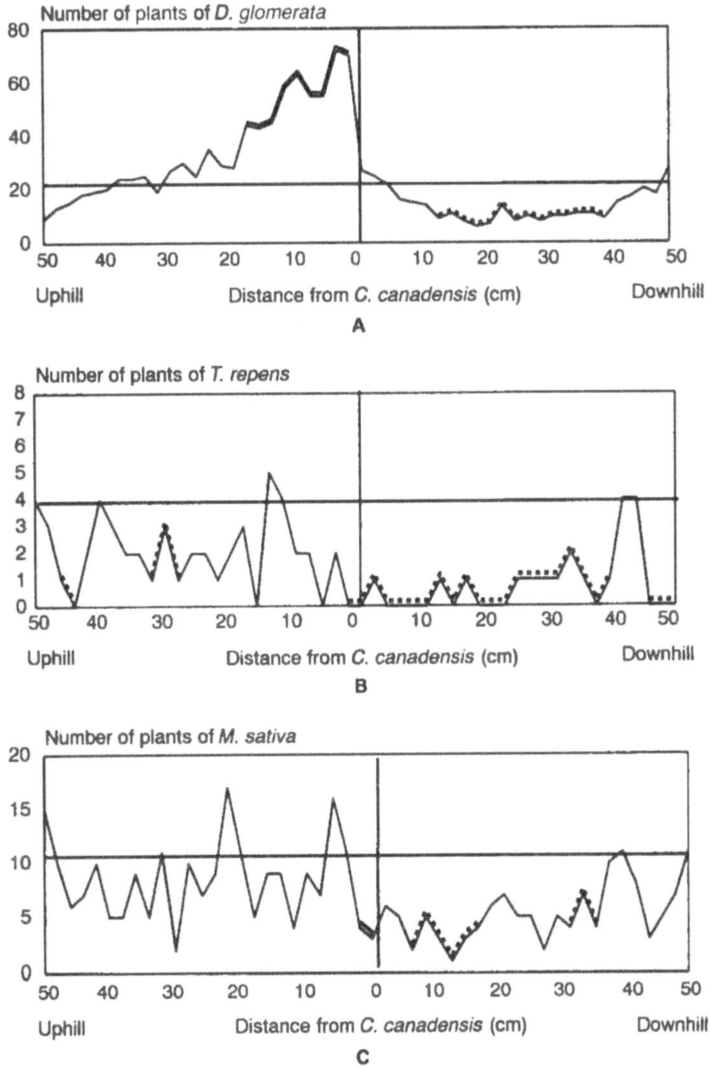

FIGURE 4 (continued)

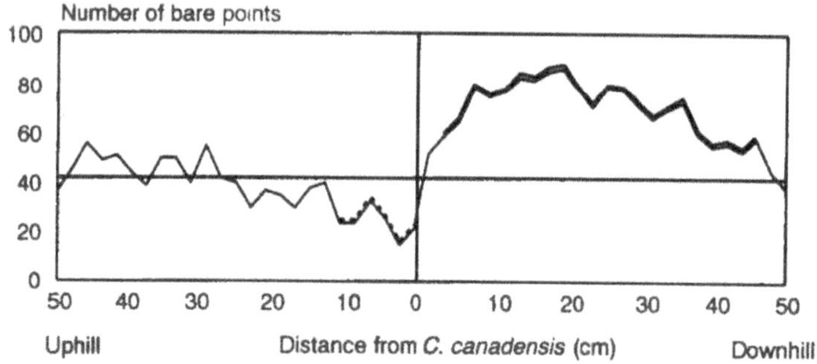

D

TABLE 1. Semiquantitative effect of phenolic compounds from *Capsicum annuum* on soil microorganisms population nine days after allelochemicals addition. (– represents decrease in microorganisms population, + means increase in microorganisms population, 0 indicates no variation in microorganisms population).

	0.25 μM	0.50 μM	0.75 μM
Dinitrogen fixers	– –	– –	– –
Proteolitic	+ + +	+ + +	+ + +
Ammonifiers	0	0	0
Denitrifiers	+	+	+
Nitrosomonas	– – –	– – –	– – –
Nitrobacter	– – –	– – –	– –
Amilolitic	0	0	0
Algae	– –	– – –	– – –
Bacteria	+	0	0
Fungi	+++	+++	+++
Cellulolitic	0	+	++

recommended in an integrated management approach to crop growth. In some agronomic strategies valuable compounds have been lost to make crops seemingly more attractive or in the process to improve yield or fruit quality. The use of man-made herbicides must then compensate for the disappearance of the chemicals. Wink (1993) also pointed that the development of new natural pesticides is only one aspect of the changes that are required to develop new and safe agricultural practices. All these aspects should be integrated and allelopathy would be another element in the whole process.

PHYSIOLOGICAL ASPECTS

The phytochemical mode of action in an agroecosystem includes all areas of interaction between an allelochemical and the crop or weed species Finally, the process that leads to plant interaction induces changes in the plant metabolism. According to Cobb (1992), this physiological effect needs a precise site of action where the location of a molecular interaction, which triggers a series of secondary events, leads to the developmental changes in the target plant. Therefore, it is necessary to know the real amount of phytochemical that the plant can absorb to establish the effective concentration that interferes with the target plant. In this sense, our research work in Spain focused on establishing the presence of known allelochemical compounds in the crop and later to quantify the concentration of these in the soil solution. We developed an extraction method that was able to avoid the breakdown of binding forms of allelochemicals from humus, that are not available at any time to the target plant (González, Souto, and Reigosa, 1997).

Using the above approach our group demonstrated that the phenolic profile in the crop-soil which is available to the plant, is much lower than that in the pepper plant. The physiological effect of these phenols on the target plant is very different to that expected after laboratory bioassays. The great amount of physico-chemical and biological interactions that happens in the first layer of the soil reduces and transforms the phytochemicals released from the pepper plants. Hence, the potential activity of the donor crop is different than would be expected if the phenol profile of the pepper plant was considered alone.

The interference that phytochemicals produce on target species is variable and is dependent on the age of the plant and the environmental conditions. In addition, it was demonstrated that the allelochemicals released by pepper plants in the crop-system, or extracted from leaves of sunflower show different effects which relate to the metabolic parameter measured (González, Souto, and Reigosa, 1993, 1997; Macías, Varela, and Torres, 1993; Macías et al., 1996; Souto, Reigosa, and González, 1995). It has also been shown that the effect brought about by continuous dilutions of the allelochemicals is not related to a lineal effect on the physiological parameter tested. It is really difficult to explain these findings. The response seems to be affected by multiple ecophysiological factors, but the biochemical explanation of this behavior needs to evaluate the possibility that the same compound may act on more than one protein or enzyme (Macías, Galindo, and Massanet, 1992).

They postulated that these compounds could have different sites of action, with different values of best activity concentrations for each one. So, the non-linear effect observed would correspond to the combination of the different activity curves for each site of action. From these results, it is difficult to find the effective concentration that injures weeds in agroecosystems but more effort must be devoted to understanding the physiology of both crops and weeds.

Diversity within and among living organisms is a biological imperative in evolutionary terms. Thus, vast phytochemical diversity between different plant species was expected. But, great variability between varieties of the same species was unexpected. Nine varieties of pepper (González, Souto, and Reigosa, 1993) and two varieties of sunflower (Macías, Varela, and Torres, 1993; Macías et al., 1996) that are widely cultivated in Spain showed different phytochemical profile between varieties.

To know the biological activity of these compounds would involve a great variety of laboratory bioassays and the success rate would probably be low, if the activity being sought is not ecologically relevant to the organism (Berenbaum and Zangerl, 1996). There is no real reason to expect that pepper or sunflower plants produce compounds that would be toxic to other organisms or biological systems not related with its ecological history, but it could be possible. On the other hand, it is more likely that some of the compounds found in those different varieties can interfere in some specific way at any time on the weeds associated with the crop. The phytochemical diversity improves the possibility of choosing the most suitable variety to control weeds under an integrated and sustainable weed management system.

RESEARCH PROGRESS

Since plants have physico-chemical adaptations for producing a wide spectrum of secondary metabolites, some of which are active allelopathic agents, the natural products are an attractive source of potential leads to new agrochemicals (Macías et al., 1996). This is the hypothesis that drives the agrochemical research work of the group lead by F. Macías in Southern Spain. They had been looking for natural products from sunflower that could be released to the environment to improve crop. It was hypothesized that these related compounds were effective against plant pathogenic bacteria and fungi, insects and weeds that usually cause significant crop losses in sunflower production in Spain. It was hoped that they could synthesize these compounds to be used as bioherbicides. The group had isolated, elucidated the structure and had carried out allelopathic bioassays of different sesquiterpenes lactones with eudesmanolide, melampolide, *cis,cis*-germacranolide, *trans,trans*-germacranolide, melampolide, heliangolide, annuithrin, helivypolide and guaianolide skeletons (Aguilar et al., 1988; Collado et al., 1985 a,b,c; Macías and Fischer, 1992; Macías, Simonet, and Esteban, 1994; Massanet et al., 1983).

In order to obtain information about the specific structural requirements needed for the biological activity of these allelochemicals in sunflower-agrosystem, the F. Macías team isolated allelochemicals from two cultivars of *Helianthus annuus* L. Through aqueous serial dilutions, they identified the compounds' effects on various species. By continuing with this approach, they hope to improve our understanding of the mode of action and allelopathic activity (Macías, 1995).

The contribution of the regulatory activity on germination and plant growth of sesquiterpene lactones has been related to the presence of a α-methylene-γ-lactone moiety. Those studies confirmed that the activity is clearly affected by the conformation of the molecules tested and the accessibility of groups that can be alkylated. To improve the formulation of new herbicides they had demonstrated that the effect of these compounds on shoot and root length could be a clear indication that the major requirement in the activity of eudesmanolides is the conformation that these compounds can adopt, rather than the functionalization they may posses. Otherwise, the behavior of some melampolides and *cis,cis*-germacranolides on the germination of different species had led them to propose that in "non-rigid" systems the effect depends both on the presence of alkylable groups (such as, α-methylene-γ-lactone) and the conformation of the molecule. The latter can determine the accessibility for alkylation (Macías, Galindo, and Massanet, 1992). So, it may be more reasonable to relate the presence of the α-methylene-γ-lactone functionalization with the germination responses than with shoot and root growth responses, at least in conformationally "less-flexible" systems.

The findings of this group (Macías, Varela, and Torres, 1993) suggest that the guaianolides found in sunflower leaves var. SH-222® are likely to be involved in the allelopathic action in the crop-system with a certain specificity to dicotyledon species (Figure 5).

The extraction of the fresh leaf aqueous extract of the variety VYP® of the same species contained different germacranolides that possessed an α-methylene-γ-lactone moiety. Therefore, the different profiles of allelopathic activity (Figure 6) were attributed to the presence of a second or a third receptor site for alkylation, such as α,β-unsaturated carbonyl systems as well as to conformational changes due to the different functionalization. They reported that those effects fundamentally influence root and shoot growth, more than germination.

These phenomena were demonstrated for many researches into plant-chemical-relationship (allelopathic interactions) in the laboratory. Our research strategy on allelopathy in agroecosystems has been derived from a

FIGURE 5. Effect of four annuolides (concentration 10^{-5} M) from sunflower var. SH-222® on the germination, radicle and shoot length of *Lactuca sativa* (A) and *Hordeum vulgare* (B).

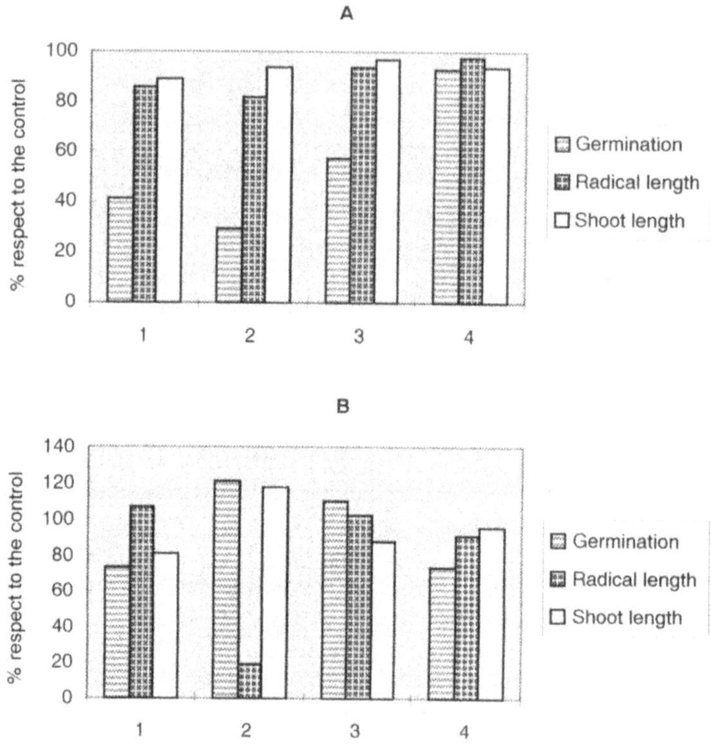

sequence of bioassays carried out in the laboratory. They followed a logical progression through a set of experiments of increasing complexity, approaching practical field conditions. In the first stage we looked for allelochemicals in different crops or weeds species or varieties. The second stage consisted of investigations on interactions of those allelochemicals with secondary physiological effects in seeds or seedlings. The third stage was on allelochemicals reactions within the crop soil. This led automatically to the work on soil microorganisms. In the fourth stage of the progression, we examined primary physiological effects on target species. This provided and is still providing both theoretical and experimental support for work with natural herbicides. Entering the results of laboratory studies into integrated crop models, and then testing the laboratory predictions against field experiments can determine progress. Using this approach we have already demonstrated a real

FIGURE 6. Effect of four lactones (concentration 10^{-5} M) from sunflower var. VYP® on the germination radicle and shoot length of *Lactuca sativa* (A) and *Hordeum vulgare* (B).

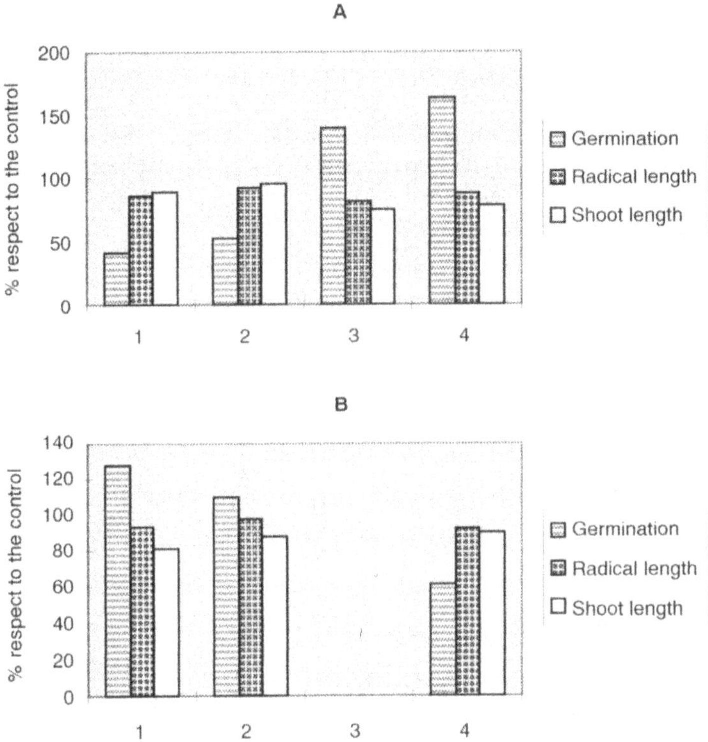

effect in the field, showing the relationship between weeds and the associated flora in a sloping field. The weed was in the center of the transept and an association index was calculated, which showed the probability of interaction between two species (Sanchez, 1996). The analysis of the field data has showed allelopathic effects (Figures 3 and 4).

Knowing the allelopathic potential of different varieties of pepper plants (González, Souto, and Reigosa, 1993) we looked for the chemicals that are involved in this phenomenon. A wide family of phenolic compounds was checked in the plant and in extracts of shoots and roots of pepper plants which had been decomposing in soil and up to ten individual low molecular weight phenolic compounds were found in the samples. The different varieties of peppers that were studied over the growing season did not show a regular behavior in terms of phenolic profile. This was due to the fact that the plants are living materials and are, therefore subject to physiological changes that are function of the relation between origin of variety, development stage, and environment. Local studies are necessary to improve an integrated pepper management strategy. The results also showed that there was no correlation between phenolic compounds of pepper determined and measured one day after beginning of decomposition and the phenolic profile of pepper plants before they were embedded in the soil (Figure 7). This reflects the real and not the potential concentration that is available for the plants in the agroecosystem. Our results had confirmed that allelopathic interference exists between the different pepper varieties and lettuce-target species (González, Souto, and Reigosa, 1997).

FIGURE 7. Concentration of phenolic compounds determined from *Capsicum annuum* leaves (A) and soil with leaves in decomposition process (B). 1: Caffeic acid, 2: Chlorogenic acid, 3: *p*-Coumaric acid, 4: Ferulic acid, 5: *p*-Hydroxybenzaldehyde, 6: *p*-Hydroxybenzoic acid, 7: Protocatechuic acid, 8: Syringic acid, 9: Vanillic acid.

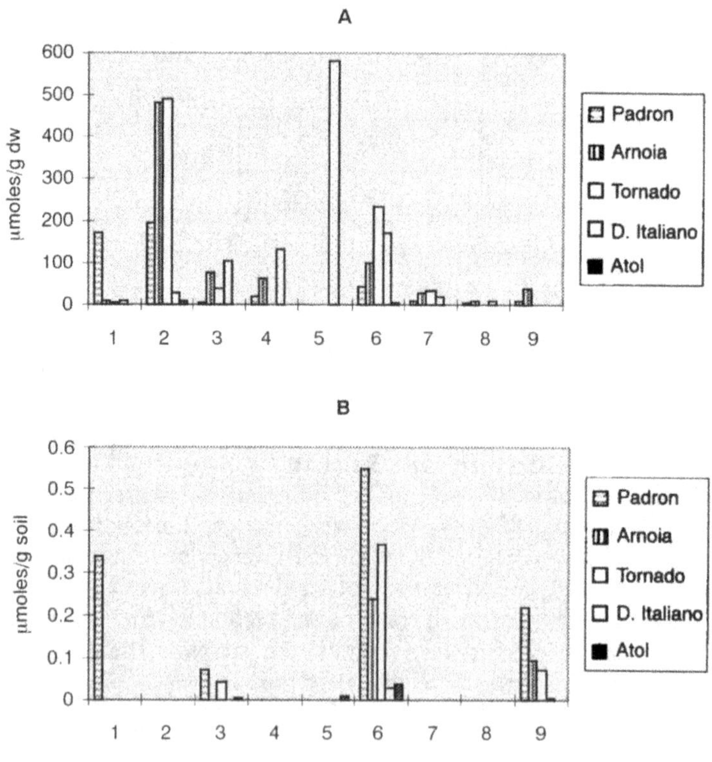

In order to assess the capacity of these phenolic compounds found in pepper plants against common weeds in the Northwest of Spain, we carried out different experiments which showed that the effect on weeds was not homogeneous (Figure 8). While the higher concentration bioassay inhibited both germination and growth, the rest of the concentrations showed a variable and non-linear range of effects which depended on the weed, the phenolic concentration assayed and the measured physiological parameter (Reigosa, González, and Souto, 1999).

Many of the pepper phenols and sesquiterpenes lactones isolated or studied to date (at least those used in the bioassays carried out by the groups leaded by Macías and Reigosa) showed different levels of activity against plants at biologically realistic concentrations. A majority of phenols and

FIGURE 8. Effects of six phenolic compounds (concentration 10^{-2} M) from *Capsicum annuum* on the germination and growth of weed species: *Amaranthus retroflexus* (A) and *Rumex crispus* (B). 1: Gallic acid, 2: Ferulic acid, 3: Vanillic acid, 4: *p*-Coumaric acid, 5: *p*-Hydroxybenzoic acid, and 6: Vanillin.

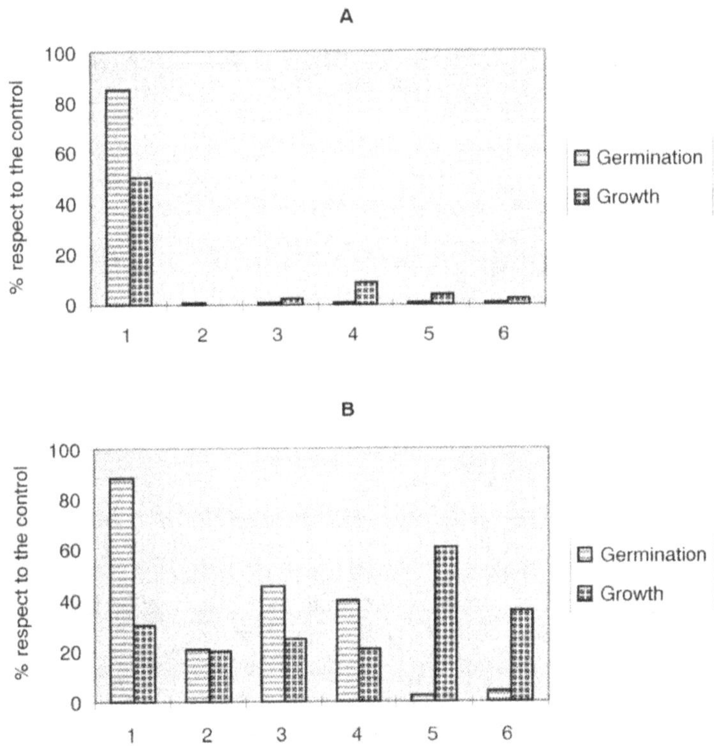

sesquiterpenes lactones tested appear to have some inhibitory activity on some target plant. More importantly, in an integrated crop management system, the vast majority of these compounds have not yet been widely screened against other organisms concurrent at the crop ecosystem, e.g., plant pathogens and other microorganism. We must also acknowledge that our understanding of the role of these compounds in the natural context is almost non-existent but it gives us an excellent opportunity to investigate further their potential in the field. From current state of knowledge, we cannot predict the bio-herbicide activity of individual phenols or novel sesquiterpene lactones with any degree of confidence on pepper or sunflower agrosystems.

FUTURE PROSPECTS AND CONCLUSIONS

Weed-control allelochemicals in the future will be used extensively within integrated crop management strategies. In such strategies, the harvest crop interacts with phytochemicals such as plant-derived anti-growth, or anti-germination. There are three ways in which they interact: employing donor-allelochemicals cultivars, employing resistant cultivars to allelochemicals released from weeds or using natural chemicals from plants which interfere with the crop-weed system. Thus, the plants to be harvested are protected and the biological compounds reduce the weed population. Conditions or the varieties selected may be modified to benefit the allelochemical action. However, for full exploitation of an allelopathy strategy in the crop field, it is necessary to define much more extensively the components of the crop-weed relationship. Special analytical methods should be developed to understand the phytochemical behavior in the agroecosystem.

Pepper and sunflower have revealed the presence of a wealth of secondary plant compounds and that, as our experiments demonstrated, probably play a pivotal role in crop-weed defense. We are just beginning to understand how crop plants communicate with other plants via chemical signals, which are often very specific. This knowledge should lead to major improvements in agriculture in Spain. But the journals are full of new promising natural products with expected commercial success. It is clear that the introduction of natural products and its production route '*is an evolutionary and not a revolutionary process*' (Van Beek and Breteler, 1993). Agricultural management changes occur much more slowly than anticipated from scientific developments. The changes are always driven by economic reasons.

We believe that these findings in relation to allelopathy in Spain will allow us to exploit allelopathy as a weed management strategy by isolating superior weed suppressing crop varieties (e.g., pepper and sunflower).

REFERENCES

Aguilar, J.M., I.G. Collado, G.M. Macias, J.M.G. Molinillo, F. Rodrigues-Luis, F.R. Fronczek, and S.F. Watkins. (1988). Sesquiterpene lactones from *Artemisia lanata*. *Phytochemistry* 27: 2229-2223.

Berenbaum, M.R. (1985). Brementown revisited: Interactions among allelochemicals in plants. In *Recent Advances in Phytochemistry. Vol. 19. Chemically Mediated Interactions Between Plants and Other Organisms*, eds. G.A. Cooper-Driver, T. Swain, and E.S. Conn. New York: Plenum Press, pp. 139-169.

Berenbaum, M.R. and A.R. Zangerl. (1996). Phytochemical diversity: Adaptation or random variation? In *Recent Advances in Phytochemistry. Vol. 30. Phytochemical Diversity and Redundancy in Ecological Interactions*, eds. J.T. Romeo, J.A. Saunders, and P. Barbosa. New York: Plenum Press, pp. 1-24.

Carballeira, A., E. Carral, and M.J. Reigosa. (1988). Asymmetric small-scale distribution and allelopathy: Interaction between *Rumex obtusifolius* L. and meadow species. *Journal of Chemical Ecology* 14: 1773-1784.

Carral, E., M.J. Reigosa, and A. Carballeira. (1988). *Rumex obtusifolius* L.: Release of allelochemical agents and their influence on small-scale spatial distribution of meadow species. *Journal of Chemical Ecology* 14: 1761-1772.

Cobb, A. (1992). *Herbicides and Plant Physiology*. London: Chapman & Hall.

Collado, I.G., F.A. Macias, G.M. Massanet, and F. Rodriguez-Luis. (1985a). Guaianolides from *Centaurea canariensis*. *Phytochemistry* 24: 912-913.

Collado, I.G., F.A. Macias, G.M. Massanet, and F. Rodriguez-Luis. (1985b). Flavonoids from *Centaurea clementei*. *Journal of Natural Products* 48: 819-822.

Collado, I.G., F.A. Macias, G.M. Massanet, F. Rodriguez-Luis, J. Salva, and C. Vergara. (1985c). Mariolin, a germacranolide from *Anacyclis radiatus*. *Phytochemistry* 24: 2447-2448.

Gonzalez, L., X.C. Souto, and M.J. Reigosa. (1993). Potential of different pepper cultivars (*Capsicum annuum* L.) as weed controllers by allelopathic agents in Galicia (Northwest Spain). In *Proceedings of 8th EWRS Symposium "Quantitative Approaches in Weed and Herbicide Research and their Practical Application."* Braunschweig, pp. 143-150.

Gonzalez, L., X.C. Souto, and M.J. Reigosa. (1997). Weed control by *Capsicum annuum*. *Allelopathy Journal* 4: 101-110.

Isman, M.B., H. Matsuura, and S. Mackinnon. (1996). Phytochemistry of the Meliaceae: So many terpenoids, so few insecticides. In *Recent Advances in Phytochemistry-Vol. 30. Phytochemical Diversity and Redundancy in Ecological Interactions*, eds. J.T. Romeo, J.A. Saunders, and P. Barbosa. New York: Plenum Press, pp. 155-178.

Macias, F.A. (1995). Allelopathy in the search for natural herbicide models. In *Allelopathy: Organisms, Processes and Applications*, eds. Inderjit, K.M.M. Dakshini, and F.A. Einhellig. ACS Symposium Series 582. Washington, DC: American Chemical Society, pp. 310-329.

Macias, F.A., A. Torres, J.M.G. Molinillo, R.M. Varela, and D. Castellano. (1996). Potential allelopathic sesquiterpene lactones from sunflower leaves. *Phytochemistry* 43: 1205-1215.

Macias, F.A., A.M. Simonet, and M.D. Esteban. (1994). Potential allelopathic lupane triterpenes from bioactive fractions of *Melilotus messanensis*. *Phytochemistry* 35: 1369-1379.

Macias, F.A., J.C.G. Galindo, and G.M. Massanet. (1992). Potential allelopathic activity of several sesquiterpene lactone models. *Phytochemistry* 31: 1969-1977.

Macias, F.A. and N.H. Fischer. (1992). Melampolides from *Leucocarpus pinnatifidus*. *Phytochemistry* 31: 2747-2754.

Macias, F.A., R.M. Varela, and A. Torres. (1993). Potential allelopathic guaianolides from cultivars sunflowers leaves var. SH-222. *Phytochemistry* 34: 669-674.

Macias, F.A., R.M. Varela, A. Torres, and J.M.G. Molinillo. (1996). Field crops as source of natural herbicide models: sunflowers. In *Allelopathy in Pest Management for Sustainable Agriculture*. Vol. II, eds. S.S. Narwal and P. Tauro. Jodhpur, India: Scientific Publishers, pp. 77-92.

Massanet, G.M., I.G. Collado, F.A. Macias, F. Bohlmann, and J. Jakupovic. (1983). Structural determination of Clementein, a new guaianolide isolated from *Centaurea clementei*. *Tetrahedron Letters* 24: 1641-1642.

Reigosa, M.J., L. Gonzalez, and X.C. Souto. (1999). Effect of some different phenolic compounds on weed species. *Plant Growth Regulation* 28: 83-88.

Reigosa, M.J., A. Sanchez, and L. Gonzalez. (1999). Ecophysiological approach in allelopathy. *Critical Reviews in Plant Sciences* 18: 577-608.

Souto, X.C., M.J. Reigosa, and L. Gonzalez. (1995). Effect of potential phenolic allelochemicals released by *Capsicum annuum* on the growth of some microorganisms populations. In *Proceedings 22nd Annual Meeting Plant Growth Regulator Society of America*, pp. 81-85.

Sanchez, A.M. (1996). *Efectos fisiológicos producidos por la acción de aleloquímicos*. Tesis de licenciatura, Facultade de Ciencias de Vigo, 88 pp. [Spanish].

Siqueira, J.O., M.G. Nair, R. Hammerschmidt, and G.R. Safir. (1991). Significance of phenolic compounds in plant-soil-microbial systems. *Critical Reviews in Plant Sciences* 10:63-121.

Van Beek, T.A. and H. Breteler. (1993). Introduction. In *Proceedings of the Phytochemical Society of Europe-Vol. 34. Phytochemistry and Agriculture*, eds. T.A. Van Beek and Breteler. New York: Oxford University Press, pp. XIII-XXV.

Wink, M. (1993). Production and application of phytochemicals from an agricultural perspective. In *Proceedings of the Phytochemical Society of Europe-Vol. 34. Phytochemistry and Agriculture*, eds. T.A. Van Beek and H. Breteler. New York: Oxford University Press, pp. 171-213.

Index

For Product Safety Concerns and Information please contact our EU representative GPSR@taylorandfrancis.com
Taylor & Francis Verlag GmbH, Kaufingerstraße 24, 80331 München, Germany

www.ingramcontent.com/pod-product-compliance
Lightning Source LLC
LaVergne TN
LVHW010619300926
849162LV00001B/1

* 9 7 8 1 5 6 0 2 2 0 9 0 9 *